EQUIPMENT HANDBOOK

ELECTRICAL EQUIPMENT HANDBOOK

Troubleshooting and Maintenance

Philip Kiameh

McGRAW-HILL
New York   Chicago   San Francisco   Lisbon   London   Madrid
Mexico City   Milan   New Delhi   San Juan   Seoul
Singapore   Sydney   Toronto

The McGraw·Hill Companies

Library of Congress Cataloging-in-Publication Data

Kiameh, Philip.
    Electrical equipment handbook : troubleshooting and maintenance / Philip Kiameh.
       p.   cm.
    Includes index.
    ISBN 0-07-139603-9
    1. Electrical machinery—Maintenance and repair—Handbooks, manuals, etc. I. Title.

TK2184.K48   2003
621.31'042'0288—dc21   2002044498

Copyright © 2003 by The McGraw-Hill Companies, Inc. All rights reserved. Printed in the United States of America. Except as permitted under the United States Copyright Act of 1976, no part of this publication may be reproduced or distributed in any form or by any means, or stored in a data base or retrieval system, without the prior written permission of the publisher.

2 3 4 5 6 7 8 9 0   BKM/BKM   0 9 8 7 6

ISBN 0-07-139603-9

The sponsoring editor for this book was Kenneth P. McCombs, the editing supervisor was David E. Fogarty, and the production supervisor was Pamela A. Pelton. It was set in Times Roman by Wayne A. Palmer of McGraw-Hill Professional's Hightstown, N.J., composition unit.

This book was printed on recycled, acid-free paper containing a minimum of 50% recycled, de-inked fiber.

McGraw-Hill books are available at special quantity discounts to use as premiums and sales promotions, or for use in corporate training programs. For more information, please write to the Director of Special Sales, Professional Publishing, McGraw-Hill, Two Penn Plaza, New York, NY 10121-2298. Or contact your local bookstore.

Information contained in this work has been obtained by The McGraw-Hill Companies, Inc. ("McGraw-Hill") from sources believed to be reliable. However, neither McGraw-Hill nor its authors guarantee the accuracy or completeness of any information published herein and neither McGraw-Hill nor its authors shall be responsible for any errors, omissions, or damages arising out of use of this information. This work is published with the understanding that McGraw-Hill and its authors are supplying information but are not attempting to render engineering or other professional services. If such services are required, the assistance of an appropriate professional should be sought.

To my son, Joseph

CONTENTS

Preface xiii
Acknowledgments xv

Chapter 1. Fundamentals of Electric Systems   1.1

Capacitors / 1.1
Current and Resistance / 1.4
The Magnetic Field / 1.6
Faraday's Law of Induction / 1.9
Lenz's Law / 1.11
Inductance / 1.13
Alternating Current / 1.15
Three-Phase Systems / 1.21
References / 1.24

Chapter 2. Introduction to Machinery Principles   2.1

Electric Machines and Transformers / 2.1
Common Terms and Principles / 2.1
The Magnetic Field / 2.2
Magnetic Behavior of Ferromagnetic Materials / 2.3
Faraday's Law—Induced Voltage from a Magnetic Field Changing with Time / 2.7
Core Loss Values / 2.9
Permanent Magnets / 2.9
Production of Induced Force on a Wire / 2.14
Induced Voltage on a Conductor Moving in a Magnetic Field / 2.15
References / 2.15

Chapter 3. Transformers   3.1

Importance of Transformers / 3.1
Types and Construction of Transformers / 3.1
The Ideal Transformer / 3.2
Impedance Transformation through a Transformer / 3.4
Analysis of Circuits Containing Ideal Transformers / 3.5
Theory of Operation of Real Single-Phase Transformers / 3.9
The Voltage Ratio across a Transformer / 3.10
The Magnetizing Current in a Real Transformer / 3.12
The Dot Convention / 3.14
The Equivalent Circuit of a Transformer / 3.15
The Transformer Voltage Regulation and Efficiency / 3.18
The Autotransformer / 3.22
Three-Phase Transformers / 3.23
Transformer Ratings / 3.25
Reference / 3.28

Chapter 4. Transformer Components and Maintenance — 4.1

Introduction / 4.1
Classification of Transformers / 4.1
Main Components of a Power Transformer / 4.2
Types and Features of Insulation / 4.9
Forces / 4.11
Cause of Transformer Failures / 4.11
Transformer Oil / 4.13
Gas Relay and Collection Systems / 4.22
Relief Devices / 4.24
Interconnection with the Grid / 4.24
Reference / 4.27

Chapter 5. AC Machine Fundamentals — 5.1

The Rotating Magnetic Field / 5.1
The Induced Voltage in AC Machines / 5.6
The Induced Torque in an AC Machine / 5.9
Winding Insulation in AC Machines / 5.10
AC Machine Power Flow and Losses / 5.11
Reference / 5.12

Chapter 6. Induction Motors — 6.1

Induction Motor Construction / 6.1
Basic Induction Motor Concepts / 6.1
The Equivalent Circuit of an Induction Motor / 6.6
Losses and the Power Flow Diagram / 6.9
Induction Motor Torque-Speed Characteristics / 6.9
Control of Motor Characteristics by Squirrel-Cage Rotor Design / 6.14
Starting Induction Motors / 6.17
References / 6.21

Chapter 7. Speed Control of Induction Motors — 7.1

Speed Control by Changing the Line Frequency / 7.1
Speed Control by Changing the Line Voltage / 7.3
Speed Control by Changing the Rotor Resistance / 7.5
Solid-State Induction Motor Drives / 7.5
Motor Protection / 7.5
The Induction Generator / 7.5
Induction Motor Ratings / 7.8
Reference / 7.11

Chapter 8. Maintenance of Motors — 8.1

Characteristics of Motors / 8.1
Enclosures and Cooling Methods / 8.1
Application Data / 8.3
Design Characteristics / 8.4
Insulation of AC Motors / 8.5

CONTENTS ix

Failures in Three-Phase Stator Windings / 8.6
Predictive Maintenance / 8.6
Motor Troubleshooting / 8.7
Diagnostic Testing for Motors / 8.7
Repair and Refurbishment of AC Induction Motors / 8.19
Appendix: Typical Causes of Winding Failures in Three-Phase Stators / 8.21
Reference / 8.22

Chapter 9. Power Electronics, Rectifiers, and Pulse-Width Modulation Inverters   9.1

Introduction to Power Electronics / 9.1
Power Electronics Components / 9.1
Power and Speed Comparison of Power Electronic Components / 9.7
Basic Rectifier Circuits / 9.7
Filtering Rectifier Output / 9.10
Pulse Circuits / 9.12
A Relaxation Oscillator Using a PNPN Diode / 9.13
Pulse Synchronization / 9.15
Voltage Variation by AC Phase Control / 9.16
The Effect of Inductive Loads on Phase Angle Control / 9.18
Inverters / 9.19
Reference / 9.24

Chapter 10. Variable-Speed Drives   10.1

Basic Principles of AC Variable-Speed Drivers (VSDs) / 10.1
Inverters / 10.1
Input Power Converter (Rectifier) / 10.3
DC Link Energy / 10.4
Output IGBT Inverter / 10.4
Input Sources for Regeneration or Dynamic Slowdown / 10.5
Regeneration / 10.7
PWM-2 Considerations / 10.8
Transients, Harmonics Power Factor, and Failures / 10.9
Thyristor Failures and Testing / 10.11
AC Drive Application Issues / 10.12
AC Power Factor / 10.13
IGBT Switching Transients / 10.13
Cabling Details for AC Drives / 10.16
Cable Details / 10.16
Motor Bearing Currents / 10.17
Summary of Application Rules for AC Drives / 10.20
Selection Criteria of VSDs / 10.21
Regeneration / 10.21
Maintenance / 10.23
Common Failure Modes / 10.23
Motor Application Guidelines / 10.24
Reference / 10.24

Chapter 11. Synchronous Machines   11.1

Physical Description / 11.1
Pole Pitch: Electrical Degrees / 11.2
Air Gap and Magnetic Circuit of a Synchronous Machine / 11.2

Synchronous Machine Windings / *11.4*
Field Excitation / *11.5*
No-Load and Short-Circuit Values / *11.6*
Torque Tests / *11.7*
Excitation of a Synchronous Machine / *11.9*
Machine Losses / *11.10*
Reference / *11.12*

Chapter 12. Synchronous Generators *12.1*

Synchronous Generator Construction / *12.1*
The Speed of Rotation of a Synchronous Generator / *12.3*
The Internal Generated Voltage of a Synchronous Generator / *12.3*
The Equivalent Circuit of a Synchronous Generator / *12.6*
The Phasor Diagram of a Synchronous Generator / *12.10*
Power and Torque in Synchronous Generators / *12.11*
The Synchronous Generator Operating Alone / *12.13*
Parallel Operation of AC Generators / *12.15*
Operation of Generators in Parallel with Large Power Systems / *12.21*
Synchronous Generator Ratings / *12.25*
Synchronous Generator Capability Curves / *12.26*
Short-Time Operation and Service Factor / *12.28*
Reference / *12.29*

Chapter 13. Generator Components, Auxiliaries, and Excitation *13.1*

The Rotor / *13.1*
Turbine-Generator Components: The Stator / *13.17*
Cooling Systems / *13.31*
Shaft Seals and Seal Oil Systems / *13.35*
Stator Winding Water Cooling Systems / *13.38*
Other Cooling Systems / *13.43*
Excitation / *13.43*
The Voltage Regulator / *13.49*
The Power System Stabilizer / *13.53*
Characteristics of Generator Exciter Power (GEP) Systems / *13.55*
Generator Operation / *13.55*
Reference / *13.58*

Chapter 14. Generator Main Connections *14.1*

Introduction / *14.1*
Isolated Phase Bus Bar Circulatory Currents / *14.1*
System Description / *14.3*
Reference / *14.3*

Chapter 15. Performance and Operation of Generators *15.1*

Generator Systems / *15.1*
Condition Monitoring / *15.3*
Operational Limitations / *15.7*
Fault Conditions / *15.7*
Reference / *15.16*

CONTENTS   xi

Chapter 16. Generator Surveillance and Testing   16.1

Generator Operational Checks (Surveillance and Monitoring) / 16.1
Appendix A: Generator Diagnostic Testing / 16.2
Appendix B: Mechanical Tests / 16.18

Chapter 17. Generator Inspection and Maintenance   17.1

On-Load Maintenance and Monitoring / 17.1
Off-Load Maintenance / 17.4
Generator Testing / 17.7
Reference / 17.12

Chapter 18. Generator Operational Problems and Refurbishment Options   18.1

Typical Generator Operational Problems / 18.1
Generator Rotor Reliability and Life Expectancy / 18.7
Generator Rotor Refurbishment / 18.8
Types of Insulation / 18.8
Generator Rotor Modifications, Upgrades, and Uprates / 18.9
High-Speed Balancing / 18.9
Flux Probe Test / 18.11
References / 18.11

Chapter 19. Circuit Breakers   19.1

Theory of Circuit Interruption / 19.1
Physics of Arc Phenomena / 19.1
Circuit Breaker Rating / 19.3
Conventional Circuit Breakers / 19.4
Methods for Increasing Arc Resistance / 19.4
Plain Break Type / 19.4
Magnetic Blowout Type / 19.4
Arc Splitter Type / 19.6
Application / 19.6
Oil Circuit Breakers / 19.6
Recent Developments in Circuit Breakers / 19.9

Chapter 20. Fuses   20.1

Types of Fuses / 20.1
Features of Current-Limiting Fuses / 20.4
Advantages of Fuses over Circuit Breakers / 20.6
Appendix: Electrical System Protection Considerations / 20.6

Chapter 21. Bearings and Lubrication   21.1

Types of Bearings / 21.1
Statistical Nature of Bearing Life / 21.3
Materials and Finish / 21.3

Sizes of Bearings / 21.3
Types of Roller Bearings / 21.5
Thrust Bearings / 21.7
Lubrication / 21.7
References / 21.13

Chapter 22. Used-Oil Analysis 22.1

Proper Lube Oil Sampling Technique / 22.1
Test Description and Significance / 22.1
Summary / 22.7
Reference / 22.8

Chapter 23. Vibration Analysis 23.1

The Application of Sine Waves to Vibration / 23.1
Multimass Systems / 23.1
Resonance / 23.3
Logarithms and Decibels / 23.4
The Use of Filtering / 23.5
Vibration Instrumentation / 23.5
Time Domain / 23.6
Frequency Domain / 23.8
Machinery Example / 23.9
Vibration Analysis / 23.9
Resonant Frequency / 23.12
Vibration Severity / 23.12
A Case History: Condensate Pump Misalignment / 23.12

Chapter 24. Power Station Electrical Systems and Design Requirements 24.1

Introduction / 24.1
System Requirements / 24.1
Electrical System Description / 24.3
System Performance / 24.4
Power Plant Outages and Faults / 24.15
Uninterruptible Power Supply (UPS) Systems / 24.16
DC Systems / 24.16
References / 24.19

Chapter 25. Power Station Protective Systems 25.1

Introduction / 25.1
Design Criteria / 25.1
Generator Protection / 25.8
DC Tripping Systems / 25.8
Reference / 25.14

Appendix. Frequently Asked Questions A.1

Index I.1

PREFACE

Electrical Equipment Handbook provides a comprehensive understanding of the various types of motors, variable-speed drives, transformers, generators, rectifiers, inverters, circuit breakers, fuses, bearings, vibration analysis, used oil analysis, power station electrical systems and design requirements, and power station protective systems. All the fundamentals, basic design, operating characteristics, testing, inspection, specification, selection criteria, advanced fault detection techniques, critical components, troubleshooting, as well as all maintenance issues of these equipment and systems are covered in detail. The book provides the latest in technology. It covers how these equipment and systems operate, and provides guidelines and rules that must be followed to maximize their efficiency, reliability, and longevity and minimizes their capital, operating, and maintenance costs. The reader will be able to specify, select, commission, and maintain these equipment and systems for their applications.

The book was written for all technical individuals including engineers of all disciplines, managers, technicians, operators, and maintenance personnel. All the concepts are presented in a simple and practical manner that allows the reader to understand them without relying on advanced mathematical equations. This book is a *must* for anyone who is involved in the selection, application, or maintenance of electrical equipment and systems.

Philip Kiameh

ACKNOWLEDGMENTS

Reprinted from D. Halliday, and R. Resnick, *Physics, Part Two*, 3d ed., Wiley, Hoboken, N.J., 1978, with permission from Wiley: Figs. 1.1 to 1.21, Tables 1.1 and 1.2.

Reprinted from A. Nasar, *Handbook of Electric Machines*, McGraw-Hill, New York, 1987, with permission from McGraw-Hill: Figs. 1.27 to 1.32, 2.3, 2.6 to 2.11, 11.1 to 11.9, Tables 2.1 to 2.3.

Reprinted from S.J. Chapman, *Electric Machinery Fundamentals*, 2d ed., McGraw-Hill, New York, 1991, with permission from McGraw-Hill: Figs. 2.1, 2.2, 2.4, 2.5, 2.12, 2.13, 3.1 to 3.25, 5.1 to 5.9, 6.6 to 6.17, 6.19 to 6.25, 7.1 to 7.3, 7.5 to 7.11, 9.1 to 9.17, 9.19 to 9.36, 10.6, 10.8, 12.1a, 12.3 to 12.5, 12.7 to 12.20, 12.22 to 12.34, Example 3.1..

Reprinted with permission from VA Tech Ferranti-Packard Transformers Ltd.: Figs. 4.2, 4.4, 4.5.

Reprinted from S. D. Myers, , and J. J. Kelly, *Transformer Maintenance Guide*, Transformer Maintenance, Institute, Division of S.D. Myers, Inc., Cuyahoga Falls, Ohio with permission from S.D. Myers: Figs. 4.8 to 4.12, Tables 4.2 to 4.4.

Reprinted with permission from *Failures in Three-Phase Stator Windings*, Electrical Apparatus Association, Inc., St. Louis, Mo.: Figs. 8.7 to 8.19.

Reprinted from L.R. Higgins, *Maintenance Engineering*, 5th ed., McGraw-Hill, New York, 1995, with permission from McGraw-Hill: Figs. 8.1 and 8.2, Tables 8.1 and 8.2.

Reprinted from A.C. Stevenson, *Power Converter Application Handbook*, Institute of Electrical and Electronics Engineers, New York, 1999, with permission from A.C. Stevenson: Figs. 10.1 to 10.5, 10.9 to 10.21.

Reprinted with permission from National Electrical Manufacturers Association: Table 6.1.

Reprinted with permission from General Electric: Figs. 6.2, 6.3b, 6.4, 12.1b and c, 12.6, 18.1 to 18.10.

Reprinted with permission from Siemens-Westinghouse Power Corp.: Fig. 12.1d.

Reprinted with permission from Magnetek: Figs. 6.1, 6.3a, 6.5, 6.18, 7.4, 7.12, 9.18.

Reprinted with permission from Armco: Figs. 2.6 and 2.7.

Reprinted from British Electricity International, *Modern Power Station Practice*, vol. C, *Turbines, Generators and Associated Plant*, 3d ed., Pergamon Press, U.K., 1991, with permission from Elsevier Science: Figs. 13.1 to 13.46.

Reprinted from British Electricity International, *Modern Power Station Practice*, vol. D, *Electrical Systems and Equipment*, 3d ed., Pergamon Press, Oxford, U.K., 1992, with permission from Elsevier Science: Figs. 14.1 to 14.4, 24.1 to 24.8, 25.1 to 25.7.

Reprinted from British Electricity International, *Modern Power Station Practice*, vol. G, *Station Operation and Maintenance*, 3d ed., Pergamon Press, Oxford, U.K., 1991, with permission from Elsevier Science: Figs. 15.1 to 15.9, 17.1 to 17.4, Table 15.1.

Reprinted from R.W. Smeaton, *Switchgear and Control Handbook*, McGraw-Hill, New York, 1987, with permission from McGraw-Hill: Figs. 19.1 to 19.10.

Reprinted from NOR-AM, with permission from Ferraz Shawmut: Figs. 20.1 to 20.7.

Reprinted from I. J. Karassik, *Pump Handbook*, 2d ed., McGraw-Hill, New York, 1986, with permission from McGraw-Hill: Figs. 21.16 to 21.20.

Reprinted from V.M. Faires, *Design of Machine Elements*, 4th ed., Collier-MacMillan, Toronto, 1965, with permission from Collier-MacMillan: Figs. 21.1 and 21.2, Tables 21.1 and 21.2.

Reprinted with permission from SKF Industries: Figs. 21.4, 21.5, and 21.14.

Reprinted with permission from Torrington: Fig. 21.7.

Reprinted with permission from Timken: Figs. 21.8 and 21.11.

Reprinted with permission from Aetna: Fig. 21.10.Reprinted with permission from Texaco: Figs. 21.12, 21.13, 21.1 to 21.4, 21.6 to 21.11, Table 22.1.

Reprinted with permission from Noria Corp.: Fig. 22.6.

Reprinted from E. Marshall, *Used Oil Analysis – AVital Part of Maintenance*, Texaco technical publication *Lubrication*, vol.79, no. 2, 1993, with permission from Texaco: Figs. 22.1 to 22.5, 22.7, Table 22.1.

CHAPTER 1
FUNDAMENTALS OF ELECTRIC SYSTEMS

CAPACITORS

Figure 1.1 illustrates a capacitor. It consists of two insulated conductors a and b. They carry equal and opposite charges $+q$ and $-q$, respectively. All lines of force that originate on a terminate on b. The capacitor is characterized by the following parameters:

- q, the magnitude of the charge on each conductor
- V, the potential difference between the conductors

The parameters q and V are proportional to each other in a capacitor, or $q = CV$, where C is the constant of proportionality. It is called the *capacitance* of the capacitor. The capacitance depends on the following parameters:

- Shape of the conductors
- Relative position of the conductors
- Medium that separates the conductors

The unit of capacitance is the coulomb/volt (C/V) or farad (F). Thus

$$1\ F = 1\ C/V$$

It is important to note that

$$\frac{dq}{dt} = C\frac{dV}{dt}$$

but since

$$\frac{dq}{dt} = i$$

Thus,

$$i = C\frac{dV}{dt}$$

This means that the current in a capacitor is proportional to the rate of change of the voltage with time.

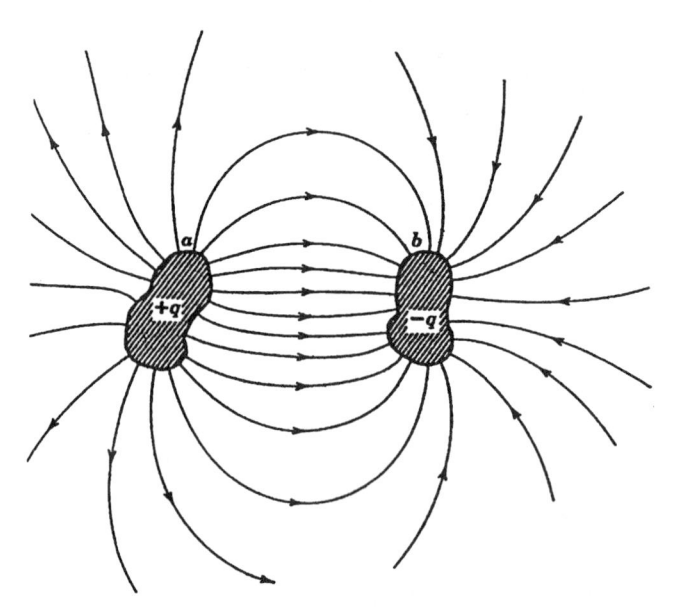

FIGURE 1.1 Two insulated conductors, totally isolated from their surroundings and carrying equal and opposite charges, form a capacitor.

In industry, the following submultiples of farad are used:

- Microfarad (1 μF = 10^{-6} F)
- Picofarad (1 pF = 10^{-12} F)

Capacitors are very useful electric devices. They are used in the following applications:

- To store energy in an electric field. The energy is stored between the conductors, which are normally called plates. The electric energy stored in the capacitor is given by

$$U_E = \frac{1}{2}\frac{q^2}{C}$$

- To reduce voltage fluctuations in electronic power supplies
- To transmit pulsed signals
- To generate or detect electromagnetic oscillations at radio frequencies
- To provide electronic time delays

Figure 1.2 illustrates a parallel-plate capacitor in which the conductors are two parallel plates of area A separated by a distance d. If each plate is connected momentarily to the terminals of the battery, a charge $+q$ will appear on one plate and a charge $-q$ on the other. If d is relatively small, the electric field **E** between the plates will be uniform.

The capacitance of a capacitor increases when a dielectric (insulation) is placed between the plates. The *dielectric constant* κ of a material is the ratio of the capacitance with

FUNDAMENTALS OF ELECTRIC SYSTEMS 1.3

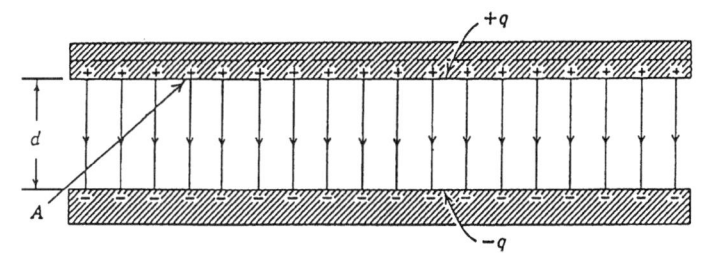

FIGURE 1.2 A parallel-plate capacitor with conductors (plates) of area A.

TABLE 1.1 Properties of Some Dielectrics*

Material	Dielectric constant	Dielectric strength,[†] kV/mm
Vacuum	1.00000	∞
Air	1.00054	0.8
Water	78	—
Paper	3.5	14
Ruby mica	5.4	160
Porcelain	6.5	4
Fused quartz	3.8	8
Pyrex glass	4.5	13
Bakelite	4.8	12
Polyethylene	2.3	50
Amber	2.7	90
Polystyrene	2.6	25
Teflon	2.1	60
Neoprene	6.9	12
Transformer oil	4.5	12
Titanium dioxide	100	6

*These properties are at approximately room temperature and for conditions such that the electric field **E** in the dielectric does not vary with time.
[†]This is the maximum potential gradient that may exist in the dielectric without the occurrence of electrical breakdown. Dielectrics are often placed between conducting plates to permit a higher potential difference to be applied between them than would be possible with air as the dielectric.

dielectric to that without dielectric. Table 1.1 illustrates the dielectric constant and dielectric strength of various materials.

The high dielectric strength of vacuum (∞, infinity) should be noted. It indicates that if two plates are separated by vacuum, the voltage difference between them can reach infinity without having flashover (arcing) between the plates. This important characteristic of vacuum has led to the development of vacuum circuit breakers, which have proved to have excellent performance in modern industry.

CURRENT AND RESISTANCE

The *electric current i* is established in a conductor when a net charge q passes through it in time t. Thus, the current is

$$i = \frac{q}{t}$$

The units for the parameters are

- *i:* amperes (A)
- *q:* coulombs (C)
- *t:* seconds (s)

The *electric field* exerts a force on the electrons to move them through the conductor. A positive charge moving in one direction has the same effect as a negative charge moving in the opposite direction. Thus, for simplicity we assume that all charge carriers are positive. We draw the current arrows in the direction that positive charges flow (Fig. 1.3).

A conductor is characterized by its resistance (symbol ⌇⌇⌇). It is defined as the voltage difference between two points divided by the current flowing through the conductor. Thus,

$$R = \frac{V}{i}$$

where V is in volts, i is in amperes, and the resistance R is in ohms (abbreviated Ω).

The current, which is the flow of charge through a conductor, is often compared to the flow of water through a pipe. The water flow occurs due to the pressure difference between the inlet and outlet of a pipe. Similarly, the charge flows through the conductor due to the voltage difference.

The resistivity ρ is a characteristic of the conductor material. It is a measure of the resistance that the material has to the current. For example, the resistivity of copper is 1.7×10^{-8} $\Omega \cdot$m; that of fused quartz is about 10^{16} $\Omega \cdot$m. Table 1.2 lists some electrical properties of common metals.

The temperature coefficient of resistivity α is given by

$$\alpha = \frac{1}{\rho} \frac{d\rho}{dT}$$

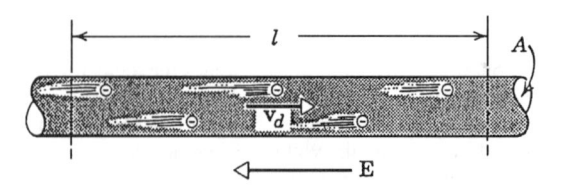

FIGURE 1.3 Electrons drift in a direction opposite to the electric field in a conductor.

FUNDAMENTALS OF ELECTRIC SYSTEMS 1.5

It represents the rate of variation of resistivity with temperature. Its units are 1/°C (or 1/°F). Conductivity (σ), is used more commonly than resistivity. It is the inverse of conductivity, given by

$$\sigma = \frac{1}{\rho}$$

The units for conductivity are $(\Omega \cdot m)^{-1}$.

Across a resistor, the voltage and current have this relationship:

$$V = iR$$

The power dissipated across the resistor (conversion of electric energy to heat) is given by

$$P = i^2 R$$

or

$$P = \frac{V^2}{R}$$

where P is in watts, i in amperes, V in volts, and R in ohms.

TABLE 1.2 Properties of Metals as Conductors

Metal	Resistivity (at 20°C), $10^{-8}\ \Omega \cdot m$	Temperature coefficient of resistivity α, per C° ($\times 10^{-5}$)*
Silver	1.6	380
Copper	1.7	390
Aluminum	2.8	390
Tungsten	5.6	450
Nickel	6.8	600
Iron	10	500
Steel	18	300
Manganin	44	1.0
Carbon[†]	3500	−50

*This quantity, defined from

$$\alpha = \frac{1}{\rho}\frac{d\rho}{dT}$$

is the fractional change in resistivity $d\rho/\rho$ per unit change in temperature. It varies with temperature, the values here referring to 20°C. For copper ($\alpha = 3.9 \times 10^{-3}$/°C) the resistivity increases by 0.39 percent for a temperature increase of 1°C near 20°C. Note that α for carbon is negative, which means that the resistivity *decreases* with increasing temperature.

[†]Carbon, not strictly a metal, is included for comparison.

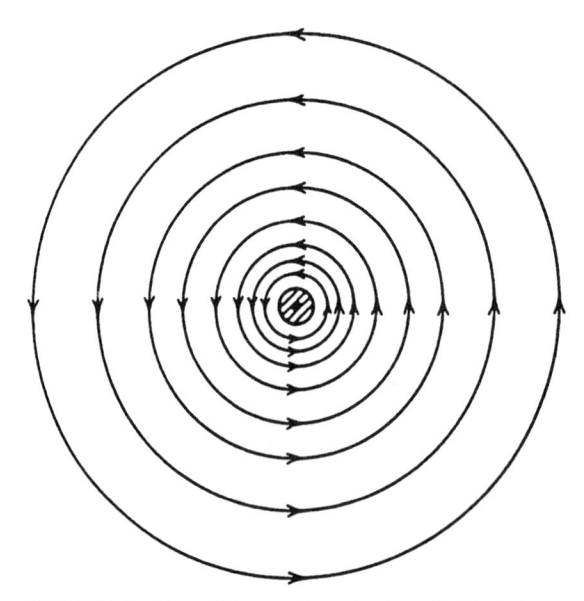

FIGURE 1.4 Lines of **B** near a long, circular cylindrical wire. A current *i*, suggested by the central dot, emerges from the page.

THE MAGNETIC FIELD

A magnetic field is defined as the space around a magnet or a current-carrying conductor. The magnetic field **B** is represented by lines of induction. Figure 1.4 illustrates the lines of induction of a magnetic field **B** near a long current-carrying conductor.

The vector of the magnetic field is related to its lines of induction in this way:

1. The direction of **B** at any point is given by the tangent to the line of induction.
2. The number of lines of induction per unit cross-sectional area (perpendicular to the lines) is proportional to the magnitude of **B**. Magnetic field **B** is large if the lines are close together, and it is small if they are far apart.

The flux Φ_B of magnetic field **B** is given by

$$\Phi_B = \int \mathbf{B} \cdot d\mathbf{S}$$

The integral is taken over the surface for which Φ_B is defined.

The magnetic field exerts a force on any charge moving through it. If q_0 is a positive charge moving at a velocity **v** in a magnetic field **B**, the force **F** acting on the charge (Fig. 1.5) is given by

$$\mathbf{F} = q_0 \mathbf{v} \times \mathbf{B}$$

The magnitude of the force **F** is given by

$$F = q_0 v B \sin \theta$$

where θ is the angle between **v** and **B**.

The force **F** will always be at a right angle to the plane formed by **v** and **B**. Thus, it will always be a sideways force. The force will disappear in these cases:

1. If the charge stops moving
2. If **v** is parallel or antiparallel to the direction of **B**

The force **F** has a maximum value if **v** is at a right angle to **B** ($\theta = 90°$).

Figure 1.6 illustrates the force created on a positive and a negative electron moving in a magnetic field **B** pointing out of the plane of the figure (symbol ⊙). The unit of **B** is the tesla (T) or weber per square meter (Wb/m^2). Thus

$$1 \text{ tesla (T)} = 1 \text{ weber/meter}^2 = \frac{1 \text{ N}}{\text{A} \cdot \text{m}}$$

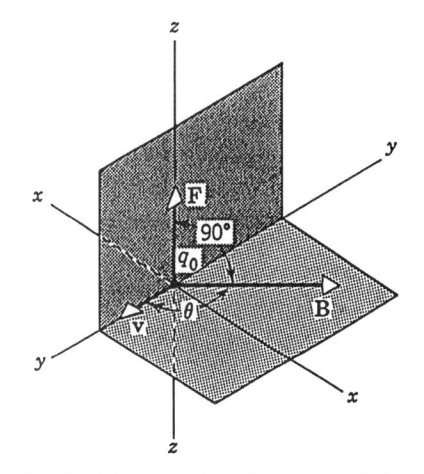

FIGURE 1.5 Illustration of $\mathbf{F} = q_0 \mathbf{v} \times \mathbf{B}$. Test charge q_0 is fired through the origin with velocity **v**.

The force acting on a current-carrying conductor placed at a right angle to a magnetic field **B** (Fig. 1.7) is given by

$$F = ilB$$

where l is the length of conductor placed in the magnetic field.

Ampère's Law

Figure 1.8 illustrates a current-carrying conductor surrounded by small magnets. If there is no current in the conductor, all the magnets will be aligned with the horizontal component

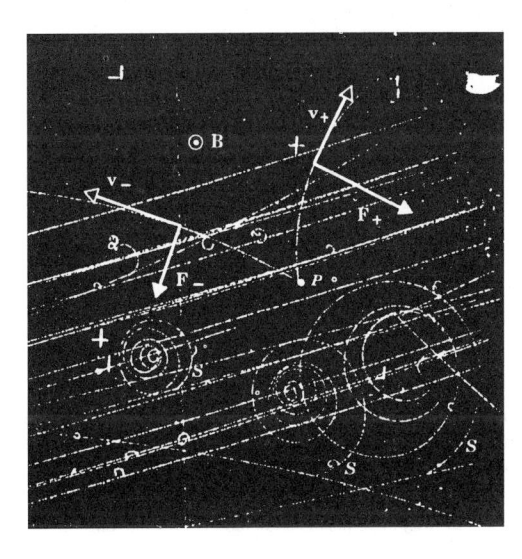

FIGURE 1.6 A *bubble chamber* is a device for rendering visible, by means of small bubbles, the tracks of charged particles that pass through the chamber. The figure shows a photograph taken with such a chamber immersed in a magnetic field **B** and exposed to radiations from a large cyclotronlike accelerator. The curved υ at point *P* is formed by a positive and a negative electron, which deflect in opposite directions in the magnetic field. The spirals *S* are tracks of three low-energy electrons.

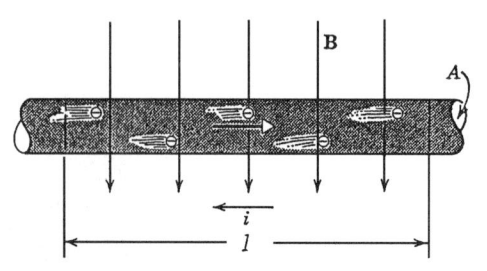

FIGURE 1.7 A wire carrying a current i is placed at right angles to a magnetic field **B**. Only the drift velocity of the electrons, not their random motion, is suggested.

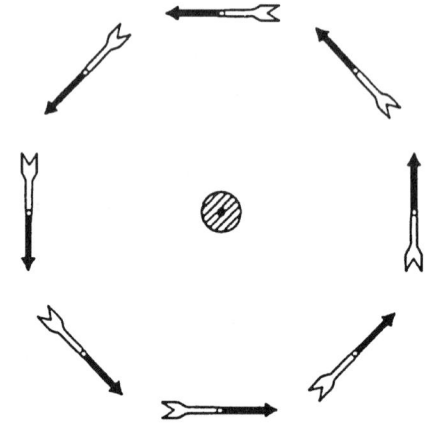

FIGURE 1.8 An array of compass needles near a central wire carrying a strong current. The black ends of the compass needles are their north poles. The central dot shows the current emerging from the page. As usual, the direction of the current is taken as the direction of flow of positive charge.

of the earth's magnetic field. When a current flows through the conductor, the orientation of the magnets suggests that the lines of induction of the magnetic field form closed circles around the conductor. This observation is reinforced by the experiment shown in Fig. 1.9. It shows a current-carrying conductor passing through the center of a horizontal glass plate with iron filings on it.

Ampère's law states that

$$\oint \mathbf{B} \cdot d\mathbf{l} = \mu_0 i$$

where **B** is the magnetic field, **l** is the length of the circumference around the wire, i is the current, μ_0 is the permeability constant ($\mu_0 = 4\pi \times 10^{-7}$ T·m/A). The integration is carried around the circumference.

If the current in the conductor shown in Fig. 1.8 is reverse direction, all the compass needles change their direction as well. Thus, the direction of **B** near a current-carrying conductor is given by the *right-hand-rule:*

> If the current is grasped by the right hand and the thumb points in the direction of the current, the fingers will curl around the wire in the direction **B**.

FUNDAMENTALS OF ELECTRIC SYSTEMS   1.9

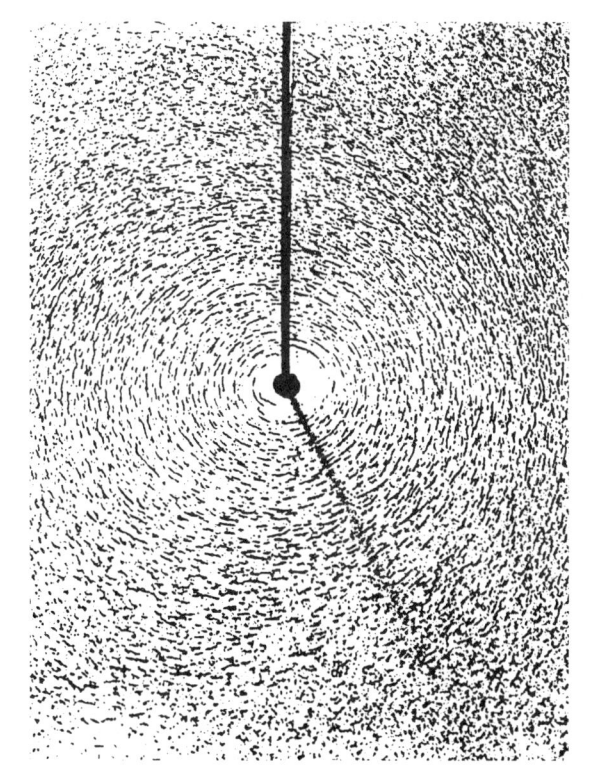

FIGURE 1.9 Iron filings around a wire carrying a strong current.

Magnetic Field in a Solenoid

A solenoid (an inductor) is a long, current-carrying conductor wound in a close-packed helix. Figure 1.10 shows a "solenoid" having widely spaced turns. The fields cancel between the wires. Inside the solenoid, **B** is parallel to the solenoid axis. Figure 1.11 shows the lines of **B** for a real solenoid. By applying Ampere's law to this solenoid, we have

$$B = \mu_0 in$$

where n is the number of turns per unit length. The flux Φ_B for the magnetic field **B** will become

$$\Phi_B = \mathbf{B} \cdot \mathbf{A}$$

FARADAY'S LAW OF INDUCTION

Faraday's law of induction is one of the basic equations of electromagnetism. Figure 1.12 shows a coil connected to a galvanometer. If a bar magnet is pushed toward the coil, the galvanometer deflects. This indicates that a current has been induced in the coil. If the magnet is held stationary with respect to the coil, the galvanometer does not deflect. If the magnet is moved away from the coil, the galvanometer deflects in the opposite direction. This indicates that the current induced in the coil is in the opposite direction.

1.10 CHAPTER ONE

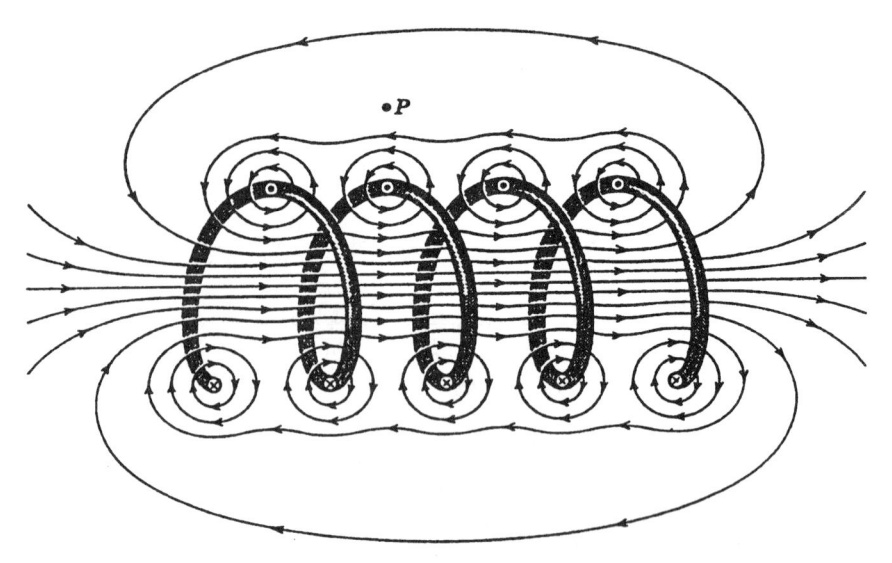

FIGURE 1.10 A loosely wound solenoid.

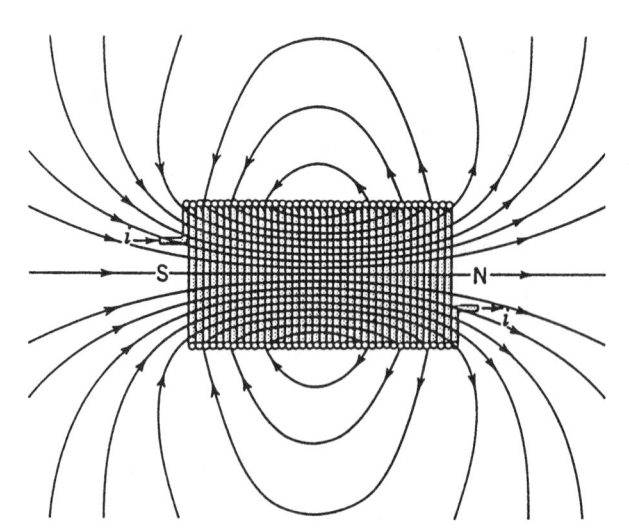

FIGURE 1.11 A solenoid of finite length. The right end, from which lines of **B** emerge, behaves as the north pole of a compass needle does. The left end behaves as the south pole.

Figure 1.13 shows another experiment in which when the switch S is closed, thus establishing a steady current in the right-hand coil, the galvanometer deflects momentarily. When the switch is opened, the galvanometer deflects again momentarily, but in the opposite direction. This experiment proves that a voltage known as an electromagnetic force (emf) is induced in the left coil when the current in the right coil changes.

FUNDAMENTALS OF ELECTRIC SYSTEMS   1.11

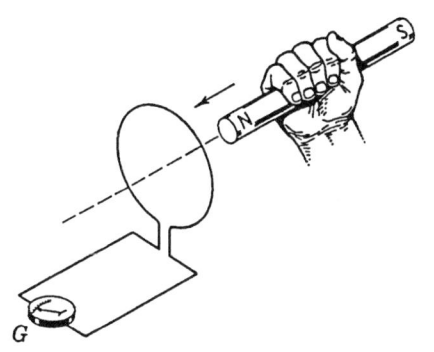

FIGURE 1.12 Galvanometer G deflects while the magnet is moving with respect to the coil. Only their relative motion counts.

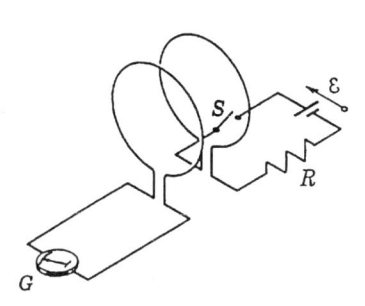

FIGURE 1.13 Galvanometer G deflects momentarily when switch S is closed or opened. No motion is involved.

Faraday's law of induction is given by

$$\mathscr{E} = -N \frac{d\Phi_B}{dt}$$

where $\mathscr{E}$ = emf for voltage
N = number of turns in coil
$d\Phi_B/dt$ = rate of change of flux with time

The minus sign will be explained by Lenz' law.

LENZ'S LAW

Lenz's law states that *the induced current will be in a direction that opposes the change that produced it.* If a magnet is pushed toward a loop as shown in Fig. 1.14, an induced current will be established in the loop. Lenz's law predicts that the current in the loop must be in a direction such that the flux established by it must oppose the change. Thus, the face of the loop toward the magnet must have the north pole. The north pole from the current loop and the north pole from the magnet will repel each other. The right-hand rule indicates that the magnetic field established by the loop should emerge from the right side of the loop. Thus, the induced current must be as shown. Lenz's law can be explained as follows: When the magnet is pushed toward the loop, this "change" induces a current. The direction of this current should oppose the "push." If the magnet is pulled away from the coil, the induced current will create the south pole on the right-hand face of the loop because this will oppose the "pull." Thus, the current must be in the opposite direction to the one shown in Fig. 1.14 to make the right-hand face a south pole. Whether the magnet is pulled or pushed, its motion will always be opposed. The force that moves the magnet will always experience a resisting force. Thus, the force moving the magnet will always be required to do work.

Figure 1.15 shows a rectangular loop of width l. One end of it has a uniform field **B** pointing at a right angle to the plane of the loop into the page ($\otimes$ indicates into the page and $\odot$ out of the page). The flux enclosed by the loop is given by

$$\Phi_B = Blx$$

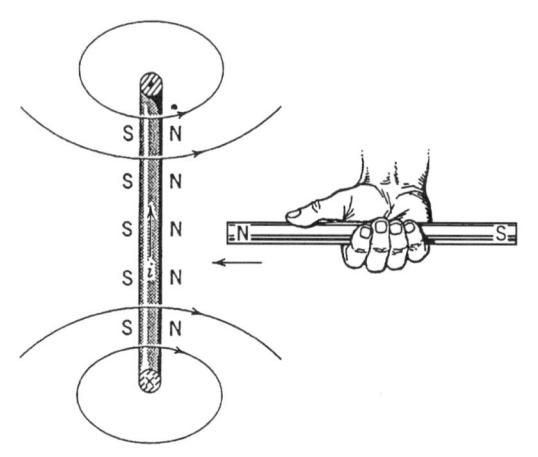

FIGURE 1.14 If we move the magnet toward the loop, the induced current points as shown, setting up a magnetic field that opposes the motion of the magnet.

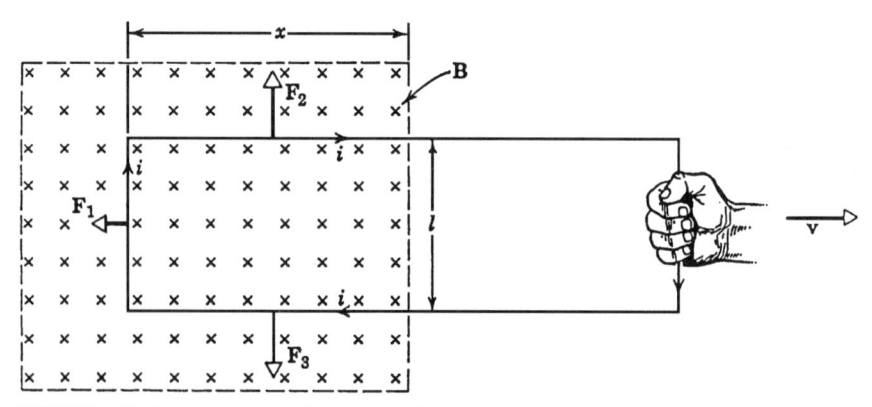

FIGURE 1.15 A rectangular loop is pulled out of a magnetic field with velocity **v**.

Faraday's law states that the induced voltage or emf $\mathcal{E}$ is given by

$$\mathcal{E} = -\frac{d\Phi_B}{dt} = -\frac{d}{dt}(Blx) = -Bl\frac{dx}{dt} = Blv$$

where $-dx/dt$ is the velocity v of the loop being pulled out of the magnetic field. The current induced in the loop is given by

$$i = \frac{\mathcal{E}}{R} = \frac{Blv}{R}$$

where R is the loop resistance. From Lenz's law, this current must be clockwise because it is opposing the *change* (the decrease in Φ_B). It establishes a magnetic field in the same direction as the external magnetic field within the loop. Forces $\mathbf{F}_2$ and $\mathbf{F}_3$ cancel each other because they are equal and in opposite directions. Force $\mathbf{F}_1$ is obtained from the equation ($\mathbf{F} = i\mathbf{l} \times \mathbf{B}$)

$$F_1 = ilB \sin 90° = \frac{B^2l^2v}{R}$$

The force pulling the loop must do a steady work given by

$$P = F_1 v = \frac{B^2l^2v^2}{R}$$

Figure 1.16 illustrates a rectangular loop of resistance R, width l, and length a being pulled at constant speed v through a magnetic field **B** of thickness d. There is no flux Φ_B when the loop is not in the field. The flux Φ_B is Bla when the loop is entirely in the field. It is Blx when the loop is entering the field. The induced voltage or emf $\mathscr{E}$ in the loop is given by

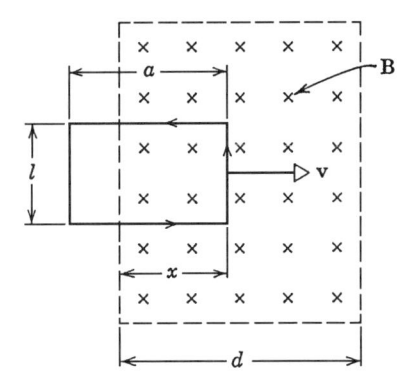

FIGURE 1.16 A rectangular loop is caused to move with a velocity **v** through a magnetic field. The position of the loop is measured by x, the distance between the effective left edge of field **B** and the right end of the loop.

$$\mathscr{E} = -\frac{d\Phi_B}{dt} = -\frac{d\Phi_B}{dx}\frac{dx}{dt} = -\frac{d\Phi_B}{dx}v$$

where $d\Phi_B/dx$ is the slope of the curve shown in Fig. 1.17a.

The voltage $\mathscr{E}(x)$ is shown in Fig. 1.17b. Lenz's law indicates that $\mathscr{E}(x)$ is counterclockwise. There is no voltage induced in the coil when it is entirely in the magnetic field because the flux Φ_B through the coil does not change with time. Figure 1.17c shows the rate P of thermal energy generation in the loop, and P is given by

$$P = \frac{\mathscr{E}^2}{R}$$

If a real magnetic field is considered, its strength will decrease from the center to the peripheries. Thus, the sharp bends and corners shown in Fig. 1.17 will be replaced by smooth curves. The voltage $\mathscr{E}$ induced in this case will be given by $\mathscr{E}_{max} \sin \omega t$ (a sine wave). This is exactly how ac voltage is induced in a real generator. Also note that the prime mover has to do significant work to rotate the generator rotor inside the stator.

INDUCTANCE

When the current in a coil changes, an induced voltage appears in that same coil. This is called *self-induction*. The voltage (electromagnetic force) induced is called *self-induced emf*. It obeys Faraday's law of induction as do any other induced emf's. For a closed-packed coil (an inductor) we have

$$N\Phi_B = Li$$

where N = number of turns of coil
 Φ_B = flux
 i = current
 L = *inductance* of the device

1.14 CHAPTER ONE

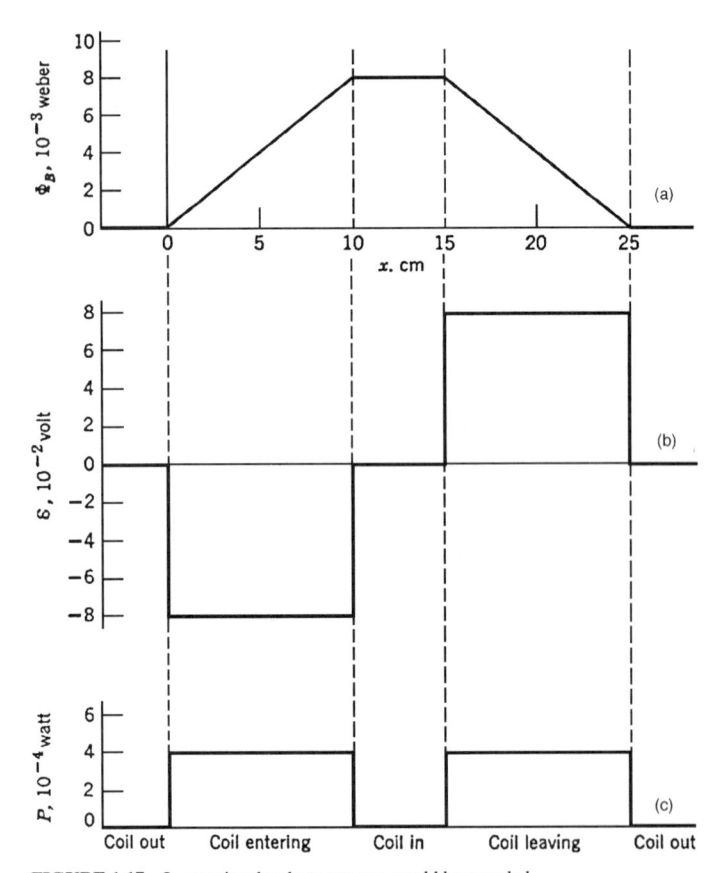

FIGURE 1.17 In practice the sharp corners would be rounded.

From Faraday's law, we can write the induced voltage (emf) as

$$\mathscr{E} = -\frac{d(N\Phi_B)}{dt} = -L\frac{di}{dt}$$

This relationship can be used for inductors of all shapes and sizes. In an inductor (symbol ⏦), L depends only on the geometry of the device. The unit of inductance is the henry (abbreviated H). It is given by

$$1 \text{ henry (H)} = \frac{1 \text{ volt} \cdot \text{second}}{\text{ampere}} \left(\frac{V \cdot s}{A}\right)$$

In an inductor, energy is stored in a magnetic field. The amount of magnetic energy stored U_B in the inductor is given by

$$U_B = \tfrac{1}{2} L i^2$$

In summary, an inductor stores energy in a magnetic field, and the capacitor stores energy in an electric field.

ALTERNATING CURRENT

An alternating current (ac) in a circuit establishes a voltage (emf) that varies with time as

$$\mathcal{E} = \mathcal{E}_m \sin \omega t$$

where $\mathcal{E}_m$ is the maximum emf and $\omega = 2\pi\nu$, where ν is the frequency measured in hertz (Hz). This type of emf is established by an ac generator in a power plant. In North America, $\nu = 60$ Hz. In western Europe and Australia it is 50 Hz. The symbol for a source of alternating emf is ⊙. This device is called an *alternating-current generator* or an ac generator. Alternating currents are essential for modern society. Power distribution systems, radio, television, satellite communication systems, computer systems, etc. would not exist without alternating voltages and currents.

The alternating current in the circuit shown in Fig. 1.18 is given by

$$i = i_m \sin(\omega t - \theta)$$

where i_m = maximum amplitude of current
ω = angular frequency of applied alternating voltage (or emf)
θ = phase angle between alternating current and alternating voltage

Let us consider each component of the circuit separately.

A Resistive Circuit

Figure 1.19a shows an alternating voltage applied across a resistor. We can write the following equations:

$$V_R = \mathcal{E}_m \sin \omega t$$

and

$$V_R = i_R R$$

or

$$i_R = \frac{\mathcal{E}_m}{R} \sin \omega t$$

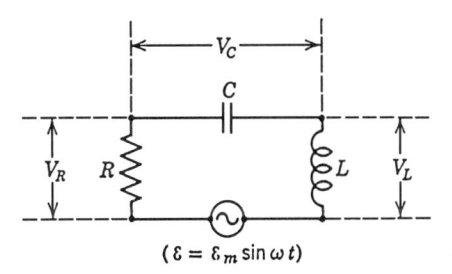

FIGURE 1.18 A single-loop *RCL* circuit contains an ac generator. Voltages V_R, V_C, and V_L are the time-varying potential differences across the resistor, the capacitor, and the inductor, respectively.

1.16 CHAPTER ONE

($\mathcal{E} = \mathcal{E}_m \sin \omega t$)
(a)

(b)

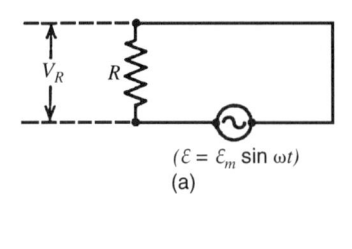

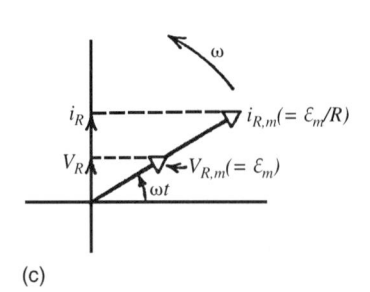

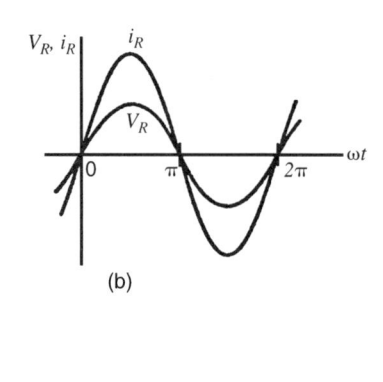

(c)

FIGURE 1.19 (a) A single-loop resistive circuit containing an ac generator. (b) The current and the potential difference across the resistor are in phase ($\theta = 0$). (c) A phasor diagram shows the same thing. The arrows on the vertical axis are *instantaneous* values.

A comparison between the previous equations shows that the time-varying (instantaneous) quantities V_R and i_R are *in phase*. This means that they reach their maximum and minimum values at the same time. They also have the same angular frequency ω. These facts are shown in Fig. 1.19b and c.

Figure 1.19c illustrates a *phasor diagram*. It is another method used to describe the situation. The phasors in this diagram are represented by open arrows. They rotate counterclockwise with an angular frequency ω about the origin. The phasors have the following properties:

1. The length of the phasor is proportional to the *maximum* value of the alternating quantity described, that is, $\mathcal{E}_m$ for V_R and $\mathcal{E}_m/R$ for i_R.
2. The projection of the phasors on the vertical axis gives the *instantaneous* values of the alternating parameter (current or voltage) described. Thus, the arrows on the vertical axis represent the instantaneous values of V_R and i_R. Since V_R and i_R are in phase, their phasors lie along the same line (Fig. 1.19c).

A Capacitive Circuit

Figure 1.20a illustrates an alternating voltage acting on a capacitor. We can write the following equations:

$$V_c = \mathcal{E}_m \sin \omega t$$

and

$$V_c = \frac{q}{C} \quad \text{(definition of } C\text{)}$$

FUNDAMENTALS OF ELECTRIC SYSTEMS   1.17

From these relationships, we have

$$q = \mathcal{E}_m C \sin \omega t$$

or

$$i_c = \frac{dq}{dt} = \omega C \mathcal{E}_m \cos \omega t$$

A comparison between these equations shows that the instantaneous values of V_c and i_c are one-quarter cycle out of phase. This is illustrated in Fig. 1.20b.

Voltage V_c lags i_c; that is, as time passes, V_c reaches its maximum after i_c does, by one-quarter cycle (90°). This is also shown clearly in the phasor diagram (Fig. 1.20c). Since the phasors rotate in counterclockwise direction, it is clear that phasor $V_{c,m}$ lags behind phasor $i_{c,m}$ by one-quarter cycle. The reason for this lag is that the capacitor stores energy in its electric field. The current goes through it before the voltage is established across it. Since the current is given by

$$i = i_m \sin(\omega t - \theta)$$

θ is the angle between V_c and i_c. In this case, it is equal to $-90°$. If we put this value of θ in the equation of current, we obtain

$$i = i_m \cos \omega t$$

This equation is in agreement with the previous equation for current that we obtained,

$$i_c = \frac{dq}{dt} = \omega C \mathcal{E}_m \cos \omega t$$

where $i_m = \omega C \mathcal{E}_m$.

Also i_c is expressed as follows:

$$i_c = \frac{\mathcal{E}_m}{x_c} \cos \omega t$$

and x_c is called the *capacitive reactance*. Its unit is the ohm (Ω). Since the maximum value of $V_c = V_{c,m}$ and the maximum value of $i_c = i_{c,m}$ we can write

$$V_{c,m} = i_{c,m} x_c$$

Voltage $V_{c,m}$ represents the maximum voltage established across the capacitor when the current is i.

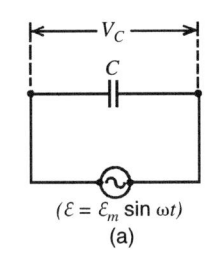

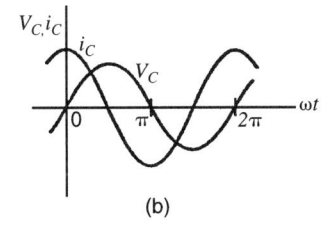

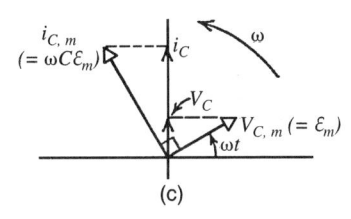

FIGURE 1.20 (a) A single-loop capacitive circuit containing an ac generator. (b) The potential difference across the capacitor lags the current by one-quarter cycle. (c) A phasor diagram shows the same thing. The arrows on the vertical axis are instantaneous values.

An Inductive Circuit

Figure 1.21a shows a circuit containing an alternating voltage acting on an inductor. We can write the following equations:

$$V_L = \mathcal{E}_m \sin \omega t$$

and

$$V_L = L \frac{di}{dt} \quad \text{(from definition of L)}$$

From these equations, we have

$$di = \frac{\mathcal{E}_m}{L} \sin \omega\, dt$$

or

$$i_L = \int di = -\frac{\mathcal{E}_m}{\omega L} \cos \omega t$$

A comparison between the instantaneous values of V_L and i_L shows that these parameters are out of phase by one-quarter cycle (90°). This is illustrated in Fig. 1.21b. It is clear that V_L leads i_L. This means that as time passes, V_L reaches its maximum before i_L does, by one-quarter cycle.

This fact is also shown in the phasor diagram of Fig. 1.21c. As the phasors rotate in the counterclockwise direction, it is clear that phasor $V_{L,m}$ leads (precedes) $i_{L,m}$ by one-quarter cycle.

The phase angle θ by which V_L leads i_L in this case is +90°. If this value is put in the current equation

$$i = i_m \sin(\omega t - \theta)$$

we obtain

$$i = -i_m \cos \omega t$$

This equation is in agreement with the previous equation of the current:

$$i_L = \int di = -\frac{\mathcal{E}_m}{\omega L} \cos \omega t$$

Again, for reasons of compactness of notation, we rewrite the equation as

$$i_L = -\frac{\mathcal{E}_m}{X_L} \cos \omega t$$

where

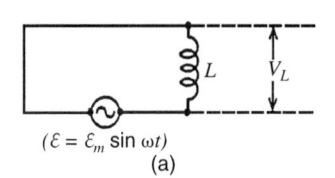

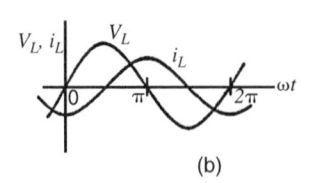

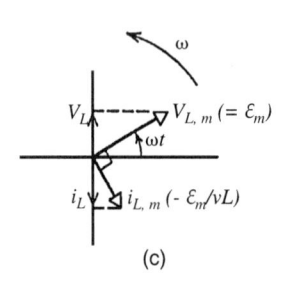

$$X_L = \omega L$$

FIGURE 1.21 (a) A single-loop inductive circuit containing an ac generator. (b) The potential difference across the inductor *leads* the current by one-quarter cycle. (c) A phasor diagram shows the same thing. The arrows on the vertical axis are instantaneous values.

FUNDAMENTALS OF ELECTRIC SYSTEMS

and X_L is called the inductive reactance. As for the capacitive reactance, the unit for X_L is the ohm. Since $\mathcal{E}_m$ is the maximum value of V_L (= $V_{L,m}$), we can write

$$V_{L,m} = i_{L,m} X_L$$

This indicated that when any alternating current of amplitude i_m and angular frequency ω exists in an inductor, the maximum voltage difference across the inductor is given by

$$V_{L,m} = i_m X_L$$

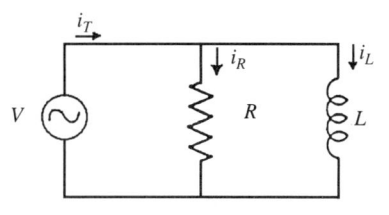

FIGURE 1.22 Circuit containing a resistor and an inductor.

Let us now examine the circuit shown in Fig. 1.22. Figure 1.23 illustrates the phasor diagram of the circuit. The total current is $\mathbf{i}_T = \mathbf{i}_R + \mathbf{i}_L$, and θ represents the angle between $\mathbf{i}_T$ and the voltage $\mathbf{V}$. It is called the *phase angle* of the system. An increase in the value of the inductance L will result in increasing the angle θ. The power factor (abbreviation PF) is defined as

$$PF = \cos \theta$$

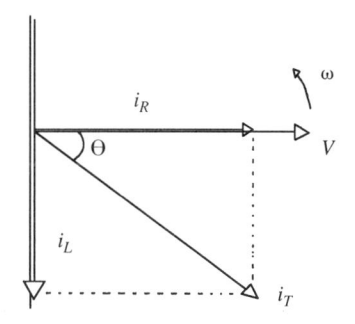

FIGURE 1.23 A phasor diagram of the circuit in Fig. 1.22.

It is a measure of the ratio of the magnitudes of $|\mathbf{i}_R|/|\mathbf{i}_T|$.

The circuit shown in Fig. 1.22 shows that the load supplied by a power plant has two natures $\mathbf{i}_R$ and $\mathbf{i}_L$. Equipment such as motors, welders, and fluorescent lights require both types of currents. However, equipment such as heaters and incandescent bulbs require the resistive current i_R only.

The power in the resistive part of the circuit is given by

$$P = Vi_R \quad \text{or} \quad P = Vi_T \cos \theta$$

This is the *real power* in the circuit. It is the energy dissipated by the resistor. This is the energy converted from electric power to heat. This power is also used to provide the mechanical power (torque × speed) in a motor. The unit of this power is watts (W) or megawatts (MW).

The power in the inductor is given by

$$Q = Vi_L \quad \text{or} \quad Q = Vi_T \sin \theta$$

This is the *reactive* or *inductive power* in the circuit. It is the power stored in the inductor in the form of a magnetic field. This power is not consumed as the real power is. It returns to the system (power plant and transmission lines) every half-cycle. It is used to create the magnetic field in the windings of the motor. The main effects of reactive power on the system are as follows:

1. The transmission line losses between the power plant and the load are proportional to $i_T^2 R_T$, where $\mathbf{i}_T = \mathbf{i}_R + \mathbf{i}_L$ and R_T is the resistance in the transmission lines. Therefore, i_L is a contributor to transmission losses.
2. The transmission lines have a specific current rating. If the inductive current i_L is high, the magnitude of i_R will be limited to a lower value. This creates a problem for the utility because its revenue is mainly based on i_R.

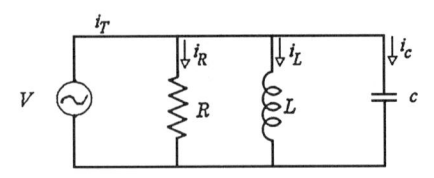

FIGURE 1.24 Addition of capacitor banks at an industry.

3. If an industry has large motors, it will require a high inductive current to magnetize these motors. This creates a localized reduction in voltage (a voltage dip) at the industry. The utility will not be able to correct for this voltage dip from the power plant. Capacitor banks are normally installed at the industry to "correct" the power factor. Figures 1.24 and 1.25 illustrate the correction in power factor.

Angle θ' is smaller than θ. Therefore, the new power factor (cos θ') is larger than the previous power factor (cos θ). Most utilities charge a penalty when the power factor drops below 0.9 to 0.92. This penalty is charged to the industry even if the power factor drops once during the month below the limit specified by the utility. Most industries use the following methods to ensure that their power factor remains above the limit specified by the utility:

a. The capacitor banks are sized to give the industry a margin above the limit specified by the utility.
b. The induction motors at the industry are started in sequence. This is done to stagger the inrush current required by each motor.

Note: The inrush current is the starting current of the induction motor. It is normally 6 to 8 times larger than the normal running current. The inrush current is mainly an inductive current. This is due to the fact that the mechanical energy (torque × speed) developed by the motor and the heat losses during the starting period of the motor are minimal (the real power provides the mechanical energy and heat losses in the motor).

c. Use synchronous motors in conjunction with induction motors. A synchronous motor is supplied by ac power to its stator. It is also supplied by direct-current (dc) power to its rotor. The dc power allows the synchronous motor to deliver reactive (inductive) power. Therefore, a synchronous motor can operate at a leading power factor, as shown in Fig. 1.26. This allows the synchronous motors to correct the power factor at the industry by compensating for the lagging power factor generated by induction motors.

The third form of power used is the apparent power. It is given by

$$S = i_T V \quad \text{where } \mathbf{i}_T = \mathbf{i}_R + \mathbf{i}_L$$

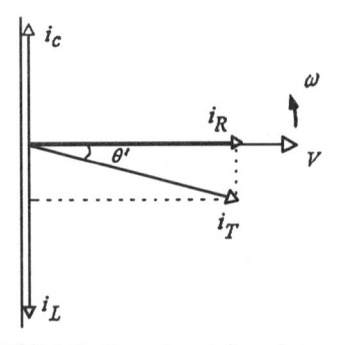

FIGURE 1.25 Correction of power factor at an industry.

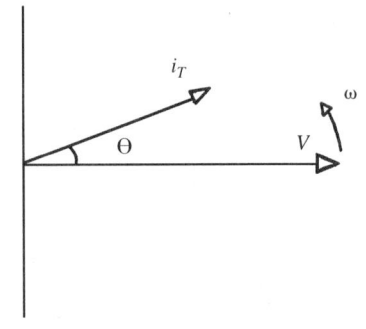

FIGURE 1.26 Phasor diagram of a synchronous motor.

FUNDAMENTALS OF ELECTRIC SYSTEMS   1.21

The unit of this power is voltamperes (VA) or megavoltamperes (MVA). This power includes the combined effect of the real power and the reactive power. All electrical equipment such as transformers, motors, and generators are rated by their apparent power. This is so because the apparent power specifies the total power (real and reactive) requirement of equipment.

THREE-PHASE SYSTEMS

Most of the transmission, distribution, and energy conversion systems having an apparent power higher than 10 kVA use three-phase circuits. The reason for this is that the *power density* (the ratio of power to weight) of a device is higher when it is a three-phase rather than a single-phase design. For example, the weight of a three-phase motor is lower than the weight of a single-phase motor having the same rating. The voltages of a three-phase system are normally given by

$$v_{a'a} = V_m \sin \omega t$$

$$v_{b'b} = V_m \sin (\omega t - 120°)$$

$$v_{c'c} = V_m \sin (\omega t - 240°)$$

where $V_m = \mathcal{E}_m$. Figure 1.27 illustrates the variations of these voltages versus time. The phasors of these voltages are

$$V_{a'a} = V \angle 0°$$

$$V_{b'b} = V \angle -120°$$

$$V_{c'c} = V \angle -240°$$

where V is the root-mean-square (rms) value of the voltage.

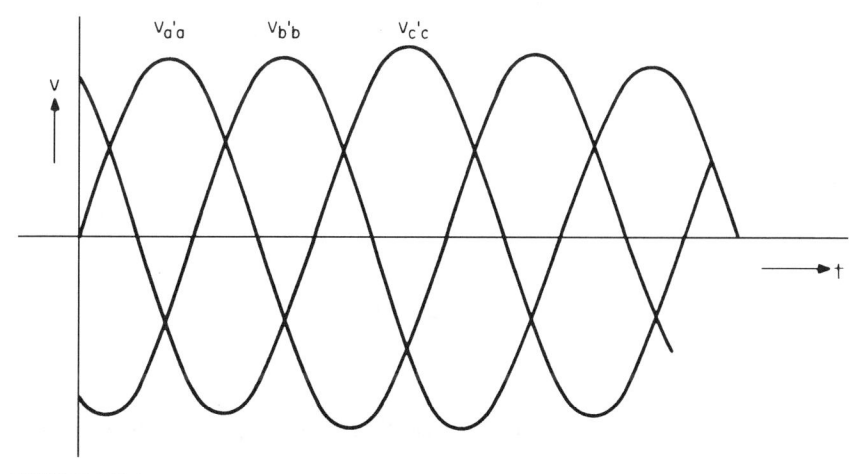

FIGURE 1.27 A system of three voltages of equal magnitude, but displaced from each other by 120°.

$V_{c'c} = V\angle -240°$

$V_{a'a} = V\angle 0°$

$V_{b'b} = V\angle -120°$

(a)

(b)

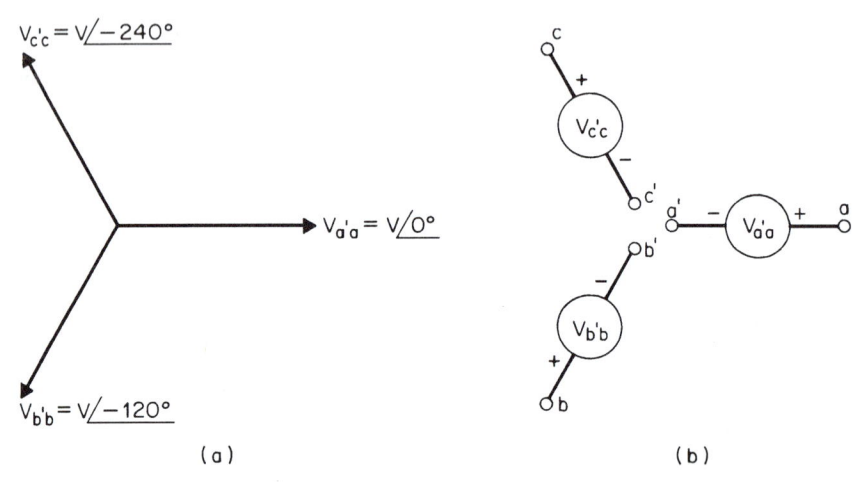

FIGURE 1.28 (*a*) Balanced three-phase phasor representation; (*b*) three-phase voltage source.

Figure 1.28a illustrates a graphical representation of the phasors. Figure 1.28b also shows the three voltage sources. When the three voltages are equal in magnitude, the system is called a *three-phase balanced system*. If the three voltages are unequal and/or the phase displacement is different from 120°, the system will be unbalanced. The phasor sum of the three voltages in a balanced system is zero.

Three-Phase Connections

The three-phase voltage sources are normally interconnected as a "wye" (Y) and a "delta" (Δ), as shown in Figs. 1.29a and b, respectively. Terminals a', b', and c' join together in the wye connection to form the neutral point O. The system becomes a *four-wire, three-phase* system when a lead is brought out from point O. In the delta connection, terminals a and b', b and c', and c and a' are joined to form the delta connection.

In the wye connection (Fig. 1.29a), the voltages across the individual phases are identified as $V_{a'a}$, $V_{b'b}$, and $V_{c'c}$. These are known as *phase voltages*. The voltages across the lines a, b, and c (or A, B, and C) are known as *line voltages*. The relationship between the line voltages and phase voltages is

$$V_l = \sqrt{3} V_p$$

Figure 1.30 illustrates the relationships between all the phase voltages and line voltages. The line currents I_l and phase currents I_p are the same in the wye connection. Thus,

$$I_l = I_p$$

In the delta connection, the line voltages V_l are the same as the phase voltages V_p. Thus,

$$V_l = V_p$$

Figure 1.31 illustrates the phasors of the phase currents and line currents in the delta-connected system. The relationship between the phase currents and line currents is given by

$$I_l = \sqrt{3} I_p$$

Power in Three-Phase Systems

The average power in a single-phase ac circuit is given by

$$P_T = V_p I_p \cos \theta_p$$

where θ_p is the power factor angle. The total power delivered in a balanced three-phase circuit is given by

$$P_T = 3 \, (V_p I_p \cos \theta_p)$$

The total power expressed in terms of line voltages and currents for a wye or delta connection is

$$P_T = \sqrt{3} \, V_l I_l \cos \theta_p$$

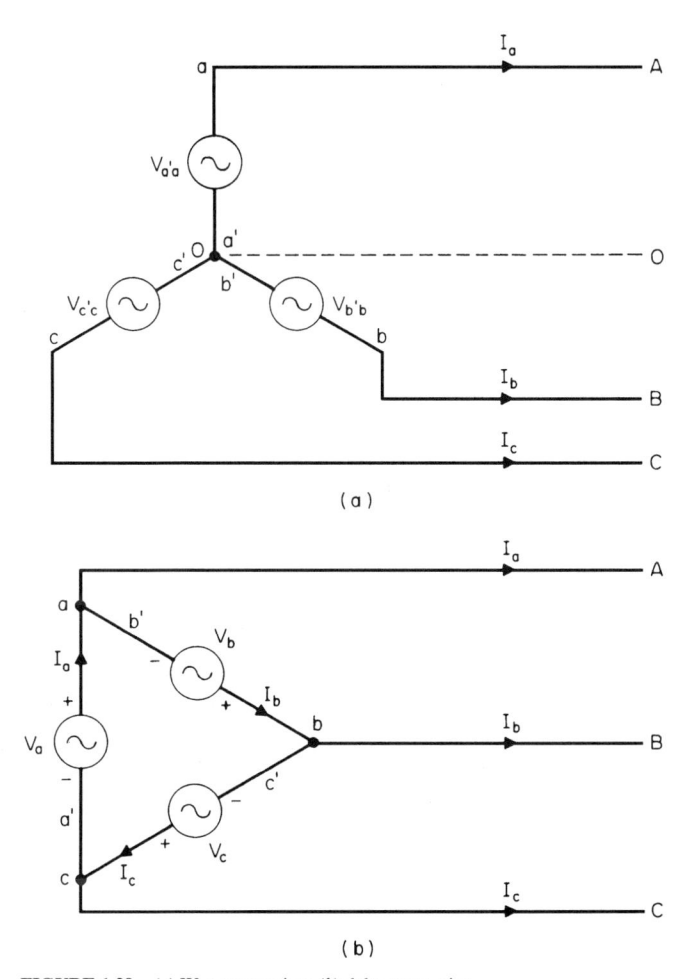

FIGURE 1.29 (*a*) Wye connection; (*b*) delta connection.

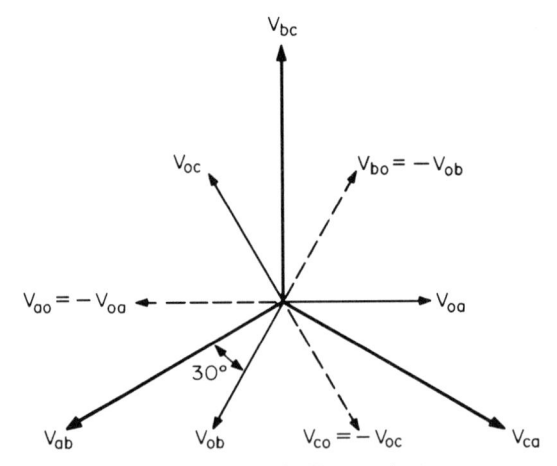

FIGURE 1.30 Voltage phases for Y connection.

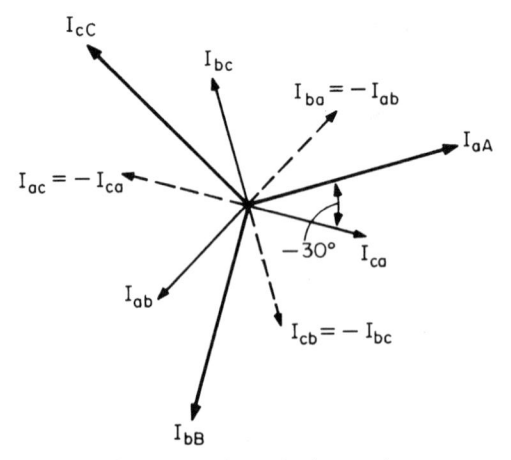

FIGURE 1.31 Current phasors for Δ connection.

Figure 1.32 illustrates a graphical representation of the instantaneous power in a three-phase system. It is clear that the instantaneous power is constant and equal to 3 times the average power. This is an important feature for three-phase motors because the constant instantaneous power eliminates torque pulsations and resulting vibrations.

REFERENCES

1. D. Halliday and R. Resnick, *Physics, Part Two,* 3d ed., Wiley, Hoboken, N.J., 1978.
2. A. S. Nasar, *Handbook of Electric Machines,* McGraw-Hill, New York, 1987.

FUNDAMENTALS OF ELECTRIC SYSTEMS 1.25

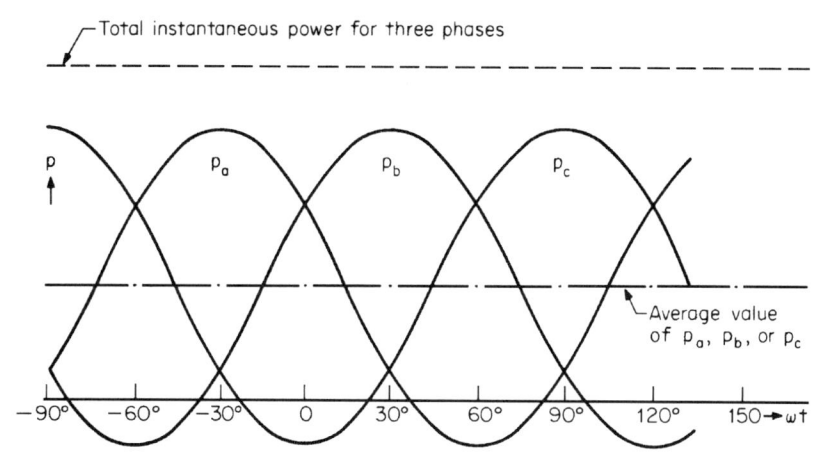

FIGURE 1.32 Power in a three-phase system.

CHAPTER 2
INTRODUCTION TO MACHINERY PRINCIPLES

ELECTRIC MACHINES AND TRANSFORMERS

An *electric machine* is a device that can convert either mechanical energy to electric energy or electric energy to mechanical energy. Such a device is called a *generator* when it converts mechanical energy to electric energy. The device is called a *motor* when it converts electric energy to mechanical energy. Since an electric machine can convert power in either direction, such a machine can be used as either a generator or a motor. Thus, all motors and generators can be used to convert energy from one form to another, using the action of a magnetic field.

A *transformer* is a device that converts ac electric energy at one voltage level to ac electric energy at another voltage level. Transformers operate on the same principles as generators and motors.

COMMON TERMS AND PRINCIPLES

- θ = angular position of an object. It is the angle at which it is oriented. It is measured from one arbitrary reference point (units: rad or deg).
- ω = angular velocity = $d\theta/dt$. It is the rate of variation of angular position with time (units: rad/s or deg/s).
- f_m = angular velocity, expressed in revolutions per second = $\omega_m/2\pi$.
- α = angular acceleration = $d\omega/dt$. It is the rate of variation of angular velocity with time (units: rad/s^2).
- τ = torque = (force applied) $\times$ (perpendicular distance). Units are newton-meters (N·m)

Newton's law of rotation:

$$\tau = J\alpha$$

where J is the moment of inertia of the rotor (units: kg·m^2).

- W = work = $T\theta$, if T is constant (units: J).
- P = power = dW/dt. It is the rate of variation of work with time (units: W):

$$P = T\omega$$

THE MAGNETIC FIELD

Energy is converted from one form to another in motors, generators, and transformers by the action of magnetic fields. These are the four basic principles that describe how magnetic fields are used in these devices:

Production of a Magnetic Field

Ampere's law is the basic law that governs the production of a magnetic field:

$$\oint \mathbf{H} \cdot d\mathbf{l} = I_{net}$$

where **H** is the magnetic field intensity produced by current I_{net}. Current I is measured in amperes and **H** in ampere-turns per meter. Figure 2.1 shows a rectangular core having a winding of N turns of wire wrapped on one leg of the core. If the core is made of ferromagnetic material (such as iron), most of the magnetic field produced by the current will remain inside the core.

Ampere's law becomes

$$Hl_c = Ni$$

where l_c is the mean path length of the core. The magnetic field intensity **H** is a measure of the "effort" that the current is putting out to establish a magnetic field. The material of the core affects the strength of the magnetic field flux produced in the core. The magnetic field intensity **H** is linked with the resulting magnetic flux density **B** within the material by

$$\mathbf{B} = \mu \mathbf{H}$$

where **H** = magnetic field intensity
   μ = magnetic *permeability* of material
   B = resulting magnetic flux density produced

Thus, the actual magnetic flux density produced in a piece of material is given by the product of two terms:

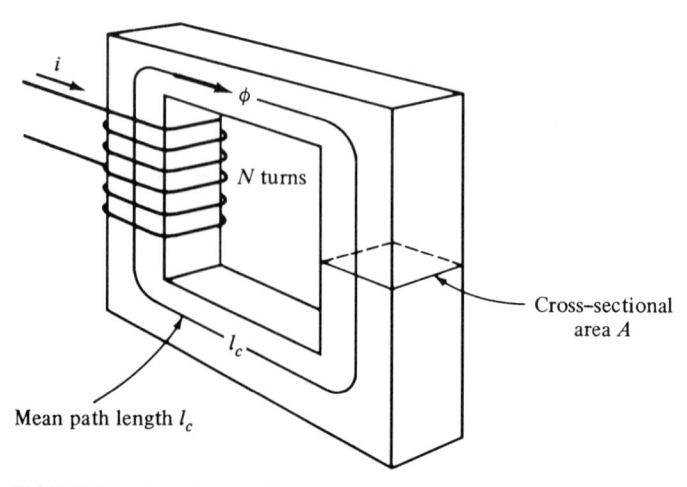

FIGURE 2.1 A simple magnetic core.

H represents effort exerted by current to establish a magnetic field
μ represents relative ease of establishing a magnetic field in a given material

In SI, the units are as follows: **H** ampere-turns per meter; μ henrys/meter (H/m); **B** webers/m², known as teslas (T). And μ_0 is the permeability of free space. Its value is

$$\mu_0 = 4\pi \times 10^{-7} \text{ H/m}$$

The relative permeability compares the magnetizability of materials. For example, in modern machines, the steels used in the cores have relative permeabilities of 2000 to 7000. Thus, for a given current, the flux established in a steel core is 2000 to 7000 times stronger than in a corresponding area of air (air has the same permeability as free space). Thus, the metals of the core in transformers, motors, and generators play an essential part in increasing and concentrating the magnetic flux in the device. The magnitude of the flux density is given by

$$B = \mu H = \frac{\mu N i}{l_c}$$

Thus, the total flux in the core in Fig. 2.1 is

$$\boxed{\phi = BA = \frac{\mu N i A}{l_c}}$$

where A is the cross-sectional area of the core.

MAGNETIC BEHAVIOR OF FERROMAGNETIC MATERIALS

The *magnetic permeability* is defined by the equation

$$\mathbf{B} = \mu \mathbf{H}$$

The permeability of ferromagnetic materials is up to 6000 times higher than the permeability of free space. However, the permeability of ferromagnetic materials is not constant. Suppose we apply a direct current to the core shown in Fig. 2.1 (starting with 0 A and increasing the current). Figure 2.2a illustrates the variation of the flux produced in the core versus the magnetomotive force. This graph is known as the *saturation curve* or *magnetization curve*. At first, a slight increase in the current (magnetomotive force) results in a significant increase in the flux. However, at a certain point, a further increase in current results in no change in the flux. The region where the curve is flat is called the *saturation region*. The core has become *saturated*. The region where the flux changes rapidly is called the *unsaturated region*. The transition region between the unsaturated region and the saturated region is called the *knee* of the curve.

Figure 2.2b illustrates the variation of magnetic flux density **B** with magnetizing intensity **H**. These are the equations:

$$H = \frac{Ni}{l_c}$$

$$B = \frac{\phi}{A}$$

It can easily be seen that the magnetizing intensity is directly proportional to the magnetomotive force, and the magnetic flux density is directly proportional to the flux. Therefore, the relationship between **B** and **H** has the same shape as the relationship between the flux and the magnetomotive force. The slope of flux-density versus the magnetizing intensity curve (Fig. 2.2c) is by definition the permeability of the core at that magnetizing intensity. The curve shows that in the unsaturated region the permeability is high and almost constant.

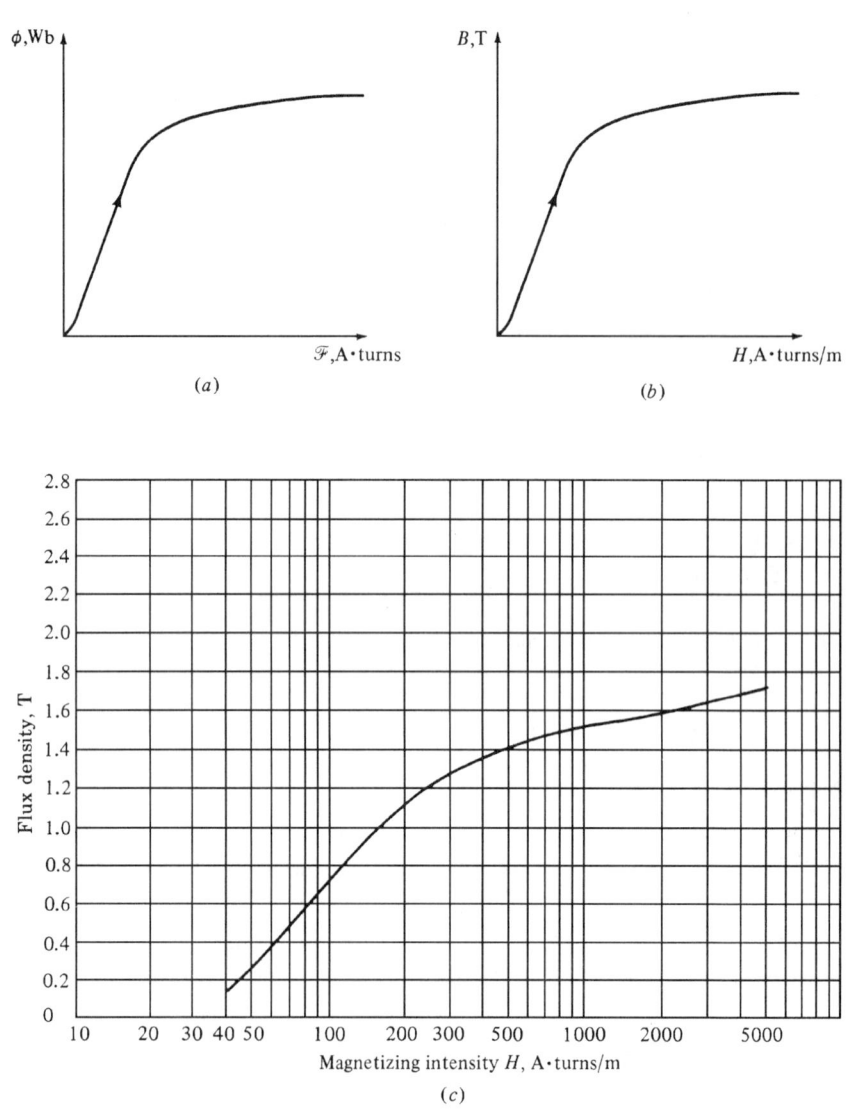

FIGURE 2.2 (a) Sketch of a dc magnetization curve for a ferromagnetic core. (b) The magnetization curve expressed in terms of flux density and magnetizing intensity. (c) A detailed magnetization curve for a typical piece of steel.

In the saturated region, the permeability drops to a very low value. Electric machines and transformers use ferromagnetic material for their cores because these materials produce much more flux than other materials.

Table 2.1 lists the characteristics of soft magnetic materials including the *Curie temperature* (or Curie point) T_c. Above this temperature a ferromagnetic material becomes paramagnetic (weakly magnetized). Figure 2.3 shows several **B-H** curves of some soft magnetic materials.

Permalloy, supermendur, and other nickel alloys have a relative permeability greater than 10^5. Only a few materials have this high permeability over a limited range of operation. The highest permeability ratio of good and poor magnetic materials over a typical operating range is 10^4.

Energy Losses in a Ferromagnetic Core

If an alternating current (Fig. 2.4*a*) is applied to the core, the flux in the core will follow path *ab* (Fig. 2.4*b*). This graph is the saturation curve shown in Fig. 2.2. However, when the current drops, the flux follows a different path from the one it took when the current increased. When the current decreases, the flux follows path *bcd*. When the current increases again, the flux follows path *bed*.

The amount of flux present in the core depends on the history of the flux in the core and the magnitude of the current applied to the windings of the core. The dependence on the history of the preceding flux and the resulting failure to retrace the flux path is called *hysteresis*. Path *bcdeb* shown in Fig. 2.4 is called a *hysteresis loop*.

Notice that if a magnetomotive force is applied to the core and then removed, the flux will follow path *abc*. The flux does not return to zero when the magnetomotive force is removed. Instead, a magnetic field remains in the core. The magnetic field is known as *residual flux* in the core. This is the technique used for producing permanent magnets. A magnetomotive force must be applied to the core in the opposite direction to return the flux to zero. This force is called the *coercive magnetomotive force* $\mathcal{F}_c$.

To understand the cause of hysteresis, it is necessary to know the structure of the metal. There are many small regions within the metal called domains. The magnetic fields of all the atoms in each domain are pointing in the same direction. Thus, each domain within the metal acts as a small permanent magnet. These tiny domains are oriented randomly within the material. This is the reason that a piece of iron does not have a resultant flux (Fig. 2.5).

When an external magnetic field is applied to the block of iron, all the domains will line up in the direction of the field. This switching to align all the fields increases the magnetic flux in the iron. This is the reason why iron has a much higher permeability than air.

When all the atoms and domains of the iron line up with the external field, a further increase in the magnetomotive force will not be able to increase the flux. At this point, the iron has become saturated with flux. The core has reached the saturation region of the magnetization curve (Fig. 2.2).

The cause of hysteresis is that when the external magnetic field is removed, the domains do not become completely random again. This is so because energy is required to turn the atoms in the domains. Originally, the external magnetic field provided energy to align the domains. When the field is removed, there is no source of energy to rotate the domains. The piece of iron has now become a permanent magnet.

Some of the domains will remain aligned until an external source of energy is supplied to change them. A large mechanical shock and heating are examples of external energy that can change the alignment of the domains. This is the reason why permanent magnets lose their magnetism when hit with a hammer or heated.

TABLE 2.1 Characteristics of Soft Magnetic Materials

Trade name	Principal alloys	Saturation flux density, T	H at B_{sat}, A/m	Amplitude permeability max. μ_m	Coercive force H_c, A/m	Electrical resistivity, $\mu\Omega\cdot cm$	Curie temperature, °C
48NI	48% Ni	1.25	80	200,000		65	
Monimax	48% Ni	1.35	6,360	100,000	4.0	65	398
High Perm 49	49% Ni	1.1	80			48	
Satmumetal	Ni, Cu	1.5	32	240,000		45	398
Permalloy (sheet)	Ni, Mo	0.8	400	100,000	1.6	55	454
Moly Permalloy (powder)	Ni, Mo	0.7	15,900	125			
Deltamax	50% Ni	1.4	25	200,000	8	45	499
M-19	Si	2.0	40,000	10,000	28	47	
Silectron	Si	1.95	8,000	20,000	40	50	732
Oriented T	Si	1.6	175	30,000		47	
Oriented M-5	Si	2.0	11,900		26	48	746
Ingot iron	None	2.15	55,000		80	10.7	
Supermendur	49% Co, V	2.4	15,900	80,000	8	26	932
Vanadium Permendur	49% Co, V	2.3	12,700	4,900	92	40	925
Hyperco 27	27% Co	2.36	70,000	2,800	198	19	
Flake iron	Carbonal power	≈0.8	5,200	5–130		10^5–10^{15}	
Ferrotron (powder)	Mo, Ni	(Linear)	(Linear)	5–25		10^{16}	
Ferrite	Mg, Zn	0.39	1,115	3,400	13	10^7	135
Ferrite	Mn, Zn	0.453	1,590	10,000	6.3	3×10^7	190
Ferrite	Ni, Zn	0.22	2,000	160	318	10^9	500
Ferrite	Ni, Al	0.28	6,360	400	143		500
Ferrite	Mg, Mn	0.37	2,000	4,000	30	1.8×10^8	210

INTRODUCTION TO MACHINERY PRINCIPLES 2.7

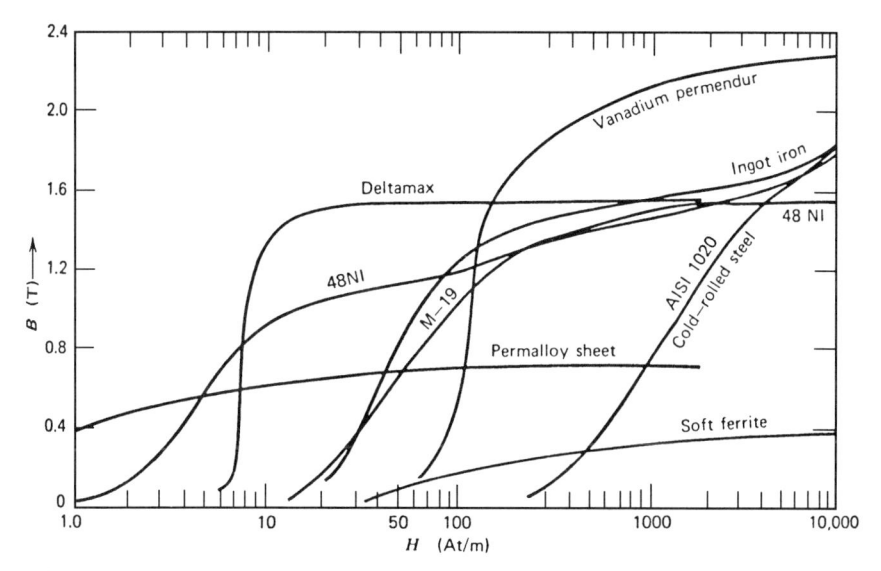

FIGURE 2.3 *B-H* curves of selected soft magnetic materials.

Energy is lost in all iron cores due to the fact that energy is required to turn the domains. The energy required to reorient the domains during each cycle of the alternating current is called the hysteresis loss in the iron core. The area enclosed in the hysteresis loop is directly proportional to the energy lost in a given ac cycle (Fig. 2.4).

FARADAY'S LAW—INDUCED VOLTAGE FROM A MAGNETIC FIELD CHANGING WITH TIME

Faraday's law states that if a flux passes through a turn of a coil of wire, a voltage will be induced in the turn of wire that is directly proportional to the *rate of change* of the flux with time. The equation is

$$e_{ind} = -\frac{d\phi}{dt}$$

where e_{ind} is the voltage induced in the turn of the coil and ϕ is the flux passing through it. If the coil has N turns and if a flux ϕ passes through them all, then the voltage induced across the whole coil is

$$e_{ind} = -N\frac{d\phi}{dt}$$

where e_{ind} = voltage induced in coil
N = number of turns of wire in coil
ϕ = flux passing through coil

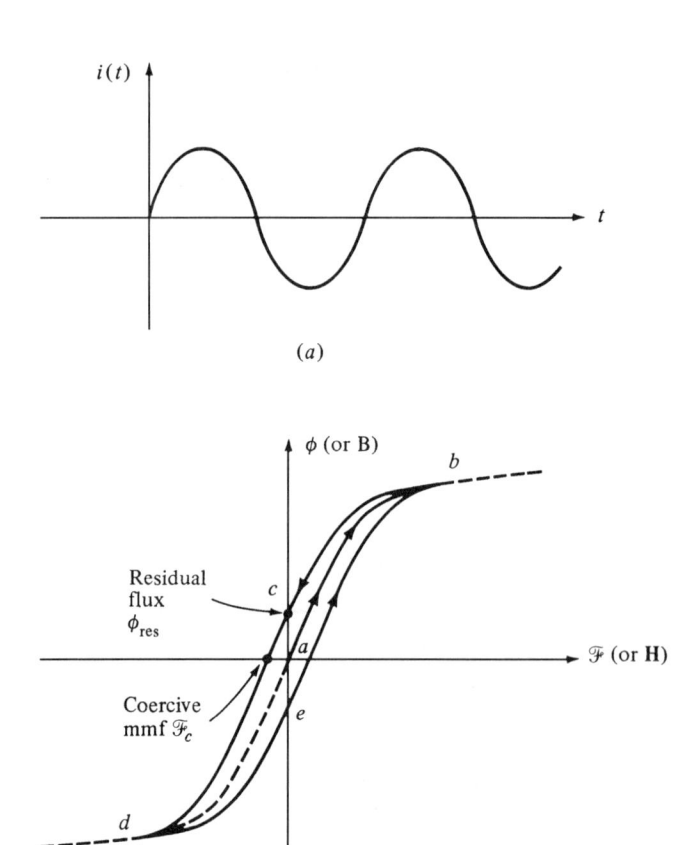

FIGURE 2.4 The hysteresis loop traced out by the flux in a core when the current $i(t)$ is applied to it.

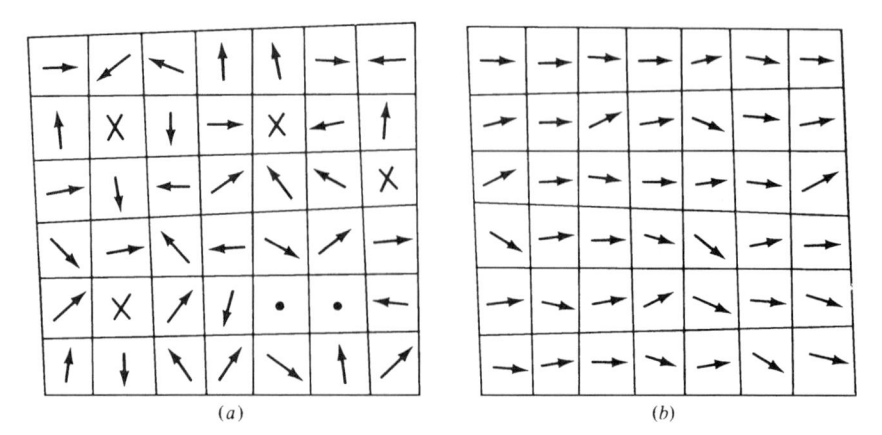

FIGURE 2.5 (*a*) Magnetic domains oriented randomly. (*b*) Magnetic domains lined up in the presence of an external magnetic field.

Based on Faraday's law, a flux changing with time induces a voltage within a ferromagnetic core in a similar manner as it would in a wire wrapped around the core. These voltages can generate swirls of current inside the core. They are similar to the eddies seen at the edges of a river. They are called eddy currents. Energy is dissipated by these flowing eddy currents. The lost energy heats the iron core. Eddy current losses are proportional to the length of the paths they follow within the core. For this reason, all ferromagnetic cores subjected to alternating fluxes are made of many small strips, or *laminations*. The strips are insulated on both sides to reduce the paths of the eddy currents. The strips are oriented in a parallel direction to the magnetic flux. The eddy current losses have the following characteristics:

- They are proportional to the square of the lamination thickness.
- They are inversely proportional to the electrical resistivity of the material.

The thickness of the laminations is between 0.5 and 5 mm in power equipment and between 0.01 and 0.5 mm in electronic equipment. The volume of a material increases when it is laminated. The *stacking factor* is the ratio of the actual volume of the magnetic material to its total volume after it has been laminated. This is an important variable for accurately calculating the flux densities in magnetic materials. Table 2.2 lists the typical stacking factors for different lamination thicknesses. Since hysteresis losses and eddy current losses occur in the core, their sum is called *core losses.*

CORE LOSS VALUES

Figure 2.6 shows the core loss data for M-15, which is a 3 percent silicon steel. This magnetic material is used in many transformers and small motors. Figure 2.7a and b shows the core loss data for a nickel alloy widely used in electronics equipment (48 NI) and a ferrite material, respectively.

PERMANENT MAGNETS

Permanent magnets are a common excitation source for rotating machines. The performance of a permanent magnet depends on how the magnet is installed in the machine and whether it was magnetized before or after installation. Most permanent magnets, except for the new neodymium-iron-boron magnet, are not machinable. They must be used in the machine as obtained from the manufacturer. Table 2.3 lists the main characteristics of common permanent magnets.

TABLE 2.2 Stacking Factors for Laminated Cores

Lamination thickness, mm	Stacking factor
0.0127	0.50
0.0254	0.75
0.0508	0.85
0.1–0.25	0.90
0.27–0.36	0.95

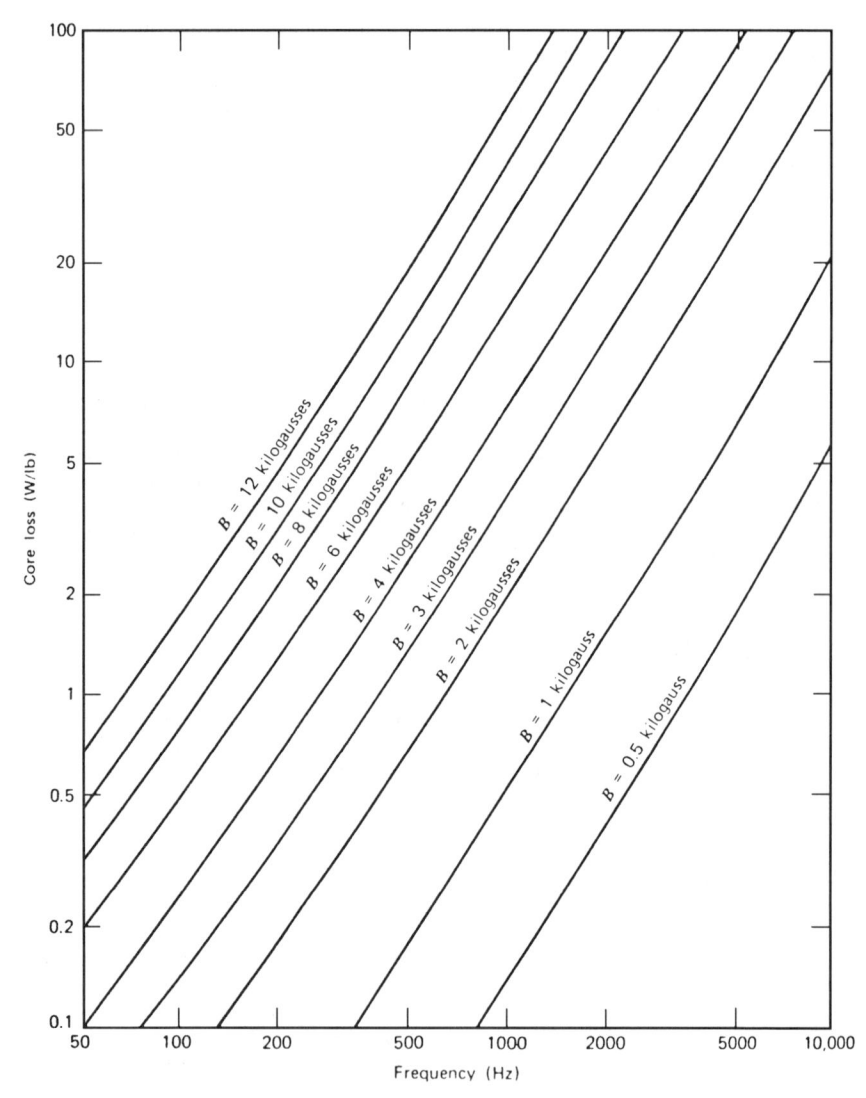

FIGURE 2.6 Core loss for nonoriented silicon steel 0.019-in-thick lamination. (*Courtesy of Armco Steel Corporation.*)

Figure 2.8 illustrates the demagnetization curve which is a portion of the hysteresis loop of alnico V. The *coercive force* H_c (the intersection of the curve with the horizontal H axis) represents the ability of the metal to withstand demagnetization from external magnetic sources. A second curve known as the *energy product* is often shown on this figure. It is the product of **B** and **H** plotted as a function of **H**. It represents the energy stored in the permanent magnet.

Figure 2.9 illustrates the **B-H** characteristics of several alnico permanent magnets. The characteristics of several ferrite magnets are shown in Fig. 2.10. The neodymium-iron-boron (NdFeB) permanent magnets are superior to most permanent magnets.

INTRODUCTION TO MACHINERY PRINCIPLES 2.11

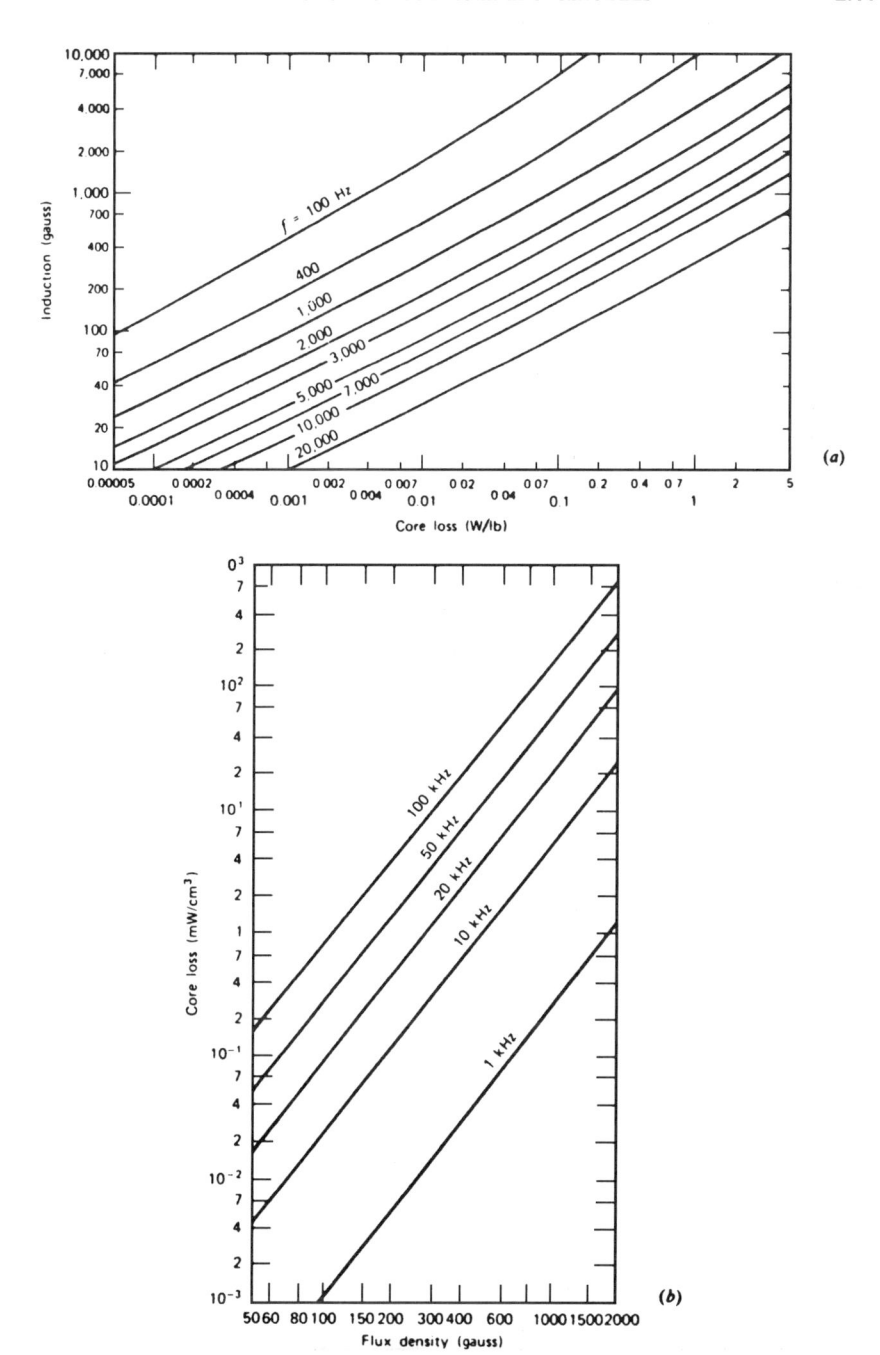

FIGURE 2.7 (*a*) Core loss for typical 48 percent nickel alloy 4 mils thick. (*Courtesy of Armco Steel Corporation.*) (*b*) Core loss for Mn-Zn ferrites.

TABLE 2.3 Characteristics of Permanent Magnets

Type	Residual flux density B_r, G	Coercive force H_c, Oe	Maximum energy product, G·Oe × 10^6	Average recoil permeability
1% Carbon steel	9,000	50	0.18	
3^1/2% Chrome steel	9,500	65	0.29	35
36% Cobalt steel	9,300	230	0.94	10
Alnico I	7,000	440	1.4	6.8
Alnico IV	5,500	730	1.3	4.1
Alnico V	12,500	640	5.25	3.8
Alnico VI	10,500	790	3.8	4.9
Alnico VIII	7,800	1,650	5.0	—
Cunife	5,600	570	1.75	1.4
Cunico	3,400	710	0.85	3.0
Vicalloy 2	9,050	415	2.3	—
Platinum-cobalt	6,200	4,100	8.2	1.1
Barium ferrite-isotropic	2,200	1,825	1.0	1.15
Oriented type A	3,850	2,000	3.5	1.05
Oriented type B	3,300	3,000	2.6	1.06
Strontium ferrite				
Oriented type A	4,000	2,220	3.7	1.05
Oriented type B	3,550	3,150	3.0	1.05
Rare earth—cobalt	8,600	8,000	18.0	1.05
NdFeB	11,200	8,500	30.0	1.05

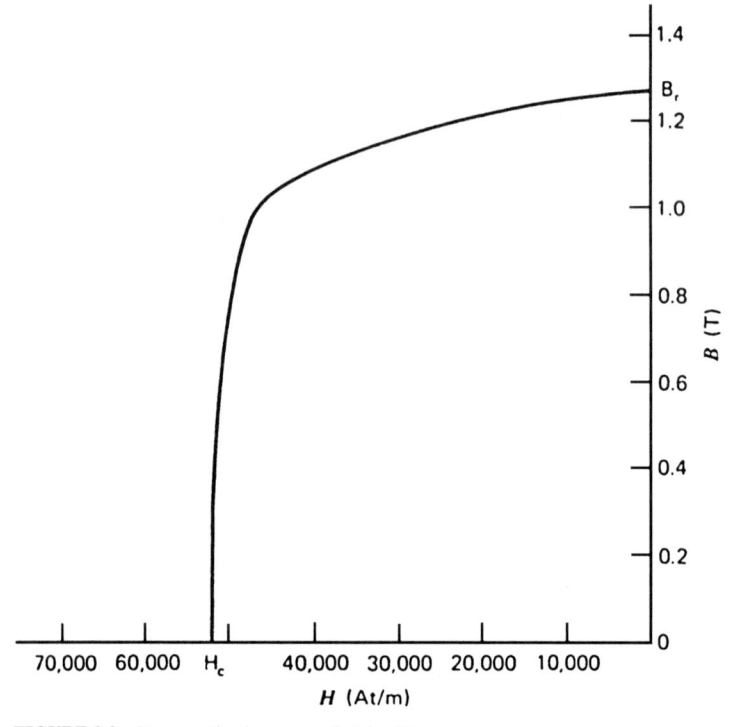

FIGURE 2.8 Demagnetization curve of alnico V.

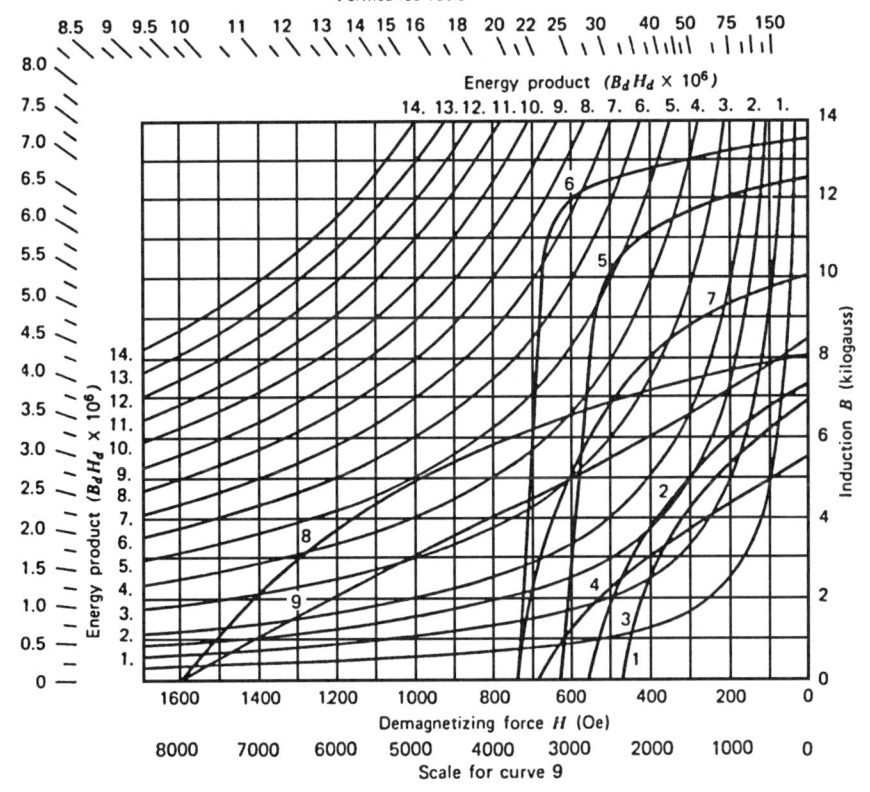

FIGURE 2.9 Demagnetization and energy product curves for alnicos I to VIII. *Key:* 1, alnico I; 2, alnico II; 3, alnico III; 4, alnico IV; 5, alnico V; 6, alnico VI; 7, alnico VII; 8, alnico VIII; 9, rare earth-cobalt.

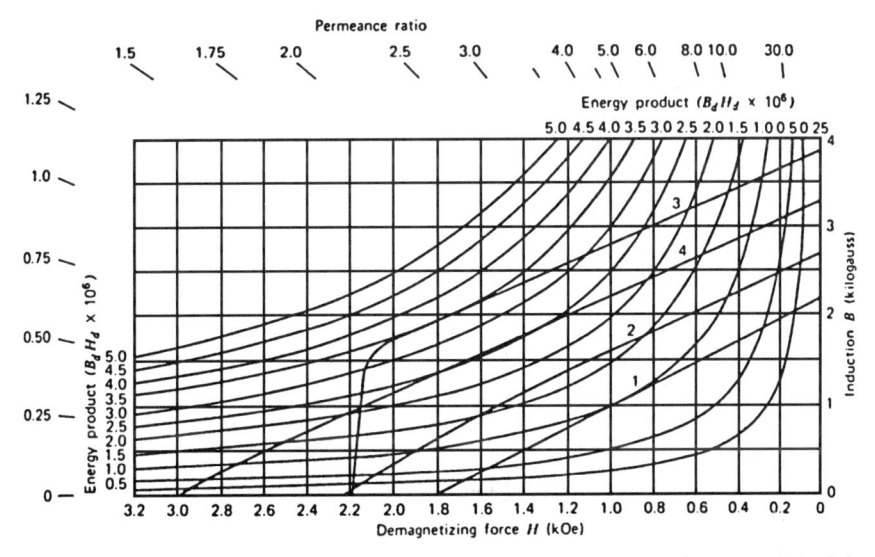

FIGURE 2.10 Demagnetization and energy product curves for Indox ceramic magnets. *Key:* 1, Index I; 2, Index II; 3, Index V; and 4, Index VI-A.

They also have a lower cost than samarium-cobalt (SmCo) magnets. Their machining characteristics, strength, and hardness are similar to those of iron and steel. Figure 2.11 shows a comparison of the NdFeB magnet characteristics with those of other common magnets. The *energy product* [product of **B** in gauss (G) and **H** in oersteds (Oe)] and the permeance ratio (ratio of **B/H**) are also shown on these figures. Permanent magnets are most efficient when operated at conditions that result in maximum energy product. The permeance ratios are useful in designing magnetic circuits. The flux density $\mathbf{B}_d$ and field intensity $\mathbf{H}_d$ are used to designate the coordinates of the demagnetization curve.

PRODUCTION OF INDUCED FORCE ON A WIRE

A magnetic field induces a force on a current-carrying conductor within the field (Fig. 2.12). The force induced on the conductor is given by

$$\mathbf{F} = i\,(\mathbf{l} \times \mathbf{B})$$

The direction of the force is given by the right-hand rule. If the index finger of the right hand points in the direction of vector **l**, and the middle finger points in the direction of the flux density vector **B**, the thumb will point in the direction of the resultant force on the wire. The magnitude of the force is

$$F = ilB \sin \theta$$

where θ is the angle between vector **l** and vector **B**.

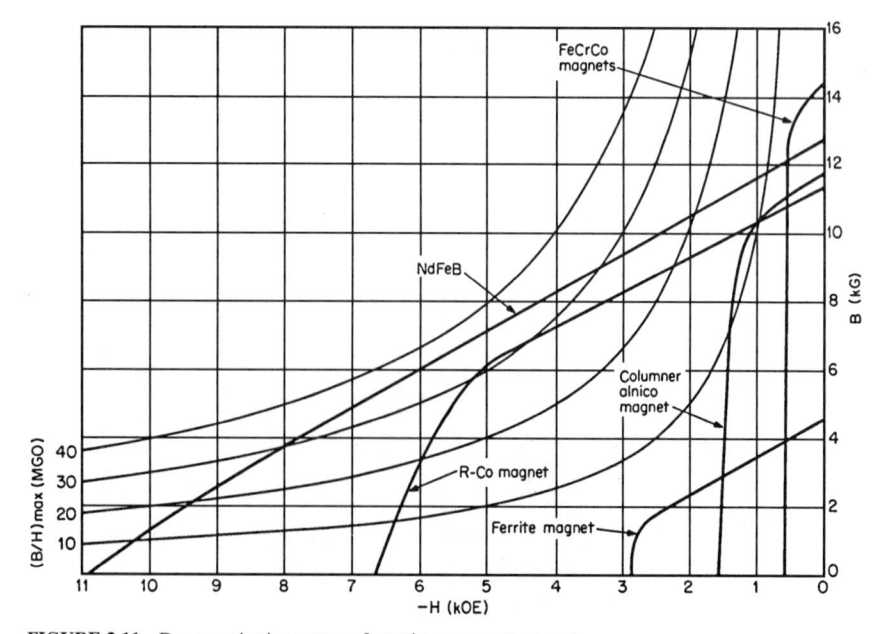

FIGURE 2.11 Demagnetization curves of certain permanent magnets.

INTRODUCTION TO MACHINERY PRINCIPLES 2.15

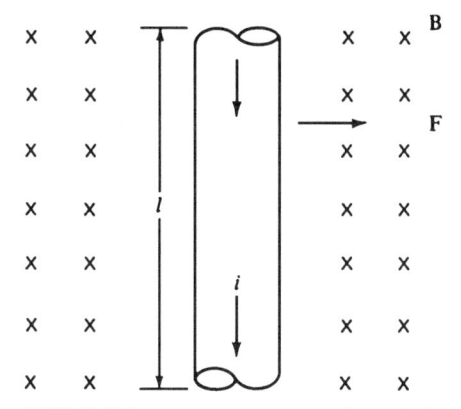

FIGURE 2.12 A current-carrying conductor in the presence of a magnetic field.

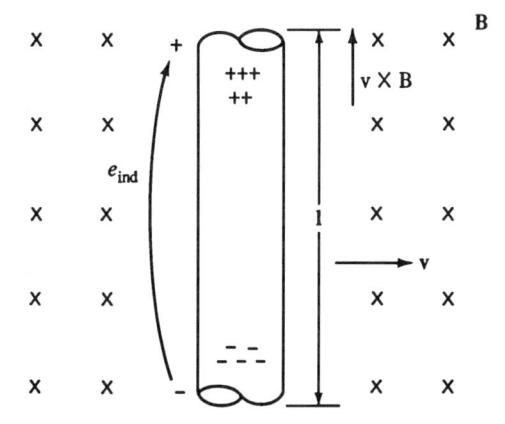

FIGURE 2.13 A conductor moving in the presence of a magnetic field.

INDUCED VOLTAGE ON A CONDUCTOR MOVING IN A MAGNETIC FIELD

A magnetic field induces a voltage on a conductor moving in the field (Fig. 2.13). The induced voltage in the conductor is given by

$$e_{ind} = (\mathbf{v} \times \mathbf{B}) \cdot \mathbf{l}$$

where $\mathbf{v}$ = velocity of conductor
$\mathbf{B}$ = magnetic flux density
$\mathbf{l}$ = length of conductor in magnetic field

REFERENCES

1. S. J. Chapman, *Electric Machinery Fundamentals*, 2d. ed., McGraw-Hill, New York, 1991.
2. A. S. Nasar, *Handbook of Electric Machines*, McGraw-Hill, New York, 1987.

CHAPTER 3
TRANSFORMERS

A transformer is a device that uses the action of a magnetic field to change ac electric energy at one voltage level to ac electric energy at another voltage level. It consists of a ferromagnetic core with two or more coils wrapped around it. The common magnetic flux within the core is the only connection between the coils. The source of ac electric power is connected to one of the transformer windings. The second winding supplies power to loads. The winding connected to the power source is called the *primary winding* or *input winding*. The winding connected to the loads is called the *secondary winding* or *output winding*.

IMPORTANCE OF TRANSFORMERS

When a transformer steps up the voltage level of a circuit, it decreases the current because the power remains constant. Therefore, ac power can be generated at one central station. The voltage is stepped up for transmission over long distances at very low losses. The voltage is stepped down again for final use. Since the transmission losses are proportional to the square of the current, raising the voltage by a factor of 10 will reduce the transmission losses by a factor of 100. Also, when the voltage is increased by a factor of 10, the current is decreased by a factor of 10. This allows the use of much thinner conductors to transmit power.

In modern power stations, power is generated at 12 to 25 kV. Transformers step up the voltage to 110 to 1000 kV for transmission over long distances at very low losses. Transformers then step it down to 12 to 34.5 kV for local distribution and then permit power to be used in homes and industry at 120 V.

TYPES AND CONSTRUCTION OF TRANSFORMERS

The function of a transformer is to convert ac power from a voltage level to another voltage level at the same frequency. The core of a transformer is constructed from thin laminations electrically isolated from each other to reduce eddy current losses (Fig. 3.1).

The primary and secondary windings are wrapped one on top of the other around the core with the low-voltage winding innermost. This arrangement serves two purposes:

1. The problem of insulating the high-voltage winding from the core is simplified.
2. It reduces the leakage flux compared to if the windings were separated by a distance on the core.

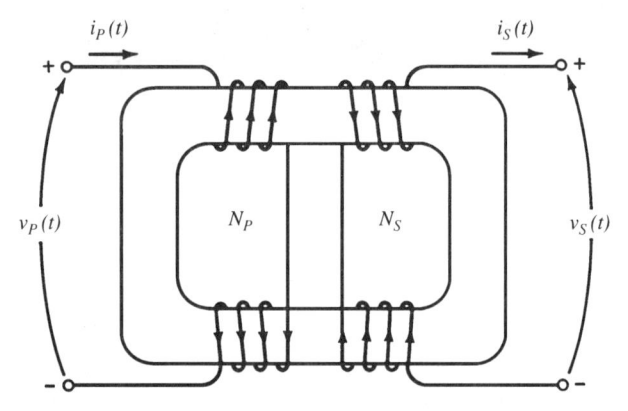

FIGURE 3.1 Core-form transformer construction.

The transformer that steps up the output of a generator to transmission levels (110+ kV) is called the *unit transformer*. The transformer that steps the voltage down from transmission levels to distribution levels (2.3–34.5 kV) is called a *substation transformer*. The transformer that steps down the distribution voltage to the final voltage at which the power is used (110, 208, 220 V, etc.) is called a *distribution transformer*.

There are also two special-purpose transformers used with electric machinery and power systems. The first is used to sample a high voltage and produce a low secondary voltage proportional to it (*potential transformers*). The potential transformer is designed to handle only a very small current. A *current transformer* is designed to give a secondary current much smaller than its primary current.

THE IDEAL TRANSFORMER

An ideal transformer does not have any losses (Fig. 3.2). The voltages and currents are related by these equations:

$$\frac{v_p(t)}{v_s(t)} = \frac{N_P}{N_S} = a$$

$$N_P i_p(t) = N_S i_s(t)$$

$$\frac{i_p(t)}{i_s(t)} = \frac{1}{a}$$

The equations of the phasor quantities are

$$\frac{\mathbf{V}_P}{\mathbf{V}_S} = a$$

TRANSFORMERS 3.3

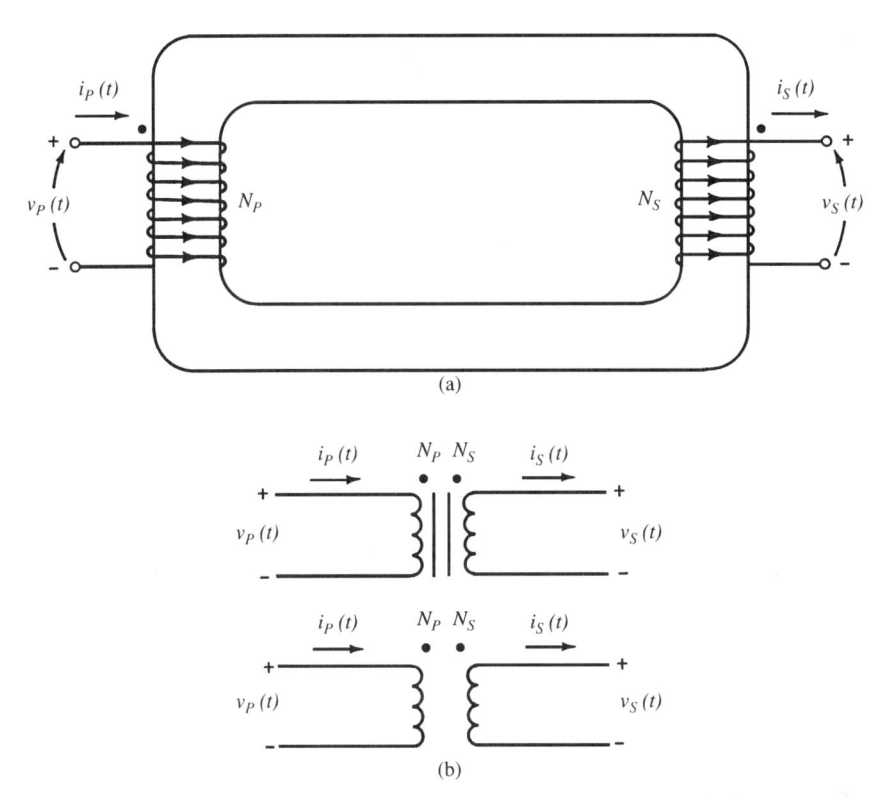

FIGURE 3.2 (a) Sketch of an ideal transformer. (b) Schematic symbols of a transformer.

$$\frac{\mathbf{I}_P}{\mathbf{I}_S} = \frac{1}{a}$$

The phase of angle $\mathbf{V}_P$ is the same as the angle of $\mathbf{V}_S$, and the phase angle of $\mathbf{I}_P$ is the same as the phase angle of $\mathbf{I}_S$.

Power in an Ideal Transformer

The power given to the transformer by the primary circuit is

$$P_{in} = V_P I_P \cos \theta_P$$

where θ_P is the angle between the primary voltage and current. The power supplied by the secondary side of the transformer to its loads is

$$P_{out} = V_S I_S \cos \theta_S$$

where θ_S is the angle between the secondary voltage and current.

An ideal transformer does not affect the voltage and power angle, $\theta_P = \theta_S = \theta$. The primary and secondary windings of an ideal transformer have the *same power factor*. The power out of a transformer is

$$\boxed{P_{out} = V_S I_S \cos \theta_S}$$

Applying the turns-ratio equations gives

$$V_S = \frac{V_P}{a} \quad \text{and} \quad I_S = aI_P$$

so

$$P_{out} = \frac{V_P}{a} aI_P \cos \theta$$

$$\boxed{P_{out} = V_P I_P \cos \theta = P_{in}}$$

Therefore, *the output power of an ideal transformer is equal to its input power.*

The same relationship is applicable to the reactive power Q and apparent power S:

$$\boxed{Q_{in} = V_P I_P \sin \theta = V_S I_S \sin \theta = Q_{out}}$$

$$\boxed{S_{in} = V_P I_P = V_S I_S = S_{out}}$$

IMPEDANCE TRANSFORMATION THROUGH A TRANSFORMER

The *impedance* of a device is defined as the ratio of the phasor voltage across it to the phasor current flowing through it.

$$Z_L = \frac{\mathbf{V}_L}{\mathbf{I}_L}$$

Since a transformer changes the current and voltage levels, it also changes the impedance of an element. The impedance of the load shown in Fig. 3.3b is

$$Z_L = \frac{\mathbf{V}_S}{\mathbf{I}_S}$$

The primary circuit apparent impedance is

$$Z'_L = \frac{\mathbf{V}_P}{\mathbf{I}_P}$$

Since the primary voltage and current can be expressed as

$$V_P = aV_S \qquad I_P = \frac{I_S}{a}$$

the apparent impedance of the primary is

$$Z'_L = \frac{V_P}{I_P} = \frac{aV_S}{I_S/a} = a^2 \frac{V_S}{I_S}$$

$$\boxed{Z'_L = a^2 Z_L}$$

It is possible to match the magnitude of load impedance to a source impedance by simply selecting the proper turns ratio of a transformer.

ANALYSIS OF CIRCUITS CONTAINING IDEAL TRANSFORMERS

The easiest way to analyze a circuit containing an ideal transformer is by replacing the portion of the circuit on one side of the transformer by an equivalent circuit with the same terminal characteristics. After substitution of the equivalent circuit, the new circuit (without a

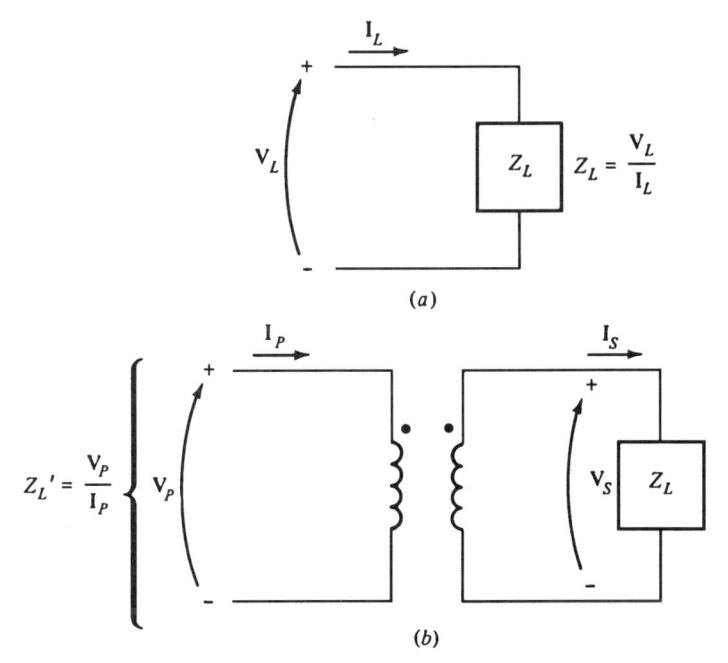

FIGURE 3.3 (*a*) Definition of impedance. (*b*) Impedance scaling through a transformer.

transformer present) can be solved for its voltages and currents. The process of replacing one side of a transformer by its equivalent at the second side's voltage level is known as *reflecting* or *referring* the first side of the transformer to the second side.

The solution for circuits containing ideal transformers is shown in Example 3.1.

EXAMPLE 3.1 A single-phase power system consists of a 480-V 60-Hz generator supplying a load $Z_{load} = 4 + j3$ Ω through a transmission impedance $Z_{line} = 0.18 + j0.24$ Ω. Answer the following questions about this system.

1. If the power system is exactly as described above (Fig. 3.4a), what will be the voltage at the load? What will the transmission line losses be?
2. Suppose a 1:10 step-up transformer is placed at the generator end of the transmission line and a 10:1 step-down transformer is placed at the load end of the line (Fig. 3.4b). What will the load voltage be now? What will the transmission losses be now?

Solution

1. Figure 3.4a shows the power system without transformers. Here $I_G = I_{line} = I_{load}$. The line current in this system is given by

$$I_{line} = \frac{V}{Z_{line} + Z_{load}}$$

$$= \frac{480 \angle 0° \text{ V}}{(0.18 \text{ Ω} + j0.24 \text{ Ω}) + (4 \text{ Ω} + j3 \text{ Ω})}$$

$$= \frac{480 \angle 0°}{4.18 + j3.24}$$

$$= \frac{480 \angle 0°}{5.29 \angle 37.8°}$$

$$= 90.8 \angle -37.8° \text{ A}$$

Therefore the load voltage is

$$V_{load} = I_{line} Z_{load}$$

$$= (90.8 \angle -37.8° \text{ A})(4 \text{ Ω} + j3 \text{ Ω})$$

$$= (90.8 \angle -37.8° \text{ A})(5 \angle 36.9° \text{ Ω})$$

$$= 454 \angle -0.9° \text{ V}$$

and the line losses are

$$P_{loss} = (I_{line})^2 R_{line}$$

$$= (90.8 \text{ A})^2 (0.18 \text{ Ω})$$

$$= 1484 \text{ W}$$

2. Figure 3.4b shows the power system with the transformers. To analyze this system, it is necessary to convert it to a common voltage level. This is done in two steps:

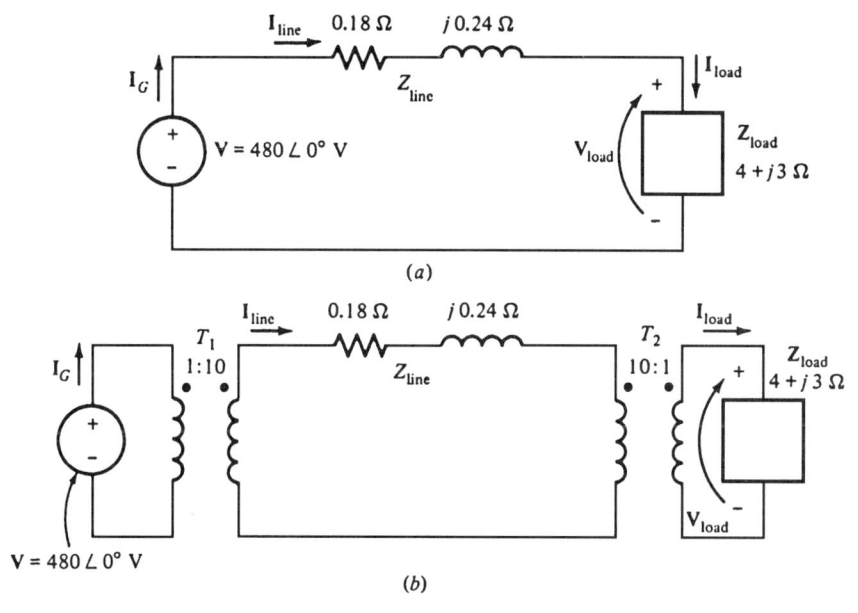

FIGURE 3.4 The power system of Example 3.1 (*a*) without and (*b*) with transformers at the ends of the transmission line.

a. Eliminate transformer T_2 by referring the load over to the transmission line's voltage level.
b. Eliminate transformer T_1 by referring the transmission line's elements and the equivalent load at the transmission line's voltage over to the source side.

The value of the load's impedance when reflected to the transmission system's voltage is

$$Z'_{load} = a^2 Z_{load}$$
$$= \left(\frac{10}{1}\right)^2 (4\,\Omega + j3\,\Omega)$$
$$= 400 + j300\,\Omega$$

The total impedance at the transmission line level is now

$$Z_{eq} = Z_{line} + Z'_{load}$$
$$= 400.18 + j300.24\,\Omega = 500.3\,\angle 36.88°\,\Omega$$

This equivalent circuit is shown in Fig. 3.5*a*. The total impedance at the transmission line level ($Z_{line} + Z'_{load}$) is now reflected in across T_1 source's voltage level:

$$Z'_{eq} = a^2 Z_{eq}$$
$$= a^2 (Z_{line} + Z'_{load})$$
$$= (\tfrac{1}{10})^2 [(0.18 + j0.24) + (400 + j300)]\,\Omega$$

$$= [(0.0018 + j0.0024) + (4 + j3)] \: \Omega$$

$$= 5.003 \: \angle 36.88° \: \Omega$$

Notice that $Z''_{load} = 4 + j3 \: \Omega$ and $Z'_{line} = 0.0018 + j0.0024 \: \Omega$. The resulting equivalent circuit is shown in Fig. 3.5b. The generator's current is

$$\mathbf{I}_G = \frac{480 \: \angle 0° \: \text{V}}{5.003 \: \angle 36.88° \: \Omega} = 95.94 \: \angle -36.88° \: \text{A}$$

Knowing the current $\mathbf{I}_G$, we can now work back and find $\mathbf{I}_{line}$ and $\mathbf{I}_{load}$. Working back through T_1, we get

$$N_{P_1} \mathbf{I}_G = N_{S_1} \mathbf{I}_{line}$$

$$\mathbf{I}_{line} = \frac{N_{P_1}}{N_{S_1}} \mathbf{I}_G$$

$$= \tfrac{1}{10}(95.94 \: \angle -36.88° \: \text{A})$$

$$= 9.594 \: \angle -36.88° \: \text{A}$$

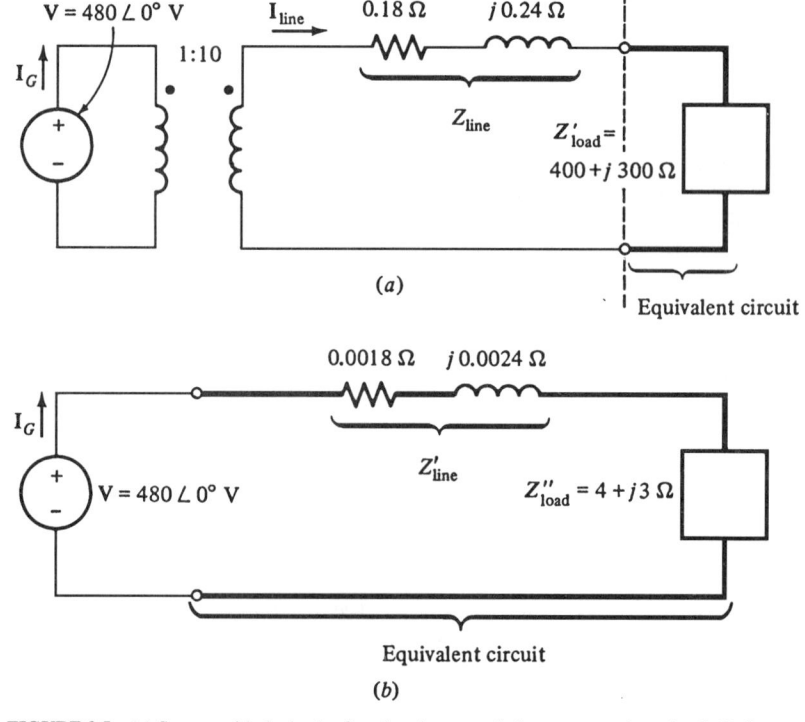

FIGURE 3.5 (a) System with the load referred to the transmission system voltage level. (b) System with the load and transmission line referred to the generator's voltage level.

Working back through T_2 gives

$$N_{P_2} \mathbf{I}_{\text{line}} = N_{S_2} \mathbf{I}_{\text{load}}$$

$$\mathbf{I}_{\text{load}} = \frac{N_{P_2}}{N_{S_2}} \mathbf{I}_{\text{line}}$$

$$= \frac{10}{1} (9.594 \angle -36.88° \text{ A})$$

$$= 95.94 \angle -36.88° \text{ A}$$

It is now possible to answer the questions originally asked. The load voltage is given by

$$\mathbf{V}_{\text{load}} = \mathbf{I}_{\text{load}} Z_{\text{load}}$$

$$= (95.94 \angle -36.88° \text{ A})(5 \angle 36.87° \text{ }\Omega)$$

$$= 479.7 \angle -0.01° \text{ V}$$

and the losses are given by

$$P_{\text{loss}} = (I_{\text{line}})^2 R_{\text{line}}$$

$$= (9.594 \text{ A})^2 (0.18 \text{ }\Omega)$$

$$= 16.7 \text{ W}$$

Notice that by stepping up the transmission voltage of the power system, the transmission losses have been reduced by a factor of 90. Also, the voltage at the load dropped significantly in the system with transformers compared to the system without transformers.

THEORY OF OPERATION OF REAL SINGLE-PHASE TRANSFORMERS

Figure 3.6 illustrates a transformer consisting of two conductors wrapped around a transformer core. Faraday's law describes the basis of transformer operation:

$$e_{\text{ind}} = \frac{d\lambda}{dt}$$

where λ is the flux linkage in the coil across which the voltage is being induced. Flux linkage λ is the sum of the flux passing through each turn in the coil added over all coil turns:

$$\lambda = \sum_{i=1}^{n} \phi_i$$

The total flux linkage through a coil is not $N\phi$ (N is the number of turns). This is so because the flux passing through each turn of a coil is slightly different from the flux in the

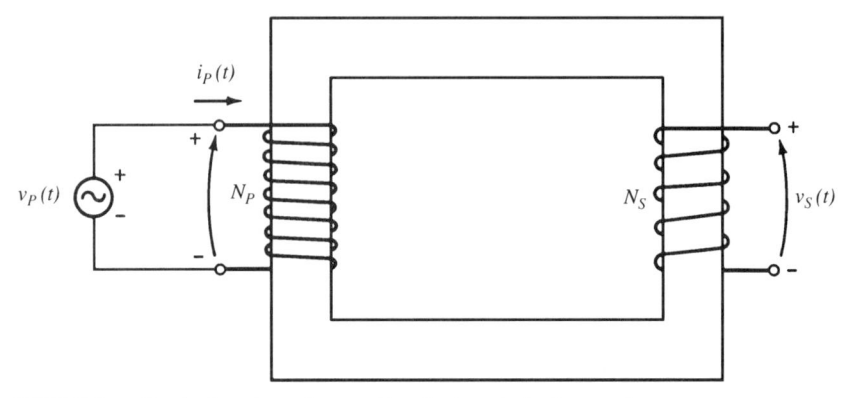

FIGURE 3.6 Sketch of a real transformer with no load attached to its secondary.

neighboring turns. An average flux per turn is defined. If λ is the total flux linkage in all the turns of the coils, the *average flux per turn* is

$$\overline{\phi} = \frac{\lambda}{N}$$

where N is the number of turns.

Faraday's law can be written as

$$e_{ind} = N \frac{d\overline{\phi}}{dt}$$

THE VOLTAGE RATIO ACROSS A TRANSFORMER

The average flux present in the primary winding of a transformer is

$$\overline{\phi} = \frac{1}{N_P} \int v_P(t)\, dt$$

The effect of this flux on the secondary coil of the transformer depends on how much of the flux reaches the secondary coil. Only a portion of the flux produced in the primary coil reaches the secondary coil. Some of the flux lines pass through the surrounding air instead of through the iron core (Fig. 3.7). The *leakage flux* is the portion of the flux that passes through one of the coils but not the other.

The flux in the primary coil can be divided into two components: a *mutual flux*, which remains in the core and links both coils (windings), and a small *leakage flux*, which passes through the primary winding and returns through air, bypassing the secondary winding:

$$\overline{\phi}_P = \phi_M + \phi_{LP}$$

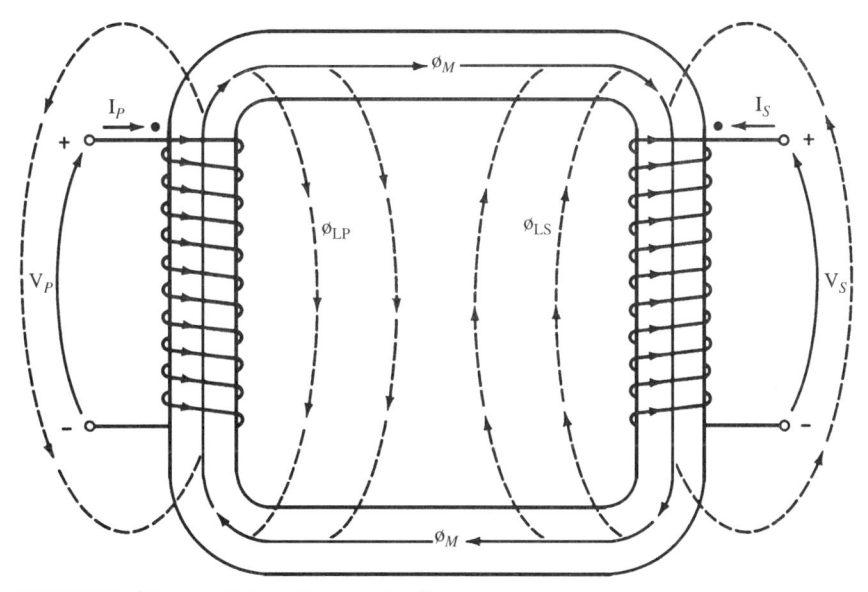

FIGURE 3.7 Mutual and leakage fluxes in a transformer core.

where $\overline{\phi}_P$ = total average primary flux
ϕ_M = flux component linking both primary and secondary coils
ϕ_{LP} = primary leakage flux

Similarly, the flux in the secondary winding is divided between the mutual and leakage fluxes which pass through the secondary winding and return through air, bypassing the primary winding:

$$\overline{\phi}_S = \phi_M + \phi_{LS}$$

where $\overline{\phi}_S$ = total average secondary flux
ϕ_M = flux component linking both primary and secondary coils
ϕ_{LS} = secondary leakage flux

The primary circuit can be expressed as

$$v_P(t) = N_P \frac{d\overline{\phi}_P}{dt}$$

$$= N_P \frac{d\phi_M}{dt} + N_P \frac{d\phi_{LP}}{dt}$$

The first term is $e_P(t)$, and the second is $e_{LP}(t)$.

$$v_P(t) = e_P(t) + e_{LP}(t)$$

The voltage in the secondary coil is expressed as

$$v_S(t) = N_S \frac{d\overline{\phi}_S}{dt}$$

$$= N_S \frac{d\phi_M}{dt} + N_S \frac{d\phi_{LS}}{dt}$$

$$= e_S(t) + e_{LS}(t)$$

The mutual flux is

$$\frac{e_P(t)}{N_P} = \frac{d\phi_M}{dt} = \frac{e_S(t)}{N_S}$$

Therefore,

$$\boxed{\frac{e_P(t)}{e_S(t)} = \frac{N_P}{N_S} = a}$$

This equation indicates that the ratio of the *primary voltage caused by the mutual flux to the secondary voltage caused by the mutual flux is equal to the turns ratio of the transformer*.

In well-designed transformers, $\phi_M \gg \phi_{LP}$ and $\phi_M \gg \phi_{LS}$; the ratio of the total voltage on the primary to the total voltage on the secondary is approximately

$$\frac{v_P(t)}{v_S(t)} \approx \frac{N_P}{N_S} = a$$

THE MAGNETIZING CURRENT IN A REAL TRANSFORMER

When the ac power source is connected to a transformer, a current flows in the primary winding, *even when the secondary winding is open-circuited*. This is the current required to produce the flux in ferromagnetic core. It consists of these components:

1. *The magnetization current i_m*. The current required to produce the flux in the core of the transformer.
2. *The core-loss current I_{h+e}*. The current required to make up for the hysteresis and eddy current losses.

Figure 3.8a illustrates the magnetization curve of a transformer core. The magnetization current can be determined if the flux in the core is known. If we assume that the leakage flux is negligible, the average flux in the core will be given by

$$\overline{\phi} = \frac{1}{N_P} \int v_P(t)\, dt$$

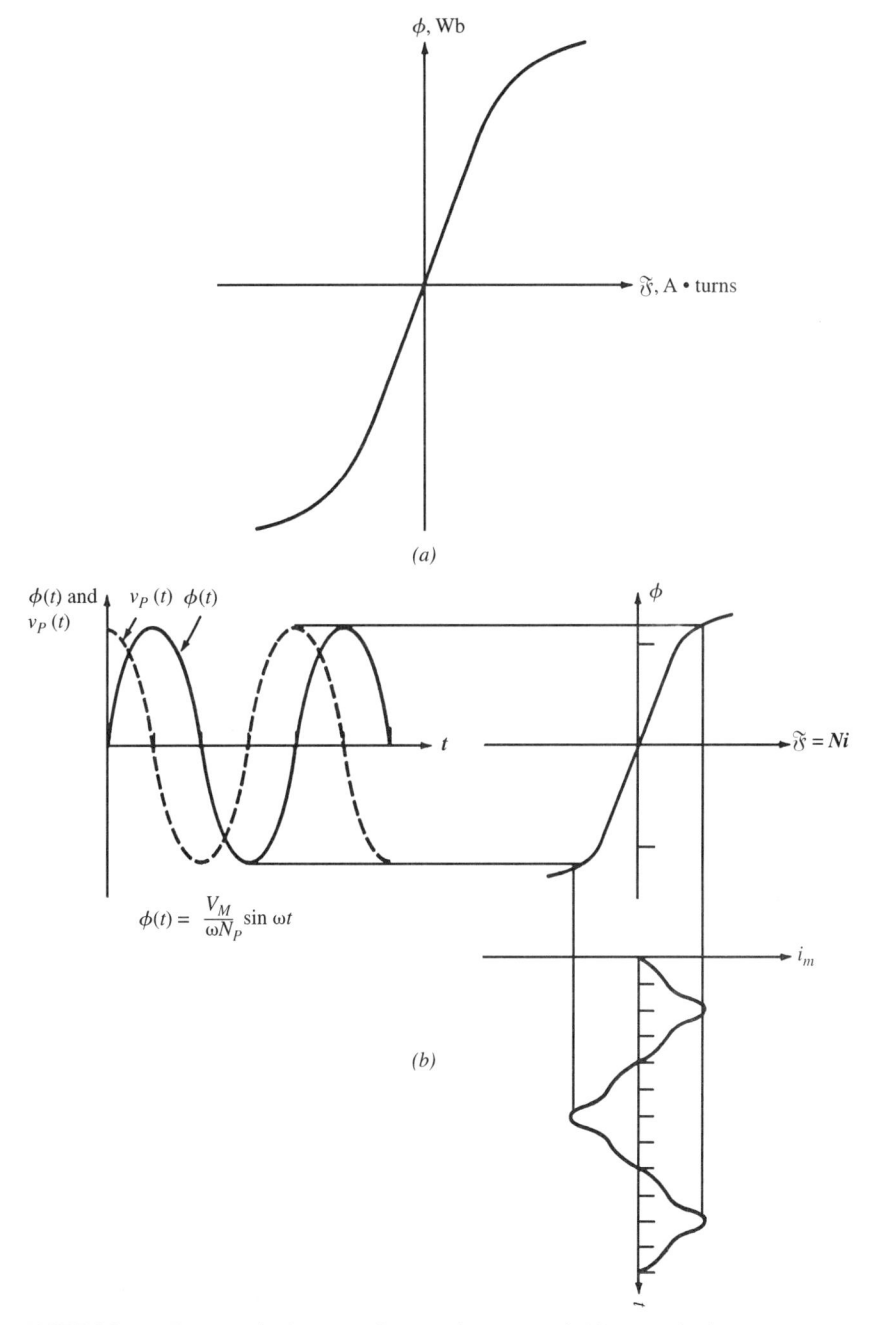

FIGURE 3.8 (*a*) The magnetization curve of the transformer core. (*b*) The magnetization current caused by the flux in the transformer core.

If the primary voltage is given by $v_p(t) = V_M \cos \omega t$ V, the resulting flux will be

$$\overline{\phi} = \frac{1}{N_P} \int V_M \cos \omega t \, dt$$

$$= \frac{V_M}{\omega N_P} \sin \omega t \quad \text{Wb}$$

Figure 3.8*b* illustrates the variations of the current required to produce the flux in the core. The following observations can be made:

1. The magnetization current is not sinusoidal. It has a higher-frequency component due to magnetic saturation in the transformer core.
2. In the saturation region, a large increase in magnetizing current is required to provide a slight increase in the flux.
3. The fundamental component of the magnetization current lags the applied voltage by 90°.
4. The higher-frequency (harmonic) component of the magnetization current increases as the core is driven into saturation.

The second component of the no-load current in the transformer is required to supply the core losses. The largest eddy current losses occur when the flux passes through 0 Wb because these losses are proportional to $d\phi/dt$. The hysteresis losses are also the highest when the flux passes through zero. Therefore, the greatest core loss occurs when the flux goes through zero. Figure 3.9 illustrates the variations in the total current required to make up for core losses.

The total no-load current is known as the *excitation current*. It is given by

$$i_{ex} = i_m + i_{h+e}$$

Figure 3.10 illustrates the total excitation current in a transformer.

THE DOT CONVENTION

Figure 3.11 illustrates a load supplied from a transformer. The dots on the windings of the transformer help to determine the polarity of the voltages and currents in the core without performing a physical examination. The dot convention states that *a positive magnetomotive force $\mathcal{F}$ is produced when the current flows into the dotted end of the winding*. A negative magnetomotive force is produced when the current flows into the undotted end of the winding. Therefore, the magnetomotive forces will be subtracted if one current flows into the dotted end of a winding and the second flows out of the dotted end.

The primary current shown in Fig. 3.11 produces a positive magnetomotive force $\mathcal{F}_P = N_P i_P$. A negative magnetomotive force $\mathcal{F}_S = -N_S i_S$ is produced by the secondary current. Therefore, the net magnetomotive force in the core is given by

$$\mathcal{F}_{net} = N_P i_P - N_S i_S$$

In well-designed transformers, the net magnetomotive force is negligible. Therefore,

$$\boxed{N_P i_P \approx N_S i_S}$$

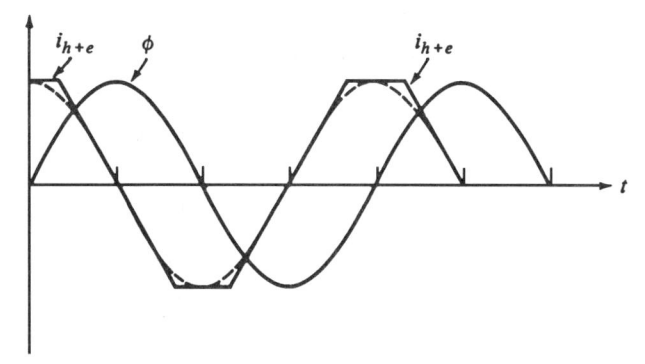

FIGURE 3.9 The core-loss current in a transformer.

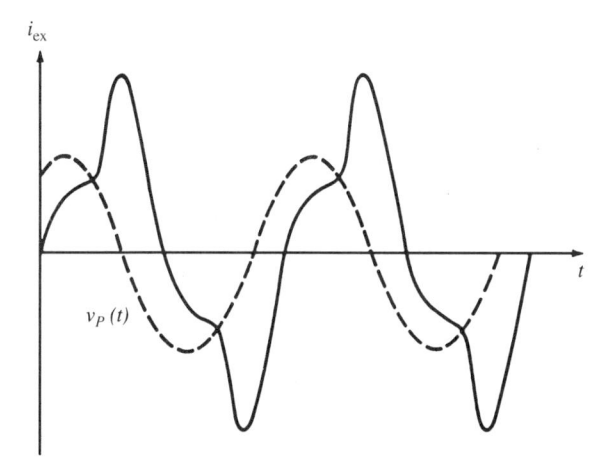

FIGURE 3.10 The total excitation current in a transformer.

$$\frac{i_P}{i_S} \approx \frac{N_S}{N_P} = \frac{1}{a}$$

The current then must flow into one dotted end and out of the other to eliminate the net magnetomotive force in the core.

THE EQUIVALENT CIRCUIT OF A TRANSFORMER

Any accurate model of transformer behavior should show the losses that occur in real transformers. The major items that should be considered in such a model are

1. *Copper losses.* These are the resistive losses in the primary and secondary windings of the transformer. They are proportional to the square of the current in the windings.
2. *Eddy current losses.* These are the resistive heating losses in the core of the transformer. They are proportional to the square of the voltage applied to the transformer.

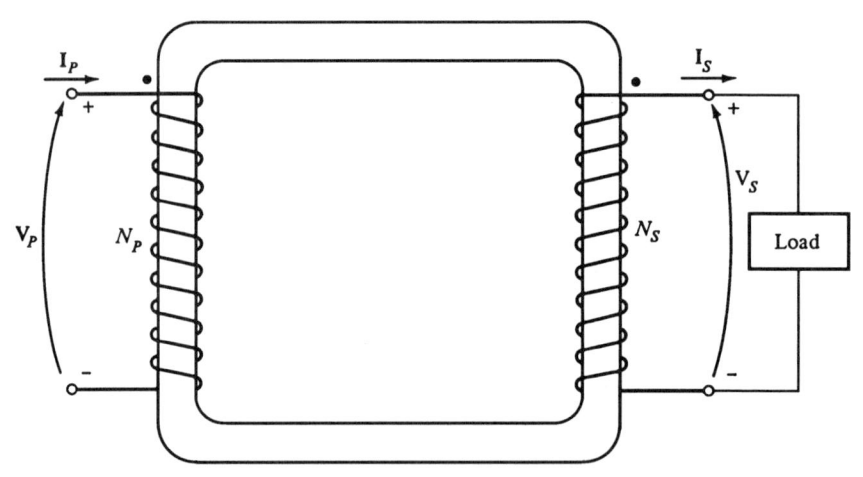

FIGURE 3.11 A real transformer with a load connected to its secondary.

3. *Hysteresis losses.* These are associated with the rearrangement of the magnetic domains in the core during each half-cycle. They are a nonlinear function of the voltage applied to the transformer.
4. *Leakage flux.* These fluxes ϕ_{LP} and ϕ_{LS} pass through only one winding, escaping the core. A self-inductance in the primary and secondary windings is associated with these escaped fluxes. The effects of these inductances must be included in any transformer model.

The copper losses can be modeled by placing resistors R_P and R_S in the primary and secondary circuits, respectively.

A voltage e_{LP} is produced by the leakage flux in the primary windings ϕ_{LP}. It is given by

$$e_{LP}(t) = N_P \frac{d\phi_{LP}}{dt}$$

Similarly, in the secondary windings, the voltage produced is

$$e_{LS}(t) = N_S \frac{d\phi_{LS}}{dt}$$

Since the leakage flux passes through air and air has a *constant* reluctance much higher than the reluctance of the core, the flux ϕ_{LP} is proportional to the current in the windings:

$$\phi_{LP} = (\mathcal{P}N_P)i_P$$

$$\phi_{LS} = (\mathcal{P}N_S)i_S$$

where $\mathcal{P}$ = permeance of flux path
N_P = number of turns on primary coil
N_S = number of turns on secondary coil

By substituting these equations into the previous ones, we get

$$e_{\text{LP}}(t) = N_P \frac{d}{dt}(\mathcal{P}N_P) i_P = N_P^2 \mathcal{P} \frac{di_P}{dt}$$

$$e_{\text{LS}}(t) = N_S \frac{d}{dt}(\mathcal{P}N_S) i_S = N_S^2 \mathcal{P} \frac{di_S}{dt}$$

By lumping the constants together, we get

$$e_{\text{LP}}(t) = L_P \frac{di_P}{dt}$$

$$e_{\text{LS}}(t) = L_S \frac{di_S}{dt}$$

where L_P and L_S are the self-inductances of the primary and secondary windings, respectively. Therefore, the leakage flux will be modeled as an inductor.

The magnetization current I_m is proportional (in the unsaturated region) to the voltage applied to the core but lags the applied voltage by 90°. Hence, it can be modeled as a reactance X_m connected across the primary voltage source.

The core-loss current I_{h+e} is proportional to the voltage applied to the core and in phase with it. Hence, it can be modeled as a resistance R_c connected across the primary voltage source. Figure 3.12 illustrates the equivalent circuit of a real transformer.

Although Fig. 3.12 represents an accurate model of a transformer, it is not a very useful one. The entire circuit is normally converted to an equivalent circuit at a single voltage level. This equivalent circuit is referred to its primary or secondary side (Fig. 3.13).

Approximate Equivalent Circuits of a Transformer

In practical engineering applications, the exact transformer model is more complex than necessary in order to get good results. Since the excitation branch has a very small current compared to the load current of the transformers, a simplified equivalent circuit is produced.

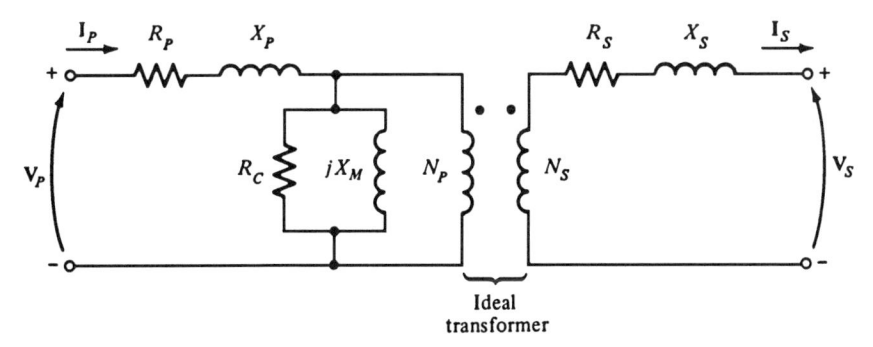

FIGURE 3.12 The model of a real transformer.

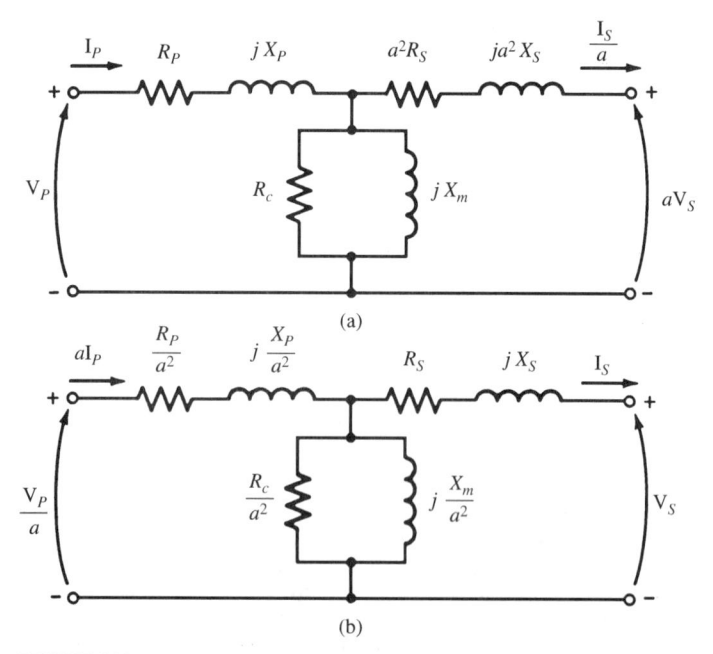

FIGURE 3.13 (*a*) The transformer model referred to its primary voltage level. (*b*) The transformer model referred to its secondary voltage level.

The excitation branch is moved to the front of the transformer, and the primary and secondary impedances are added, creating equivalent circuits (Fig. 3.14*a* and *b*). In some applications, the excitation branch is neglected entirely without causing serious error (Fig. 3.14*c* and *d*).

THE TRANSFORMER VOLTAGE REGULATION AND EFFICIENCY

The transformer's output voltage varies with the load even when the input voltage remains constant. This is due to the fact that a real transformer has impedances within it. The *voltage regulation* (VR) is used to compare the voltage variations in transformers. The *full-load voltage regulation* is a parameter that compares the transformer's output voltage at no load with the output voltage at full load.

It is given by

$$\text{VR} = \frac{V_{S,\text{nl}} - V_{S,\text{fl}}}{V_{S,\text{fl}}} \times 100\%$$

Since the secondary voltage at no load is given by

$$V_S = \frac{V_P}{a}$$

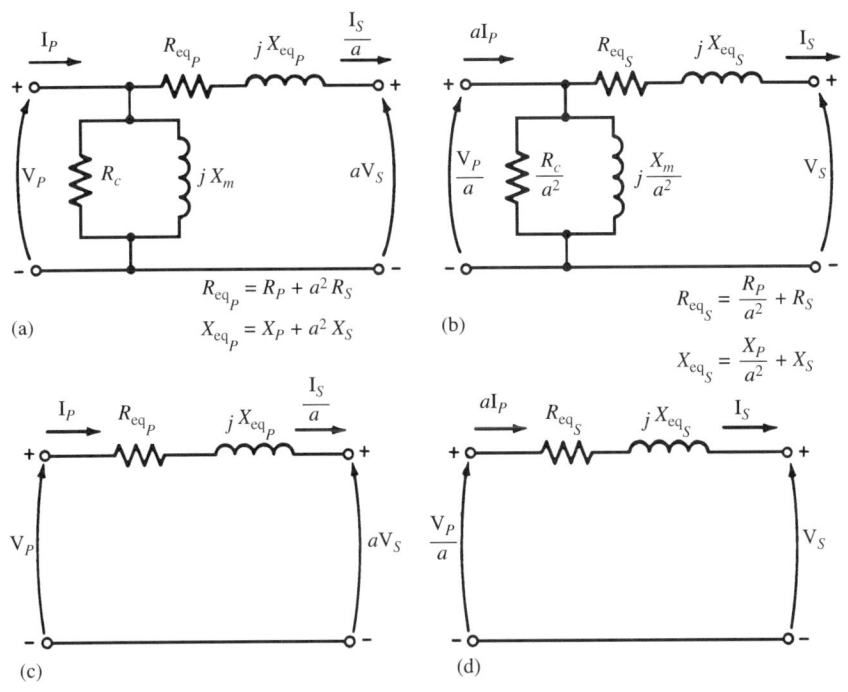

FIGURE 3.14 Approximate transformer models: (*a*) Referred to the primary side; (*b*) referred to the secondary side; (*c*) with no excitation branch, referred to the primary side; (*d*) with no excitation branch, referred to the secondary side.

where a is the turns ratio of the transformer, then

$$\text{VR} = \frac{V_P/a - V_{S,\text{fl}}}{V_{S,\text{fl}}} \times 100\%$$

Most transformer applications limit the VR to a small value. In an ideal transformer VR = 0 percent.

The Transformer Phasor Diagram

Figure 3.13 illustrates a simplified equivalent circuit for a transformer. The excitation branch (R_c and X_m) has a negligible effect on the voltage regulation of the transformer. The secondary voltage of the transformer depends on the magnitude of the series impedances within it and on the phase angle of the current flowing through it.

The primary voltage of the transformer is given by

$$\frac{\mathbf{V}_P}{a} = \mathbf{V}_S + R_{eq}\mathbf{I}_S + jX_{eq}\mathbf{I}_S$$

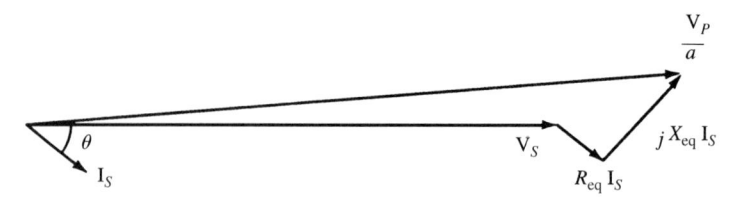

FIGURE 3.15 Phasor diagram of a transformer operating at a lagging power factor.

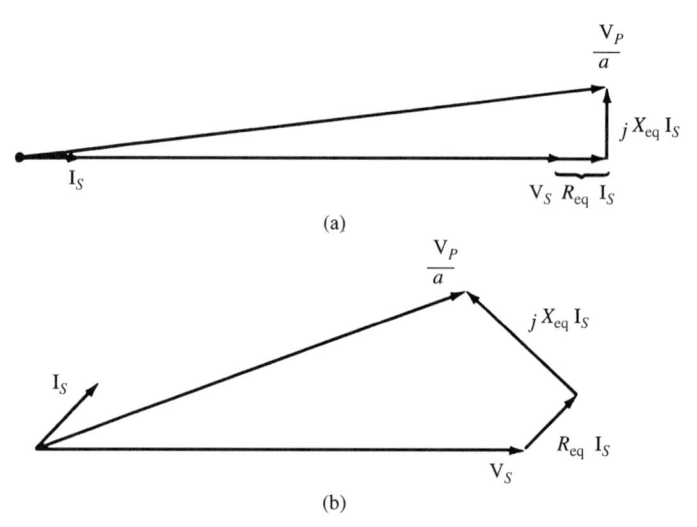

FIGURE 3.16 Phasor diagram of a transformer operating at (*a*) unity and (*b*) leading power factor.

Figure 3.15 illustrates a phasor diagram of a transformer operating at a lagging power factor. It is easy to notice that $\mathbf{V}_p/a > \mathbf{V}_S$. Thus, the voltage regulation of a transformer with lagging loads is greater than zero.

Figure 3.16*a* illustrates a phasor diagram of a transformer having a power factor of 1. The secondary voltage is again lower than the primary voltage. Thus, the voltage regulation is greater than zero. However, in this case, the voltage regulation is smaller than it was in the previous case (for a lagging current). The secondary voltage will actually be larger if the secondary current is leading (Fig. 3.16*b*). In this case, the transformer will have a *negative* voltage regulation.

Simplified Voltage Regulation

Figure 3.17 illustrates a phasor diagram of a typical transformer operating with a lagging load. The primary voltage is given approximately by

$$\frac{V_P}{a} = V_S + R_{eq}I_S \cos\theta + X_{eq}I_S \sin\theta$$

TRANSFORMERS

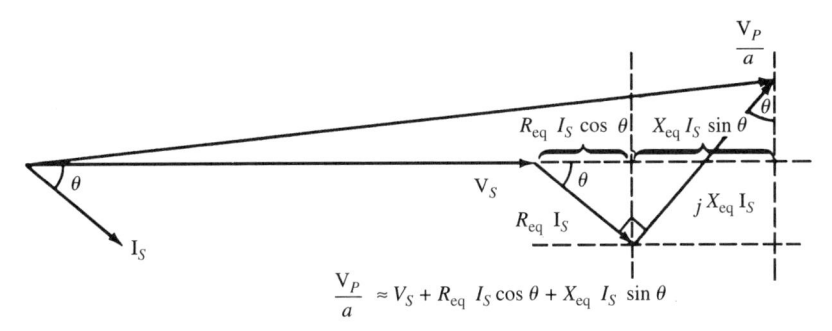

FIGURE 3.17 Derivation of the approximate equation V_p/a.

In the figure: $\dfrac{V_P}{a} \approx V_S + R_{eq} I_S \cos\theta + X_{eq} I_S \sin\theta$

The voltage regulation can be determined by using the V_p/a term calculated from the equation above in the voltage regulation equation:

$$\text{VR} = \frac{V_P/a - V_{S,\text{fl}}}{V_{S,\text{fl}}} \times 100\%$$

Transformer Efficiency

The efficiency of a device is normally given by

$$\eta = \frac{P_{\text{out}}}{P_{\text{in}}} \times 100\%$$

$$= \frac{P_{\text{out}}}{P_{\text{out}} + P_{\text{loss}}} \times 100\%$$

Since the output power of the transformer is expressed by

$$P_{\text{out}} = V_S I_S \cos\theta$$

the efficiency of the transformer is given by

$$\eta = \frac{V_S I_S \cos\theta}{P_{\text{Cu}} + P_{\text{core}} + V_S I_S \cos\theta} \times 100\%$$

where P_{Cu} = copper losses and P_{core} = core losses.

Transformer Taps and Voltage Regulators

Since the transformer's output voltage varies with the load, most distribution transformers use *taps* in the windings to change the turns ratio of the transformer. Most applications have four taps with spacings of 2.5 percent of full voltage between them in addition to the nominal setting. This arrangement permits a 5 percent adjustment above or below the nominal voltage rating of the transformer.

The taps are adjusted normally when the transformer is deenergized. However, in some applications, the transformer output voltage varies significantly with the load. These variations could be caused by large line impedance (possibly due to long distance) between the generator and the load. Since normal loads require a steady voltage, these applications use a *tap changing under load* (TCUL) *transformer* or *voltage regulator*. This transformer has the ability to change the taps on power. A voltage regulator is a TCUL transformer that detects the voltage and changes the taps automatically to maintain constant system voltage.

THE AUTOTRANSFORMER

In some applications, small adjustments in voltage are required. For example, the voltage may need to be increased from 110 to115 V or from 13.2 to 13.6 kV. These changes could be necessary to accommodate reductions in voltage that occur in power systems far from the generators. In these cases, it would be very expensive to have two full windings in the transformer. The *autotransformer* is used in these applications instead of the conventional transformer.

Figure 3.18 illustrates a diagram of a step-up transformer. The two windings are arranged in a conventional manner in Fig. 3.18a. Figure 3.18b illustrates the connection of the two windings in an autotransformer. The two windings are now connected in series. The turns ratio of the autotransformer determines the relationship between the voltages on the first and second windings. *The output voltage of the transformer is the sum of the voltages on the first and second windings.* The first winding in the autotransformer is called the *common winding*. Its voltage appears on both sides of the transformer. The second winding is normally smaller than the first. It is called the *series winding*.

Figure 3.19 illustrates a step-down autotransformer. The input voltage is the sum of the voltages on the series and common windings. The output voltage is only the voltage on the common winding.

The terminology used for autotransformers is different from that for conventional *common voltage* transformers. The voltage on the common coil is V_C and the *common current* I_C is the current through the coil. The voltage on the series coil is called the *series voltage*

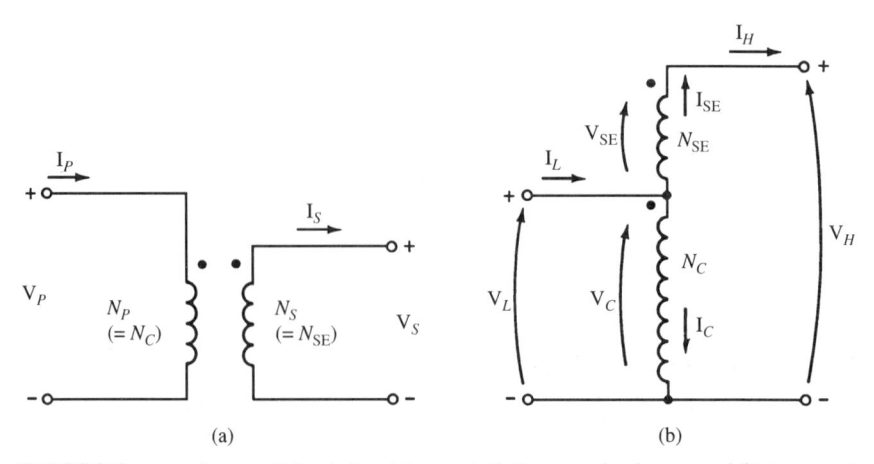

FIGURE 3.18 A transformer with its windings (*a*) connected in the conventional manner and (*b*) reconnected as an autotransformer.

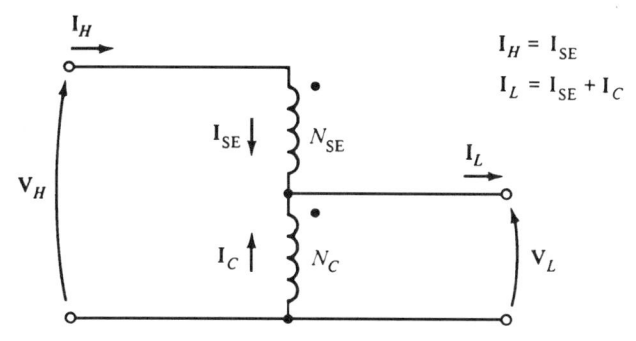

FIGURE 3.19 A step-down connection.

V_{SE}. The current in this coil is called the *series current* I_{SE}. The current and voltage on the high-voltage side of the transformer are called I_H and V_H, respectively. The corresponding parameters on the low-voltage side are called I_L and V_L, respectively. From Fig. 3.18b, the voltages and currents in the coils are related by:

$$\frac{V_C}{V_{SE}} = \frac{N_C}{N_{SE}}$$

$$N_C I_C = N_{SE} I_{SE}$$

The voltages in the coils are given by

$$V_L = V_C$$

$$V_H = V_C + V_{SE}$$

The currents in the coils are given by

$$I_L = I_C + I_{SE}$$

$$I_H = I_{SE}$$

The ratio of low and high voltage is given by

$$\boxed{\frac{V_L}{V_H} = \frac{N_C}{N_{SE} + N_C}}$$

The ratio of low to high current is given by

$$\boxed{\frac{I_L}{I_H} = \frac{N_{SE} + N_C}{N_C}}$$

THREE-PHASE TRANSFORMERS

Most of the transformers used in power generation and distribution systems are three-phase. These transformers can be constructed from three single-phase transformers connected in a three-phase bank (Fig. 3.20). They can also be constructed by wrapping three sets of

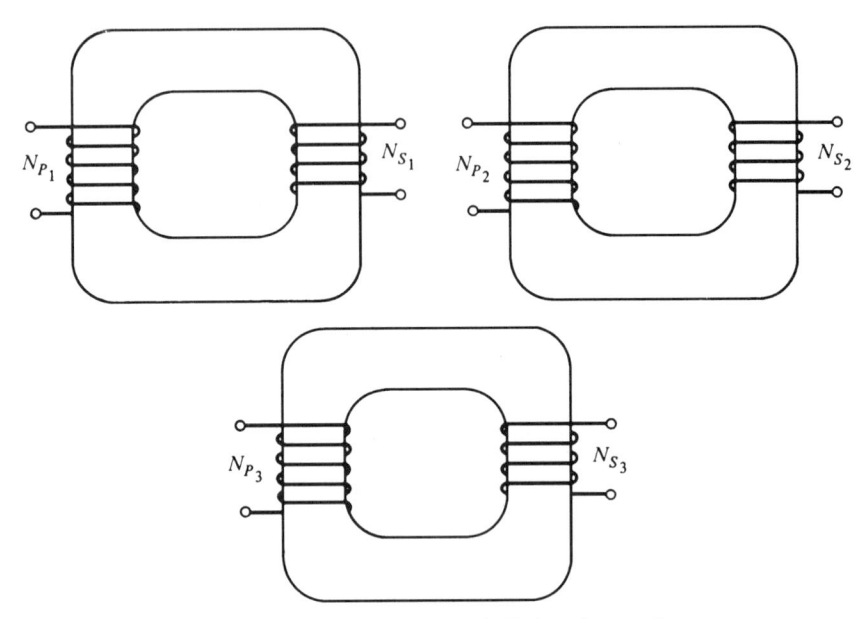

FIGURE 3.20 A three-phase transformer bank composed of independent transformers.

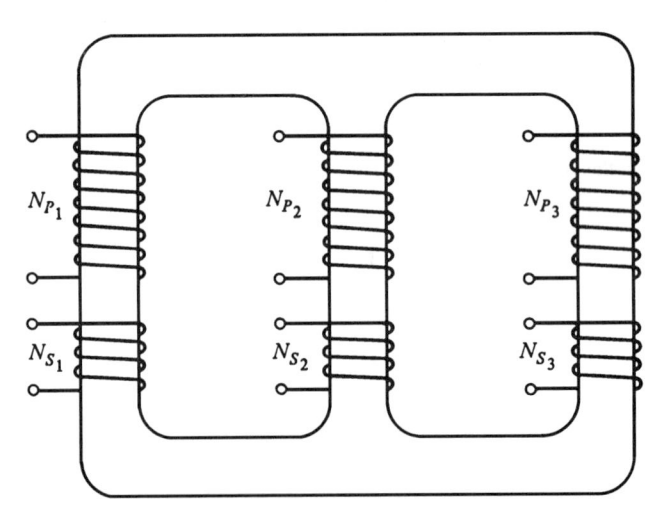

FIGURE 3.21 A three-phase transformer wound on a single three-legged core.

windings on a common core (Fig. 3.21). The second alternative is preferred because the transformer is cheaper, smaller, lighter, and slightly more efficient.

Three-Phase Transformer Connections

The primary and secondary phases of any three-phase transformer are independently connected as a wye (Y) or delta (Δ). The possible connections of a three-phase transformer are

1. Wye-wye (Y-Y)
2. Wye-delta (Y-Δ)
3. Delta-wye (Δ-Y)
4. Delta-delta (Δ-Δ)

Figure 3.22 illustrates a Y-Y connection.

TRANSFORMER RATINGS

Transformers have ratings for apparent power, voltage, current, and frequency.

Voltage and Frequency Ratings of a Transformer

The voltage rating of a transformer is used to protect the winding insulation from failure due to high voltage applied to it. It also has a second purpose associated with the

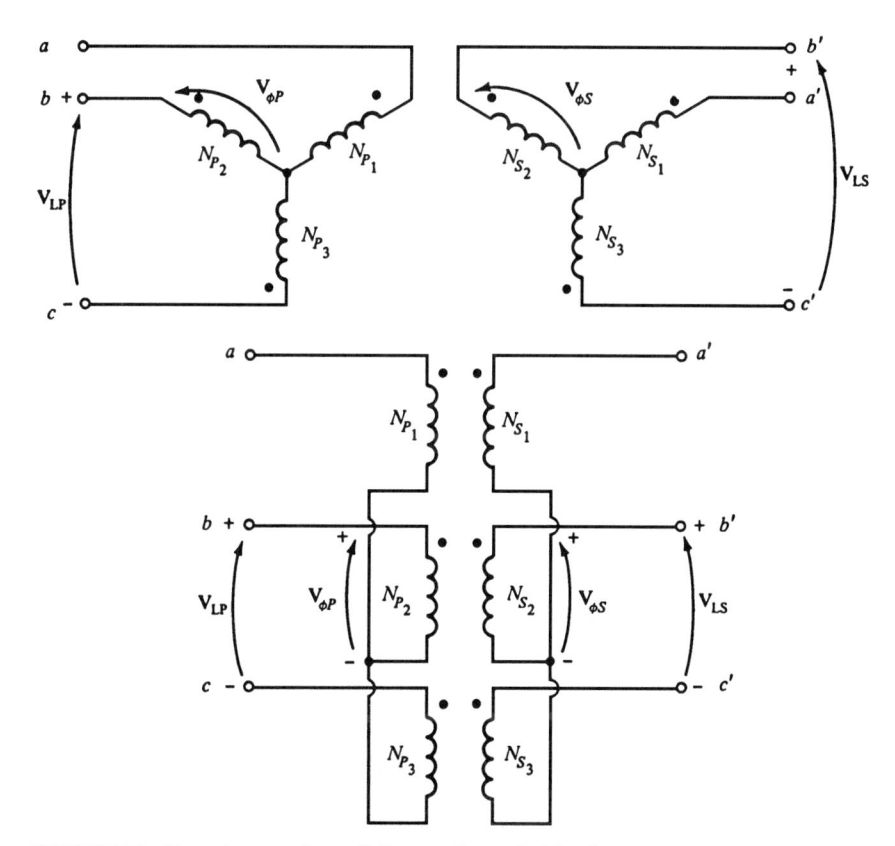

FIGURE 3.22 Three-phase transformer Y-Y connections and wiring diagram.

magnetization current of the transformer. Figure 3.8a illustrates the magnetization curve of a transformer. When a voltage $\upsilon(t) = V_M \sin \omega t$ V is applied to the primary winding of the transformer, the flux will be

$$\phi(t) = -\frac{V_M}{\omega N_P} \cos \omega t$$

A 5 percent increase in the applied voltage $\upsilon(t)$ will increase the flux in the core by 5 percent. However, in the saturation region, a 5 percent increase in flux requires a much larger increase in the magnetization current than 5 percent (Fig. 3.23). Thus, at a specified voltage, this high magnetization current starts to damage the insulation. The rated voltage is determined by the maximum acceptable magnetization current.

Since the flux is proportional to V/ω, a transformer designed for 60-Hz operation can be operated on 50 Hz if the applied voltage is reduced by one-sixth. This is known as *derating* the transformer. Similarly, a transformer designed for 50-Hz operation can have 20 percent higher voltage when operated at 60 Hz if its insulation system can withstand the higher voltage.

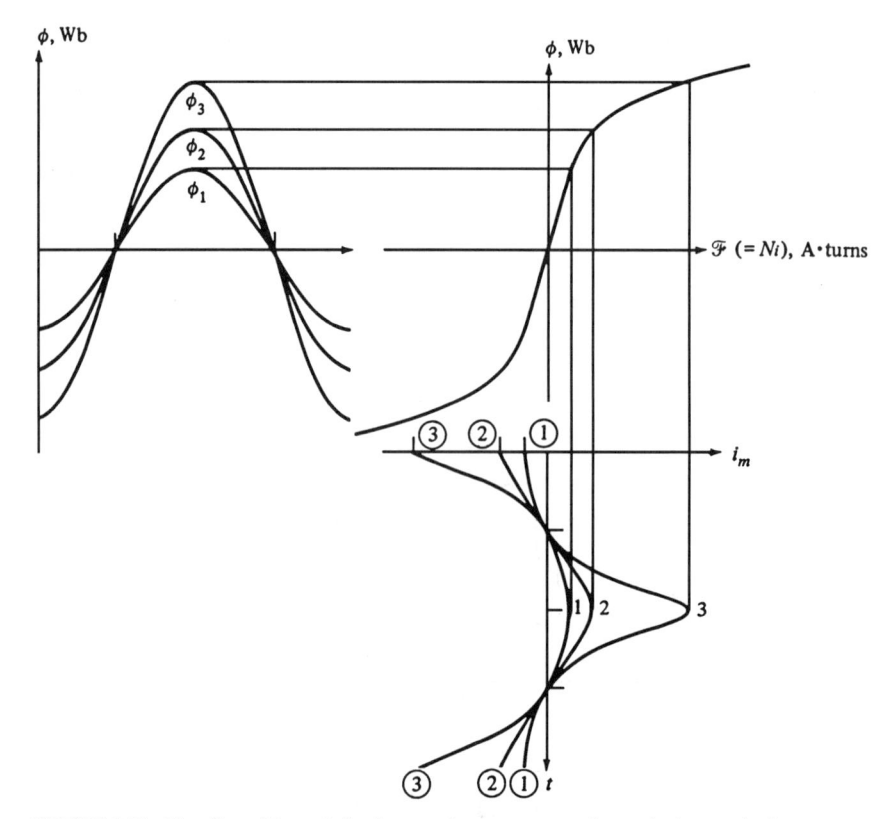

FIGURE 3.23 The effect of the peak flux in a transformer core upon the required magnetization current.

Apparent Power Rating of a Transformer

The apparent power rating of a transformer is used in conjunction with the voltage rating to determine the maximum allowable current in the transformer. Since the heat losses are proportional to i^2R, the current must be limited to prevent a significant reduction in the life of the insulation due to overheating.

If the transformer is operated at a lower frequency than normal (say, 50 Hz rather than 60 Hz) and its applied voltage is reduced as discussed above, its apparent power rating must be reduced by an equal amount. This is done to prevent overheating.

Inrush Current

The inrush current to the transformer is around 12 times the normal operating current. It usually lasts 8 to 10 cycles. The transformer and the power system connected to it must be able to withstand these high currents.

Transformer Nameplate

Figure 3.24 illustrates a typical nameplate for a distribution transformer. The information on the nameplate includes

- Rated voltage
- Rated apparent power

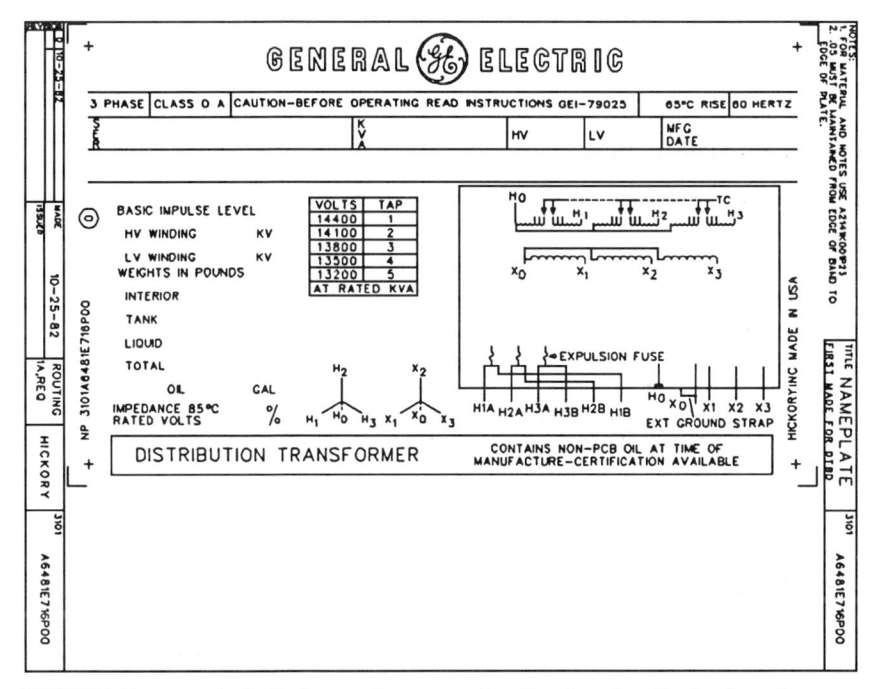

FIGURE 3.24 A sample distribution transformer nameplate. Note the ratings listed: voltage, frequency, apparent power, and tap settings. (*Courtesy of General Electric Company.*)

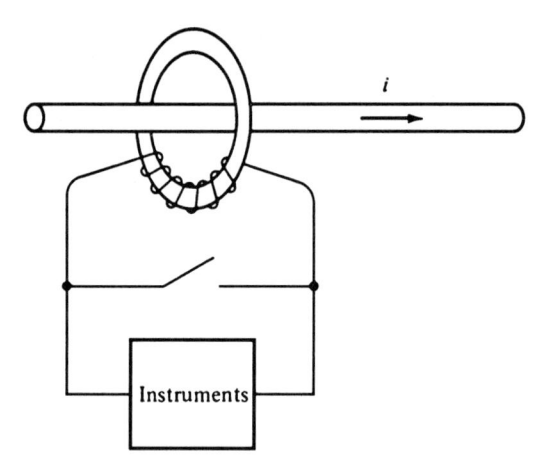

FIGURE 3.25 Sketch of a current transformer.

- Rated frequency
- Transformer impedance
- Voltage rating for each tap on the transformer
- Wiring schematic of the transformer
- Transformer type designation
- References to operating instructions

Instrument Transformers

Potential transformers and current transformers are two special-purpose transformers used for taking measurements.

Potential transformer is used to sample the voltage of a power system. It has high voltage on the primary side and low voltage on the secondary. Its power rating is low. These transformers are manufactured with different *accuracy classes*.

Current transformer is used to sample the current in a line (Fig. 3.25). Its secondary winding is wrapped around a ferromagnetic ring while the single primary line goes through the center of the ring. A typical rating of a current transformer includes 600 : 5 and 1000 : 5. Extremely high voltages will appear across the secondary terminals of a current transformer if the terminals are open. The current transformer should always be kept short-circuited to prevent dangerous high voltages from appearing at the secondary terminals.

REFERENCE

1. S. J. Chapman, *Electric Machinery Fundamentals,* 2d ed., McGraw-Hill, New York, 1991.

CHAPTER 4
TRANSFORMER COMPONENTS AND MAINTENANCE

INTRODUCTION

Electric power is generated most economically at 14 to 25 kV. System loads such as motors, lights, etc., require a voltage source of 440, 220, and 110 V. Transformers are needed to change the voltage level. They are also used for these reasons:

1. To reduce transmission losses between power plants and the load by stepping up the voltage. (When the voltage is stepped up, the current will be stepped down. This results in reduction of transmission losses which are proportional to the square of the current.)
2. Reduction in the diameter of the transmission line (amount of copper or aluminum) due to a reduction in the current flowing in the line.

Typical transmission voltages are 13.2, 22, 66, 230, 345, and 500 kV. In general, the longer the distance, the higher the voltage used.

CLASSIFICATION OF TRANSFORMERS

The two types of transformers are air-cooled (dry type) and oil-filled. The transformer rating increases with improved cooling methods. These are the typical ratings for various types of transformers:

Dry Transformers

The two types of dry transformers are self-air-cooled and forced-air-cooled. The heat is removed by natural convection in self-air-cooled transformers. In forced-air-cooled, it is removed by blowers. The rating of dry-type transformers used to be limited to less than 1 MVA. However, modern technology pushed this rating to 20 MVA. This was mainly due to improvements in the quality of insulation and mechanisms of heat removal from the transformer.

Oil-Immersed Transformers

In this type of transformer, the windings and core are immersed in oil. The main types of oil-immersed transformers are

- Oil-immersed self-cooled (heat is removed by natural convection of the oil through radiators)
- Oil-immersed cooled by forced air (heat is removed by blowers blowing air on radiators)
- Oil-immersed cooled by water (the oil is cooled by an oil-water heat exchanger)

The lowest rating of oil-type transformers is around 750 kVA. Since modern dry-type transformers are being manufactured up to a rating of 20 MVA, they are replacing oil-type transformers. The main reason is that oil-immersed transformers constitute a fire hazard, and they are very hard to maintain. Dry-type transformers are preferred in most industries.

The rating of oil-immersed cooled-by-water transformers is normally higher than 100 MVA. However, they could be used for transformers having a rating as low as 10 MVA if the transformer is feeding a rectifier. Harmonics (deformation in the sine wave of current and voltage) are normally generated in this application, causing significant heat generation, that necessitate cooling through a heat exchanger.

Most failures in transformers are caused by erosion of the insulating materials. Analysis of a transformer's oil can provide trends and early warning signs of premature failure. Figure 4.1 illustrates a basic electric power system from the utility to the consumer.

MAIN COMPONENTS OF A POWER TRANSFORMER

Figure 4.2 illustrates the outline drawing of a transformer. Figure 4.3 illustrates a cutaway view of an oil-filled transformer. The main components of the transformer are as follows:

1. *Concrete base.* This is to provide support for the transformer. It must be leveled and fire-resistant.
2. *Core.* It provides a route for the magnetic flux and supports the low-voltage and high-voltage windings.
3. *Low-voltage winding.* It has fewer turns compared with the high-voltage windings. Its conductor has a large diameter because it carries more current.

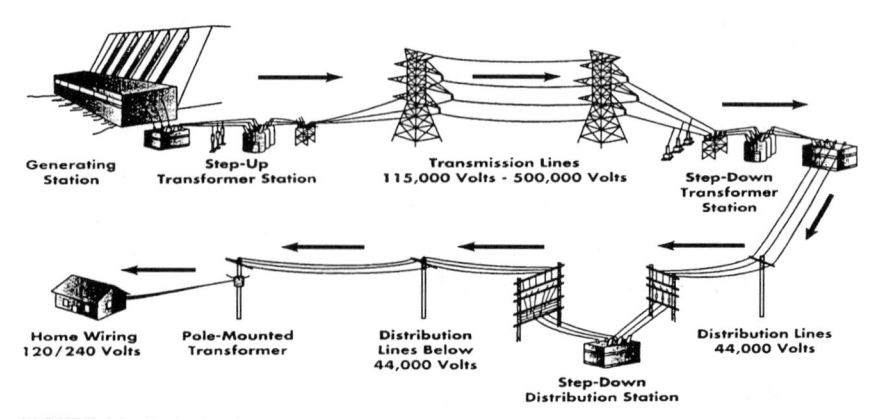

FIGURE 4.1 Basic electric power system.

TRANSFORMER COMPONENTS AND MAINTENANCE 4.3

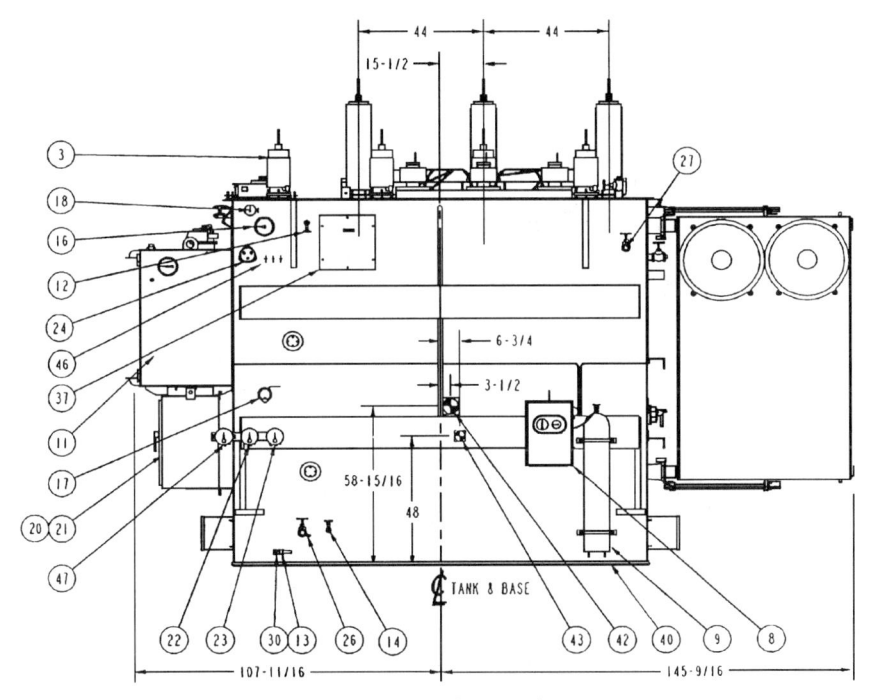

Legend for Outline Drawing

1. 3 H. V. bushings; 69 kV, 350 kV B.I.L. 400 amps, C/W 1.50"-12 silver-plated stud. 2" usable thread length. LAPP Cat #B-88726-70. Internal bushing lead is drawlead connected. Shipping weight per bushing is 139 lb.
2. 3 L. V. bushings; 25 kV, 150 kV B.I.L. 2000 amps, C/W 1.50"-12 silver-plated stud. 2.5" usable thread length. LAPP Cat # B-88723-70. Internal lead is bottom connected.
3. 1 L. V. neutral bushings; 25 kV, 150 kV B.I.L. 2000 amps, C/W 1.50"-12 silver-plated stud. 2.5" usable thread length. LAPP Cat # B-88723-70. Internal lead is bottom connected.
4. 3 H. V. lightning arresters, 60 kV, 48 kV MCOV. Ohio brass polymer housed, station class PVN 314048-3001. Shipping weight of arrester is 53.6 lb.
5. H. V. lightning arrester supports C/W seismic bracing. Shipping weight of supports is 304 lb.
6. 3 L. V. lightning arresters, 9 kV, 7.65 kV MCOV. Ohio brass polymer housed, station class PVN 314008-3001. Shipping weight of arrester is 21.3 lb.
7. L. V. lightning arrester supports C/W seismic bracing. Shipping weight of supports is 138 lb.
8. Nitrogen preservation system, ABB Type RNB, in a hinged, lockable door compartment C/W regulators, cylinder pressure gauge, empty cylinder alarm switch, transformer tank pressure/vacuum gauge, high- & low-tank-pressure alarm switches & tank gas space sampling valve. Connected to a nitrogen cylinder.
9. Nitrogen cylinder.
10. Deenergized tap changer handle C/W position indicator & provision for padlocking.
11. Load tap changer, Reinhausen Type RMV-II, C/W Qualitrol Self-Sealing Pressure Relief Device #208-60U, C/W deflector, alarm contacts & semaphore, liquid level indicator C/W contacts, 1½" NPT drain valve with sampler, dehydrating breather, vacuum interrupters and monitoring system. (See wiring diagram.)
12. 2 - 1" globe valves located approximately 6' below cover.
13. Ground connectors, Anderson # SW11-025-B.
14. 1" globe valve and plug.
15. 2 - Multiaxis impact recorders (one at each end). To be returned to Ferranti-Packard.
16. Magnetic liquid level indicator C/W contacts. (See wiring diagram.)

FIGURE 4.2 Transformer outline drawing, Los Medanos Energy Center (*Courtesy of VA Tech Ferranti-Packard Transformers.*)

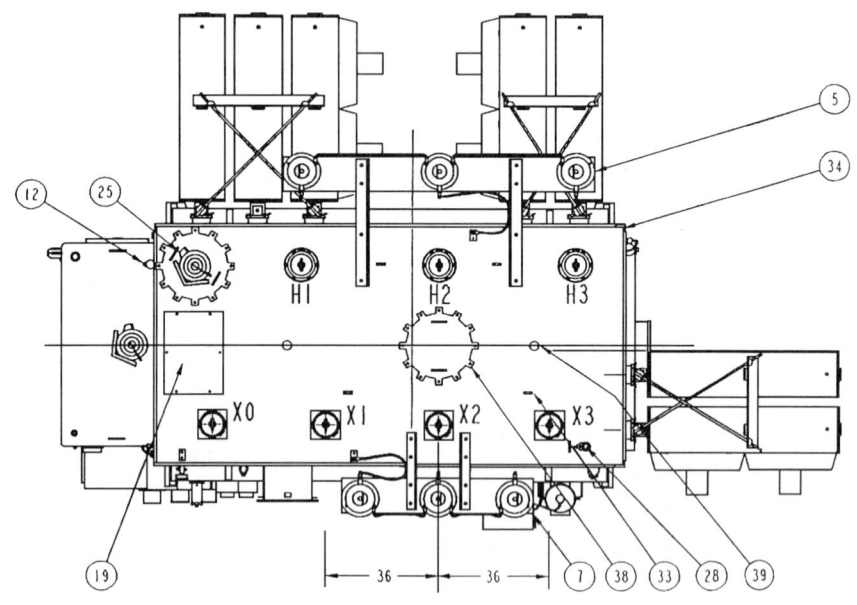

4.4 CHAPTER FOUR

Legend for Outline Drawing (*Cont.*)

17. Rapid pressure rise relay C/W contacts. Located under oil. (See wiring diagram.)
18. Pressure vacuum gauge C/W bleeder.
19. C. T. outlet box for current transformer leads continued to control cabinet.
20. Weatherproof control cabinet C/W LTC motor drive and C/W 7"× 21¼" opening(s) in bottom.
21. Nameplate.
22. Liquid temperature indicator C/W contacts. (See wiring diagram.)
23. Winding temperature indicator C/W Contacts. (See wiring diagram.)
24. Thermowell for liquid temperature indicator and winding temperature indicator.
25. Qualitrol Self-Sealing Pressure Relief Device #208-60U C/W deflector, alarm contacts & semaphore. (See wiring diagram.)
26. 2" NPT globe valve for draining C/W sampler.
27. 2" NPT top filter press globe valve.
28. 2"NPT globe valve located on tank cover for vacuum filling.
29. 2" NPT globe valve (for F.P. Use).
30. Copper-faced ground pads.
31. 4 - Transformer lifting lugs, each C/W a 2½" dia. hole. Use slings with minimum leg length of 12 ft.
32. 4 - Transformer jacking steps.
33. Cover lifting lugs.
34. Tank cover (welded on).
35. 7 Detachable cooling radiators C/W shutoff valves, lifting lugs, 1" NPT top vent C/W brass plug and 1" NPT bottom drain plug. Shipping weight per radiator is 1336 lb.
36. 9 Cooling fans C/W guards, 2 stages. (See wiring diagram.) Shipping weight per fan is 35 lb.
37. Accessible core ground(s), inside box C/W 5-kV bushings.
38. 2 - 18" dia. manholes.
39. Safety harness anchor rod (for F.P. use).
40. Slab base suitable for skidding and rolling in both directions.
41. 2 sets of seismic radiator bracing per radiator bank (1 at top & 1 at bottom). Shipping weight per set is 41 lb.
42. Center of gravity—operation.

FIGURE 4.2 (*Continued*) Transformer outline drawing, Los Medanos Energy Center (*Courtesy of VA Tech Ferranti-Packard Transformers.*)

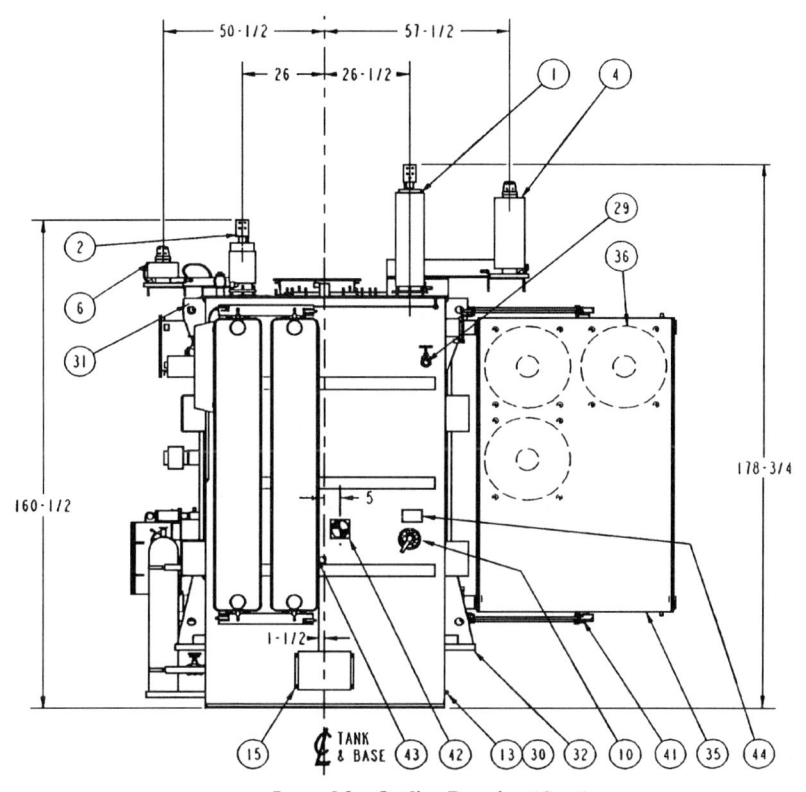

TRANSFORMER COMPONENTS AND MAINTENANCE 4.5

Legend for Outline Drawing (*Cont.*)

43. Center of gravity—shipping.
44. Warning plate—deenergized tap changer.
45. Bushing terminal connectors on HV, LV & neutral bushings are SEFCOR #SNFT-44-4B-SND (tin plated).
46. Liquid cover mark (+++).
47. Thermal load gauge center. (See wiring diagram.)

Notes:
- [Bracketed] Items in legend are removed for shipping.
- Oil is removed for shipping.

Transformer rating data	Estimaed mass of components		
Oil-filled transformer	Core & windings		44,855 lb
3-phase 60 Hz 55°/65°C rise	Tank & fittings		24,257 lb
Type: OA / FA / FA	Removable pads		9,353 lb
MVA: 18 / 24 / 30		U.S. Gal.	
H.V. - 67000 delta	Liquid (trans)	4969	36,111 lb
L.V. - 12470 wye	Liquid (rads)	492	3,574 lb
Tank characteristics	Liquid (LTC)	280	2,032 lb
Tank strength: ±14.7 P.S.I.	Total liquid	5741	41,717 lb
Tank size: 73 × 145 × 133-1/2"	**Total mass**		120,183 lb
Paint: ANSI 70 gray	**Shipping mass**		72,000 lb

FIGURE 4.2 (*Continued*) Transformer outline drawing, Los Medanos Energy Center (*Courtesy of VA Tech Ferranti-Packard Transformers.*)

4.6 CHAPTER FOUR

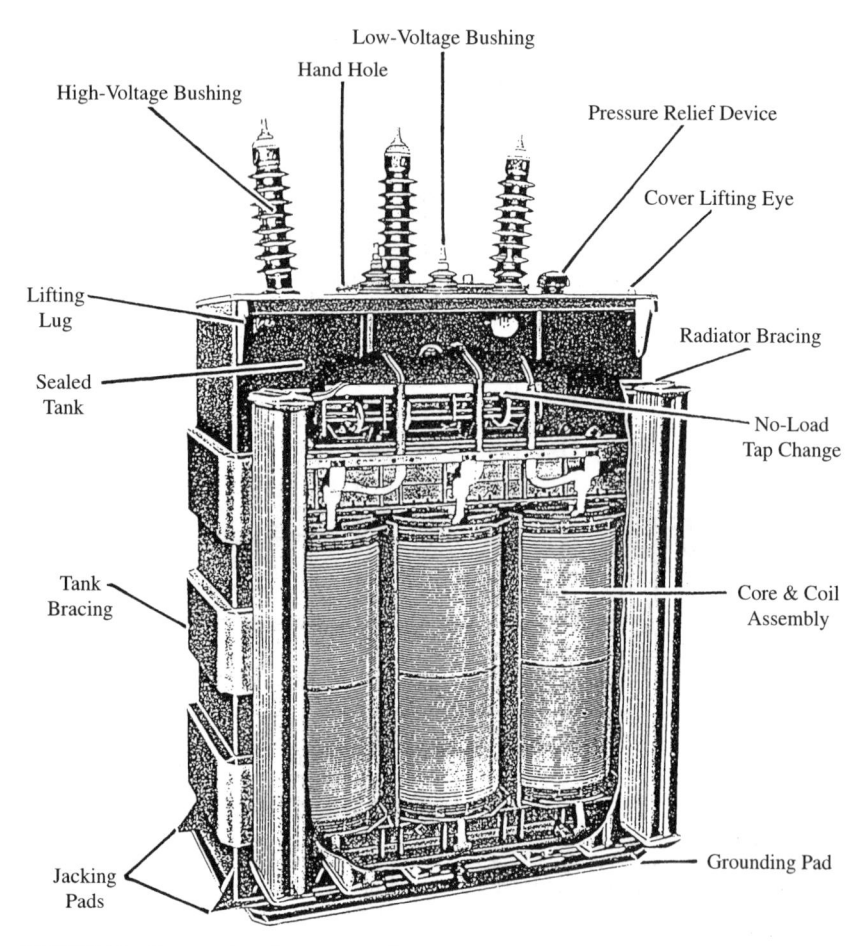

FIGURE 4.3 Cutaway view of a power transformer.

4. *High-voltage winding.* It has a larger number of turns, and its conductor has a smaller diameter than the low-voltage winding conductor. The high-voltage winding is usually wound around the low-voltage winding (only cooling ducts and insulation separate the windings). This is done to minimize the voltage stress on the core insulation.
5. *Tank.* It houses the windings, core, and oil. It must be strong enough to withstand the gas pressures and electromagnetic forces that could develop when a fault occurs.
6. *Oil.* It is a good-quality mineral oil. It provides insulation between the windings, core, and transformer tank. It also removes the heat generated. The oil is specially refined, and it must be free from impurities such as water, inorganic acid, alkali, sulfur, and vegetable and mineral oil.

Note: Some transformers use Askarel, which is nonflammable insulating and cooling medium, instead of insulating oil. It gives good fire protection, which is a significant advantage when the transformer is located inside a building. However, Askarel contains polychlorinated biphenals (PCBs). They have been linked to cancer-causing substances. Therefore, they have been banned. Most industries are now in the process of replacing transformers containing Askarel with dry-type transformers or transformers containing mineral oil.

7. *Thermometer.* It measures the oil temperature and initiates an alarm when the temperature exceeds the alarm set point.
8. *Low/high-voltage bushing.* This ceramic bushing carries the low/high-voltage conductor and insulates it from the tank. The high-voltage bushing is usually filled with oil to enhance the heat removal capability.
9. *Low/High-voltage connection.* It connects the low/high-voltage conductor to the circuit.
10. *Conservator tank.* It contains oil and has the capability of absorbing the swell of the oil when it becomes hot.
11. *Gas detector relay.* Electrical faults are characterized by the formation of gas. In the early stages of the fault, small quantities of gas are liberated. The amount of gas formed increases with time, and a violent explosion could occur. Most of the damage and expense can be saved if the fault can be discovered and corrected in its early stages. The gas detector relay detects gas buildup in the tank. Any gas generation inside the transformer passes through it. It has two parts. One part detects large rate of gas production which could be caused by a major fault (such as hydrogen, acetylene, and carbon monoxide). The second part detects the slow accumulation of air and gases which are released from the oil when it gets warm or from minor arcs. When the gas level reaches a predetermined amount, an alarm is annunciated. The relay has also the capability to trip the transformer in case of a serious fault which would result in a sudden rush of oil or gas through it.
12. *Explosion vent.* It prevents the buildup of high pressure in the tank due to gas formation (caused by oil disintegration) when a fault occurs. A relief diaphragm at the end of the explosion vent ruptures when the gas pressure reaches a certain predetermined value, to relieve the pressure to atmosphere.
13. *Oil level.* The oil level changes with the temperature of the oil. The core and winding must always be immersed in oil to ensure they are adequately cooled and insulated.
14. *Sight glass.* It indicates oil level.
15. *Breather.* This allows air movement to the conservator upon swell and shrinkage of the oil. It has a silica gel (air dryer) to remove moisture from the air entering the conservator tank. The transformer oil must be kept dry always. A water content of 8 parts per million (ppm) in the oil will reduce the dielectric strength to a dangerous level. Oxygen has also adverse consequences on the oil (it oxidizes it, resulting in the formation of sludge). Some transformers have a seal of inert gas (nitrogen) between the oil and atmosphere to prevent oxygen and moisture from reaching the oil.
16. *Radiator.* It is a heat exchanger which cools the oil by natural circulation.
17. *Ground connection.* The tank is connected to ground (earth) to ensure the safety of maintenance personnel.

Transformer Core

The core is made from thousands of laminations of grain-oriented steel. The thickness of a lamination is around 12 mils. Each lamination is coated with a thin layer of insulating material. The windings have a circular cross section. They are normally concentric. The lowest-voltage winding is placed next to the core to reduce the voltage stress. There is a layer of insulation between the low-voltage winding and the core. There are also oil ducts and insulating barriers between the coils. The core is grounded at one point. The ground connection is normally accessible externally for test purposes.

Windings

The windings must be able to withstand the large mechanical forces created by a short circuit. The winding insulation must be able to withstand the highest operating temperature without excessive degradation. The cooling fluid must be able to flow freely through spaces between the windings to remove the heat. The windings are arranged concentrically. The highest voltage is located on the outside.

Nitrogen Demand System

Many transformers are equipped with an automatic nitrogen demand system. It regulates the pressure in the transformer during the thermal cycles of the oil (swell and shrinkage of the oil). Nitrogen is used as a buffer gas between the oil and the air. Its purpose is to keep outside air (containing water vapor) from contacting the oil. The water vapor has devastating effects on the oil. The dew point of nitrogen is less than $-50°C$ to ensure that it is very dry. Most units have an alarm indicating low nitrogen pressure in the cylinder.

Conservator Tank with Air Cell

The conservator air cell preservation system is an expansion tank located above the main transformer tank. It provides a head of oil so that the transformer tank is filled always. The quality of oil in the transformer is preserved by sealing it from contaminants such as moisture and oxygen (Fig. 4.4).

Air is drawn into the air cell (part 1) or expelled through the breather (part 2) as the oil level changes. The air cell prevents the transformer oil from coming in direct contact with atmospheric air. The air cell will inflate and deflate as the oil volume changes in the conservator tank.

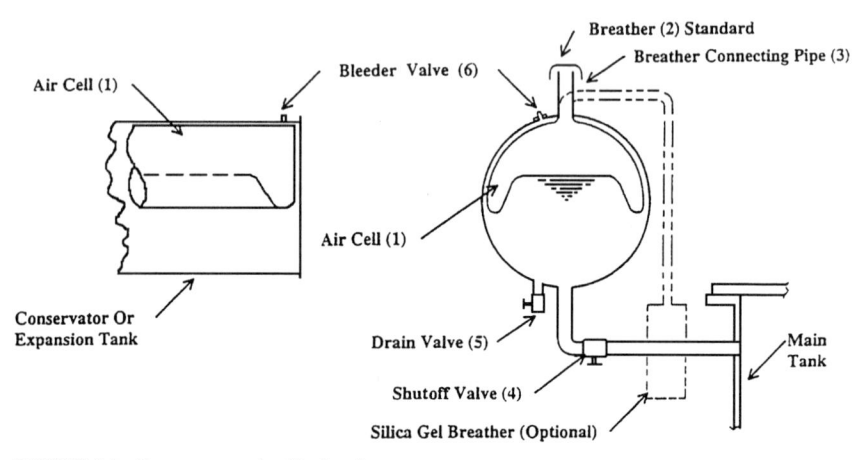

FIGURE 4.4 Conservator tank with air cell.

Current Transformers

Current transformers (CTs) are auxiliary transformers used normally for metering or operation of auxiliary equipment and relays. They could be located inside or outside the tank assembly.

Bushings

A bushing is an insulated conductor passing through a cover of a piece of enclosed electrical equipment. The bushing carries the current through it and seals the opening against weather or oil pressure. It also supports the leads and provides protection against flashover due to overvoltage or heavy contamination (Fig. 4.5).

A bushing has a voltage less than 35 kV and consists of a current-carrying conductor, an epoxy or porcelain insulator, and a mechanical assembly to hold it all together. It also has terminals on the top and bottom for connection. The bushing is filled with oil for cooling.

Tap Changers

Many transformers have on-load or off-load tap changers. They change the effective number of the windings in the transformer to maintain the secondary voltage constant.

Insulation

Figure 4.6 shows the main components of a power transformer and the major uses of Kraft paper. The winding insulation is normally made of paper. Cellulose board is also used as internal insulation in liquid-filled transformers. Cellulose insulation is impregnable with the insulating fluid of the transformer. This maintains a uniform dielectric stress throughout the transformer.

TYPES AND FEATURES OF INSULATION

1. Low-density calendered board, available in flat and formed parts in thickness up to 0.188 in.
2. Laminated low-density calendered board using dextrin resin.
3. High-density precompressed board designed for high mechanical strength and dimensional stability. The dimensional stability is important for the long life of the transformer. It ensures that the windings do not loosen over time. This insulation is used for space ducts in the transformer.
4. Laminated high-density boards using polyester resin. This material is typically used for clamping plates at each end of the transformer windings. It is completely impregnable with oil despite its high density. This ensures freedom from partial discharge.

Aramid is another type of insulation used in power transformers. It is chemically similar to nylon. It is completely impregnable with oil. It is generally known as Nomex. It has high strength, and it can withstand an operating temperature of 170°C without deterioration (cellulose insulation can only withstand 105°C). However, Nomex is very expensive

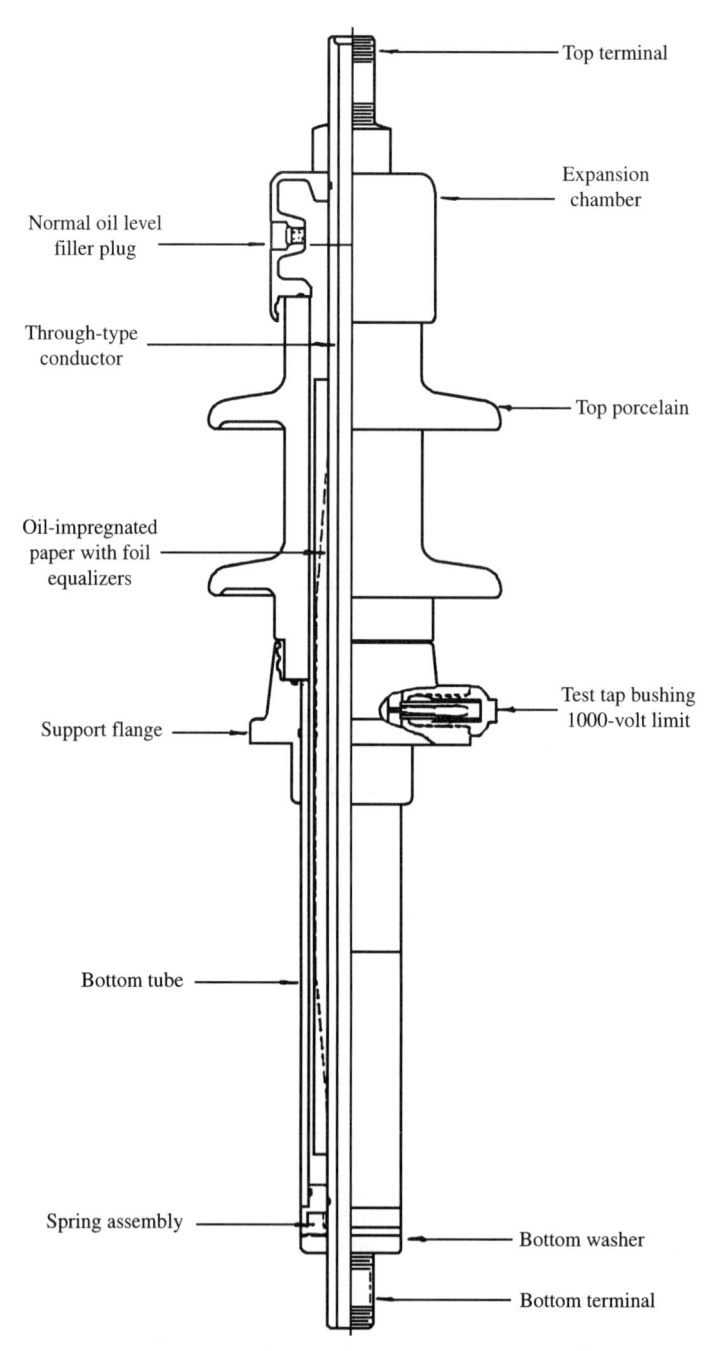

FIGURE 4.5 Typical bushing construction.

Kraft paper providing turn to turn insulation of winding coil.

Conductor tap leads wrapped in paper insulation.

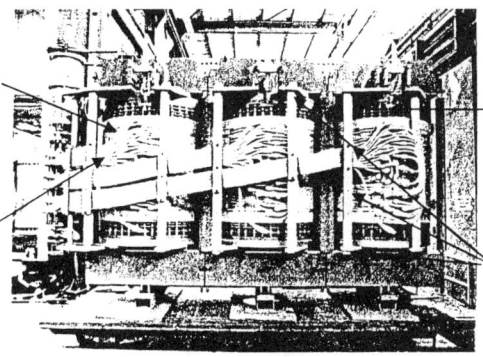

End rings using dense pressboard to aid in mechanical strength. Spaced to also provide for cooling of the inner winding(s).

Dense pressboard used for support.

FIGURE 4.6 Three-phase power transformer (75/125 MVA).

(about 10 times more than cellulose insulation). In general, Nomex is not cost-effective for use as a major insulation in a power transformer.

Reasons for Deterioration

Insulation deteriorates due to the following:

1. Heat
2. Contamination such as dirt, moisture, or oxygen
3. Electrical stress
4. Mechanical stress and strain

Insulation deterioration normally results in the loss of its mechanical properties. As the insulation weakens, it loses flexibility and becomes brittle. It would not be able to resist the mechanical stresses resulting from the magnetic forces, differential temperature expansion, and vibration. The insulation disintegrates, leading to electrical faults.

FORCES

During normal operation, the axial and radial forces between the windings are moderate. These forces become severe during a short circuit. The transformer must be able to withstand fault conditions. This includes forces 10 times higher than normal. Figure 4.7 illustrates damaged top and bottom coils in a transformer.

CAUSE OF TRANSFORMER FAILURES

Most transformers fail due to mechanical reasons. The windings are subjected to physical forces that operate in all directions. These forces can become astronomical under short-circuit conditions. For example, a 16-MVA transformer will develop a 500,000-lb vertical force

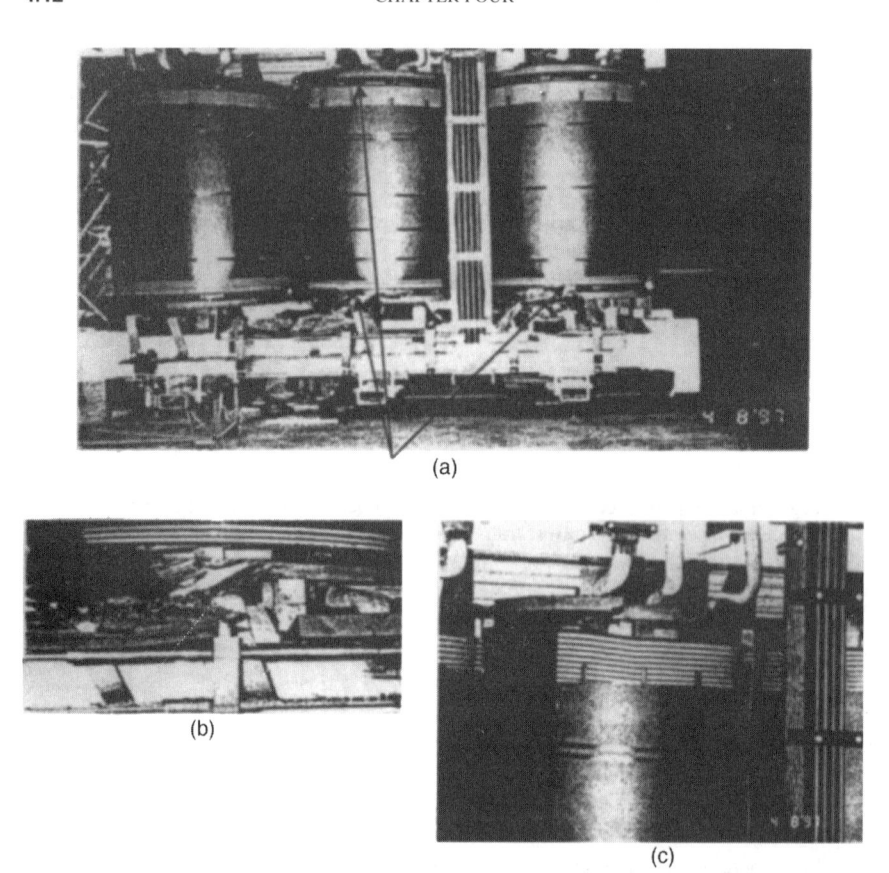

FIGURE 4.7 (*a*) Examples of damage to top and bottom coils in a transformer; (*b*) bottom view of coil; (*c*) top view of coil.

and a hoop force (horizontal) of 3,000,000 lb. Therefore, the windings must be braced to withstand these forces. If the windings are not properly braced, physical movement occurs. Short circuit is developed, leading to transformer failure. Therefore, the root cause of electrical faults is mechanical in nature.

When the Kraft paper is impregnable with a good, clean, dry oil, it becomes one of the best dielectrics known in industry. Water has devastating effects on Kraft paper. Most transformers fail due to the presence of water. The water that weakens the Kraft paper is the *microscopic* droplets formed by paper degradation and oil oxidation (Fig. 4.8). The water droplets are produced by the inner layers of paper and oil that is trapped between the coil (copper or aluminum) and the paper. The water acts as a solvent to dissolve and weaken the paper by destroying the fiber of the Kraft paper. This results in loosening the windings. The paper insulation will get abraded by the constant moving of the windings. A total failure is created by having a failure in an extremely small amount of the paper in the transformer. Adequate measures must be taken early, and promptly to protect the transformer.

The water generated in the "innards of the transformer" that causes the destruction of the unit is significantly below the level of detection through oil test procedures and/or electrical testing. Therefore, advanced deterioration would have already occurred by the time the evidence appeared in the test data. A suitable solution involves continuous dehydration of the transformer during normal operation. A less expensive solution involves servicing the transformer every 3 to 5 years.

The amount of moisture in the paper of a transformer is expressed as *percent moisture by dry weight* (%M/dw). The aging factor is controlled by controlling the %M/dw. The upper limit of 0.5 %M/dw should be specified when a transformer is selected. If this limit is not specified, the moisture content could be as high as 1.5 to 2.0 %M/dw.

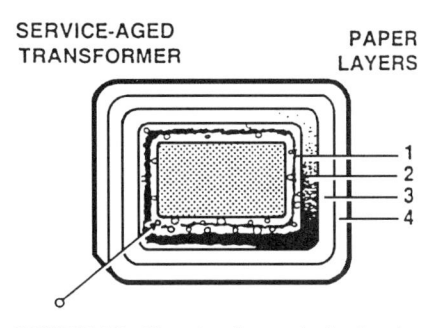

FIGURE 4.8 Water in microscopic droplets that cause the Kraft paper to dissolve from oil decay products.

Figure 4.9 illustrates the aging factor versus %M/dw. As shown, moisture weakens the Kraft paper. This weakening is measured and expressed by the *aging factor* (AF). Notice the significant increase of the AF with %M/dw. Figure 4.10 illustrates the significant decrease in transformer life expectancy with %M/dw.

Most transformers fail at the bottom due to high %M/dw. This is caused by the fact that paper has up to 3000 times higher affinity for water than oil does at lower temperatures.

TRANSFORMER OIL

The oil acts as insulation and a cooling medium for the transformer. The mineral oil used in transformers normally has these features:

1. It has a high flash point to minimize fire hazard.
2. It is nonvolatile at operating temperatures to avoid evaporation losses.
3. It has a low pour point.
4. It remains as a liquid at the lowest ambient temperatures expected.
5. It is stable and inert.
6. It resists oxidation which increases acidity and formation of sludge.

Table 4.1 lists all the oil tests and when they are normally performed. The private industry usually performs the following tests only: dielectric breakdown, neutralization number (NN), interfacial tension (IFT), specific gravity, color, and visual exam. The remaining tests may be done when necessary.

The visual examination consists of checking the color of the oil. The color varies from the clear (like water), with acidity of 0.01 to 0.03 mg KOH/g and interfacial tension of 30 to 45 dyn/cm; to black with acidity of 1.01 mg KOH/g and higher and interfacial tension of 6 to 9 dyn/cm. The acceptable conditions are clear to amber (acidity: 0.15 mg KOH/g; interfacial tension: 22 dyn/cm).

Testing Transformer Insulating Oil

The functions of the oil in the transformer are to

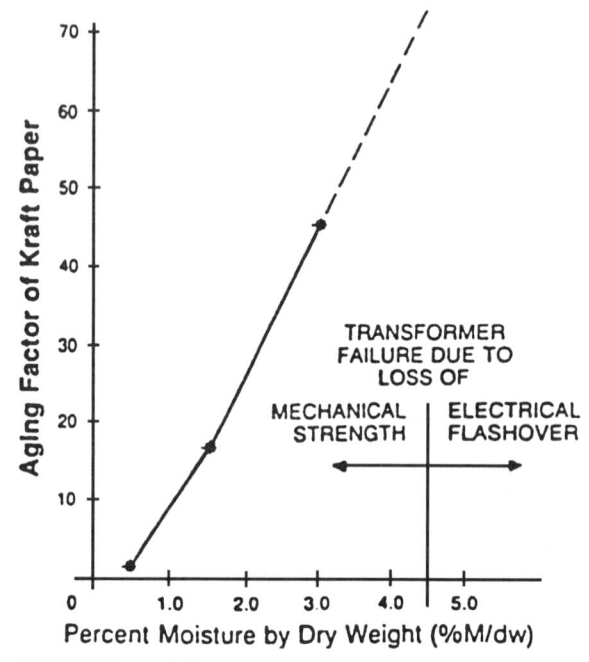

FIGURE 4.9 Variation of aging factor with percent moisture divided by dry weight.

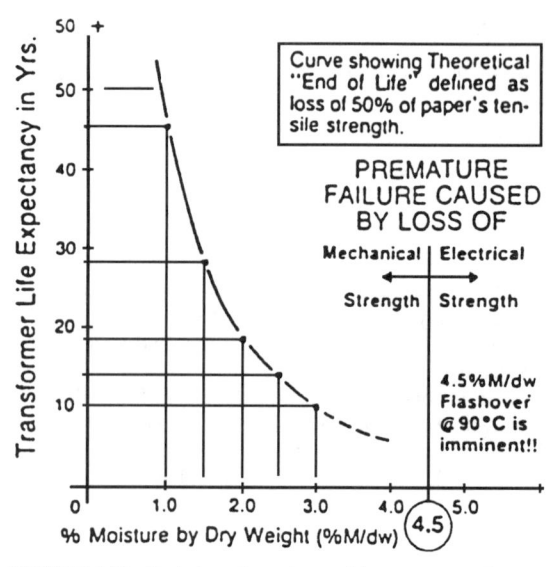

FIGURE 4.10 Variation of transformer life expectancy with percent moisture divided by dry weight.

TABLE 4.1 Insulating Oil Tests Available

ASTM test method	Test name	Units	Used oil	New oil	Aging analysis	Fault analysis
D877/D1816	Dielectric breakdown	kV	♦	♦		
D974	Neutralization number	mg KOH/g	♦	♦		
D924	Power factor @ 100°C	%		♦		
D924	Power factor @ 25°C	%	♦			
D971	Interfacial tension	dyn/cm	♦	♦		
D1500	Color	Scale 0.5 to 8.0	♦	♦		
D1298	Specific gravity	@15°C	♦	♦		
	PCB	ppm	♦	♦		
	Water content	ppm	♦	♦		
D97	Pour point	°C	♦	♦		
D1935	Steam emulsion	s		♦		
D92	Flash point	°C		♦		
D445	Viscosity	°C		♦		
D2440	Stability test			♦		
	Carbon content	ppm (g/kg)				♦
	Metal particles					♦
	Oxidation inhibitors				♦	
	Furans	ppb			♦	
	Degree of polymerization	Units P_v			♦	

1. Provide insulation
2. Provide efficient cooling
3. Protect the windings and core from chemical attack
4. Prevent the buildup of sludge in the transformer

The condition of the oil in the transformer determines the transformer life. Annual testing is the *minimum* requirement to ensure acceptable dielectric strength for the oil. These are the benefits of annual testing:

1. It indicates the internal condition of the transformer. It detects the presence of sludge. The sludge must be purged before it can precipitate on the windings and other surfaces inside the transformer.
2. It indicates any deteriorating trends. The naphthenic insulating oils used in transformers today have been used for more than 50 years. The deteriorating trend of these oils is well known. Many data are available to permit the comparison between normal and abnormal oil.
3. It prevents a forced outage. Incipient problems are detected early. Corrective action is scheduled with minimal disruption. The tests that are required for transformer oil are listed in Table 4.2. The results of one test only cannot indicate the condition of the oil. The true condition of the oil is obtained by considering the combined results of the eight tests together (especially the first four tests).

The ASTM D-877 (flat disk electrode) is the classical test for determining the dielectric strength of the oil. It will detect contaminants such as free water, dirt, or other conducting particles. However, it will not detect the presence of dissolved water, acid, or sludge.

TABLE 4.2 The Eight Most Important ASTM Tests for In-Service Transformer Oil

ASTM test method	Criteria for evaluating test results	Information provided by test
D-877, Dielectric breakdown strength	New oil should not break down at 30 kV or below	Free water present in oil
D-974, Neutralization (or acid) number (NN)	Milligrams of potassium hydroxide required to neutralize 1 g of oil (0.03 or less of new oil)	Acid present in oil
D-971, Interfacial tension (IFT)	Dynes per centimeter (40 or higher for new oil)	Sludge present in oil
D-1524, Color	Compared against color index scale of 0.5 (new oil) to 8.0 worst case	Marked change from one year to next indicates a problem
D-1298, Specific gravity	Specific gravity of new oil is approximately 0.875	Provides a quick check
D-1524, Visual evaluation of transparency/opacity	Good oil is clear and sparkling, not cloudy	Cloudiness indicates presence of moisture or other contaminants
D-1698, Sediment	None/slight/moderate/heavy	Indicates deterioration and/or contamination of oil
D-924, Power factor	Power factor of new oil is 0.05 or less	Reveals presence of moisture, resin varnishes, or other products of oxidation in oil, or of foreign contaminants such as motor oil or fuel oil

Note: For comparative purposes, specifications of new insulating oil can be obtained from ASTM D-3487-77, *Standard Specifications for Mineral Insulating Oil in Electrical Apparatus*, available from the American Society for Testing and Materials, 1916 Race St., Philadelphia, PA 19103.

Causes of Deterioration

Oxygen, heat, and moisture have adverse effects on oil. Oxygen is derived from the air that entered the transformer and from the transformer oil. Oxygen is liberated in some cases by the effect of heat on cellulose insulation. The natural oxygen inhibitors in new insulating oil deplete gradually with time. Thus, the oxidation rate of the oil increases steadily while the oil is in service.

Pure hydrocarbons do not oxidize easily under normal conditions. The American Society for Testing and Materials (ASTM) has established that oil oxidation results generally from a process that starts when oxygen combines with unstable hydrocarbon impurities. The metals in the transformer act as a catalyst for this combination. Acids, peroxides, alcohols, and ketones are the product of oxidation. The oxidation process results in continuous detrimental action on the insulating materials of the transformer. Sludge will eventually form. Greater damage will be caused by sludge formation. This is due to the inability of the sludge to circulate and remove the heat buildup.

The most important tests of the eight ASTM tests listed in Table 4.2 are neutralization number (NN) and the D-971 for interfacial tension (IFT). This is so because these tests deal directly with the acid content and the presence of sludge. The two tests provide a quantitative description of the condition of the oil.

The Neutralization Number Test

The NN test (ASTM D-974) determines the acid content of the oil. An oil sample of known quantity is titrated with the base potassium hydroxide (KOH) until the acid in the oil has been neutralized. The NN is expressed as the amount of KOH in milligrams required to neutralize 1 g of oil. A high NN indicates high acid content.

The acid formation in the transformer begins as soon as the oil is placed in service. Figure 4.11 illustrates the increase of NN with time and temperature. Electromechanical vibration, mechanical shock, and especially heat will accelerate the normal deterioration of the oil. Even minute amounts of water will enhance the oxidation process and the formation of acids. The copper or copper alloys in the windings will act as a catalyst for this reaction.

The Interfacial Tension Test

The ASTM D-971 test for interfacial tension determines the concentration of sludge. In this test, a platinum ring is drawn through the interface between distilled water and the oil sample. A delicate balance (Cenco DuNuoy Tensiometer) is used to draw the ring. The test results are expressed in dynes per centimeter. This test gives good indication about the presence of oil decay products. The IFT of new oil is more than 40 dyn/cm. The IFT of badly deteriorating oil is less than 18 dyn/cm.

The Myers Index Number

A high IFT indicates that the oil is relatively sludge-free. Therefore, it will be purer than an oil with low IFT. Conversely, when the oil has a high acid content and bad deterioration, it

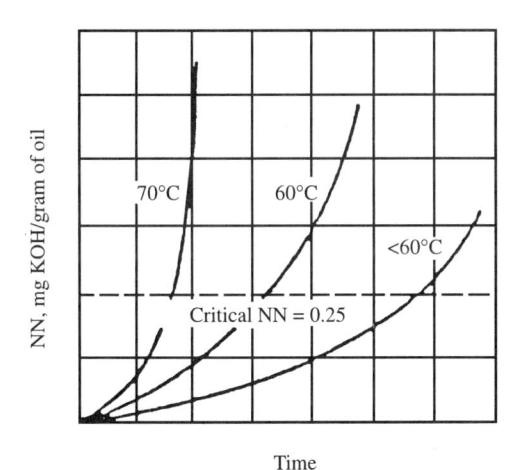

FIGURE 4.11 Typical pattern of increase in NN as a function of time is exhibited by curve for transformer oil operated at 60°C. The exponential rise in NN at the critical point results from the catalytic action of acids and the depletion of the oxidation inhibitors. Heat is the greatest accelerator of oil deterioration; deterioration is most marked above 60°C. Beyond 60°C, the rate of deterioration approximately doubles for each 10°C increase.

will have a high neutralization number. An excellent indicator of the oil condition is obtained by dividing the IFT by the NN. This ratio is known as the *Myers index number* (MIN) or *oil quality index number*. For example, the MIN for a new oil is around 1500:

$$\text{MIN} = \frac{\text{IFT}}{\text{NN}}$$

$$= \frac{45.0 \text{ (typical new oil)}}{0.03 \text{ (typical new oil)}} = 1500$$

The Transformer Oil Classification System

The seven classifications of the transformer oil are presented in Table 4.3. They are based on the oil's ability to perform its four intended functions (cooling, insulation, protection against chemical attack, and prevention of sludge buildup).

The oil in category 1 can perform efficiently all four functions. It does not require attention other than periodic testing.

Preventive maintenance is required for the oils in categories 2 and 3. The decrease of IFT to 27.0 indicates the start of sludge in solution. At this stage, the oil is less than ideal, but does not require immediate attention.

Oil in category 3 (marginal) is not providing adequate cooling and winding protection. At this stage, fatty acids have begun to coat the winding insulation. The sludge has also begun to build up in the insulation voids. Numerous studies of transformer failures for oils in categories 2 and 3 revealed sludge in the voids of the insulation system. This is the reason for not deferring transformer maintenance when they contain marginal oil.

Transformers containing oils in categories 4, 5, and 6 should be serviced promptly. Sludge has already been deposited on most of the winding and core in category 4. Considerable insulation shrinkage and blockage of cooling vents would have occurred in categories 5 and 6 (extremely bad) due to oil deterioration. The transformer should be replaced if the oil is found in category 7.

There is some overlap of MIN ranges in the first three categories. This overlap is due to the fact that the oil should meet the criteria for both minimum IFT and maximum NN for the category, in addition to falling within the range given for MIN.

Table 4.4 summarizes the results of a study on 500 transformers to determine the relationship between IFT, NN, and the sludge content. The transformers have been selected from different industrial environments. All 500 transformers experienced visible sludge buildup when the NN exceeded 0.6 mg KOH/g, or IFT dropped below 14.0 dyn/cm.

Figure 4.12 illustrates the relationship between NN, IFT, and the condition of transformer oil. If a sickeningly sweet odor is detected while the oil sample is taken, additional

TABLE 4.3 Transformer Oil Classifications

Oil condition	NN	IFT	Color	MIN
1. Good oils	0.00–0.10	30.0–45.0	Pale yellow	300–1500
2. Proposition A oils	0.05–0.10	27.1–29.9	Yellow	271–600
3. Marginal oils	0.11–0.15	24.0–27.0	Bright yellow	160–318
4. Bad oils	0.16–0.40	18.0–23.9	Amber	45–159
5. Very bad oils	0.41–0.65	14.0–17.9	Brown	22–44
6. Extremely bad oils	0.66–1.50	9.0–13.9	Dark brown	6–21
7. Oils in a disastrous condition	1.51 or more		Black	

TABLE 4.4 Correlation of Neutralization Number and Interfacial Tension to Transformer Sludging*

Neutralization number, mg KOH/g of oil	Number of units found sludged
0.00–0.10	0
0.11–0.20	190
0.21–0.60	360
0.60 or higher	500

Interfacial tension, dyn/cm	Number of units found sludged
24 or higher	0
22–24	150
20–22	165
18–20	175
16–18	345
14–16	425
14 or less	500

*Study conducted by ASTM from 1946 to 1957 on 500 transformers that had been in service for some time. Study is described in ASTM Special Publication No. 218, *Evaluation of Laboratory Tests as indicators of the Service Life of Uninhibited Electrical Insulating Oils* (1957).

tests are needed. Degradation products released by arcing can cause this odor. However, the odor of arc-over products can be masked by the odor of acid if the oil has a high acid content. The oil should be tested by gas chromotography if there is a reason to suspect that dissolved combustible gases are present.

Methods of Dealing with Bad Oil

There are two options when tests indicate the oil is in proposition A (second classification) range or worse: replacement or reclamation of the oil. The reclamation process involves these steps:

1. Dissolve the sludge on the internals of the transformer.
2. Purge the sludge from the transformer.
3. Filter the sludge from the oil to restore it to "like new" condition.

The sludge deposited on the internal components of the transformer will not be removed when the oil is replaced. Simple replacement will only put new oil into a contaminated container. There are also handling and disposal problems with the replacement option.

The same process and same equipment are used to reclaim the oil regardless of the degree of deterioration. Reclamation is relatively simple if the oil is in the proposition A range. If the oil has deteriorated further, a more extensive treatment will be required.

The reclamation process is performed in a *closed loop*. Special equipment is used to continuously heat the oil, filter it through absorbent beds, and recirculate it in the transformer. The heated oil is maintained at its aniline point (82°C) during the process. The hot oil acts as a strong solvent for decay products. The sludge is removed usually by 6 to 10 recirculation cycles. Twenty recirculations may be required to desludge the transformer if it is badly sludged (oil with NN > 0.3 and IFT < 18).

[Figure: Plot of Interfacial tension (dynes/cm) on y-axis (0 to 40) versus Neutralization number, mg KOH/gram of oil on x-axis (0.2 to 1.5), showing regions labeled "Good oil", "Prop A", "Marginal oil", "Bad oil", "Very bad oil", "Extremely bad oil", and "Oil in disastrous condition". Points A and B are marked as outliers.]

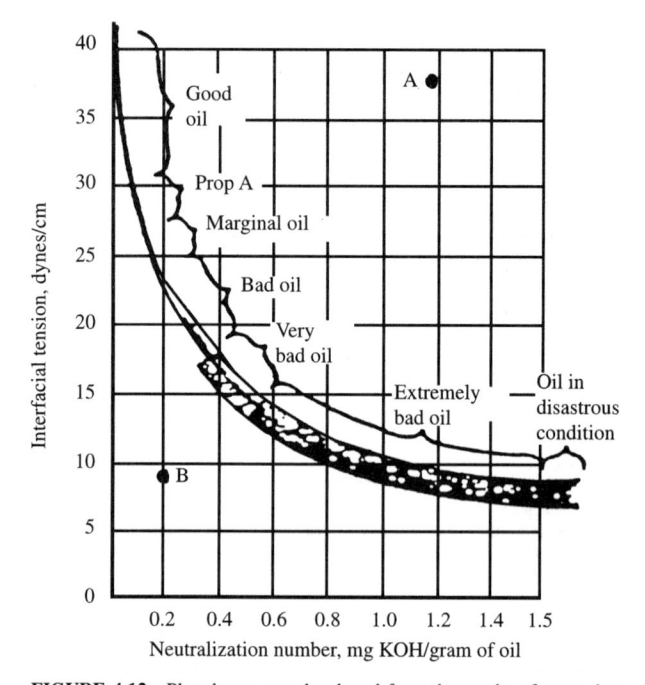

FIGURE 4.12 Plot shown was developed from the results of more than 10,000 IFT/NN tests. If a plot of IFT versus NN for a given oil sample does not fall within the range shown on either side of the median line (as in the case of points A and B), further investigation is in order. Additional tests (see Table 4.2) should be conducted. The results should be evaluated in combination to get a true picture of the condition of both the oil and the transformer.

Many separate reclamation/desludging treatments are needed if the transformer is in extremely sludged condition (NN > 1.5). It is recommended to consider replacing the transformer at this stage. Reclamation/desludging should be performed before NN exceeds 1.5 and IFT drops below 24.0, to ensure adequate operation and minimum deterioration of the insulation. The oil can be reclaimed/desludged while the transformer is online (energized and in service) by properly equipped maintenance contractors. Therefore, there is no reason to allow the oil to deteriorate until an expensive treatment is required.

It is essential to have adequate cleanliness and quality control while obtaining the test results. Dirty containers or laboratory equipment should not be used for the oil sample. It is very important to protect the sample from light, air, and moisture. The sample should be taken quickly, sealed in a container, and tested within hours. It must not be exposed to sunlight. Aging will accelerate if the sample is exposed to sunlight. The results of the sample could be misleading in this case.

Gas-in-Oil

The analysis of dissolved combustible gases in transformers is highly indicative of possible trouble. All transformers will develop a certain amount of gases over their lifetime. The two principal causes of gas formation within an operating transformer are thermal and electrical disturbances. Heat losses from the conductors produce gases from decomposition of the oil and solid insulation. Gases are also generated from the decomposition of oil and insulation exposed to arc temperatures. The following paragraph was extracted from the IEEE Guide for the interpretation of gases generated in oil-immersed transformers:

> The detection of certain gases generated in an oil filled transformer is frequently the first available indication of a malfunction that may eventually lead to failure if not corrected. Arcing, corona discharge, low-energy sparking, severe overloading, pump motor failure, and overheating in the insulation system are some of the possible mechanisms. These conditions occurring singly, or as several simultaneous events, can result in decomposition of the insulating materials and the formation of some gases. In fact, it is possible for some transformers to operate throughout their useful life with substantial quantities of combustible gas present. Operating a transformer with large quantities of combustible gas present is not a normal occurrence but it does happen, usually after some degree of investigation and an evaluation of possible risk.

The generated gases in the transformer can be found dissolved in the oil, in the gas blanket above the oil, or in gas-collecting devices. If an abnormal condition is detected, an evaluation is required to determine the amount of generated gases and their continuing rate of generation. When the composition of the generated gases is determined, some indication of the source of the gases and the kind of insulation involved will be gained.

There are many techniques for detecting and measuring gases. However, the interpretation of the significance of these gases is not a science at present. It is an art subjected to variability. It is difficult to establish a consensus due to the variability of acceptable gas limits. The main reason for not developing an exact science for fault interpretation is the lack of correlation between the fault-identifying gases with faults found in actual transformers. Table 4.5 provides a general description of the various fault types with associated developing gases.

TABLE 4.5 A General Description of the Various Fault Types with Associated Developing Gases

Fault type	Description
Arcing	Arcing is the most severe of all fault processes. Large amounts of hydrogen and acetylene are produced, with minor quantities of methane and ethylene. Arcing occurs through high-current and high-temperature conditions. Carbon dioxide and carbon monoxide may also be formed if the fault involved cellulose. In some instances, the oil may become carbonized.
Corona	Corona is a low-energy electric fault. Low-energy electric discharges produce hydrogen and methane, with small quantities of ethane and ethylene. Comparable amounts of carbon monoxide and carbon dioxide may result from discharges in cellulose.
Sparking	Sparking occurs as an intermittent high-voltage flashover without high current. Increased levels of methane and ethane are detected without concurrent increases in acetylene, ethylene, or hydrogen.
Overheating	Decomposition products include ethylene and methane, together with smaller quantities of hydrogen and ethane. Traces of acetylene may be formed if the fault is severe or involves electrical contacts.
Overheated cellulose	Large quantities of carbon dioxide and carbon monoxide are evolved from overheated cellulose. Hydrocarbon gases, such as methane and ethylene, will be formed if the fault involved an oil-impregnated structure. A furanic compound and/or degree of polymerization analysis may be performed to further assess the condition of the insulating paper.

GAS RELAY AND COLLECTION SYSTEMS

Introduction

Most utilities and large industries have a gas relay mounted on their power transformers. As mentioned earlier, gases are generated by the chemical and electrical phenomena associated with the development of faults in oil-filled transformers. A significant amount of gas is normally formed in the early stage of the fault.

The gases generated and the air expelled from the oil by the fault rise to the top of the equipment and get collected in a gas relay. Figure 4.13 illustrates a gas collection system. It ensures that gases trapped in various pockets are allowed to escape and travel to the gas relay.

Gas Relay

The gas detector relay (Fig. 4.14) gives an early indication of faults in oil-filled transformers. There are two types of faults:

1. Minor faults that result in a slow evolution of gases. These faults may result from the following:
   - Local heating
   - Defective insulation structures

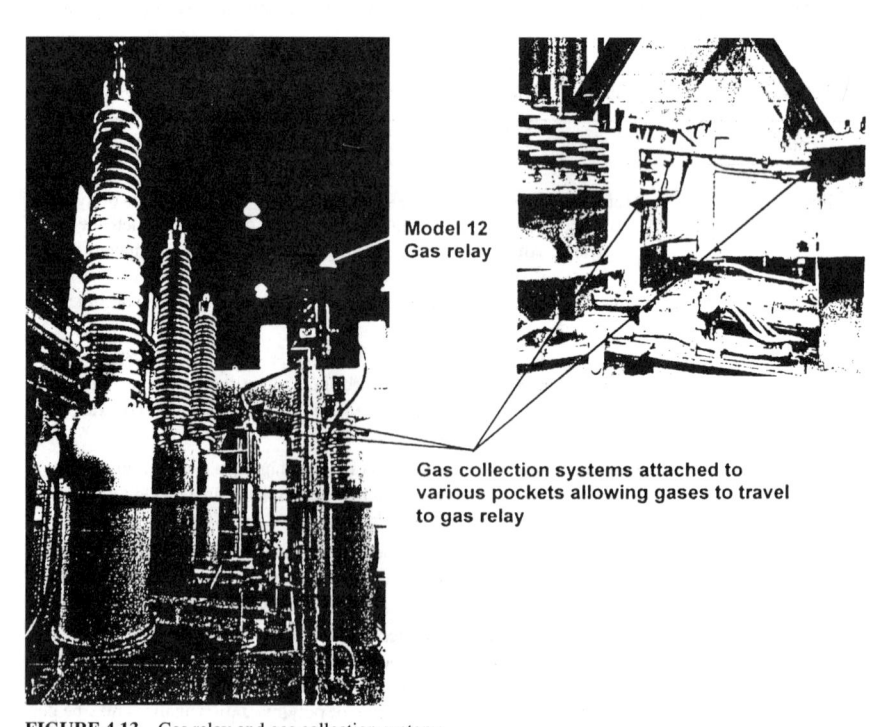

FIGURE 4.13 Gas relay and gas collection systems.

TRANSFORMER COMPONENTS AND MAINTENANCE   4.23

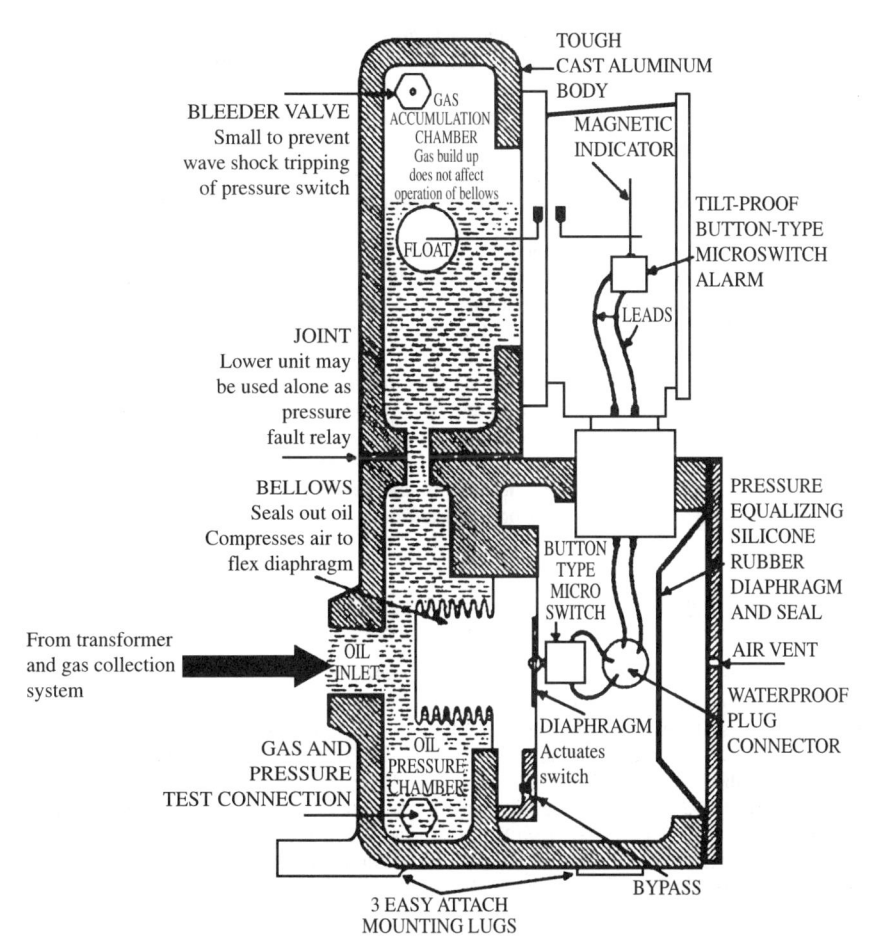

FIGURE 4.14 Cross-sectional view of CGE model 12 gas relay.

- Improperly brazed joints
- Loose contacts
- Ground faults
- Short-circuit turns
- Opening or interruption of a phase current
- Burnout of core iron
- Release of air from oil
- Leakage of air into the transformer

2. Major faults that result in a sudden increase in pressure. These faults are normally caused by flashover between parts. The relay will detect both types of faults.

The relay has two sections:

- A gas accumulation chamber is located at the top of the relay. It consists of an oil chamber with a gas bleeder needle valve. A float in the oil chamber operates a magnetic oil gauge with an alarm switch.
- A pressure chamber is located at the bottom of the relay. It has two parts. An oil chamber is seen at the rear of the relay. The chamber is connected to the transformer by a pipe entering the back of the relay. A test valve is located at the base. It is used for making operation checks. Sensitive brass bellows separate the first section from the second.

There is an air chamber in the front. It contains stops for the bellows to prevent overtravel, a flexible diaphragm, and a microswitch. When the bellows move, they compress the air behind the diaphragm. This action actuates the microswitch which is fastened to the diaphragm. When arcing occurs in the transformer, it causes a rapid evolution of gas in the oil. A pressure wave is generated through the oil. This wave will reach the relay through the pipe. It will compress the flexible bellows. The air in the chamber is compressed by the displacement of the bellows. Since the air cannot pass quickly through the bypass valve, it forces the flexible diaphragm to close the contact of the trip switch. This action disconnects the transformer.

It is essential to find the fault when the pressure contact trips the transformer. The fault should be corrected before the transformer is put back in service.

RELIEF DEVICES

Very high pressure is generated following an electrical fault under oil. These pressures could easily burst the steel tank if they are not relieved. An explosion vent was used until the early 1970s. It consists of a large-diameter pipe extending slightly above the conservator tank of the transformer and curved in the direction of the ground. A diaphragm (made of glass usually) is installed at the curved end. It ruptures at a relatively low pressure, releasing any force buildup inside the transformer. Since the early 1970s, a self-resetting pressure relief vent was installed on transformers. When the pressure in the transformer reaches a predetermined level following a fault, it forces the seal open under spring pressure. This relieves the pressure to atmosphere.

INTERCONNECTION WITH THE GRID

Figures 4.15 and 4.16 illustrate interconnections of power plants with the grid. The transformer that connects the plant with the grid is normally called the *main output transformer* (MOT) or generator step-up transformer. The transformer that connects the output of the generator with the plant itself (feeding power back to the plant) is called the *unit service transformer* (UST) or *station auxiliary transformer*. The rating of the UST is normally about 6 to 7 percent of the MOT. The transformer that connects the grid to the station loads (allowing power to be fed back to the unit) is called the *station service transformer* (SST). The design philosophy for supplying power to the loads in the plant varies. Some power plants supply one-half of their loads through the UST and the other half through the SST. However, modern power plants have opted to supply all their power through the UST. They use the incoming power through the SST as backup. In the plants where the plant loads are supplied equally from the UST and SST (Fig. 4.16), the tie breakers connecting the buses inside the plant close when the unit is disconnected from the grid. This is done to ensure that the loads inside the plant continue to be supplied with power when the unit is taken off line.

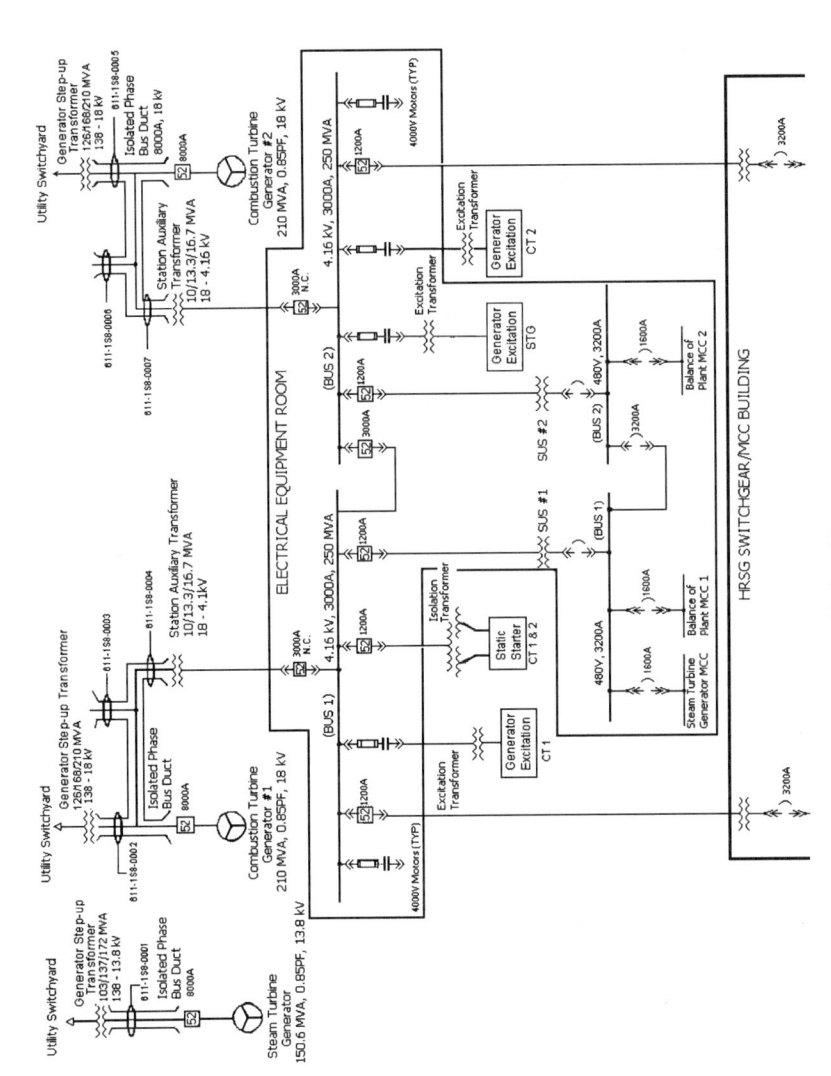

FIGURE 4.15 Interconnection of combined cycles with the grid.

4.26 CHAPTER FOUR

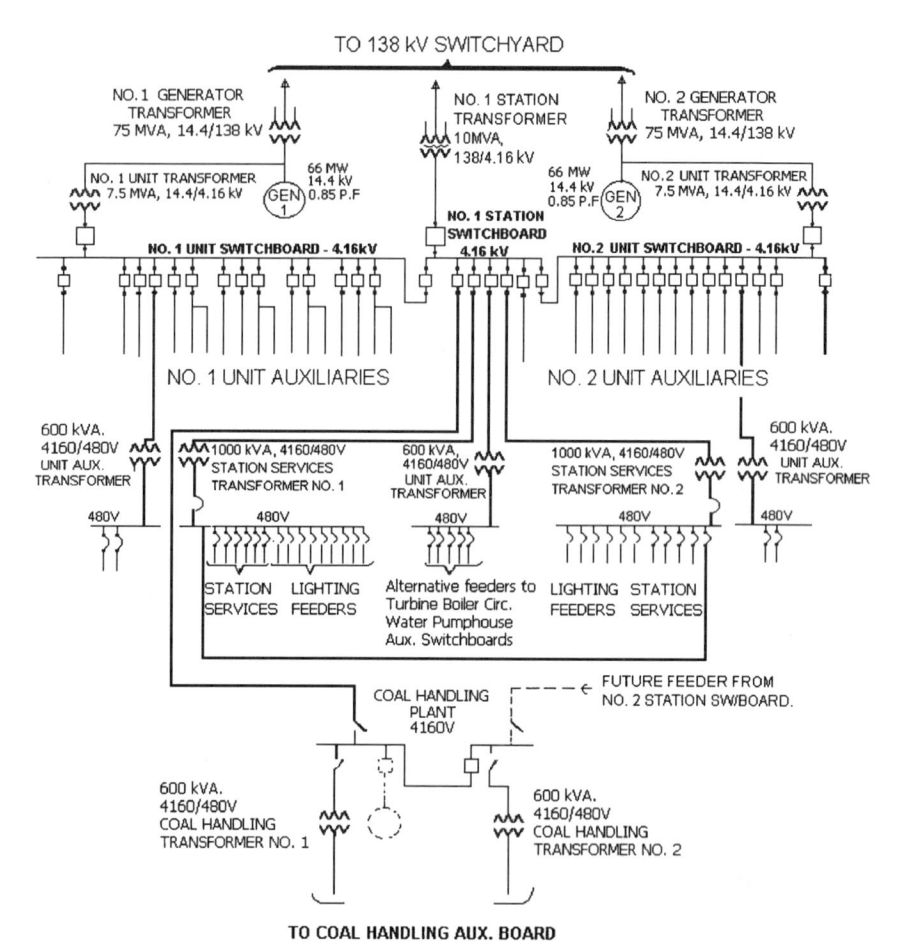

FIGURE 4.16 Power station ac single-line diagram.

Some plants install the generator breaker before the line feeding power from the output of the generator back into the plant. In this design, the plant loads are supplied from another source (e.g., the grid) when the breaker opens. Other plants install the generator breaker after the line feeding power from the output of the generator back into the plant. When the generator breaker opens (due to a load rejection), a significant load shedding occurs (about 94 percent of the load is taken off within a fraction of a second). This results in a significant reduction in countertorque on the generator shaft (about 94 percent reduction). Since the driving torque from the turbine has not changed, the turbine-generator shaft will accelerate. The governing system will limit the overspeed normally to 8 percent. However, since the generator is still supplying the plant loads at higher frequency (up to 8 percent higher), the synchronous speed of the motors inside the plant will increase. The currents pulled by these motors from the power supply will increase. This results in tripping of these motors in some plants on overload.

Transformers are used inside the plant to provide power to the loads requiring lower voltage. Some steam power plants use separate gas turbines as backup power (standby generators). They feed the 4-kV and 600-V buses inside the plant when the normal power from the UST and SST is lost. This type of power is considered more reliable than the power supplied from the UST or SST. Most plants use large battery banks to supply emergency loads (e.g., turbine emergency dc lube oil pump, generator stator emergency dc water cooling pump, and generator emergency dc seal oil pump). Some plants use inverters to supply emergency ac power from the battery bank to plant loads. The reliability of power supplies in descending order is as follows:

1. Power from the battery banks (dc or ac through inverters) (most reliable)
2. Standby generators
3. Power through UST or SST (least reliable).

REFERENCE

1. Stanley D. Myers and J. J. Kelly, Transformer Maintenance Institute, Division of S. D. Meyers, Inc., Cuyahoga Falls, Ohio.

CHAPTER 5
AC MACHINE FUNDAMENTALS

AC machines are motors that convert ac electric energy to mechanical energy and generators that convert mechanical energy to ac electric energy. The two major classes of ac machines are synchronous and induction machines. The field current of synchronous machines (motors and generators) is supplied by a separate dc power source while the field current of induction machines is supplied by magnetic induction (transformer action) into the field windings.

AC machines differ from dc machines by having their *armature windings* almost always located on the stator while their *field windings* are located on the rotor. A set of three-phase ac voltages is induced into the stator armature windings of an ac machine by the rotating magnetic field from the rotor field windings (generator action). Conversely, a set of three-phase currents flowing in the stator armature windings produces a rotating magnetic field within the stator. This magnetic field interacts with the rotor magnetic field to produce the torque in the machine (motor action).

THE ROTATING MAGNETIC FIELD

The main principle of ac machine operation is this: A three-phase set of currents, flowing in an armature windings, each of equal magnitude and differing in phase by 120°, produces a rotating magnetic field of constant magnitude.

The stator shown in Fig. 5.1 has three coils, each 120° apart.
The currents flowing in the stator are given by

$$i_{aa'}(t) = I_M \sin \omega t \quad A$$
$$i_{bb'}(t) = I_M \sin (\omega t - 120°) \quad A$$
$$i_{cc'}(t) = I_M \sin (\omega t - 240°) \quad A$$

The resulting magnetic flux densities are

$$\mathbf{B}_{aa'}(t) = B_M \sin \omega t \angle 0° \quad \text{Wb/m}^2$$
$$\mathbf{B}_{bb'}(t) = B_M \sin (\omega t - 120°) \angle 120° \quad \text{Wb/m}^2$$
$$\mathbf{B}_{cc'}(t) = B_M \sin (\omega t - 240°) \angle 240° \quad \text{Wb/m}^2$$

The directions of these fluxes are given by the right-hand rule. When the fingers of the right hand curl in the direction of the current in a coil, the thumb points in the direction of the resulting magnetic flux density.

5.2   CHAPTER FIVE

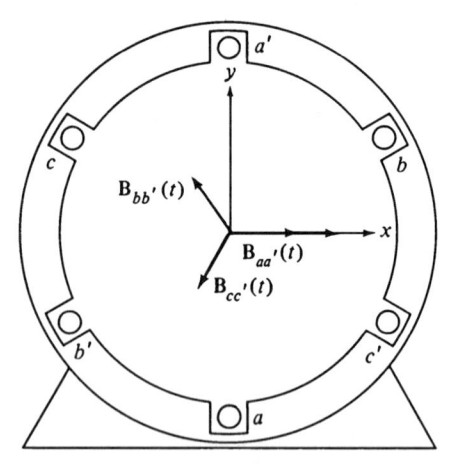

FIGURE 5.1 A simple three-phase stator. Currents in this stator are assumed positive if they flow into the unprimed and out of the primed ends of the coils.

An examination of the currents and their corresponding magnetic flux densities at specific times is used to determine the resulting net magnetic flux density. For example, at time $\omega t = 0°$, the magnetic field from coil aa' will be

$$\mathbf{B}_{aa'} = 0$$

The magnetic field from coil bb' will be

$$\mathbf{B}_{bb'} = B_M \sin(-120°) \angle 120°$$

and the magnetic field from coil cc' will be

$$\mathbf{B}_{cc'} = B_M \sin(-240°) \angle 240°$$

The total magnetic field from all three coils added together will be

$$\mathbf{B}_{net} = \mathbf{B}_{aa'} + \mathbf{B}_{bb'} + \mathbf{B}_{cc'}$$

$$= 0 + \left(-\frac{\sqrt{3}}{2} B_M\right) \angle 120° - \frac{\sqrt{3}}{2} B_M \angle 240°$$

$$= 1.5 B_M \angle -90°$$

As another example, look at the magnetic field at time $\omega t = 90°$. At that time, the currents are

$$i_{aa'} = I_M \sin 90° \quad \text{A}$$

$$i_{bb'} = I_M \sin(-30°) \quad \text{A}$$

$$i_{cc'} = I_M \sin(-150°) \quad \text{A}$$

and the magnetic fields are

$$\mathbf{B}_{aa'} = B_M \angle 0°$$
$$\mathbf{B}_{bb'} = -0.5 B_M \angle 120°$$
$$\mathbf{B}_{cc'} = -0.5 B_M \angle 240°$$

The resulting net magnetic field is

$$\mathbf{B}_{net} = B_M \angle 0° + (-0.5) B_M \angle 120° + (-0.5) B_M \angle 240°$$
$$= 1.5 B_M \angle 0°$$

The resulting magnetic flux is shown in Fig. 5.2. Notice that the direction of the magnetic flux has changed, but its magnitude remained constant. The magnetic flux is rotating counterclockwise while its magnitude remained constant.

Proof of the Rotating Magnetic Flux Concept

At any time t, the magnetic flux has the same magnitude $1.5 B_M$. It continues to rotate at angular velocity ω. A proof of this concept is presented in Ref. 1.

The Relationship between Electrical Frequency and the Speed of Magnetic Field Rotation

Figure 5.3 illustrates that the rotating magnetic field in the stator can be represented as a north and a south pole. The flux leaves the stator at the north pole and enters the stator at the south pole. The magnetic poles complete one complete revolution around the stator surface for each electrical cycle of the applied current. Therefore, the mechanical angular speed of rotation in revolutions per second is equal to the electrical frequency in hertz:

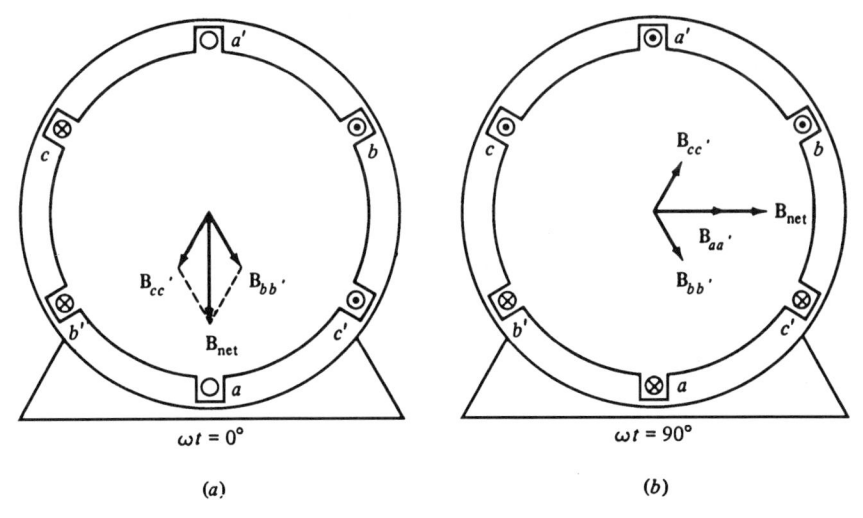

FIGURE 5.2 (a) The vector magnetic field in a stator at time $\omega t = 0°$. (b) The vector magnetic field in a stator at time $\omega t = 90°$.

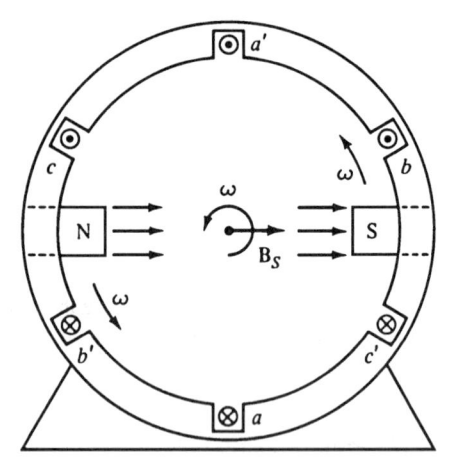

FIGURE 5.3 The rotating magnetic field in a stator represented as moving north and south stator poles.

$$f_e = f_m \quad \text{two poles}$$
$$\omega_e = \omega_m \quad \text{two poles}$$

where f_m and ω_m are the mechanical speed of rotation in revolutions per second and radians per second, respectively. Both f_e and ω_e are the electrical frequency (speed) in hertz and radians per second, respectively.

The windings on the two-pole stator shown in Fig. 5.1 occur in the order (taken counterclockwise)

$$a-c'-b-a'-c-b$$

If this pattern is repeated twice within the stator, the pattern of windings becomes

$$a-c'-b-a'-c-b'-a-c'-b-a'-c-b'$$

Figure 5.4 illustrates the *two* north poles and *two* south poles that are produced in the stator when a three-phase set of currents is applied to the stator.

In this stator, the pole moves around half the stator surface in one electrical cycle. Since the mechanical motion is 180° for a complete electrical cycle (360°), the electrical angle θ_e is related to the mechanical angle θ_m by

$$\theta_e = 2\theta_m \quad \text{four poles}$$

Therefore, for a four-pole stator, the electrical frequency is double the mechanical frequency of rotation:

$$f_e = 2f_m \quad \text{four poles}$$
$$\omega_e = 2\omega_m \quad \text{four poles}$$

In general, if P is the number of magnetic poles on the stator, then there are $P/2$ repetitions of the windings. The electrical and mechanical quantities of the machine are related by

AC MACHINE FUNDAMENTALS 5.5

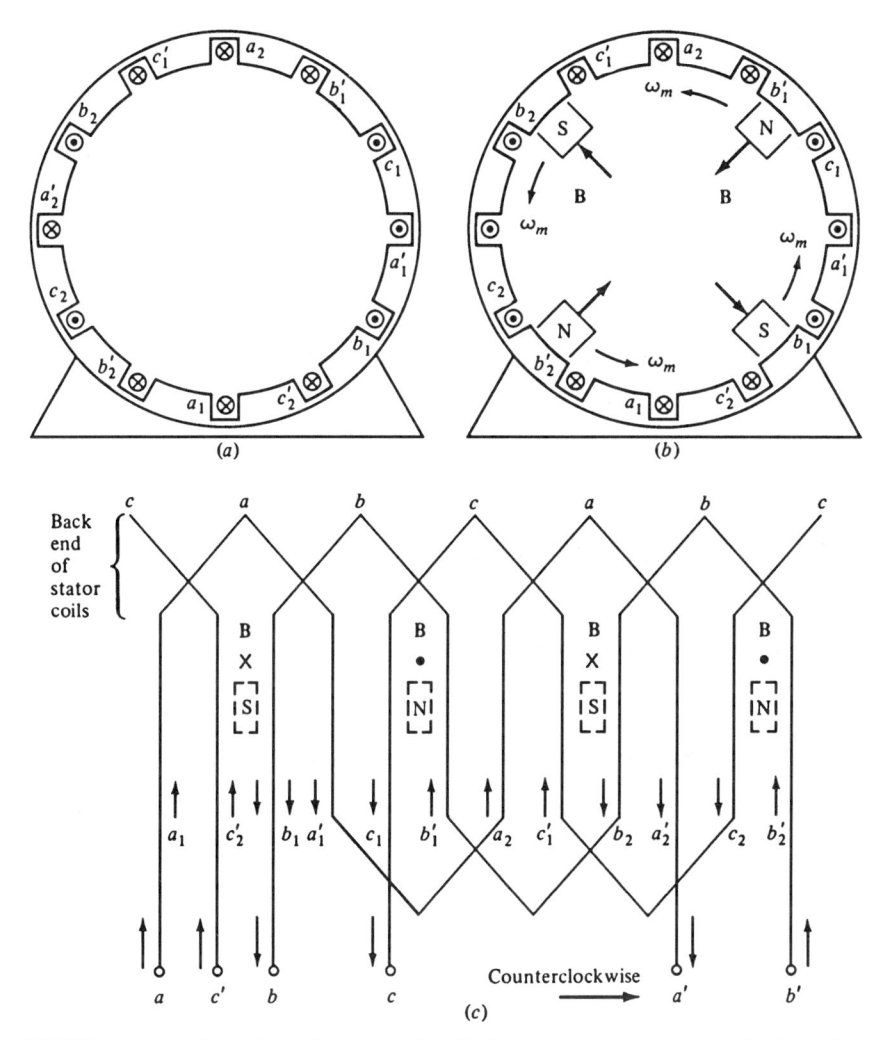

FIGURE 5.4 (a) A simple four-pole stator winding. (b) The resulting stator magnetic poles. Notice that there are moving poles of alternating polarity every 90° around the stator surface. (c) A winding diagram of the stator as seen from its inner surface, showing how the stator currents produce north and south magnetic poles.

$$\theta_e = \frac{P}{2} \theta_m$$

$$f_e = \frac{P}{2} f_m$$

$$\omega_e = \frac{P}{2}\omega_m$$

Since the mechanical frequency $f_m = n_m/60$, the electrical frequency in hertz is related to the mechanical speed of the magnetic fields in revolutions per minute by

$$f_e = \frac{n_m P}{120}$$

Reversing the Direction of the Magnetic Field Rotation

The direction of the magnetic field's rotation is reversed when the current in any two of three coils is swapped. Therefore, it is possible to reverse the direction of rotation of an ac motor by just switching any two of the three phases (Ref. 1).

THE INDUCED VOLTAGE IN AC MACHINES

Just as a rotating magnetic field can be produced by three-phase set of currents in a stator, a three-phase set of voltages in the coils of a stator can be produced by a rotating magnetic field.

The Induced Voltage in a Coil on a Two-Pole Stator

Figure 5.5 illustrates a *stationary* coil with a *rotating* magnetic field moving in its center. The induced voltage in a wire is given by

$$e_{ind} = (\mathbf{v} \times \mathbf{B}) \cdot \mathbf{l}$$

where $\mathbf{v}$ = velocity of wire *relative to magnetic field*
$\mathbf{B}$ = magnetic flux density of field
$\mathbf{l}$ = length of wire

This equation was derived for a *wire moving within a stationary magnetic field*. In ac machines, the magnetic field is moving, and the wire is stationary.

Figure 5.6 illustrates the velocities and vector magnetic field from the point of view of a moving wire and a stationary magnetic field. The voltages induced in the sides of the coil are

1. *Segment ab*. The angle between $\mathbf{v}$ and $\mathbf{B}$ in segment bc is $180° - \theta$, while the quantity $\mathbf{v} \times \mathbf{B}$ is in the direction of $\mathbf{l}$, so

$$e_{ba} = (\mathbf{v} \times \mathbf{B}) \cdot \mathbf{l}$$
$$= vBl \sin(180° - \theta) \quad \text{directed } \textit{into } \text{page}$$

The direction of e_{ba} is given by the right-hand rule. By trigonometric identity, $\sin(180° - \theta) = \sin\theta$. So

AC MACHINE FUNDAMENTALS 5.7

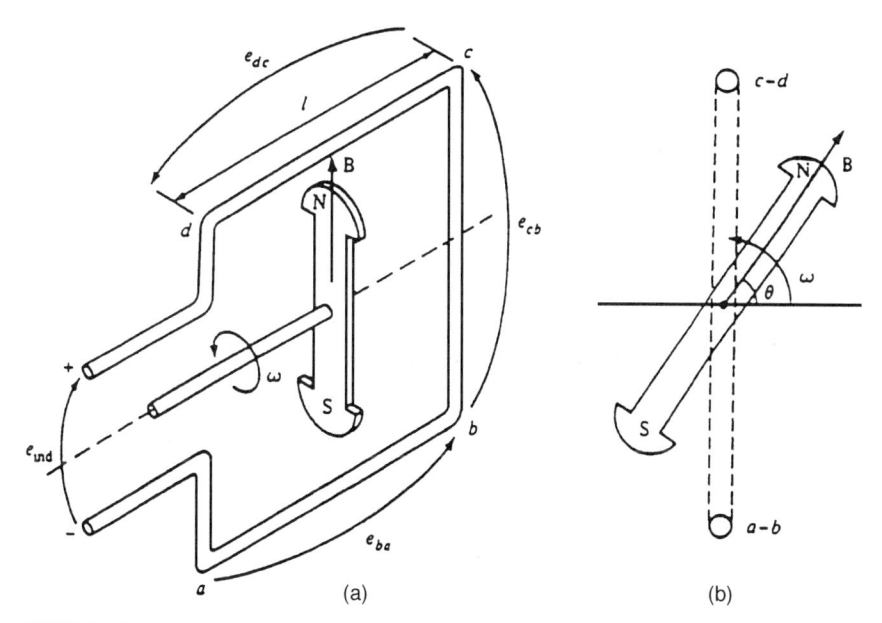

FIGURE 5.5 A rotating magnetic field inside a fixed coil: (*a*) Perspective view; (*b*) end view.

$$e_{ba} = vBl \sin \theta$$

2. *Segment bc.* The voltage on segment *bc* is zero, since the vector quantity **v** × **B** is perpendicular to **l**.

$$e_{cb} = (\mathbf{v} \times \mathbf{B}) \cdot \mathbf{l}$$
$$= 0$$

3. *Segment cd.* The angle between v and B in segment *cd* is θ, while the quantity **v** × **B** is in the direction of **l**. So

$$e_{dc} = (\mathbf{v} \times \mathbf{B}) \cdot \mathbf{l}$$
$$= vBl \sin \theta \quad \text{directed } out\ of \text{ page}$$

4. *Segment da.* The voltage on segment *da* is zero, for the same reason as in segment *bc*:

$$e_{ad} = 0$$

The total voltage induced within a single-turn coil is given by

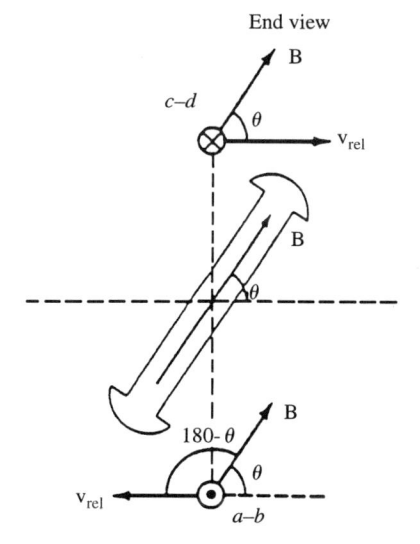

FIGURE 5.6 The magnetic fields and velocities of the coil sides as seen from a frame of reference in which the magnetic field is stationary.

$$e_{ind} = 2vBl \sin \theta$$

Since angle $\theta = \omega_e t$, the induced voltage can be rewritten as

$$e_{ind} = 2vBl \sin \omega_e t$$

Since the cross-sectional area A of the turn is $2rl$ and the velocity of the end conductors is given by $v = r\omega_m$, the equation can be rewritten as

$$e_{ind} = 2(r\omega_m)Bl \sin \omega_e t$$
$$= (2rl)B\omega_m \sin \omega_e t$$
$$= AB\omega_m \sin \omega_e t$$

The maximum flux passing through the coil is $\phi = AB$. For a two-pole stator $\omega_m = \omega_e = \omega$, the induced voltage is

$$\boxed{e_{ind} = \phi\omega \sin \omega t}$$

This equation describes the voltage induced in a single-turn coil; if the coil (phase) has N_c turns of wire in it, the total induced voltage will be

$$\boxed{e_{ind} = N_c \phi\omega \sin \omega t}$$

The Induced Voltage in a Three-Phase Set of Coils

Figure 5.7 illustrates three coils each of N_c turns placed around the rotor magnetic field. The voltage induced in each has the same magnitude but differs in phase by 120°. The resulting voltages in the three phases are

$$e_{aa'}(t) = N_c \phi\omega \sin \omega t \quad \text{V}$$
$$e_{bb'}(t) = N_c \phi\omega \sin (\omega t - 120°) \quad \text{V}$$
$$e_{cc'}(t) = N_c \phi\omega \sin (\omega t - 240°) \quad \text{V}$$

Therefore, a set of three-phase currents generates a rotating uniform magnetic field within the stator of the machine, and a uniform magnetic field induces a set of three-phase voltages in such a stator.

The RMS Voltage in a Three-Phase Stator

The peak voltage in any phase is

$$E_{max} = N_c \phi\omega$$

Since $\omega = 2\pi f$, the rms voltage in any phase is

$$E_A = \frac{2\pi}{\sqrt{2}} N_c \phi f$$

AC MACHINE FUNDAMENTALS

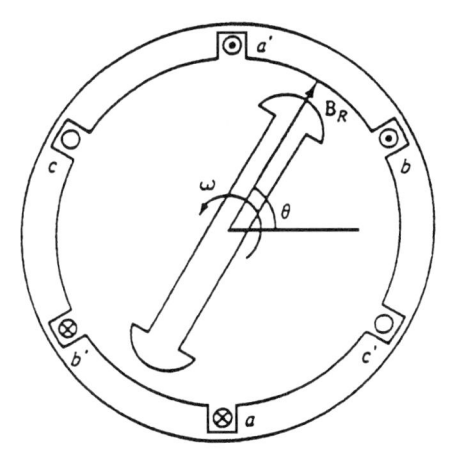

FIGURE 5.7 The production of three-phase voltages from three coils spaced 120° apart.

$$E = \sqrt{2}\pi N_c \phi f$$

The rms voltage at the terminals of the machine depends on whether the stator is Y- or Δ-connected. If the machine is Y-connected, the terminal voltage is 3 times E_A. In Δ-connected machines, the terminal voltage is the same as E_A.

THE INDUCED TORQUE IN AN AC MACHINE

During normal operation of ac machines (motors and generators), there are two magnetic fields: a magnetic field from the rotor and another from the stator. A torque is induced in the machine due to the interaction of the two magnetic fields.

A synchronous machine is illustrated in Fig. 5.8. A magnetic flux density $\mathbf{B}_R$ is produced by the rotor, and a magnetic flux density $\mathbf{B}_S$ is produced by the stator. The induced torque in a machine (motors and generators) is given by

$$\tau_{ind} = k\mathbf{B}_R \times \mathbf{B}_S$$

$$\tau_{ind} = kB_R B_S \sin \gamma$$

where τ_{ind} = induced torque in machine
 $\mathbf{B}_R$ = rotor flux density
 $\mathbf{B}_S$ = stator flux density
 γ = angle between $\mathbf{B}_R$ and $\mathbf{B}_S$

The net magnetic field in the machine is the vector sum of the fields from the stator and rotor

$$\mathbf{B}_{net} = \mathbf{B}_R + \mathbf{B}_S$$

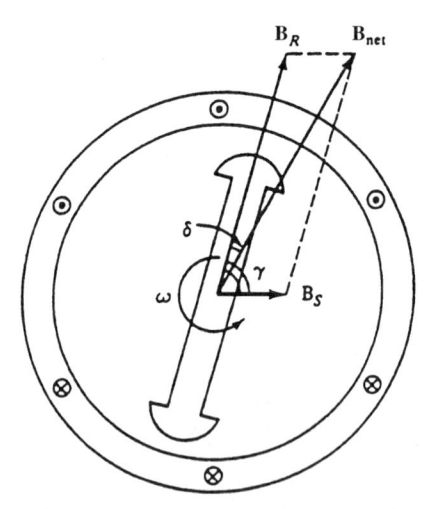

FIGURE 5.8 A simplified synchronous machine showing its rotor and stator magnetic fields.

The induced torque can be expressed as

$$\tau_{ind} = k\mathbf{B}_R \times \mathbf{B}_{net}$$

The magnitude of the torque is

$$\tau_{ind} = kB_R B_{net} \sin \delta$$

The magnetic fields of the synchronous machine shown in Fig. 5.8, are rotating in a counterclockwise direction. What is the direction of the induced torque on the rotor of the machine?

By applying the right-hand rule to the equation of the induced torque, we see that the induced torque is clockwise. It is opposing the direction of rotation of the rotor. Therefore, this machine is working as a generator.

WINDING INSULATION IN AC MACHINES

In ac machine design, one of the most critical parts is the insulation of the windings. When the insulation breaks down, the machine shorts out. The repair of machines with shorted insulation is expensive and sometimes impossible.

The temperature of the windings should be limited to prevent the insulation from breaking down due to overheating. This can be done by providing circulation of cool air over the windings. The continuous power supplied by the machine is usually limited by the maximum temperature of the windings. The increase in temperature usually degrades the insulation, causing it to fail by another cause such as shock, vibration, or electrical stress. A rule of thumb indicates that the life of an ac machine is halved for a temperature rise of 10 percent above the rated temperature of the windings.

The temperature limits of machine insulation have been standardized by the National Electrical Manufacturers Association (NEMA). A series of insulation system classes have been defined. Each insulation system class specifies the maximum temperature rise allowed for the insulation. The most common NEMA insulation classes for ac motors are B, F, and H. Each class has a higher permissible winding temperature than the one before it. For example, the temperature rise above ambient of the armature windings in continuously operating induction machines is limited to 80°C for class B, 105°C for class F, and 125°C for class H insulation.

Similar standards have been defined by the International Electrotechnical Commission (IEC) and by other national standards organizations.

AC MACHINE POWER FLOW AND LOSSES

A power flow diagram is a convenient tool to analyze ac machines. Figure 5.9 illustrates the power flow diagram of an ac generator and an ac motor.
The losses in ac machines are

1. Rotor and stator copper (I^2R) losses
2. Core losses

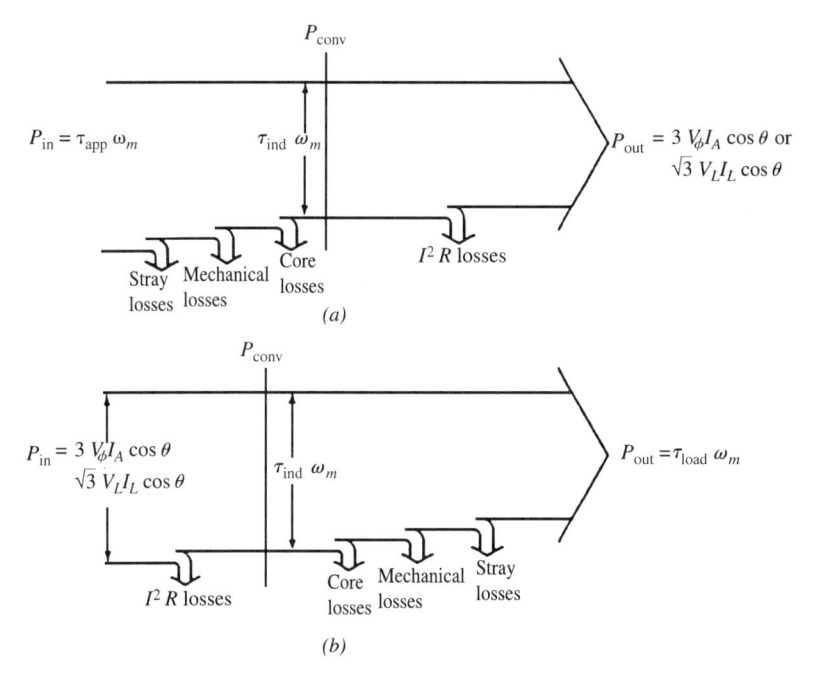

FIGURE 5.9 (a) The power flow diagram of a three-phase ac generator. (b) The power flow diagram of a three-phase ac motor.

3. Mechanical losses
4. Stray losses

The stator copper losses in ac machines are the heat losses from the conductors of the stator. They are given by

$$P_{\text{SCL}} = 3I_A^2 R_A$$

where I_A is the current flowing in each armature phase and R_A is the resistance of the conductor in each armature phase. The rotor copper losses are given by

$$P_{\text{RCL}} = I_F^2 R_F$$

The mechanical losses are caused by bearing friction and windage effects while the core losses are caused by hysteresis and eddy currents. These losses are called the *no-load* rotational losses of the machine.

All the input power at no load is used to overcome these losses. Therefore, these losses can be obtained by measuring the power to the stator at no load.

Stray load losses are all miscellaneous losses that do not fall into one of these categories. They are taken by convention as 1 percent of the output power of the machine. The overall efficiency of an ac machine is defined as the useful power output to the total input power:

$$\eta = \frac{P_{\text{out}}}{P_{\text{in}}} \times 100\%$$

REFERENCE

1. S. J. Chapman, *Electric Machinery Fundamentals*, 2d ed., McGraw-Hill, New York, 1991.

CHAPTER 6
INDUCTION MOTORS

In induction machines, the rotor voltage (which produces the rotor current and the rotor magnetic field) is not physically connected by wires to the rotor windings—it is *induced* in the rotor. The main advantage of induction motors is that there is no need for *dc field current* to run the machine. An induction machine can be used as a motor or a generator. However, it has many disadvantages as a generator.

INDUCTION MOTOR CONSTRUCTION

Figure 6.1 illustrates a typical two-pole stator for an induction motor. The two main types of rotors are *squirrel-cage* and *wound* rotors. Figures 6.2 and 6.3 illustrate squirrel-cage induction motor rotors.

The rotor consists of a series of conducting bars installed into slots carved in the face in the rotor. These bars are shorted at both ends by shorting rings. This design is known as a squirrel-cage rotor. The second type is known as a wound rotor. A *wound rotor* (Figs. 6.4 and 6.5) has three phase windings that are mirror images to the stator windings.

The three rotor phases are usually Y-connected. Slip rings on the rotor shaft tie the ends of the three rotor wires. Brushes riding on the slip rings short the rotor windings.

The rotor currents are accessible. They can be examined, and extra resistance can be added to the rotor circuit. This is a significant advantage of this design because the torque-speed characteristic of the motor can be modified.

BASIC INDUCTION MOTOR CONCEPTS

Figure 6.6 illustrates a squirrel-cage induction motor. A set of three-phase currents is flowing in the stator. A magnetic field $\mathbf{B}_S$ is produced. It rotates in a counterclockwise direction. The rotational speed of the magnetic field is given by

$$n_{sync} = \frac{120 f_e}{P}$$

where f_e is the electrical frequency in hertz and P is the number of poles in the machine. The rotating magnetic field $\mathbf{B}_S$ crosses the rotor bars and induces a voltage in them.

FIGURE 6.1 The stator of a typical induction motor, showing the stator windings. (*Courtesy of MagneTek, Inc.*)

The induced voltage in a given rotor bar is given by

$$e_{ind} = (\mathbf{v} \times \mathbf{B}) \cdot \mathbf{l}$$

where $\mathbf{v}$ = velocity of rotor bars *relative to magnetic field*
$\mathbf{B}$ = magnetic stator flux density
$\mathbf{l}$ = length of rotor bar

The voltage in a rotor bar is induced by the *relative* motion of the rotor compared to the magnetic field. The velocity of the upper rotor bars relative to the magnetic field is to the right. Therefore, the induced voltage in the upper bars is out of the page, and the induced voltage in the lower bars is into the page.

The current is flowing out of the upper bars and into the lower bars. However, the peak rotor current lags behind the peak rotor voltage due to the inductive nature of the rotor assembly. A rotor magnetic field $\mathbf{B}_R$ is produced by the current flowing in the rotor. Since the induced torque is given by

$$\tau_{ind} = k\mathbf{B}_R \times \mathbf{B}_S$$

the resulting torque is counterclockwise. The rotor accelerates in this direction.

The Concept of Rotor Slip

The speed of the rotor *relative* to the magnetic fields determines the voltage induced in the rotor. The relative speed is used because the behavior of the motor depends on the voltage and current in the rotor.

The two terms used to define the relative motion between the rotor and the magnetic fields are the *slip speed* and the *slip*. The *slip speed* is the difference between synchronous speed and rotor speed:

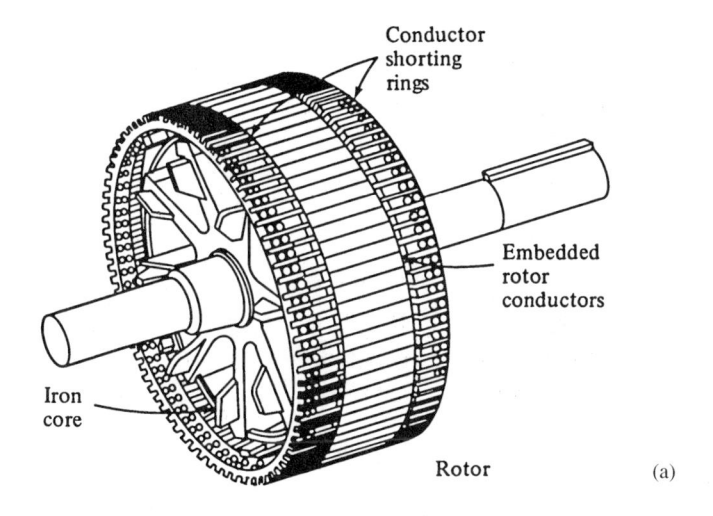

FIGURE 6.2 (a) Sketch of squirrel-cage rotor. (b) A typical squirrel-cage rotor. (*Courtesy of General Electric Company.*)

$$n_{\text{slip}} = n_{\text{sync}} - n_m$$

where n_{slip} = slip speed of machine
n_{sync} = speed of magnetic fields
n_m = mechanical shaft speed of rotor

The second term used to describe the relative motion is the slip. The *slip* is defined as

$$s = \frac{n_{\text{slip}}}{n_{\text{sync}}} \times 100\%$$

$$s = \frac{n_{\text{sync}} - n_m}{n_{\text{sync}}} \times 100\%$$

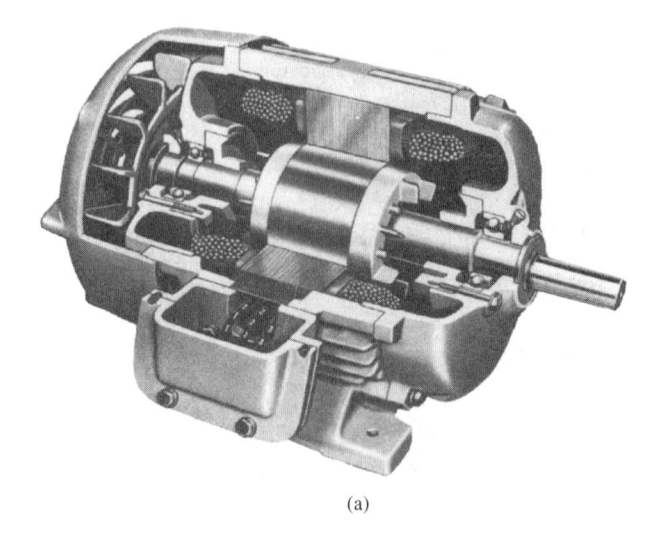

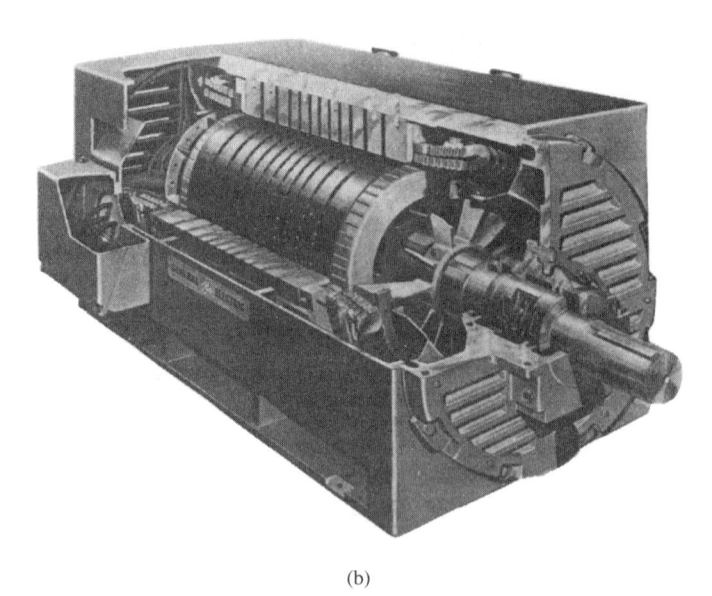

FIGURE 6.3 (*a*) Cutaway diagram of a typical small squirrel-cage induction motor. (*Courtesy of MagneTek, Inc.*) (*b*) Cutaway diagram of a typical large squirrel-cage induction motor. (*Courtesy of General Electric Company.*)

When the rotor turns at synchronous speed, $s = 0$. When the rotor is stationary, $s = 1$. All motor speeds fall between these two limits.

The mechanical speed of the rotor shaft can be expressed in terms of the synchronous speed and slip as

$$n_m = (1 - s)n_{\text{sync}}$$

(a)

(b)

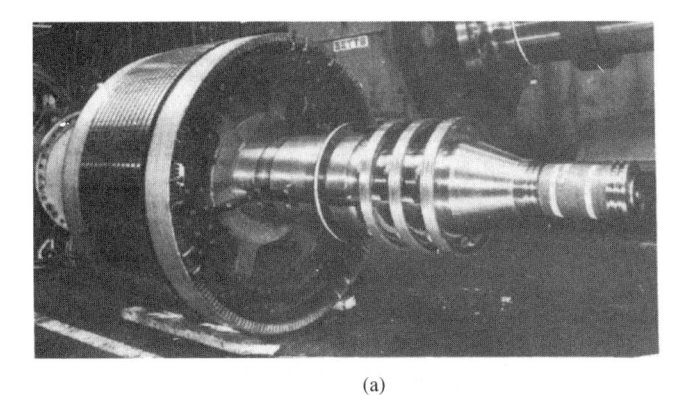

FIGURE 6.4 Typical wound rotors for induction motors. Notice the slip rings and the bars connecting the rotor windings to the slip rings. (*Courtesy of General Electric Company.*)

$$\omega_m = (1 - s)\omega_{sync}$$

The Electrical Frequency of the Rotor

Induction motors have been called *rotating transformers* because they work by inducing voltages and currents in the rotor. The primary (stator) induces a voltage in the secondary (rotor), but the secondary frequency is not necessarily the same as the primary frequency. If the rotor is locked, it will have the same frequency as the stator. If the rotor turns at synchronous speed, the frequency of the rotor will be equal to zero. For any speed in between,

$$f_r = \frac{n_{sync} - n_m}{n_{sync}} f_e$$

FIGURE 6.5 Cutaway diagram of a wound-rotor induction motor. Notice the brushes and slip rings. (*Courtesy of MagneTek, Inc.*)

Therefore,

$$f_r = \frac{P}{120}(n_{\text{sync}} - n_m)$$

THE EQUIVALENT CIRCUIT OF AN INDUCTION MOTOR

It is possible to derive the equivalent circuit of an induction motor from the knowledge of transformers. Figure 6.7 illustrates the equivalent circuit, representing the operation of an induction motor. The effective turns ratio a_{eff} couples the primary internal stator voltage $\mathbf{E}_1$ to the secondary $\mathbf{E}_R$. A current flow in the shorted rotor (or secondary) is produced by $\mathbf{E}_R$.

The Rotor Circuit Model

In induction motors, the higher the relative motion between the rotor and the stator magnetic fields, the higher the resulting rotor voltage. The relative motion is largest when the rotor is stationary. This is called the *locked-* or *blocked-rotor* condition. The induced voltage in the rotor is at maximum during this condition. When the rotor moves at the same speed as the stator magnetic field (no relative motion), the induced voltage in the rotor is zero.

If the induced rotor voltage at locked-rotor conditions is $\mathbf{E}_{R_0}$, the induced voltage at any slip is

$$\mathbf{E}_R = s\mathbf{E}_{R_0}$$

The rotor has a resistance and a reactance. Its resistance R_R is constant independent of slip while the rotor reactance depends on the slip.

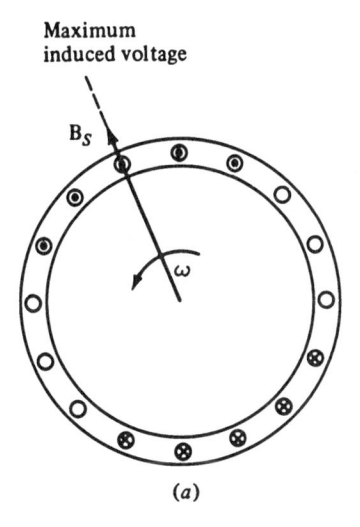

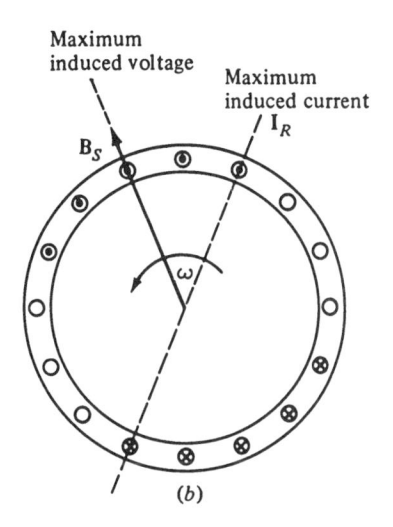

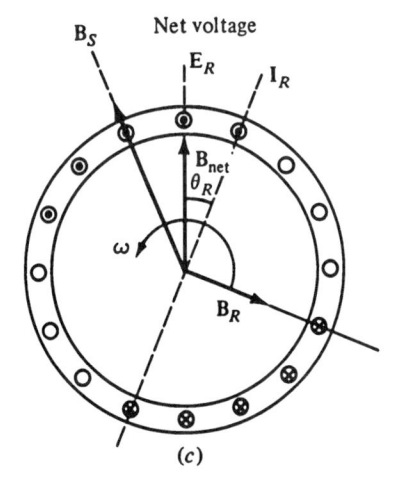

FIGURE 6.6 The development of induced torque in an induction motor. (*a*) The rotating stator field $\mathbf{B}_S$ induces a voltage in the rotor bars. (*b*) The rotor voltage produces a rotor current flow, which lags behind the voltage because of the inductance of the rotor. (*c*) The rotor current produces a rotor magnetic field $\mathbf{B}_R$ lagging 90° behind itself, and $\mathbf{B}_R$ interacts with $\mathbf{B}_{net}$ to produce a counterclockwise torque in the machine.

The reactance of a rotor depends on the rotor inductance and the frequency of the voltage and current in the rotor. If the rotor inductance is L_R, the rotor reactance is given by

$$X_R = \omega_r L_R = 2\pi f_r L_R$$

Since $f_r = sf_e$, the rotor reactance becomes

$$X_R = 2\pi s f_e L_R$$
$$= s(2\pi f_e L_R)$$
$$= s X_{R_0}$$

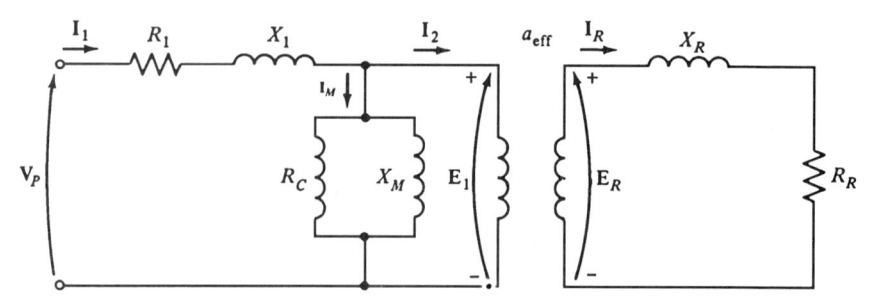

FIGURE 6.7 The transformer model of an induction motor, with rotor and stator connected by an ideal transformer of turns ratio a_{eff}.

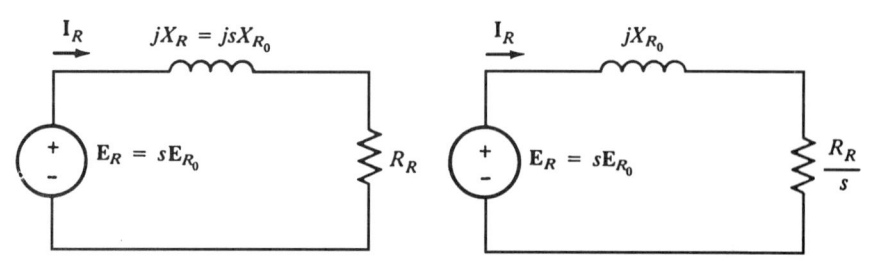

FIGURE 6.8 The rotor circuit model of an induction motor.

FIGURE 6.9 The rotor circuit model with all the frequency (slip) effects concentrated in resistor R_R.

The rotor equivalent circuit is shown in Fig. 6.8. The rotor current can be found as

$$\mathbf{I}_R = \frac{\mathbf{E}_R}{R_R + jX_R}$$

$$\boxed{\mathbf{I}_R = \frac{s\mathbf{E}_{R_0}}{R_R + jsX_{R_0}}}$$

The voltage supplied $\mathbf{E}_{R_0}$ can be treated as constant. The final rotor equivalent circuit is shown in Fig. 6.9. The equivalent impedance containing all the effects of varying rotor slip is

$$Z_{R,\text{eq}} = \frac{R_R}{s} + jX_{R_0}$$

Figure 6.10 illustrates the variation of rotor current with speed.

At high slips (rotor speed is much lower than normal operating speed or synchronous speed), X_{R_0} is much larger than R_R/s. The rotor current approaches a *steady-state* value. At low slips (rotor speed is near normal operating speed or synchronous speed), the resistive term R_R/s is much larger X_{R_0}. The rotor resistance is the dominant term, and the rotor current vary *linearly with slip*.

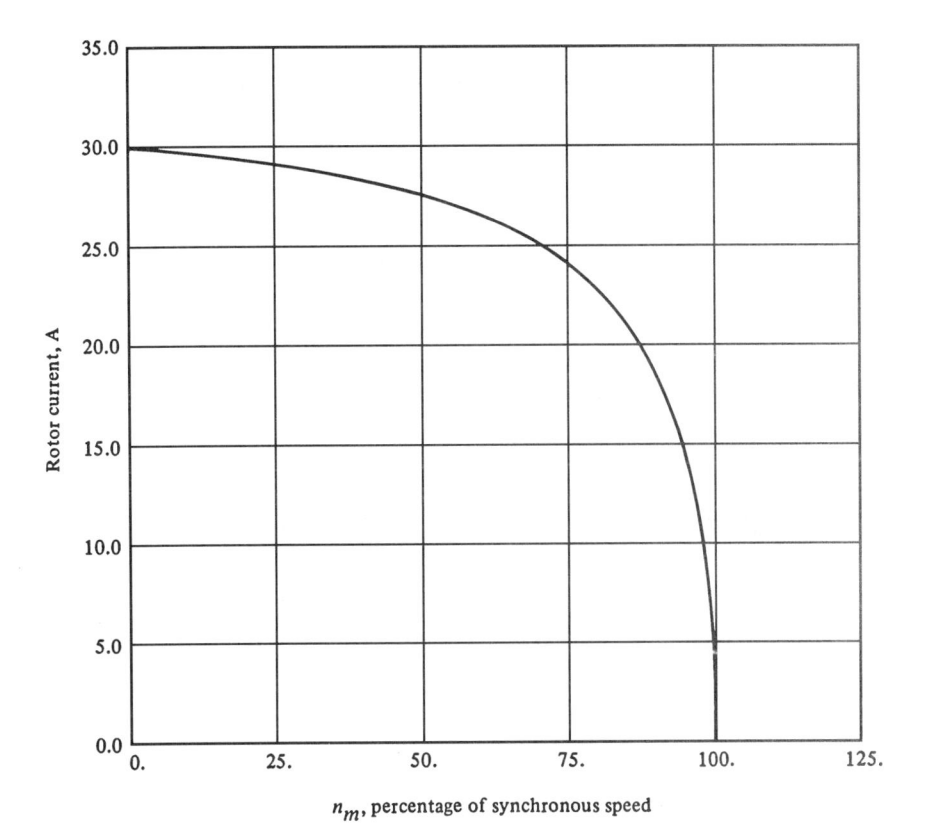

FIGURE 6.10 Rotor current as a function of rotor speed.

LOSSES AND THE POWER FLOW DIAGRAM

Induction motors have been described as rotating transformers. The input is a three-phase system of voltages and currents. The secondary windings of the motor (the rotor) are shorted out. Figure 6.11 illustrates the relationship between the input electric power and the output mechanical power. The stator copper losses are I^2R in the stator windings. The core losses include the hysteresis and eddy currents losses.

INDUCTION MOTOR TORQUE-SPEED CHARACTERISTICS

Figure 6.12a illustrates a squirrel-cage rotor of an induction motor that is operating at no load (near synchronous speed). The magnetization current $\mathbf{I}_M$ flowing in the motor's equivalent circuit (Fig. 6.7) creates the net magnetic field $\mathbf{B}_{net}$. Current $\mathbf{I}_M$ and hence $\mathbf{B}_{net}$ are proportional to $\mathbf{E}_1$. Since $\mathbf{E}_1$ remains constant with the changes in load, then $\mathbf{I}_M$ and $\mathbf{B}_{net}$ remain constant also. At no load (Fig. 6.12a), the rotor slip (the relative motion between the rotor

6.10 CHAPTER SIX

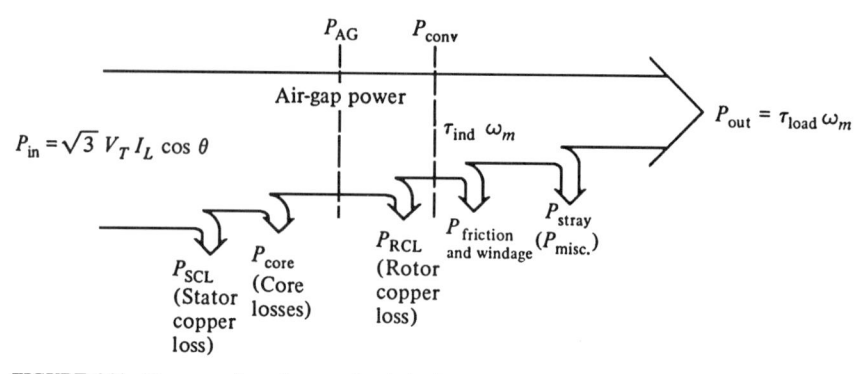

FIGURE 6.11 The power flow diagram of an induction motor.

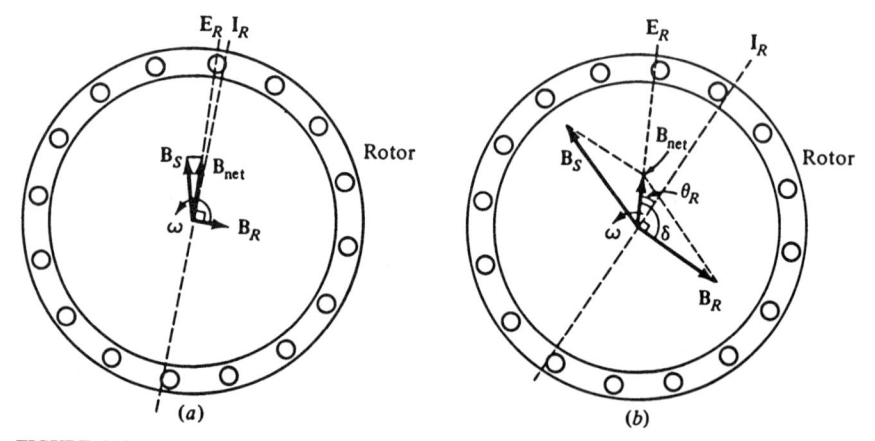

FIGURE 6.12 (*a*) The magnetic fields in an induction motor under light loads. (*b*) The magnetic fields in an induction motor under heavy loads.

and the magnetic fields) and the rotor frequency are very small. Since the relative motion is small, the induced voltage in the rotor bars E_R is very small and the resulting current I_R is small. Also, since the rotor frequency is small ($f_r = sf_e$), the reactance of the rotor ($X_R = sX_{R_0}$) is negligible, and the maximum rotor current I_R is almost in phase with the rotor voltage E_R.

The induced torque in this region is small (just enough to overcome the motor's rotational losses) because the rotor magnetic field is quite small. When the motor is loaded down (Fig. 6.12*b*), the slip increases and the rotor speed falls. Now there is more relative motion between the rotor and the magnetic fields because the rotor speed is slower. Higher rotor voltage E_R is now produced because of the higher relative motion. This in turn produces a larger rotor current I_R.

Since the induced torque is given by

$$\tau_{ind} = kB_R B_{net} \sin \delta$$

the resulting torque-speed characteristic is shown in Fig. 6.13. The torque-speed curve is divided into three regions. The first is the *low-slip region*. In this region, the motor slip

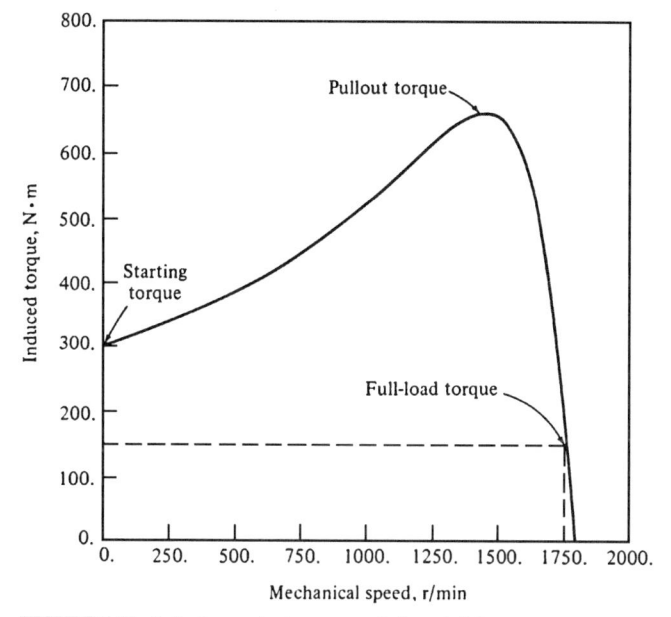

FIGURE 6.13 Induction motor torque-speed characteristic curve.

and mechanical speed change linearly with the load. The second region is the *moderate-slip region*. The peak torque (the pullout torque) of the motor occurs in this region. The third region is the *high-slip region*. In this region, the induced torque decreases with increasing load. Typically, the pullout torque is about 200 to 250 percent of the rated full-load torque of the machine. The starting torque (at zero speed) is about 150 percent of the full-load torque.

Comments on the Induction Motor Torque-Speed Curve

1. There is no induced torque at synchronous speed.
2. The torque-speed curve is linear between no load and full load. In this range, the rotor reactance is much smaller than the rotor resistance, so the rotor current, magnetic field, and induced torque varies linearly with slip.
3. There is a maximum possible torque that the motor cannot exceed. This torque is called the *pullout torque* or *breakdown torque* and is 2 to 3 times the rated full torque of the motor.
4. The starting torque is about 150 percent of the full-load torque. The motor can start carrying any load that it normally handles at full power.
5. The motor torque is proportional to the square of the applied voltage.
6. If the motor turns backward, the induced torque will stop the rotor quickly and will try to rotate it in the opposite direction. Since switching any two of the stator phases will reverse the direction of magnetic field rotation, this fact can be used to stop motors quickly. This technique is known as *plugging*.
7. If the motor is driven beyond synchronous speed, it will operate as an induction generator (Fig. 6.14).
8. The variation of power converted to mechanical form ($P_{conv} = \tau_{ind}\omega_m$) is shown in Fig. 6.15.

6.12 CHAPTER SIX

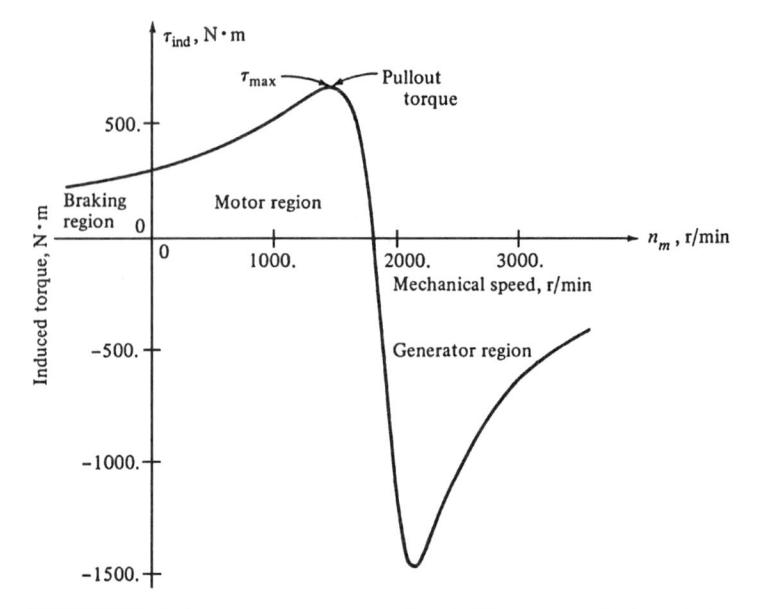

FIGURE 6.14 Induction motor torque-speed characteristic curve, showing the extended operating ranges (braking region and generator region).

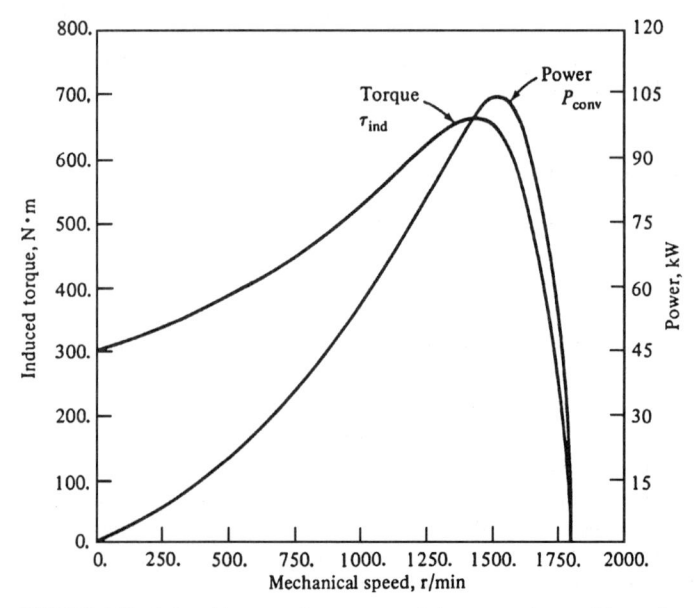

FIGURE 6.15 Induced torque and power converted versus motor speed in revolutions per minute for a four-pole induction motor.

Variation of the Torque-Speed Characteristics

Figure 6.16 illustrates the variation of the torque-speed characteristic of a wound rotor induction motor. Recall that the resistance of the rotor circuit can be changed because the rotor circuit is brought out to the stator through slip rings. As the rotor resistance increases, the pullout speed of the motor decreases, but the maximum torque remains constant. The advantage of this characteristic of wound rotor induction motors is the ability to start very heavy loads.

The maximum torque can be adjusted to occur at starting conditions by inserting a high resistance. Once the load starts to turn, the extra resistance can be removed from the circuit, and the maximum torque will shift up to near synchronous speed for normal operation. If the rotor is designed with high resistance, then the starting torque is high, but the slip is also high during normal operation. However, the higher the slip during normal operation, the smaller the fraction of power converted to mechanical power, and the lower the efficiency. A motor with a high rotor resistance has a high starting torque and poor efficiency during normal operation.

If the rotor resistance is low, then the starting torque is low and the starting current is high. However, the efficiency is high during normal operation. A compromise between high starting torque and good efficiency is needed.

A wound rotor induction motor can be used to provide high starting torque during start-up by inserting extra resistance. The extra resistance can be removed during normal operation to increase the efficiency. However, wound rotor induction motors are more expensive, require more maintenance, and have a more complex automatic control circuit than squirrel-cage induction motors. Also, wound rotor induction motors cannot be used in hazardous and explosive environments because completely sealed motors are needed. Figure 6.17 illustrates the desired motor characteristic.

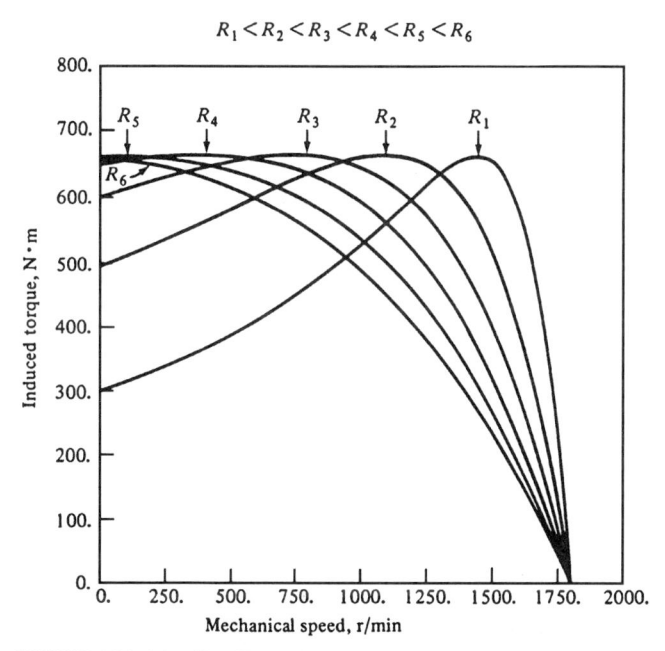

FIGURE 6.16 The effect of varying rotor resistance on the torque-speed characteristic of a wound-rotor induction motor.

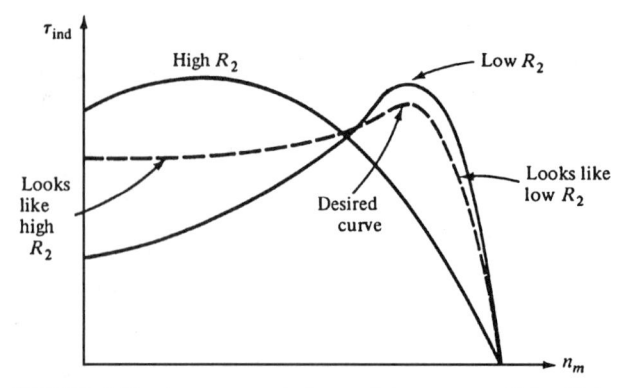

FIGURE 6.17 A torque-speed characteristic curve combining high-resistance effects at low speeds (high slip) with low-resistance effects at high speed (low slip).

CONTROL OF MOTOR CHARACTERISTICS BY SQUIRREL-CAGE ROTOR DESIGN

The reactance X_2 in the equivalent circuit of an induction motor represents the rotor's leakage reactance. This is the reactance due to the rotor flux lines that do not couple with the stator windings. In general, the farther away a rotor bar is from the stator, the greater its leakage reactance because a smaller percentage of the bar's flux reaches the stator. Therefore, if the bars of a squirrel-cage motor are placed near the surface of the rotor, the leakage reactance X_2 will be small. If the bars are placed deeper into the rotor, X_2 will be larger.

Figure 6.18a illustrates rotor lamination showing the cross section of the bars in the rotor. The bars are large and placed near the surface of the rotor. This design has a low resistance (due to the large cross section of the bars) and a low leakage reactance X_2 (due to the bar's location near the surface).

The slip at pullout torque S_{max}, the starting torque τ_{start}, and the converted power to mechanical form are given by

$$S_{max} = \frac{R_2}{\sqrt{R_{Th}^2 + (X_{Th} + X_2)^2}}$$

$$\tau_{start} = \frac{3V_{Th}^2 R_2}{\omega_{sync}[(R_{Th} + R_2)^2 + (X_{Th} + X_2)^2]}$$

$$P_{conv} = (1 - s)P_{AG}$$

where V_{Th}, R_{Th}, and X_{Th} are the Thevenin equivalent of the portion of the circuit to the left of the X's in Fig. 6.19. Reference 1 provides details on the Thevenin theorem and the derivation of these equations.

Therefore, the pullout torque will be near synchronous speed due to the low rotor resistance, and the efficiency of the motor will be high. However, the starting torque of the

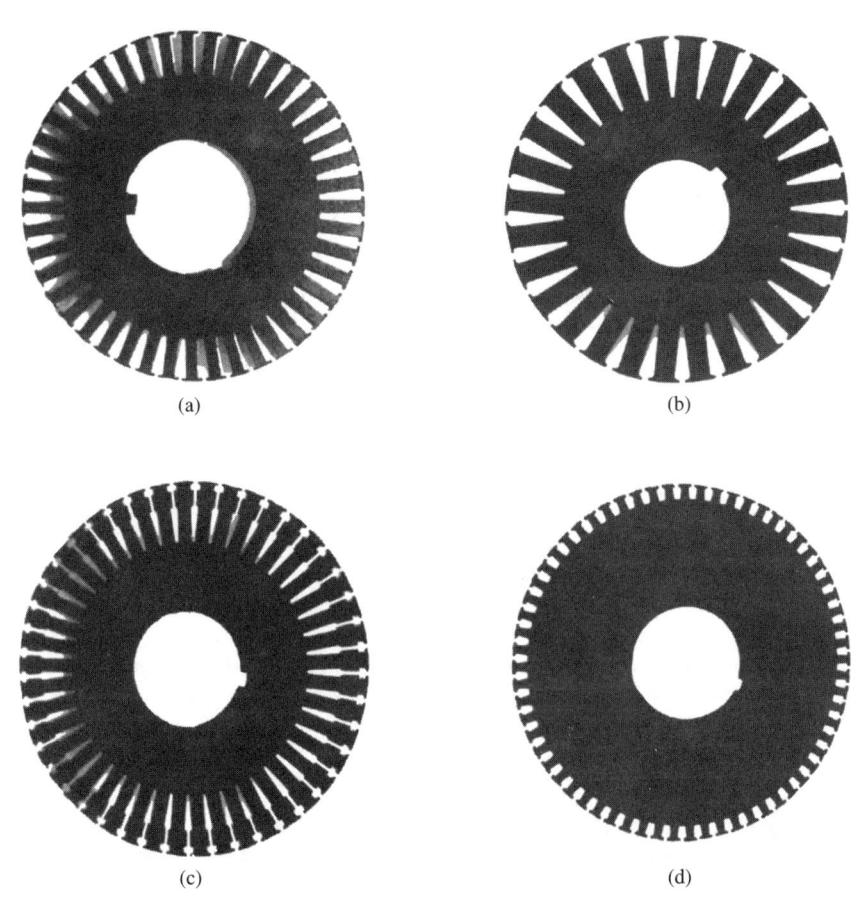

FIGURE 6.18 Laminations from typical squirrel-cage induction motor rotors, showing the cross section of the rotor bars: (*a*) NEMA design class A—large bars near the surface. (*b*) NEMA design class B—large, deep rotor bars. (*c*) NEMA design class C—double-cage rotor design. (*d*) NEMA design class D—small bars near the surface. (*Courtesy of MagneTek, Inc.*)

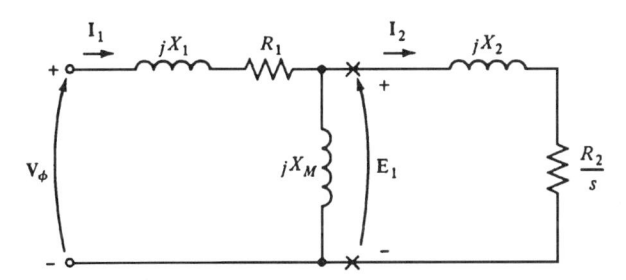

FIGURE 6.19 Per-phase equivalent circuit of an induction motor.

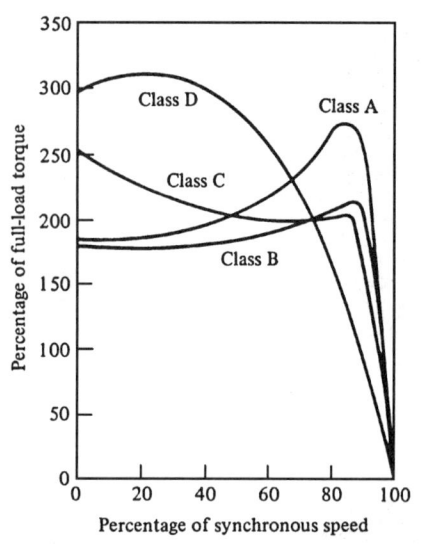

FIGURE 6.20 Typical torque-speed curves for different rotor designs.

motor will be small (low R_2), and the starting current will be high. This design is known as the National Electrical Manufacturers Association (NEMA) class A. This is a typical induction motor. Figure 6.20 illustrates its torque-speed characteristics. Figure 6.18d illustrates a rotor cross section with small bars placed near the surface. The rotor resistance is high due to the small cross-sectional area of the bars. The leakage reactance of the rotor is small because the bars are placed near the stator.

The pullout torque of this motor occurs at high slip, and the starting torque is high due to the large resistance of the rotor. This type of motors is called NEMA design class D. Figure 6.20 illustrates its torque-speed characteristic.

Deep Bar and Double-Cage Rotor Designs

This design has a *variable* rotor resistance that can combine high starting torque and low starting current (class D) with low normal operating slip and high efficiency (class A). The double-cage rotors use deep rotor bars, as illustrated in Fig. 6.21. Figure 6.21a shows the current flowing through the upper part of a deep rotor bar. The leakage inductance is small in this region due to the tight coupling between the rotor and the stator. Figure 6.21b shows the current flowing in the bottom of the bar. In this case, the leakage inductance is higher. Hence, the flux is loosely linked to the stator. Since the rotor bars are connected in parallel, they represent a series of parallel electric circuits (Fig. 6.21c). The upper ones have a smaller inductance than the lower ones.

During normal operation (low slip), the frequency of the rotor is very small. The reactances of all the parallel bars are small compared to their resistances. The rotor resistance is small due to the large cross-sectional area. This results in high efficiency at low slip.

During starting conditions (high slip), the resistances are small compared to the reactances. The current is forced to flow in the bars located near the stator due to their low reactances. Hence, the rotor resistance is higher than before due to the smaller *effective*

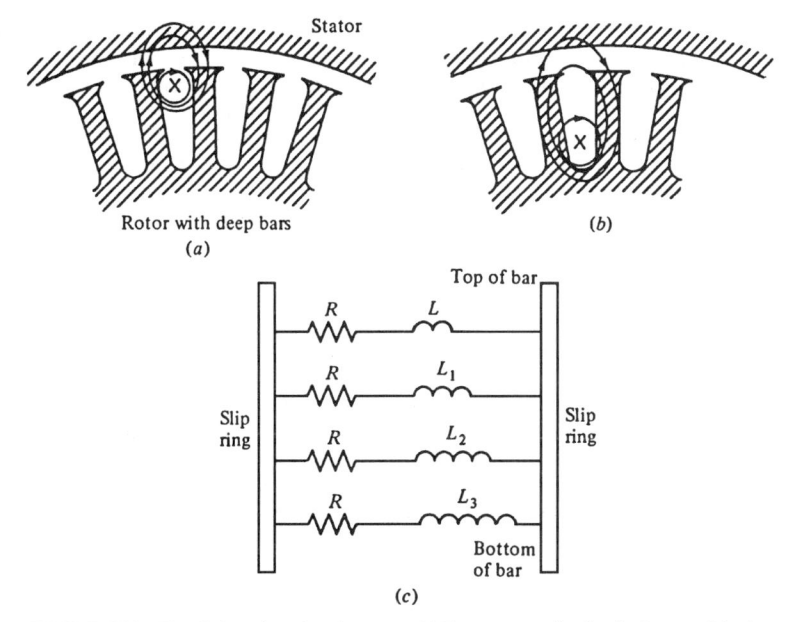

FIGURE 6.21 Flux linkage in a deep bar rotor. (*a*) For a current flowing in the top of the bar, the flux is tightly linked to the stator, and leakage inductance is small. (*b*) For a current flowing in the bottom of the bar, the flux is loosely linked to the stator, and leakage inductance is large. (*c*) The resulting equivalent circuit of the rotor bar as a function of depth in the rotor.

cross section. Since the rotor resistance is high during starting conditions, the starting torque is relatively higher and the starting current is relatively lower than for a class A motor. This design is known as class B (Fig. 6.20).

Figure 6.18*c* illustrates a cross-sectional view of a double-cage rotor. The inner bars are large (low resistance). They are buried deeply in the rotor. The upper bars are small (low resistance). They are located near the surface of the rotor. This design is similar to the deep bar rotor, but the difference between the low- and high-slip operation is even more exaggerated.

During starting conditions, the rotor resistance is high because the small bars are effective only. Hence, the starting torque is high. However, during normal operation, both bars are effective, resulting in low resistance. Double-cage rotors of this type are used to produce NEMA class B and C characteristics (Fig. 6.20). The main disadvantage of double-cage rotors is that they are more expensive than squirrel-cage rotors.

STARTING INDUCTION MOTORS

Since the inrush current of induction motors can exceed 8 times the normal operating current, *across-the-line starting* may not be acceptable because it may cause a dip in the power system voltage. In wound rotor induction motors, extra resistance can be inserted in the rotor circuit during starting. This results in an increase in the starting torque and a reduction in the starting current. The starting current of a squirrel-cage induction motor can vary widely depending on the motor's rated power and the effective resistance of the rotor at starting conditions.

A starting *code letter* has been established to estimate the rotor current when the motor is starting. This code letter sets the limit on the magnitude of the current that the motor can draw at starting conditions. These limits are given in terms of the starting apparent power of the motor as a function of its horsepower rating. Table 6.1 shows the starting kVA per horsepower for each code letter. The NEMA code letters indicate the starting kilovoltamperes per horsepower of rating for a motor. Each code letter extends up to, but does not include, the lower bound of the next higher class.

The starting current of an induction motor can be determined by

$$I_L = \frac{(\text{rated horsepower})(\text{code letter factor})}{\sqrt{3}(\text{rated voltage})}$$

The starting current can be reduced if necessary, by a starting circuit. However, this results in reduction of the starting torque of the motor. The starting current of an induction motor can be reduced by inserting extra inductors or resistors in the power line during starting conditions. Alternatively, the terminal voltage of the motor can be reduced during starting by using an autotransformer to step it down.

Figure 6.22 illustrates a typical reduced-voltage starting circuit using autotransformers. During start-up, contacts 1 and 3 are shut. The voltage supplied to the motor is lower than

TABLE 6.1 NEMA Code Letters

Nominal code letter	Locked rotor, kVA/hp	Nominal code letter	Locked rotor, kVA/hp
A	0–3.15	L	9.00–10.00
B	3.15–3.55	M	10.00–11.20
C	3.55–4.00	N	11.20–12.50
D	4.00–4.50	P	12.50–14.00
E	4.50–5.00	R	14.00–16.00
F	5.00–5.60	S	16.00–18.00
G	5.60–6.30	T	18.00–20.00
H	6.30–7.10	U	20.00–22.40
J	7.10–8.00	V	22.40 and up
K	8.00–9.00		

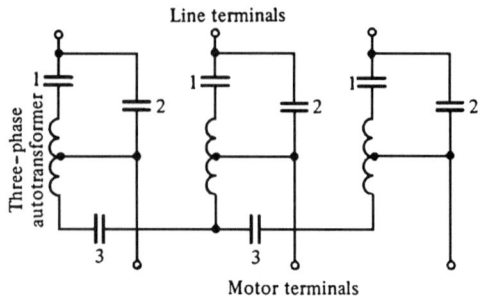

Starting sequence:
(*a*) Close 1 and 3
(*b*) Open 1 and 3
(*c*) Close 2

FIGURE 6.22 An autotransformer starter for an induction motor.

normal. When the motor approaches the operating speed, those contacts are opened, and contact 2 is shut, putting the full line voltage across the motor.

When the starting current is reduced proportionally to the decrease in terminal voltage, the starting torque decreases proportionally to the *square* of the applied voltage. Hence, the reduction in starting current is limited when the motor is coupled to a load.

Induction Motor Starting Circuits

Figure 6.23 illustrates a typical full-voltage or across-the-line induction motor starter circuit. The meanings of the symbols are explained in Fig. 6.24. When the start button is

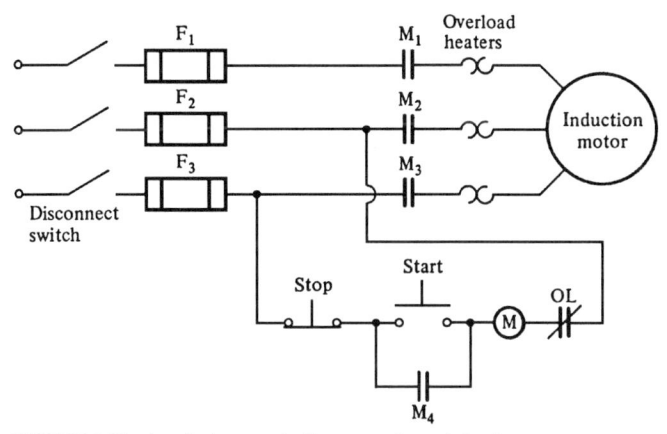

FIGURE 6.23 A typical across-the-line starter for an induction motor.

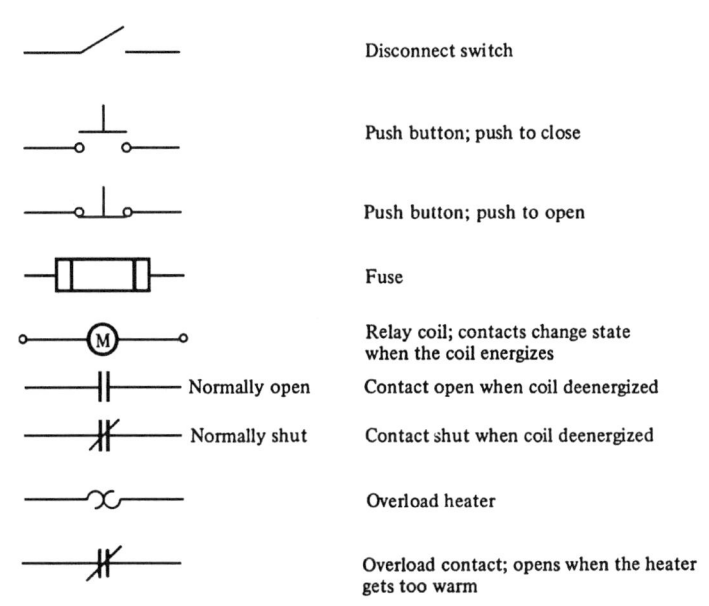

	Disconnect switch
	Push button; push to close
	Push button; push to open
	Fuse
	Relay coil; contacts change state when the coil energizes
	Normally open — Contact open when coil deenergized
	Normally shut — Contact shut when coil deenergized
	Overload heater
	Overload contact; opens when the heater gets too warm

FIGURE 6.24 Typical components found in induction motor control circuits.

6.20 CHAPTER SIX

pressed, the coil of relay (or contactor) M is energized, causing normally open contacts M_1, M_2, and M_3 to close. When these contacts close, power is applied to the motor. Also, contact M_4 closes, shorting out the starting switch, allowing the operator to release the switch while relay M remains energized. When the stop button is pressed, relay M deenergizes, stopping the motor.

The built-in protective features in this type of magnetic motor starter circuit are

1. Short-circuit protection
2. Overload protection
3. Undervoltage protection

Fuses F_1, F_2, and F_3 provide *short-circuit protection* for the motor. When a short circuit occurs, the current increases instantly to many times the rated value. The fuses blow, disconnecting the motor from the power supply to prevent it from burning up.

However, these fuses are designed to tolerate currents many times greater than the full-load current. They would *not* burn up during normal starting due to inrush current. Therefore, if a short circuit occurs through a high resistance and/or excessive loads, it will not be cleared by the fuses.

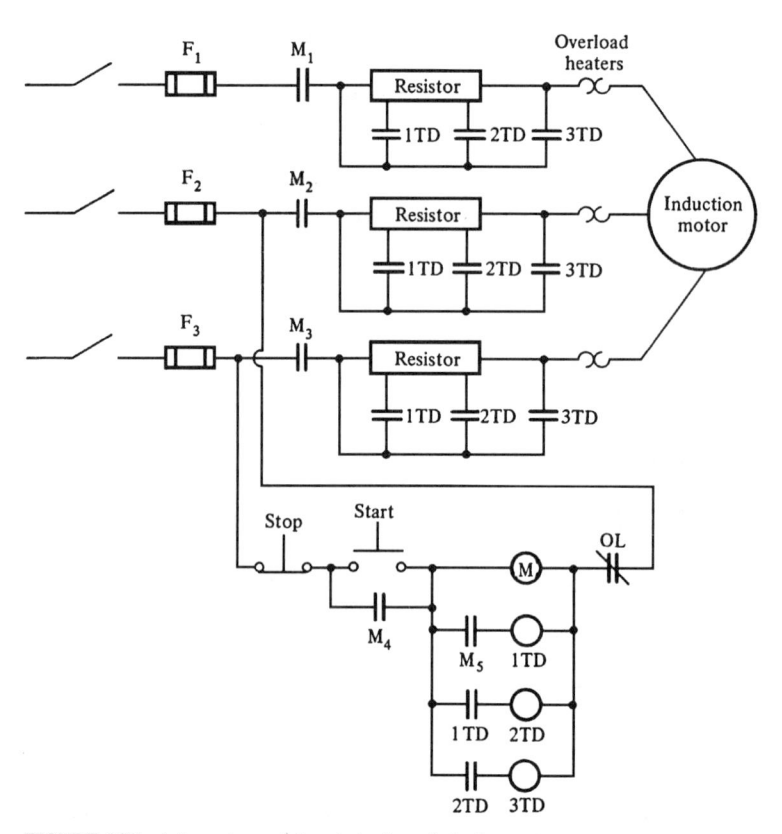

FIGURE 6.25 A three-step resistive starter for an induction motor.

The devices labeled OL provide the *overload protection* for the motor. They consist of two parts: the overload heater element and overload contacts. During normal operation, the overload contacts are closed. When the temperature of the heater elements increases to a set limit, the OL contacts open. This deenergizes the M relay opening the M contacts and isolating the motor from the power supply. If the motor is overloaded for an extended period of time, damage will occur due to the excessive heating caused by the high currents. The inrush current would not affect the heater elements because of its short duration. The heater elements will operate when there is a high current for an extended period, isolating the motor from the power supply.

The controller provides the *undervoltage protection*. The control power for the M relay comes from directly across the incoming lines to the motor. If the voltage level drops beyond a certain level, the voltage applied to the M relay will also fall, resulting in the deenergization of the relay.

Figure 6.25 illustrates an induction motor starting circuit with resistors to reduce the starting current. Relays 1TD, 2TD, and 3TD are on-time delay relays. When these relays are energized, there is a set time delay before their contacts close.

When the start button is pushed, the M relay gets energized. Its contacts close, allowing power to reach the motor. Since the contacts for the 1TD, 2TD, and 3TD relays are all open, the current is forced to flow through the full starting resistor, reducing the starting current. The 1TD relay gets energized as soon as the contacts of the M relay close. However, there is a set time delay before the contacts of the 1TD relay close. During this time, the motor would have gained some speed, and the starting current decreased.

After the specified time delay, the contacts of the 1TD relay close, cutting out the first part of the starting resistance and simultaneously causing relay 2TD to energize. After another delay, the contacts for relay 2TD close, cutting out the second part of the starting resistor and causing relay 3TD to energize. Finally, the contacts of relay 3TD close, causing the resistor to be bypassed.

The values of the resistor and time delays can be selected to prevent the starting current from becoming dangerously large while allowing enough current to ensure adequate acceleration to the normal operating speed.

REFERENCE

1. S. J. Chapman, *Electric Machinery Fundamentals*, 2d ed., McGraw-Hill, New York, 1991.

CHAPTER 7
SPEED CONTROL OF INDUCTION MOTORS

Until the advent of solid-state drives, induction motors were not used in many applications requiring speed control. The normal operating range of an induction motor is within less than 5 percent slip. At larger slip, the efficiency of the motor will drop significantly because the rotor copper losses are directly proportional to the slip of the motor ($P_{RCL} = sP_{AG}$). The speed of an induction motor can be controlled by varying the synchronous speed or the slip for a given load. The synchronous speed can be varied by changing the electrical frequency or the number of poles. The slip can be changed by varying the rotor resistance or terminal voltage.

SPEED CONTROL BY CHANGING THE LINE FREQUENCY

The rate of rotation of the stator magnetic field depends on the electrical frequency. The no-load point on the torque-speed curve changes with the frequency (Fig. 7.1). The *base speed* is the synchronous speed at rated conditions.

The speed of the motor can be adjusted by using variable frequency control. A variable frequency induction motor drive can control the speed from 5 percent of the base load to twice the base speed.

There are limits on the voltage and torque as the frequency is varied to ensure safe operation. When the speed is being reduced below the base speed, the terminal voltage to the stator should be decreased linearly with decreasing stator frequency. This process is called *derating*. If the motor is not derated, the steel in the core will saturate and large magnetization current will flow in the machine.

The flux in the core of an induction motor is given by *Faraday's law:*

$$v(t) = N \frac{d\phi}{dt}$$

Solving for the flux ϕ gives

$$\phi = \frac{1}{N} \int v(t)\, dt$$

$$= \frac{1}{N} \int V_M \sin \omega t \, dt$$

$$\boxed{\phi = \frac{V_M}{\omega N} \cos \omega t}$$

If the electrical frequency decreases by 10 percent while the voltage remains constant, the flux in the core will increase by 10 percent. The magnetization current will also increase by 10 percent in the unsaturated region of the motor's magnetization curve. The magnetization current will increase by much more than 10 percent in the saturated region.

Since induction motors are designed to operate near saturation, the increase in flux due to the decrease in frequency will cause a large magnetization current to flow. The stator

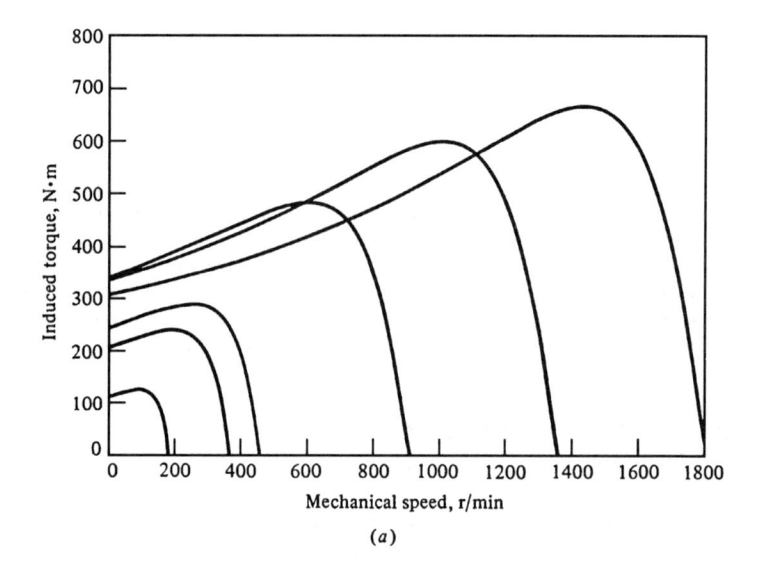

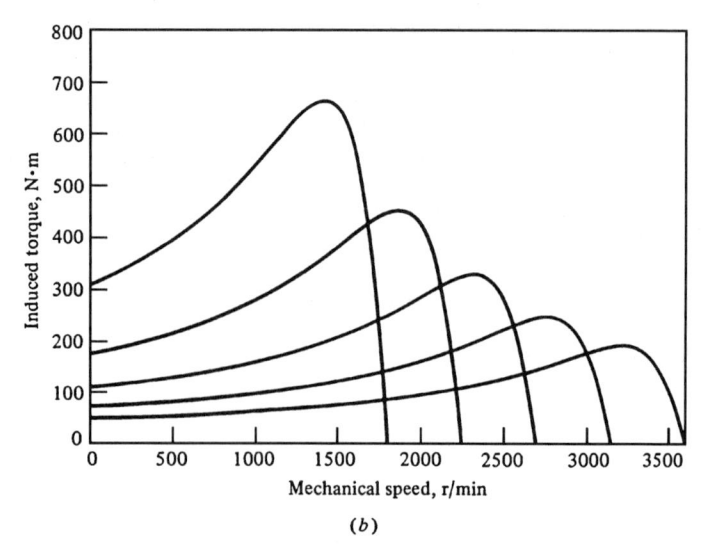

FIGURE 7.1 Variable-frequency speed control in an induction motor: (*a*) The family of torque-speed characteristic curves for speeds below base speed, assuming that the line voltage is derated linearly with frequency. (*b*) The family of torque-speed characteristic curves for speeds above base speed, assuming that the line voltage is held constant.

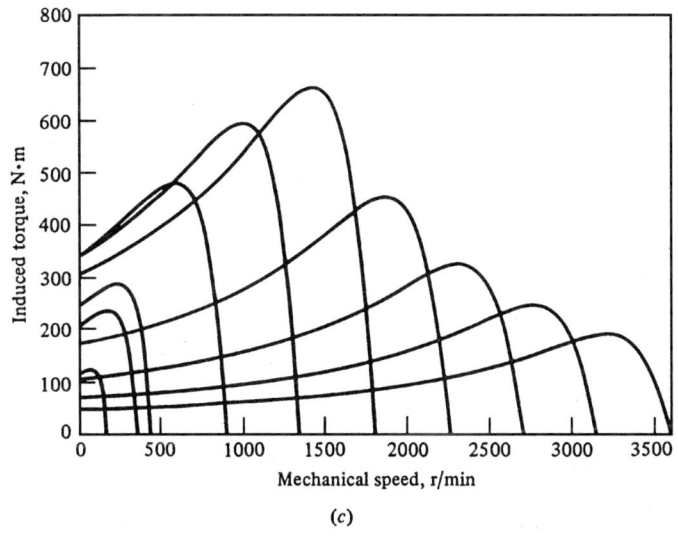

FIGURE 7.1 (*Continued*) (*c*) The torque-speed characteristic curves for all frequencies.

voltage is usually decreased in direct proportion to the decrease in frequency to avoid large magnetization currents.

The flux in the motor remains approximately constant when the voltage is decreased with frequency. Since the power supplied to the motor is given by

$$P = \sqrt{3} V_L I_L \cos \theta$$

the maximum power rating must decrease linearly with decreasing voltage to protect the stator from overheating.

Figure 7.1*a* illustrates a family of torque-speed characteristic curves for speeds below the base speed. The stator voltage was assumed to vary linearly with frequency.

When the frequency applied to the motor exceeds the rated frequency, the stator voltage is held constant at the rated value. Although the applied voltage can be raised above the rated value without reaching saturation, it is limited to the rated voltage. This is done to protect the winding insulation of the motor. As the frequency increases while the voltage remains constant, the resulting flux and the maximum torque will decrease with it.

Figure 7.1*b* shows a family of torque-speed characteristic curves for speeds higher than the base speed, assuming that the stator voltage is held constant. Figure 7.1*c* shows a family of torque-speed characteristic curves for speeds higher and lower than the base speed, assuming that the stator voltage is varied linearly with frequency below base speed and is held constant at rated value above base speed (the rated speed for the motor shown in Fig. 7.1 is 1800 r/min). Changing the line frequency with solid-state motor drives has become the preferred method for induction motor speed control.

SPEED CONTROL BY CHANGING THE LINE VOLTAGE

Since the torque developed by the induction motor is proportional to the square of the applied voltage, the speed of the motor can be controlled within a limited range by varying the line voltage as shown in Fig. 7.2.

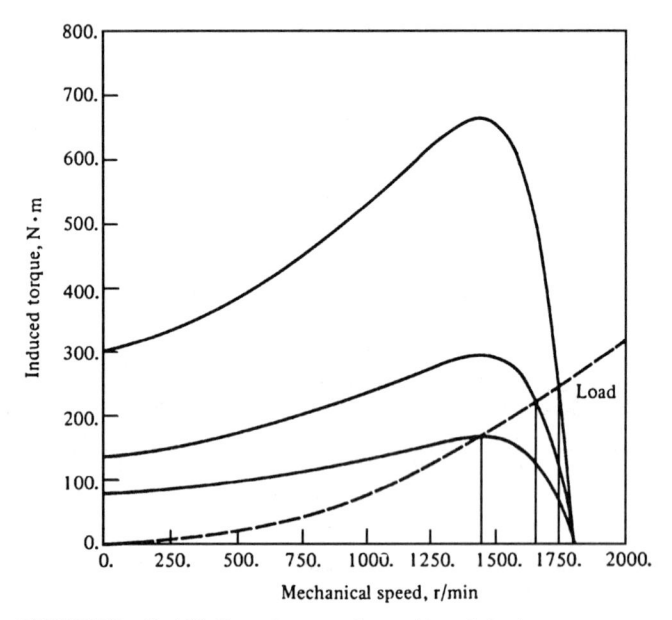

FIGURE 7.2 Variable-line-voltage speed control in an induction motor.

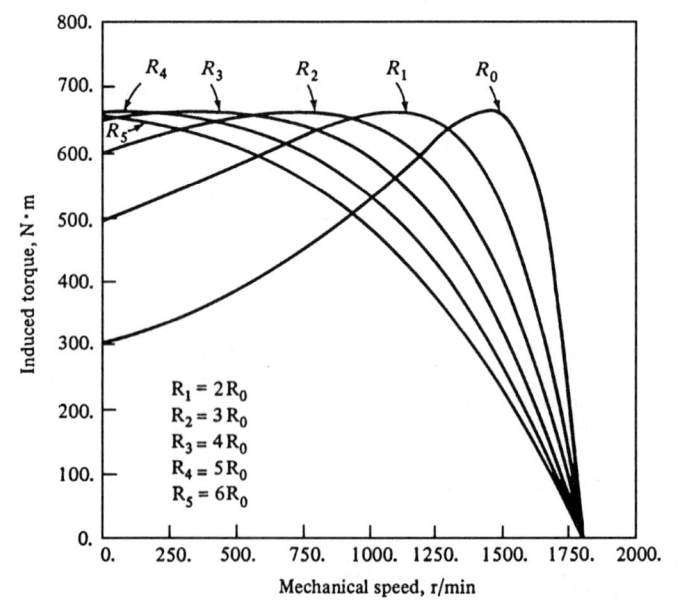

FIGURE 7.3 Speed control by varying the rotor resistance of a wound rotor induction motor.

SPEED CONTROL BY CHANGING THE ROTOR RESISTANCE

The shape of the torque-speed curve of wound rotor induction motors can be changed by inserting extra resistances into the rotor circuit, as shown in Fig. 7.3. However, inserting additional resistances into the rotor circuit will reduce the efficiency of the motor significantly. This method is usually used for short periods.

SOLID-STATE INDUCTION MOTOR DRIVES

The solid-state variable frequency induction motor drive is the preferred method for speed control. A typical drive is shown in Fig. 7.4. The drive is very flexible. Its input can be single-phase or three-phase; 50 or 60 Hz; and any voltage in the range of 208 to 230 V. The output is a three-phase voltage whose frequency can vary in the range of 0 to 120 Hz and whose voltage can vary in the range of 0 to the rated voltage of the motor. The control of the output voltage and frequency is achieved by using the pulse-width modulation (PWM) technique.

The output frequency and output voltage can be controlled independently. Figure 7.5 illustrates how the drive controls the output frequency while the root-mean-square (rms) voltage is maintained at a constant level. Figure 7.6 illustrates how the drive controls the rms voltage while maintaining the frequency at a constant value.

FIGURE 7.4 A typical solid-state variable-frequency induction motor drive. (*Courtesy of MagneTek Drives and Systems.*)

MOTOR PROTECTION

The induction motor drive has a variety of features for protecting the motor. The drive can detect and trip the motor under any of the following conditions:

1. An overload (excessive steady-state currents)
2. Excessive instantaneous currents
3. Overvoltage
4. Undervoltage

THE INDUCTION GENERATOR

Figure 7.7 illustrates the torque-speed characteristic of an induction machine. It shows clearly that if an induction motor is driven at a speed higher than the synchronous speed by

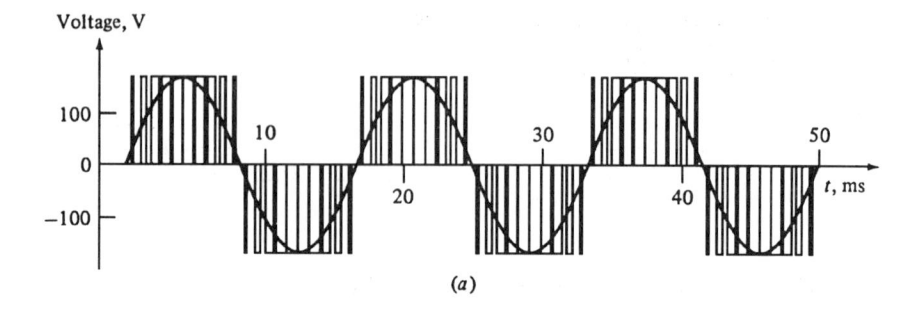

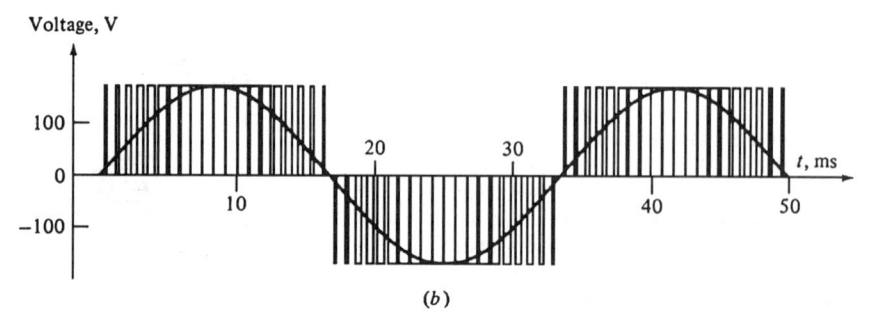

FIGURE 7.5 Variable frequency control with a PWM waveform: (a) 60-Hz 120-V PWM waveform; (b) 30-Hz 120-V PWM waveform.

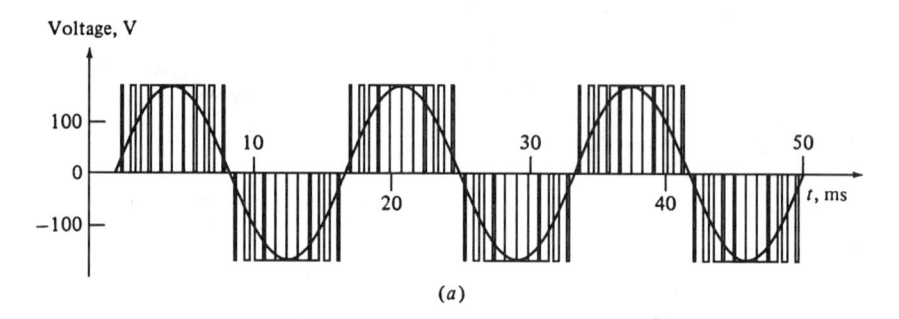

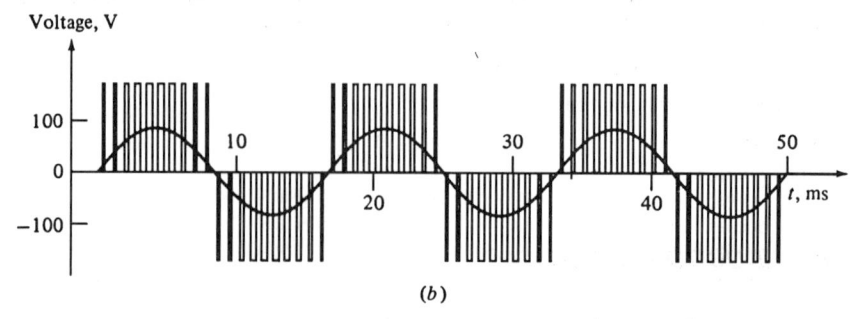

FIGURE 7.6 Variable voltage control with a PWM waveform: (a) 60-Hz 120-V PWM waveform; (b) 60-Hz 60-V PWM waveform.

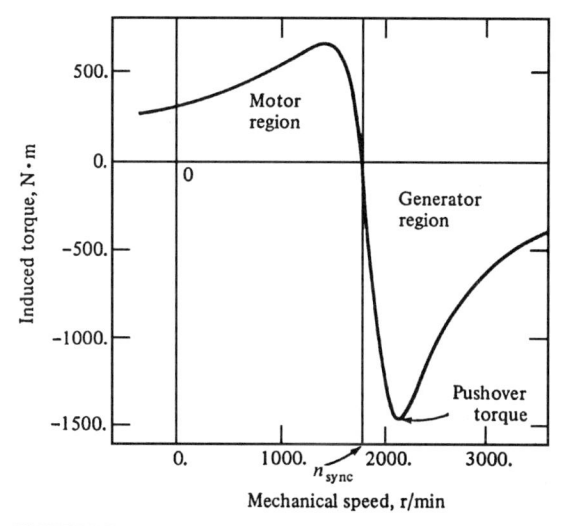

FIGURE 7.7 The torque-speed characteristic of an induction machine, showing the generator region of operation. Note the pushover torque.

a prime mover, the direction of the induced torque will reverse and it will act as a generator. As the torque applied to the shaft increases, the power generated increases. However, there is a maximum possible induced torque in the generator region of operation (*pushover torque*). If the actual torque is higher than the pushover torque, the machine will overspeed.

An induction machine operating as a generator has severe limitations. An induction generator cannot produce reactive power because it does not have a separate field circuit. In reality, it requires reactive power. An external source of reactive power must be provided to it at all times to maintain its stator magnetic field. The induction generator cannot control its own output voltage because it does not have a field circuit. The terminal voltage of the generator must be maintained by the external power system which is connected to it.

The main advantages of the induction generator are its simplicity and its ability to operate at different speeds (higher than synchronous speed). Since no sophisticated regulation is required, this generator is suitable for windmills and supplementary power sources connected to an existing power system. In these applications, the power factor correction can be provided by capacitors, and the terminal voltage can be controlled by an existing power system (the grid).

Induction Generator Operating Alone

The induction generator can operate independently of any power system, if capacitors are available to supply the reactive power required by the generator and by the load. This arrangement is shown in Fig. 7.8.

The magnetization current required by the induction machine as a function of the terminal voltage can be found by running the machine as a motor at no load and measuring its armature current. This magnetization curve is shown in Fig. 7.9*a*. Therefore, the induction generator can achieve a given voltage level if the external capacitors are supplying the magnetization current corresponding to that level. The reactive current produced by a capacitor is *directly proportional* to the voltage applied to it (straight-line relationship). Figure 7.9*b* illustrates the variation of voltage with current for a given frequency.

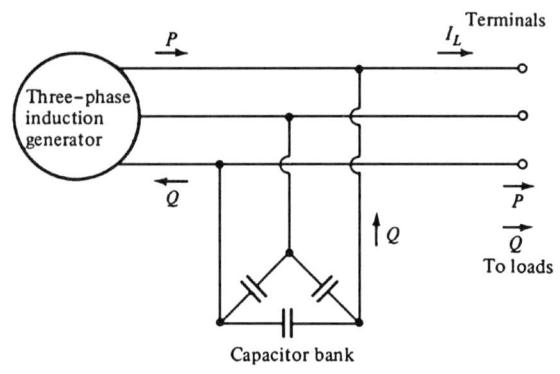

FIGURE 7.8 An induction generator operating alone with a capacitor bank to supply reactive power.

The induction generator must be flashed by momentarily running it as a motor. This is done to establish residual flux in the rotor, which is needed to start the induction generator. When the induction generator is starting, a small voltage is produced by the residual magnetism in its field circuit. A capacitive current flow is produced by the small voltage which increases the terminal voltage. The increase in terminal voltage increases the capacitive current, which increases the terminal voltage further until the voltage is fully built up.

The main disadvantage of induction generators is that their voltage varies significantly with changes in load (especially reactive load). Figure 7.10 illustrates a typical terminal voltage-current characteristic of an induction generator operating alone with a constant parallel capacitance.

The voltage collapses very rapidly when the generator is supplying inductive loads because the capacitors must supply all the reactive power needed by the load and the generator. Any reactive power diverted to the load moves the generator back along its magnetization curve. This results in a major drop in generator voltage.

A set of series capacitors is included in the power line in addition to the parallel capacitors. The capacitive reactive power increases with increasing load. This compensates for the reactive power demanded by the load.

Figure 7.11 illustrates the terminal characteristic of an induction generator with series capacitors. The frequency of the induction generator varies slightly with the load. However, this frequency variation is limited to less than 5 percent because the torque-speed characteristic is very steep in the normal operating range. This variation is acceptable in many applications such as isolated or emergency generators. The induction generator is ideal for windmills and energy recovery applications. Since most of these applications operate in parallel with the grid, the terminal voltage and frequency are controlled by the grid. Capacitors are used for power factor correction.

INDUCTION MOTOR RATINGS

Figure 7.12 shows a nameplate for a typical high-efficiency induction motor. The most important ratings are

1. Output power
2. Voltage
3. Current

SPEED CONTROL OF INDUCTION MOTORS   7.9

4. Power factor
5. Speed
6. Nominal efficiency
7. NEMA design class
8. Starting code

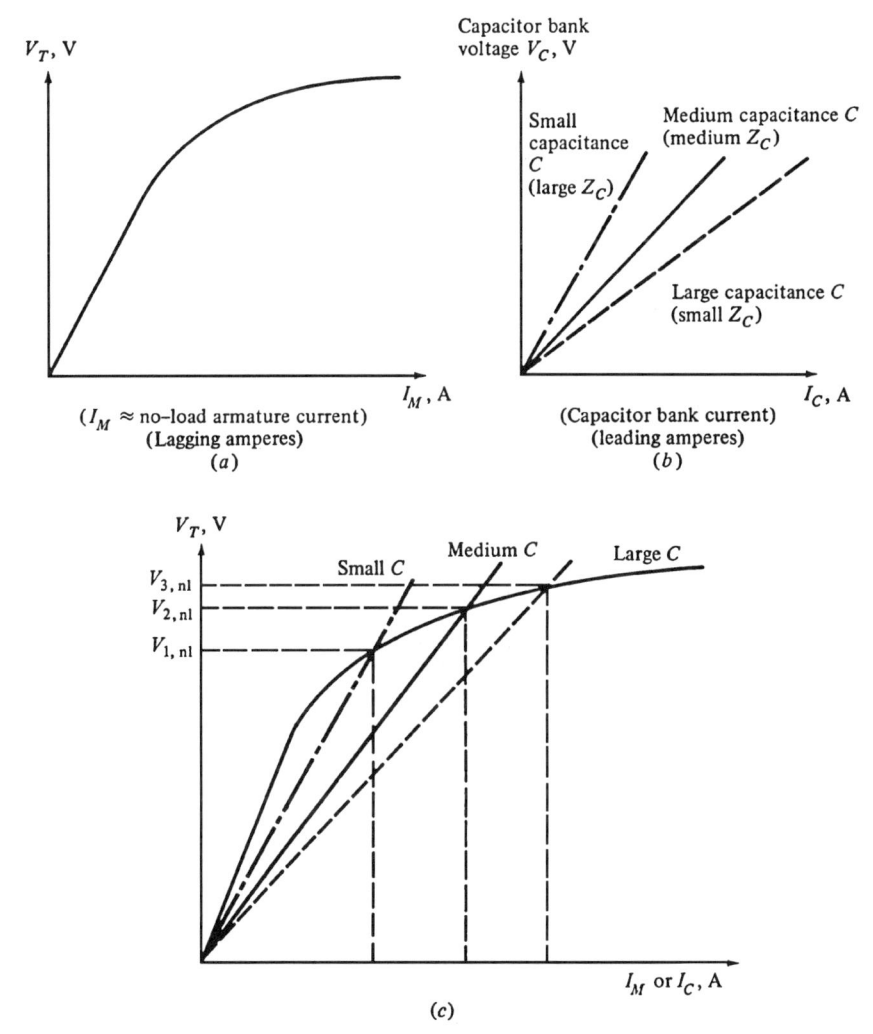

FIGURE 7.9 (a) The magnetization curve of an induction machine. It is a plot of the terminal voltage of the machine as a function of its magnetization current (which *lags* the phase voltage by approximately 90°). (b) Plot of the voltage-current characteristic of a capacitor bank. Note that the larger the capacitance, the greater its current for a given voltage. This current *leads* the phase voltage by approximately 90°. (c) The no-load terminal voltage for an isolated induction generator can be found by plotting the generator terminal characteristic and the capacitor voltage-current characteristic on a single set of axes. The intersection of the two curves is the point at which the reactive power demanded by the generator is exactly supplied by the capacitors, and this point gives the *no-load terminal voltage* of the generator.

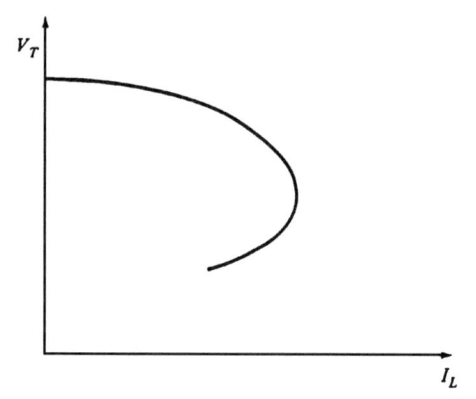

FIGURE 7.10 The terminal voltage-current characteristic of an induction generator for a load with a constant lagging power factor.

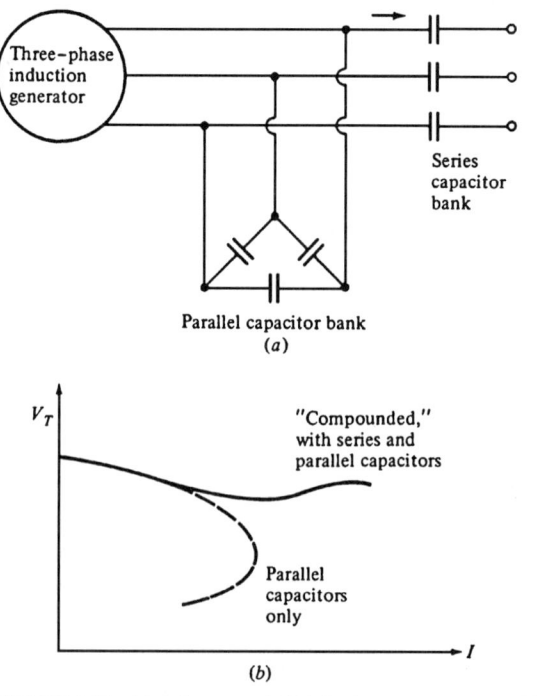

FIGURE 7.11 (a) A "compounded" induction generator, one with both "shunt" (parallel) and series capacitors. (b) The resulting voltage-current characteristic of the generator for a load with a constant lagging power factor.

SPEED CONTROL OF INDUCTION MOTORS 7.11

```
┌─────────────────────────────────────────────────────────┐
│              SPARTAN™ MOTOR                             │
├─────────────────────────────────────────────────────────┤
│ MODEL     27987J-X                                      │
├─────────────────────────────────────────────────────────┤
│ TYPE   CJ4B              │ FRAME      324TS             │
├──────────────────────────┼──────────────────────────────┤
│ VOLTS  230/460           │ °C AMB.                      │
│                          │ INS.CL.    40 B              │
├──────────────────────────┼──────────────────────────────┤
│ FRT.                     │ EXT.                         │
│ BRG.   210 SF            │ BRG.       312 SF            │
├──────────────────────────┼──────────────────────────────┤
│ SERV.                    │ OPER.                        │
│ FACT.  1.0               │ INSTR.     C-517             │
├────────────┬─────────────┼─────────────┬────────────────┤
│ PHASE  3   │ HZ   60     │ CODE  G     │ WDGS.  1       │
├────────────┴─────────────┴─────────────┴────────────────┤
│ H.P.   40                                               │
├─────────────────────────────────────────────────────────┤
│ R.P.M. 3565                                             │
├─────────────────────────────────────────────────────────┤
│ AMPS   97/48.5                                          │
├─────────────────────────────────────────────────────────┤
│ NEMA NOM. EFF.   .936                                   │
├─────────────────────────────────────────────────────────┤
│ NOM.P.F.   .827                                         │
├─────────────────────────────────────────────────────────┤
│ MIN. AIR                                                │
│ VEL.FT/MIN.                                             │
├─────────────────────────────────────────────────────────┤
│ DUTY   Cont                         │ NEMA              │
│                                     │ DESIGN  B         │
└─────────────────────────────────────────────────────────┘
```

FIGURE 7.12 The nameplate of a typical high-efficiency induction motor. (*Courtesy of MagneTek, Inc.*)

The voltage limit is based on the maximum acceptable magnetization current flow because as the voltage increases, the iron becomes more saturated and the magnetization current increases. A 60-Hz induction motor can be used on a 50-Hz power system only if the voltage rating is decreased by the same proportion as the decrease in frequency. The current limit is based on the maximum acceptable heating in the motor's windings. The power limit is determined by the combination of the voltage and current ratings with the power factor and efficiency.

REFERENCE

1. S. J. Chapman, *Electric Machinery Fundamentals*, 2d ed., McGraw-Hill, New York, 1991.

CHAPTER 8
MAINTENANCE OF MOTORS

An electric motor that has been properly maintained will contribute greatly to the continuous operation and success of a plant. Since the cost of unscheduled downtime is high, regular maintenance of motors has become one of the highest priorities in plants. A loss of a motor could result in reduction in reliability of a system, impairment of a system, or shutdown of a plant.

CHARACTERISTICS OF MOTORS

The nameplate data include the rated horsepower, speed, voltage, and service factor if different from 1.0. The service factor is a multiplier that can be applied to the horsepower of motors that are designed to handle periodic overloading.

ENCLOSURES AND COOLING METHODS

The National Electric Manufacturers Association (NEMA) classifies motors according to environmental protection and methods of cooling. The available categories include the following:

Open dripproof. Motor has ventilating openings constructed so that operation will not be affected by solid or liquid particles that strike or enter the enclosure at any angle up to 15° downward from the vertical.

Dripproof guarded. These self-ventilating motors feature louvered covers on the sides with grilles on bottom openings to prevent accidental exposure to live metal or rotating parts.

Dripproof forced ventilated. Dripproof guarded motors with forced ventilation are provided by a motor-mounted blower driven by a three-phase ac motor.

Dripproof separately ventilated. These motors provide high horsepower ratings with separate ventilation supplied by the customer.

Totally enclosed. Motors that operate in severe environments require a totally enclosed frame. These motors are nonventilated, fan-cooled, or dual-cooled, depending on the horsepower.

Totally enclosed, nonventilated. Motors are not equipped for cooling by means external to the enclosing parts, generally limited to low horsepower ratings or short-time rated machines.

Totally enclosed, fan-cooled. Motors with exterior surfaces are cooled by an external fan on the motor shaft. Cooling is dependent upon motor speed.

Totally enclosed air-over in-line. Motors with an external fan are driven by a constant-speed ac motor flange mounted to the motor fan shroud. These provide cooling independent of motor speeds. Brakes and tachometers cannot be mounted on the motor end bracket, except for specific small tachometers which can be nested between the motor bracket and fan.

Totally enclosed air-over piggyback. Motors have a top-mounted, ac motor-driven blower with shroud to direct ventilating air over the motor frame.

Totally enclosed dual-cooled air-to-air heat exchanger. Motors are cooled by circulating internal air through the heat exchanger by an ac motor-driven blower. External air circulated through the heat exchanger by another ac motor-driven blower removes heat from the circulating internal air. No free exchange of air occurs between the inside and outside of the motor.

Totally enclosed dual-cooled with air-to-water heat exchanger. This is similar to the previous description except that external circulating airflow is replaced by user-supplied water to remove heat from the heat exchanger.

Totally enclosed pipe-ventilated or totally enclosed separately ventilated. Motors are cooled by user-supplied air which is piped into the machines and ducted out of the machines by user-supplied ducts.

Explosion-proof. The enclosures of these motors are designed and constructed to withstand an explosion of a specified gas or vapor which may occur within them. It prevents the ignition of the specified gas or vapor surrounding the machines by sparks, flashes, or explosions of the specified gas or vapor which may occur within the machine casing.

Dust-ignition-proof. The enclosures of these motors are designed and constructed so as to exclude ignitable amounts of dust or amounts which might affect performance or rating. It will not permit arcs, sparks, or heat otherwise generated or liberated inside the enclosure to cause ignition of exterior accumulations or atmosphere-suspended dust on or in the vicinity of the enclosure.

Weather-protected. These motors are divided into two types, I and II. A type I motor is open, with ventilating passages constructed to minimize the ability of rain, snow, and airborne particles to come in contact with the electric parts. A type II motor includes ventilating passages at both intake and discharge in an arrangement so that high-velocity air and particles blown into the machine by storms or winds can be discharged without entering the internal ventilating passages leading directly to the motor's electric parts.

Totally enclosed, pipe-ventilated. These motors are totally enclosed except for openings so arranged that inlet and outlet pipes are connected to them for the admission and discharge of ventilating air. The air may be circulated by means integral with the motor or external to and not a part of the motor, or separately forced ventilation.

Totally enclosed water, air-cooled. A totally enclosed motor is cooled by circulating air, which, in turn, is cooled by circulating water. The motor is provided with a water-cooled heat exchanger for cooling ventilating air and with a fan or fans, integral with or separate from the rotor shaft, for circulating ventilating air.

Totally enclosed air-to-air cooled. This totally enclosed motor is cooled by circulating internal air through a heat exchanger, which is cooled by circulating external air.

APPLICATION DATA

The first step in obtaining and maintaining adequate motor performance is the proper application of motors in accordance with their service conditions, which are designated as usual and unusual, by NEMA.

Usual Service Conditions

Environment

1. Exposure to an ambient temperature in the range of 0 to 40°C (32 to 104°F) or, when water cooling is used, from 10 to 40°C (50 to 104°F)
2. Exposure to an altitude which does not exceed 3300 ft
3. Installation on a rigid mounting surface
4. Installation in areas or supplementary enclosures which do not seriously interfere with the motor's ventilation

Operating

1. V-belt drive in accordance with NEMA standard MG1-14.41 for ac motors or with NEMA Standard MG-1-14.67 for industrial dc motors
2. Flat belt, chain, and gear drives in accordance with NEMA Standard MG-1-14.07

Unusual Service Conditions

1. Exposure to
   a. Combustible, explosive, abrasive, or conducting dusts.
   b. Lint or very dirty operating conditions.
   c. Chemical fumes or flammable or explosive gases.
   d. Nuclear radiation.
   e. Salty air, steam, or oil vapor.
   f. Damp or very dry locations, radiant heat, vermin infestation, or atmosphere conducive to the growth of fungi.
   g. Abnormal shock, vibration, or mechanical loading from external sources.
   h. Abnormal axial or side thrust imposed on the motor shaft.
2. Operating where
   a. There is excessive departure from rated voltage and/or frequency.
   b. The deviation factor of the ac supply voltage exceeds 10 percent.
   c. The ac supply voltage is unbalanced by more than 1 percent.
   d. The rectifier output supplying a dc motor has current peaks unbalanced by more than 10 percent.
   e. Low noise levels are required.
3. Operations at speeds above the highest rated speed.
4. Operation in a poorly ventilated room or in an inclined position.
5. Operation where the motor is subjected to torsional impact loads, repetitive abnormal overloads, or reversing or electric braking.
6. Operation of the motor at standstill with any winding continuously energized.

TABLE 8.1 AC Motor Nameplate Voltages for Corresponding Distribution System Voltages (60 Hz)

Three-phase motors		Single-phase motors	
System voltage	Motor voltage	System voltage	Motor voltage
208	200	120	115
240	230	240	230
480	460		
600	575		
2,400	2,300		
4,160	4,000		
4,800	4,600		
6,900	6,600		
13,800	13,200		

Source: ANSI C84.1-1970, American National Standards Institute, Inc., 1430 Broadway, New York, NY 10018.

The supply voltage must be known in order to select a motor for an application. Table 8.1 shows the required voltage which should exceed the nameplate voltage by a small amount. Ideal power supplies provide constant voltage, frequency, and phasing. In reality, voltages vary from 10 percent above to 10 percent below the nominal values. The frequency is usually controlled within a tight tolerance, but it, too, can vary. Phasing, which is normally balanced when the voltage in each of the three phases is equal, can be unbalanced by a few percentage points.

DESIGN CHARACTERISTICS

NEMA standards state that motors should operate properly at rated load with a variation of up to 10 percent in the voltage. However, this does not indicate that the motor will operate at its rated performance. For example, a motor may not be able to accelerate a driven load if the voltage is reduced by 10 percent because the torque-speed curve will change. The major effects of voltage variation on motor operation are as follows:

Reduced voltage

1. Increased temperature rises
2. Reduction in starting torque
3. Reduction in maximum torque
4. Decreased starting current
5. Increased acceleration time

Increased voltage

1. Increased starting and maximum torques
2. Higher inrush current
3. Decreased power factor

Any of these conditions can shorten the effective service life of a motor.

A motor designed for 60-Hz operation can operate at 50 Hz on selected voltages at 80 or 85 percent of the 60-Hz rated horsepower. Table 8.2 illustrates the voltages required for this operation.

MAINTENANCE OF MOTORS 8.5

TABLE 8.2 Comparative Voltage Ratings for 60- and 50-Hz Operation*

	50-Hz optional voltage ratings ±5%	
Nameplate power of 60-Hz motor, V	80% of 60-Hz motor nameplate power, V	85% of 60-Hz motor nameplate power, V
230	190	200
460	380	400
575	475	500

*For 48 through 440T frames, polyphase only. The motors may operate at less than NEMA torques. Care must be taken in using for hard-to-start and hard-to-accelerate loads. It is advisable to consult the motor manufacturer when operating at frequencies other than that stated on the nameplate.

INSULATION OF AC MOTORS

The quality of the maintenance and the insulation system have a significant impact on the overall service life and performance of a motor. The main factors in increasing the life of the insulation system are to keep it clean, dry, and cool. The insulation systems (Fig. 8.1) in motors include the following:

1. *Turn insulation.* This basic wire coating is an enamel, resin, film, or film-fiber combination used to electrically insulate adjacent wire turns from one another within a coil.
2. *Phase insulation.* Sheet material is used to insulate between phases at the coil end turns.
3. *Ground insulation (slot liners).* Sheet material is used to line the stator slots and insulate the stator winding from stator iron or other structural parts.
4. *Midstick (center wedge).* Insulation is used in stator slots to separate and insulate coils from one another within the slot.
5. *Topstick (top wedge).* This is used to compact and contain the coil wires within the stator slots.
6. *Lead insulation.* Insulation materials surround lead wires.
7. *Lacing and tape.* These are used to tie lead wires in place on the stator end coils. Also they are used to tie end coils together, adding mechanical strength to the end coils and restricting their movement.
8. *Varnish.* Varnish treatment is employed to increase the resistance of the completely wound stator to environmental attack of chemicals, moisture, etc. The varnish treatment bonds the coils, connections, wedges, and stator iron into an integrated structure and improves the components' resistance to electrical and mechanical damage.

In modern motors rated above 500 hp, a *vacuum pressure impregnation* (VPI) system is used for the insulation. The benefits are a void-free sealed insulation system with high mechanical strength and thermal conductivity. The ground wall insulation of this system always contains some form of mica tape. The varnishes used in normal atmospheric dipping are substituted by a 100 percent solids resins. The stator is placed in a large vacuum pressure tank. Resins are impregnated into the stator and windings by alternately applying vacuum and pressure while the stator is fully immersed in resin. The resins are made of polyester, or epoxy. However, epoxy provides greater strength and higher resistance to abrasion or chemicals such as acids and alkalies. The VPI provides a better seal in wet and corrosive atmospheres. The absence of voids reduces the losses due to corona (partial discharge) at higher voltages.

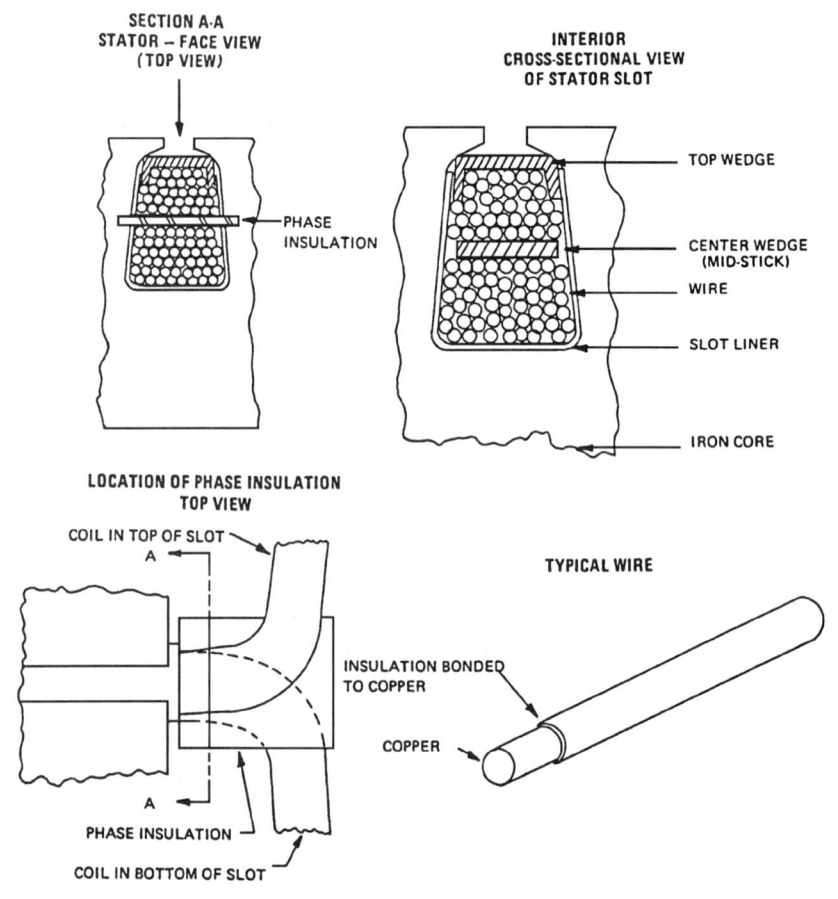

FIGURE 8.1 Typical winding insulation construction.

The motor temperature rise is the difference between the ambient temperature and the average winding temperature. The hot-spot allowance is the difference between the average winding temperature and the hottest spot in the windings.

FAILURES IN THREE-PHASE STATOR WINDINGS

The appendix at the end of this chapter illustrates 12 different failures in three-phase stator windings.

PREDICTIVE MAINTENANCE

Predictive maintenance requires the establishment of trends to monitor for vibration, temperatures, and oil contamination (wear particles) to identify potential problems. Predictive maintenance improves the performance of the motor and reduces the failure rate. This results in a reduction in the operating and maintenance costs and an increase in efficiency and reliability.

The philosophy behind predictive maintenance is to anticipate impending equipment failures by using modern technologies, and to eliminate the *root cause* before it can result in catastrophic failure. The benefits of predictive maintenance are as follows:

1. Repetitive problems are identified and eliminated.
2. Equipment installation is performed to precise standards.
3. Performance verification ensures that new and rebuilt equipment is free of defects.

Predictive maintenance technologies include

1. Vibration spectrum analysis for rotating equipment
2. Oil and wear analysis
3. Thermography for all electrical equipment

MOTOR TROUBLESHOOTING

Figure 8.2 illustrates a troubleshooting flowchart which provides logical, step-by-step methods for identifying and correcting motor problems.

Warning: The internal parts of a motor could be at line potential even when it is not rotating. Disconnect all power to the motor before performing any maintenance which could require contacting internal parts.

This is the key for Fig. 8.2 on troubleshooting ac motors.

A. Motor would not start or accelerates too slowly.

B. Motor runs noisy.

C. Motor overheats.

D. Motor bearings run hot or noisy.

DIAGNOSTIC TESTING FOR MOTORS

The following factors affect the insulation systems in motors:

- High temperature
- Environment
- Mechanical effects such as thermal expansion and contraction, vibration, electromagnetic bar forces, and motor start-up forces in the end turns
- Voltage stresses during operating and transient conditions

All these factors contribute to loss of insulation integrity and reliability.

These aging factors interact frequently to reinforce one another's effects. For example, high-temperature operation could deteriorate the insulation of a stator winding, loosen the winding bracing system, and increase vibration and erosion. At some point, high-temperature operation could lead to delamination of the core and internal discharge. This accelerates the rate of electrical aging and could lead to a winding failure. Nondestructive diagnostic tests are used to determine the condition of the insulation and the rate of electrical aging. The description of the recommended diagnostic tests for the insulation system of motors and the conditions they are designed to detect are discussed below.

PROBLEM A: Motor won't start or motor accelerates too slowly.

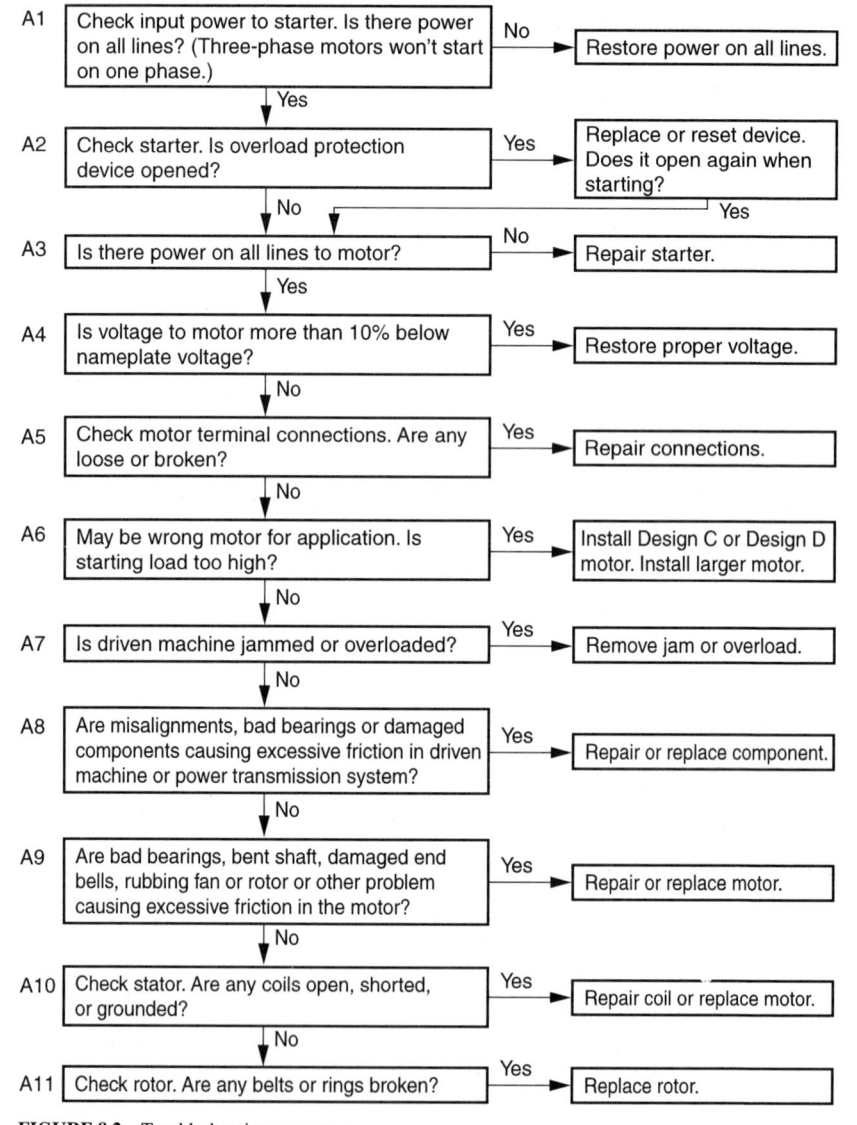

- **A1** Check input power to starter. Is there power on all lines? (Three-phase motors won't start on one phase.) — No → Restore power on all lines.
- Yes ↓
- **A2** Check starter. Is overload protection device opened? — Yes → Replace or reset device. Does it open again when starting? — Yes →
- No ↓
- **A3** Is there power on all lines to motor? — No → Repair starter.
- Yes ↓
- **A4** Is voltage to motor more than 10% below nameplate voltage? — Yes → Restore proper voltage.
- No ↓
- **A5** Check motor terminal connections. Are any loose or broken? — Yes → Repair connections.
- No ↓
- **A6** May be wrong motor for application. Is starting load too high? — Yes → Install Design C or Design D motor. Install larger motor.
- No ↓
- **A7** Is driven machine jammed or overloaded? — Yes → Remove jam or overload.
- No ↓
- **A8** Are misalignments, bad bearings or damaged components causing excessive friction in driven machine or power transmission system? — Yes → Repair or replace component.
- No ↓
- **A9** Are bad bearings, bent shaft, damaged end bells, rubbing fan or rotor or other problem causing excessive friction in the motor? — Yes → Repair or replace motor.
- No ↓
- **A10** Check stator. Are any coils open, shorted, or grounded? — Yes → Repair coil or replace motor.
- No ↓
- **A11** Check rotor. Are any belts or rings broken? — Yes → Replace rotor.

FIGURE 8.2 Troubleshooting ac motors.

Stator Insulation Tests

An electrical test is best suited to determine the condition of electrical insulation. The tests on insulation systems in electrical equipment can be divided into two categories:

1. High-potential (hipot), or voltage-withstand tests
2. Tests that measure some specific insulation property, such as resistance or dissipation factor

MAINTENANCE OF MOTORS   8.9

PROBLEM B: Motor runs noisy.

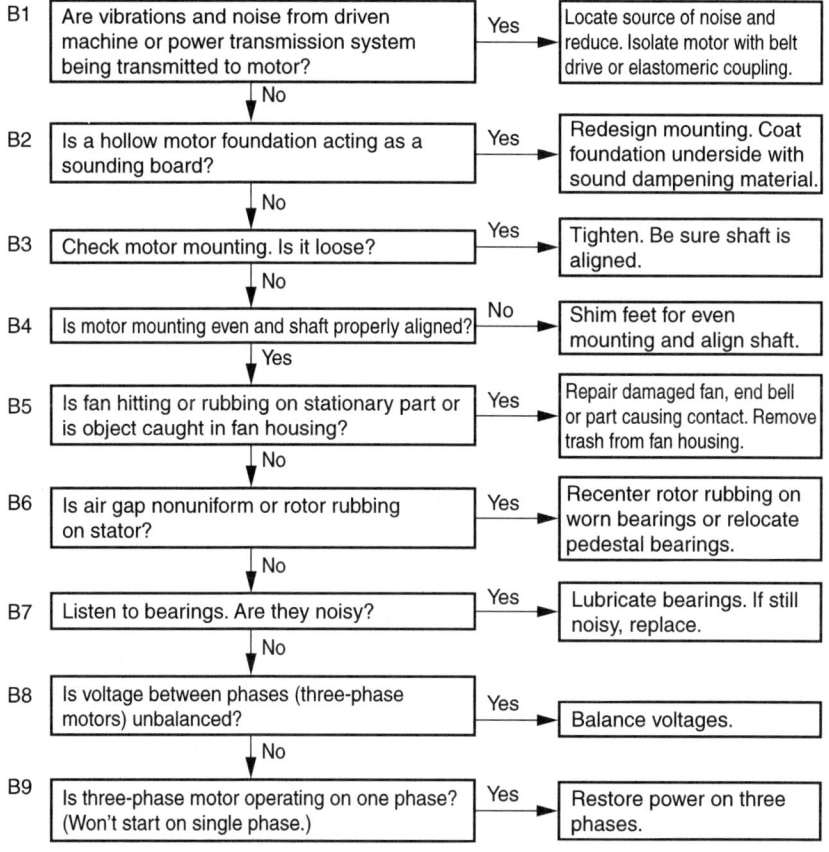

B1	Are vibrations and noise from driven machine or power transmission system being transmitted to motor?	Yes →	Locate source of noise and reduce. Isolate motor with belt drive or elastomeric coupling.

↓ No

B2	Is a hollow motor foundation acting as a sounding board?	Yes →	Redesign mounting. Coat foundation underside with sound dampening material.

↓ No

B3	Check motor mounting. Is it loose?	Yes →	Tighten. Be sure shaft is aligned.

↓ No

B4	Is motor mounting even and shaft properly aligned?	No →	Shim feet for even mounting and align shaft.

↓ Yes

B5	Is fan hitting or rubbing on stationary part or is object caught in fan housing?	Yes →	Repair damaged fan, end bell or part causing contact. Remove trash from fan housing.

↓ No

B6	Is air gap nonuniform or rotor rubbing on stator?	Yes →	Recenter rotor rubbing on worn bearings or relocate pedestal bearings.

↓ No

B7	Listen to bearings. Are they noisy?	Yes →	Lubricate bearings. If still noisy, replace.

↓ No

B8	Is voltage between phases (three-phase motors) unbalanced?	Yes →	Balance voltages.

↓ No

B9	Is three-phase motor operating on one phase? (Won't start on single phase.)	Yes →	Restore power on three phases.

FIGURE 8.2 (*Continued*) Troubleshooting ac motors.

Tests in the first category are performed at some elevated ac or dc voltage to confirm that the equipment is not in imminent danger of failure if operated at its rated voltage. Various standards give the test voltages that are appropriate to various types and classes of equipment. They confirm that the insulation has not deteriorated below a predetermined level and that the equipment will most likely survive in service for a few more years. However, they do not give a clear indication about the condition of the insulation.

The second category of electrical tests indicates the moisture content; presence of dirt; development of flaws (voids), cracks, and delamination; and other damage to the insulation.

A third category of tests includes the use of electrical or ultrasonic probes that can determine the specific location of damage in a stator winding. These tests require access to the air gap and energization of the winding from an external source. These tests are considered an aid to visual inspection.

DC Tests for Stator and Rotor Windings

These tests are sensitive indicators to the presence of dirt, moisture, and cracks. They must be performed off-line with the winding isolated from ground, as shown in Fig. 8.3.

8.10 CHAPTER EIGHT

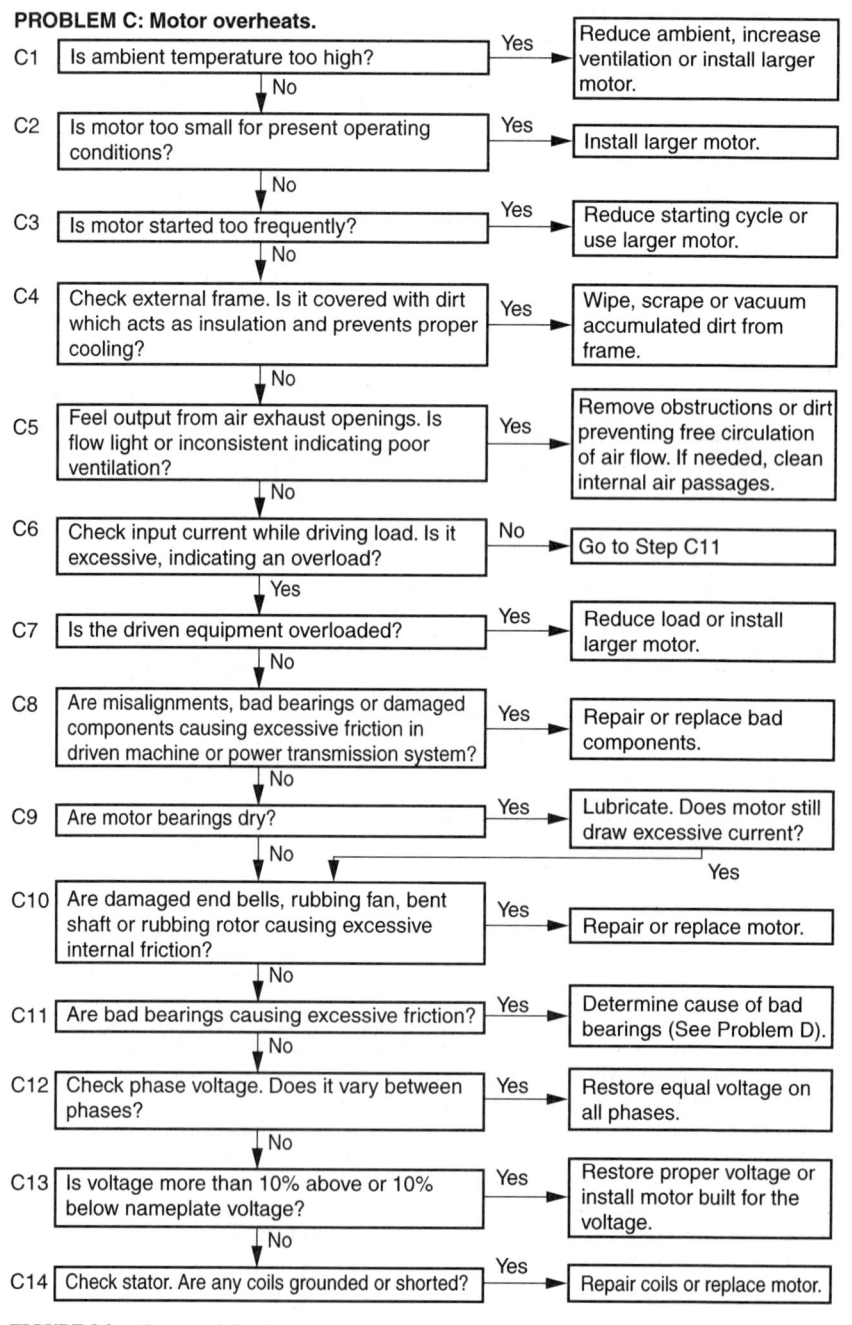

PROBLEM C: Motor overheats.

- **C1** Is ambient temperature too high? — **Yes** → Reduce ambient, increase ventilation or install larger motor.
  - No ↓
- **C2** Is motor too small for present operating conditions? — **Yes** → Install larger motor.
  - No ↓
- **C3** Is motor started too frequently? — **Yes** → Reduce starting cycle or use larger motor.
  - No ↓
- **C4** Check external frame. Is it covered with dirt which acts as insulation and prevents proper cooling? — **Yes** → Wipe, scrape or vacuum accumulated dirt from frame.
  - No ↓
- **C5** Feel output from air exhaust openings. Is flow light or inconsistent indicating poor ventilation? — **Yes** → Remove obstructions or dirt preventing free circulation of air flow. If needed, clean internal air passages.
  - No ↓
- **C6** Check input current while driving load. Is it excessive, indicating an overload? — **No** → Go to Step C11
  - Yes ↓
- **C7** Is the driven equipment overloaded? — **Yes** → Reduce load or install larger motor.
  - No ↓
- **C8** Are misalignments, bad bearings or damaged components causing excessive friction in driven machine or power transmission system? — **Yes** → Repair or replace bad components.
  - No ↓
- **C9** Are motor bearings dry? — **Yes** → Lubricate. Does motor still draw excessive current? — Yes
  - No ↓
- **C10** Are damaged end bells, rubbing fan, bent shaft or rubbing rotor causing excessive internal friction? — **Yes** → Repair or replace motor.
  - No ↓
- **C11** Are bad bearings causing excessive friction? — **Yes** → Determine cause of bad bearings (See Problem D).
  - No ↓
- **C12** Check phase voltage. Does it vary between phases? — **Yes** → Restore equal voltage on all phases.
  - No ↓
- **C13** Is voltage more than 10% above or 10% below nameplate voltage? — **Yes** → Restore proper voltage or install motor built for the voltage.
  - No ↓
- **C14** Check stator. Are any coils grounded or shorted? — **Yes** → Repair coils or replace motor.

FIGURE 8.2 (*Continued*) Troubleshooting ac motors.

MAINTENANCE OF MOTORS   8.11

PROBLEM D: Motor bearings run hot and noisy.

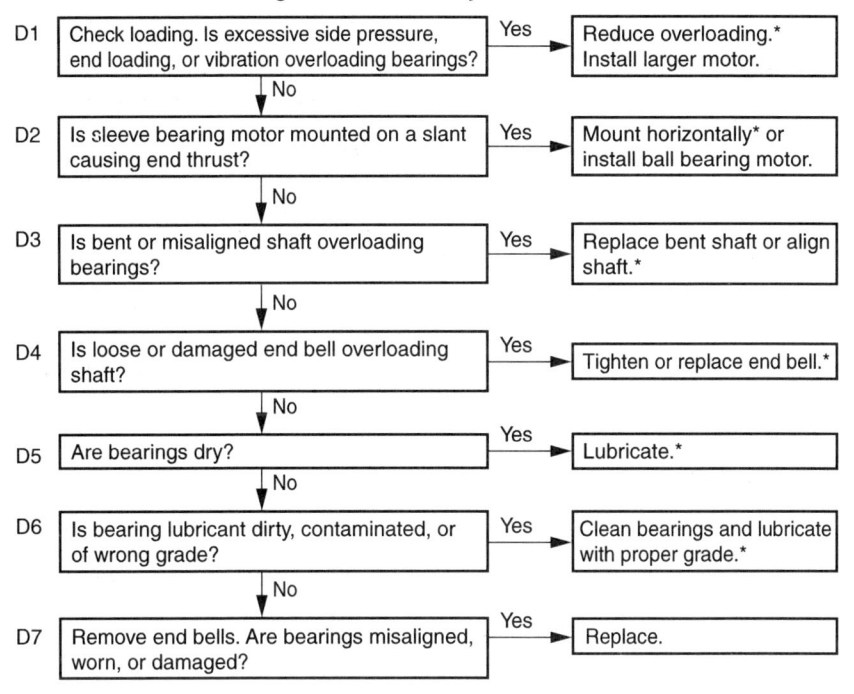

D1 Check loading. Is excessive side pressure, end loading, or vibration overloading bearings? — Yes → Reduce overloading.* Install larger motor.
↓ No

D2 Is sleeve bearing motor mounted on a slant causing end thrust? — Yes → Mount horizontally* or install ball bearing motor.
↓ No

D3 Is bent or misaligned shaft overloading bearings? — Yes → Replace bent shaft or align shaft.*
↓ No

D4 Is loose or damaged end bell overloading shaft? — Yes → Tighten or replace end bell.*
↓ No

D5 Are bearings dry? — Yes → Lubricate.*
↓ No

D6 Is bearing lubricant dirty, contaminated, or of wrong grade? — Yes → Clean bearings and lubricate with proper grade.*
↓ No

D7 Remove end bells. Are bearings misaligned, worn, or damaged? — Yes → Replace.

*BEARINGS MAY HAVE BEEN DAMAGED. IF MOTOR STILL RUNS NOISY OR HOT, REPLACE BEARINGS.

FIGURE 8.2 (*Continued*) Troubleshooting ac motors.

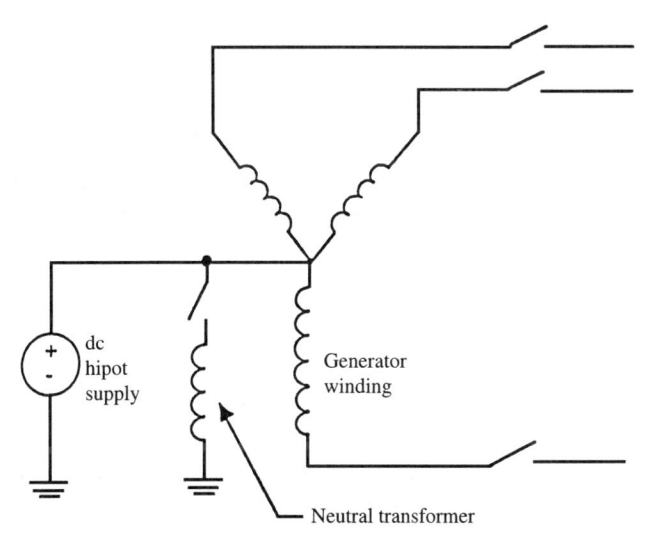

FIGURE 8.3 DC testing of a generator winding.

Suitable safety precautions should be taken when performing all high-potential tests. When high-voltage dc tests are performed on water-cooled windings, the tubes or manifolds should be dried thoroughly, to remove current leakage paths to ground and to avoid the possibility of damage by arcing between moist patches inside the insulating water tubes. For greater sensitivity, these tests can be performed on parts of the windings (phases) isolated from one another.

The charge will be retained in the insulation system for up to several hours after application of high dc voltages. Hence, the windings should be kept grounded for several hours after a high-voltage dc test to protect personnel from a shock.

Tests using dc voltages have been preferred over the ones using ac voltages for routine evaluation of large machines for two reasons:

1. The high dc voltage applied to the insulation during a test is far less damaging than high ac voltages due to the absence of partial discharges.
2. The size and weight of the dc test equipment are far less than those of the ac test equipment needed to supply the reactive power of a large winding.

Insulation Resistance and Polarization Index

The *polarization index* (PI) and insulation resistance tests indicate the presence of cracks, contamination, and moisture in the insulation. They are commonly performed on any motor and generator winding. They are suitable for stator and insulated rotor windings.

The insulation resistance is the ratio of the dc voltage applied between the winding and ground to the resultant current. When the dc voltage is applied, the following current components flow:

1. The charging current into the capacitance of the windings.
2. A polarization or absorption current due to the various molecular mechanisms in the insulation.
3. A "leakage" current between the conductors and ground (the creepage path). This component is highly dependent on the dryness of the windings.

The first two components of the current decay with time. The third component is mainly determined by the presence of moisture or a ground fault. However, it is relatively constant. Moisture is usually absorbed in the insulation and/or condensed on the end winding surfaces. If the leakage current is larger than the first two current components, then the total charging current (or insulation resistance) will not vary significantly with time. Therefore, the dryness and cleanliness of the insulation can be determined by measuring the insulation resistance after 1 min, and after 10 min. The polarization index is the ratio of the 10-min to the 1-min reading.

Test Setup and Performance

Several suppliers, such as Biddle Instruments and Genrad, offer insulation resistance meters that can determine the insulation resistance accurately by providing test voltages of 500 to 5000 V dc. For motors and generators rated 4 kV and higher, 1000 V is usually used for testing the windings of a rotor, and 5000 V is used for testing the stator windings.

To perform the test on a stator winding, the phase leads and the neutral lead (if accessible) must be isolated. The water must be drained from any water-cooled winding, and any hoses removed or dried thoroughly by establishing a vacuum (it is preferable to remove the hoses because vacuum drying is usually impossible).

MAINTENANCE OF MOTORS 8.13

The test instrument is connected between the neutral lead or one of the phase leads and the machine frame (Fig. 8.3). To test a rotor winding, the instrument should be connected between a lead from a rotor winding and the rotor steel. During the test, the test leads should be clean and dry.

Interpretation

If there is a fault or the insulation is punctured, the resistance of the insulation will approach zero. The Institute of Electrical and Electronics Engineers (IEEE) standard recommends a resistance in excess of $V_{L-L} + 1$ MΩ. If the winding is 13.8 kV, the minimum acceptable insulation resistance is 15 MΩ. This value must be considered the absolute minimum since modern machine insulation is on the order of 100 to 1000 MΩ. If the air around the machine had high humidity, the insulation resistance would be on the order of 10 MΩ.

The insulation resistance depends highly on the temperature and humidity of the winding. To monitor the changes of insulation resistance over time, it is essential to perform the test under the same humidity and temperature conditions. The insulation resistance can be corrected for changes in winding temperature. If the corrected values of the insulation resistance are decreasing over time, then there is deterioration in the insulation.

However, it is more likely that the changes in insulation resistance are caused by changes in humidity. If the windings were moist and dirty, the leakage component of the current (which is relatively constant) will predominate over the time-varying components. Hence, the total current will reach a steady value rapidly.

Therefore, the polarization index is a direct measure of the dryness and cleanliness of the insulation. The PI is high (>2) for a clean and dry winding. However, it approaches unity for a wet and dirty winding.

The insulation resistance test is a very popular diagnostic test due to its simplicity and low cost. It should be done to confirm that the winding is not wet and dirty enough to cause a failure that could have been averted by a cleaning and drying procedure. The resistance testing has a pass/fail criterion. It cannot be relied upon to predict the insulation condition, except when there is a fault in the insulation.

The high-potential tests, whether dc or ac, are destructive testing. They are not generally recommended as maintenance-type tests.

For stator windings rated 5 kV or higher, a partial discharge (PD) test, which in the past has been referred to as *corona*, should be done. The level of partial discharge should be determined because it can erode the insulation and lead to insulation aging.

DC High-Potential Testing

The dc high-potential (hipot) test is a nondestructive test used to evaluate the dielectric strength of the groundwall insulation. The voltage applied across the windings is given by

$$V_{dc\text{-}hipot} = 2V_o + 1000 \quad \text{V}$$

where V_o is the operating voltage and $V_{dc\text{-}hipot}$ is the voltage applied across the windings during the dc hipot test. The casing of the motor is maintained at ground voltage. The leakage current between the windings and the core is measured. The insulation resistance is obtained by dividing the voltage imposed across the windings by the leakage current. The test indicates that the groundwall insulation is able to withstand high voltage without being damaged.

Note that this test is different from the destructive ac and dc high-potential tests performed by the manufacturer of the motor. These tests are performed to determine the maximum voltage that the insulation of the motor can withstand. The voltage reached during these tests is much higher than the voltage recommended for the nondestructive dc high-potential test. They are performed by a qualified operator to prevent the destruction of the motor.

Surge Testing

The dc high-potential test confirms the integrity of the insulation. However, it does not indicate a failure of the insulation between the turns of the windings (interturn fault). The surge test is used to detect the early stages of insulation failures in the windings such as coil-to-coil failures, short circuits, ground, and misconnections. During the surge test, brief voltage surges (pulses) are applied across the coil. These pulses produce a momentary voltage stress between the turns of the coil. Figure 8.4 illustrates a typical response of a coil.

Each coil has a unique signature waveform, which can be displayed on the screen of the equipment during the test. The waveform obtained during the surge test is directly related to the inductance of the coil.

A surge test can detect an interturn fault due to weak insulation. If the voltage spike is greater than the dielectric strength of the interturn insulation, one or more turns could be shorted out of the circuit. The number of turns in the coil will drop, leading to a reduction in the inductance of the coil and an increase in the frequency of the waveform produced by the surge test. If the coil has an interturn fault or a phase-to-phase fault, the waveform produced during the test could become unstable. It could shift rapidly to the left and right and back to its original position (Fig. 8.5a).

A comparison is done between the surge tests performed on each of the phases. A healthy three-phase motor should have three identical phases. Therefore, the results of the surge tests performed on each of the phases should be identical. Any differences found between the three results indicate that there is a fault in the motor (Fig. 8.5a and b).

Terminal-to-Terminal Resistances (Winding Resistances)

This test involves measuring the resistance of each of the three phases. The resistance unbalance is given by

$$\text{Resistance unbalance} = \frac{\text{Maximum resistance} - \text{Minimum resistance}}{\text{Average resistance}}$$

For a healthy motor, the resistance unbalance should be less than 10 percent if the test is performed from the motor control center (MCC). It should be less than 5 percent if the

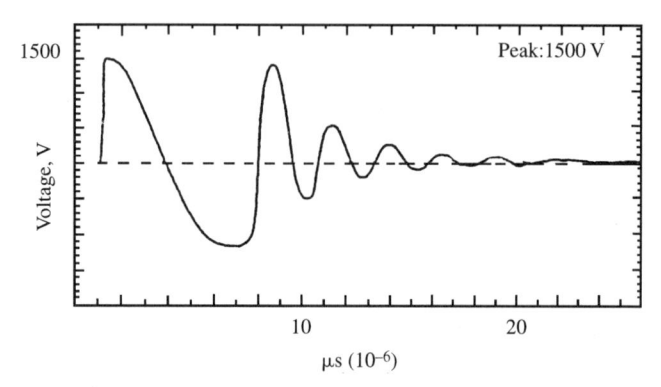

FIGURE 8.4 A typical waveform produced in a coil during surge testing.

MAINTENANCE OF MOTORS 8.15

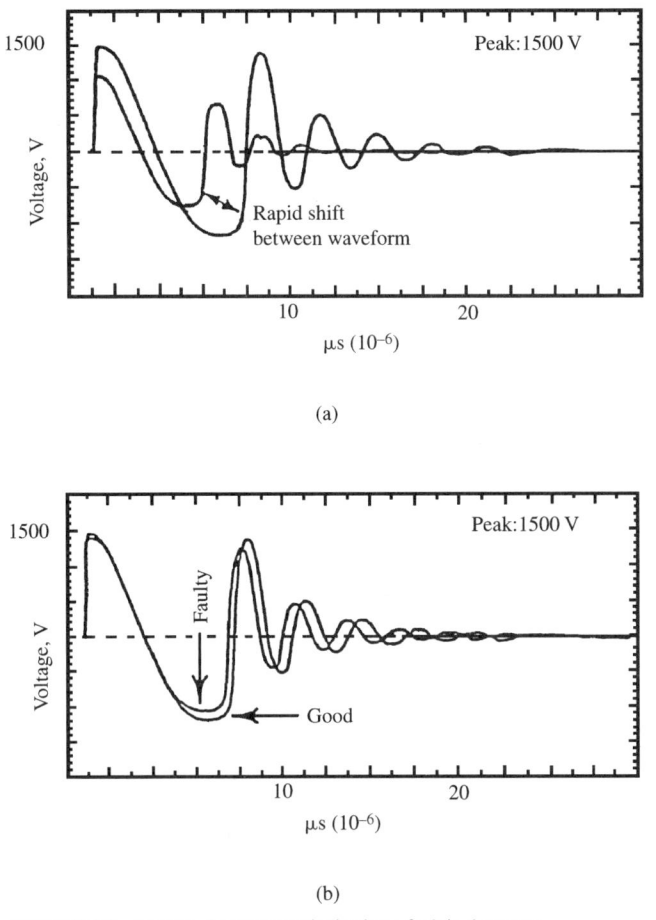

FIGURE 8.5 Results of a surge test indicating a fault in the motor.

test is performed at the motor. The larger resistance unbalance (10 percent instead of 5 percent) is allowed when the test is performed from the MCC due to the long cables between the MCC and the motor.

The voltage and current used during the test for 4-kV and 575-V motors are 5 V dc and 6 A, respectively. This test is used to detect short circuits, ground faults, phase-to-phase faults, loose connections, open circuits, dirt accumulation at connections, etc. Tables 8.3 and 8.4 list the tests required and acceptance criteria used for a 575-V and a 4-kV motor, respectively.

Tests for the Detection of Open Circuits in Induction Motor Rotor Cage Windings

Severe thermal aging and failure of rotor core lamination insulation will occur as a result of open circuits in induction motor cage windings. Failure analysis of more than 6000 induction motors confirmed that 10 percent of these failures occurred in the rotors. One-half of the rotor failures occurred in the cage winding. If a rotor winding cracks, the performance of the motor will degrade and the stator end windings could also get damaged.

TABLE 8.3 Tests and Acceptance Criteria for a 575-V Motor

Test	Test voltage	Test purpose	Acceptance criteria
Insulation resistance (IR)	1000 V dc	Determines insulation resistance	IR at 20°C $\geq$ 40 MΩ IR at 40°C $\geq$ 10 MΩ
Terminal-to-terminal resistances (winding resistances)	5 V dc at 6 A	Identifies short circuits, open circuits, ground faults, phase-to-phase faults, loose connections, dirt accumulation at connections, etc.	Resistance unbalance $\leq$ 10% if test is performed from motor control center. Resistance unbalance $\leq$ 5% if test is performed at motor. Resistance unbalance = $\dfrac{\text{Max. resistance} - \text{Min. resistance}}{\text{Average resistance}}$
DC high-potential (hipot)	Maximum voltage 2200 V	Nondestructive evaluation of dielectric strength of groundwall insulation (slot liner)	Linear variation between applied voltage and leakage current
Surge voltage	Peak voltage 2200 V	Identifies interturn faults, phase-to-phase faults, short circuits, ground faults, coil misconnections, etc.	• Stable surge waveform while voltage is increasing • Identical waveform for all three phases

TABLE 8.4 Tests and Acceptance Criteria for a 4-kV Motor

Test	Test voltage	Test purpose	Acceptance criteria
Insulation resistance (IR)	2500 V dc	Determines insulation resistance	IR at 20°C $\geq$ 1000 MΩ IR at 40°C $\geq$ 240 MΩ
Polarization index (PI)	2500 V dc	Detects contamination, dirt, and humidity in insulation	PI $\geq$ 2.0 unless insulation resistance at 1 min is > 5000 MΩ
Terminal-to-terminal resistances (winding resistances)	5 V dc at 6 A	Identifies short circuits, open circuits, ground faults, phase-to-phase faults, loose connections, dirt accumulation at connections, etc.	Resistance unbalance $\leq$ 10% if test is performed from motor control center. Resistance unbalance $\leq$ 5% if test is performed at motor Resistance unbalance = $\dfrac{\text{Max. resistance} - \text{Min. resistance}}{\text{Average resistance}}$
DC high-potential (hipot)	Maximum voltage 9000 V	Nondestructive evaluation of dielectric strength of groundwall insulation (slot liner)	Linear variation between applied voltage and leakage current
Surge voltage	Peak voltage 9000 V	Identifies interturn faults, phase-to-phase faults, short circuits, ground faults, coil misconnections, etc.	• Stable surge waveform while voltage is increasing • Identical waveform for all three phases

Large and medium-size high-speed motors are the most susceptible to cracks in the cage winding. Large, low-speed motors driving high-inertia loads such as induced-draft (ID) and forced-draft (FD) boiler fans are also susceptible to this mode of failure. The bar extensions, short-circuit rings, and joints between these components are most susceptible to cracking. Cyclic mechanical stresses occurring during start-up lead to cracking by fatigue failure. These stresses are mainly produced by centrifugal forces and differential thermal expansion.

The impedance around the rotor periphery becomes unbalanced by an open circuit in a squirrel-cage rotor winding. The amplitude of the stator current waveform will fluctuate as the magnetic field rotates relative to the rotor surface due to the open circuit. Figure 8.6 illustrates this modulating effect.

Fluctuations in shaft speed can also be caused by broken bars or short-circuit rings. In a healthy rotor, the torque varies sinusoidally at twice the slip frequency. When a bar is aligned with the center of the stator pole, it develops the maximum torque. Some sections of the cage winding will not develop any torque if the rotor has an open circuit. The shaft will slow down slightly twice during each slip cycle. The presence of an open circuit can be verified by measurement of these speed fluctuations.

The two methods described below are used to detect open circuits in rotors. The stator current fluctuation test is performed on an operating and loaded motor. The normal power supply and driven equipment should be uncoupled before performing the manual rotation test.

Stator Current Fluctuation Test

This simple online technique has been used for many years. The current in one of the phases is monitored for fluctuations at twice the slip frequency. An ammeter is observed, or a current transformer output is monitored on a strip chart recorder or oscilloscope. The results should be interpreted by an experienced operator.

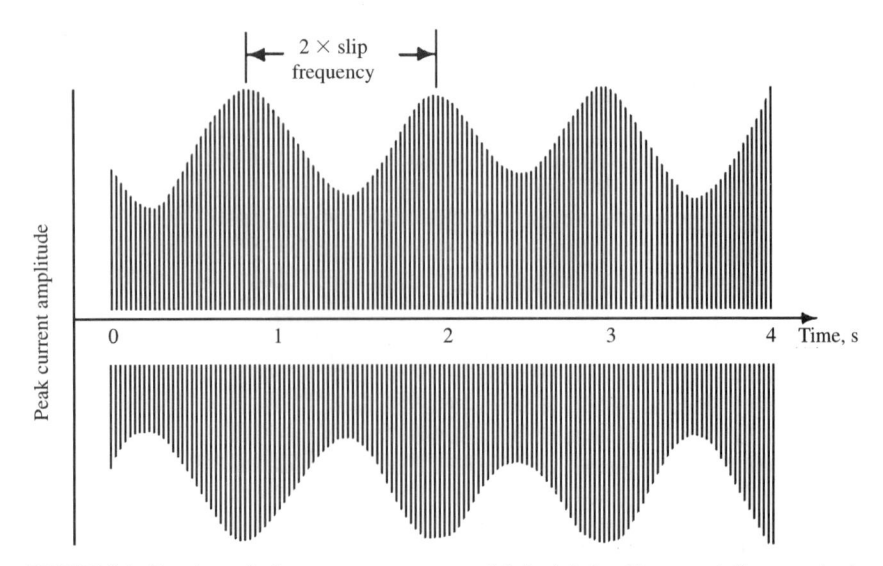

FIGURE 8.6 Two times slip frequency stator current modulation is induced by cage winding open circuit.

Manual Rotation Test

The motor is disconnected from its normal three-phase power supply for its off-line test. The driven equipment should be uncoupled unless it can be manually rotated with the motor. A single-phase ac supply is connected across two motor terminals. It has voltage rating of 10 to 25 percent of rated line-to-line volts and a kVA rating of 5 to 25 percent of rated kVA. The rotor is manually turned for one-half revolution while monitoring the variations in the current. A broken rotor cage winding is indicated by current fluctuations in excess of 10 percent. The current fluctuations can also be monitored on a strip chart recorder connected to the output of a current transformer.

It is important to note that the test time should not exceed 1 min due to rapid heating in the stator and rotor windings. The main limitation of this test is that it can only be conducted off-line. Therefore, breaks in cage windings may not be detected if they close up, giving low-resistance connections when centrifugal forces are removed.

REPAIR AND REFURBISHMENT OF AC INDUCTION MOTORS

The repair and refurbishment of an ac induction motor should include the following steps:

1. Perform a visual inspection to assess the general condition of the motor. Check for cracks, broken welds, and missing parts. Photographs are required in some cases prior to disassembly of the motor to document the motor construction and accessories.
2. Perform these tests on the stator windings and record the results:
   - Insulation resistance (IR) and polarization index (PI)
   - Winding resistances (terminal-to-terminal resistances)
3. Rotate the rotor manually, and check for any defects in the bearings and shaft.
4. Run the motor at no load. Measure and record the currents, vibration, bearing temperatures, and temperature rises.
5. Measure and record the rotor end play (axial and radial movement of the rotor in the bearings).
6. Dismantle the motor and remove the rotor.

Stator Work

7. Clean the windings, using low-pressure steam, if they are contaminated with dust, oil, or grease.
8. Dry the stator in an oven at a temperature of 105°C (220°F) for a period of 6 h.
9. Take IR and PI tests. *Note:* Following steam cleaning and drying of the windings, the results of IR and PI tests should improve.
10. Inspect the motor cable insulation for cracks, overheating, and brittleness.
11. Inspect the stator insulation for cracks, brittleness, and puffiness.
12. Inspect the slot wedges and bracing system in the stator for looseness.
13. Inspect the laminations in the stator core for looseness, damage due to rotor rubbing, localized overheating, and blockage of the vent ducts.

14. If there are no defects in the core or windings of the stator, perform these tests on the stator windings and record the results:
    - Insulation resistance and polarization index tests
    - Terminal-to-terminal resistances (winding resistances)
    - DC high-potential test
    - Surge (impulse) test

Rotor Work

15. If there is a dust, grease, or oil contamination on the rotor, clean it with low-pressure steam.
16. Inspect the rotor laminations for looseness, cracks, and damage due to rubbing with the stator and localized overheating.
17. Inspect the rotor shaft fans of the motor for cracks. If there are signs of cracks, perform dye penetrant and ultraviolet light inspection.
18. Inspect the bars and end rings of the rotor for cracks, looseness, and localized overheating. If there are signs of cracks, perform the manual rotation test as well as dye penetrant and ultraviolet light inspection.
19. Inspect the rotor shaft for cracks.
20. Mount the rotor on a lathe and measure the eccentricity of the shaft. The total indicated reading should be less than 0.0038 cm (0.0015 in).

Bearings

21. Inspect the bearings for cracks, wear, etc.

Oil and Water Heat Exchangers

If heat exchangers are used, perform the following steps:

22. Perform chemical cleaning of the heat exchanger, using a weak acid solution.
23. Inspect the heat exchanger for erosion and corrosion.
24. Perform a hydrostatic pressure test to confirm the integrity of the heat exchanger. The pressure of the water during the test should be 1.5 times the design pressure of the heat exchanger.
25. If there is evidence of corrosion or erosion, perform an eddy current inspection on the tubes of the heat exchanger. All tubes that experienced a reduction of more than 50 percent in the wall thickness should be plugged. If more than 10 percent of the tubes have experienced a reduction in wall thickness of more than 50 percent, consideration should be given to replace the heat exchanger.

Temperature Detectors

26. Perform a visual and a functional check for all resistance temperature detectors (RTDs) and thermocouples used in the winding or bearings of the motor.

MAINTENANCE OF MOTORS 8.21

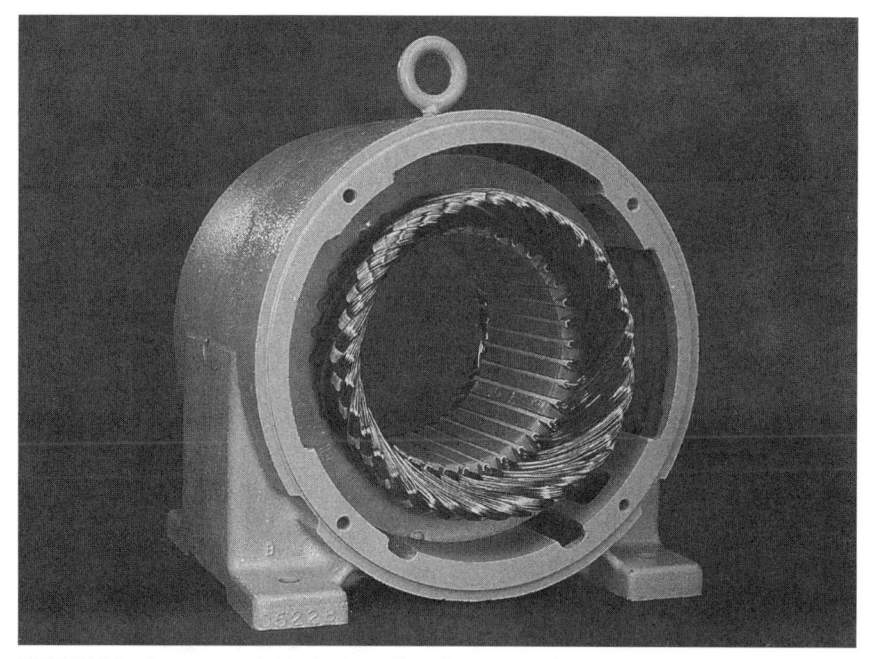

FIGURE 8.7 A new stator winding is pictured here for purposes of comparison with the winding failures shown in Figs. 8.8 to 8.19. Descriptions of the causes of failure are given.

Motor Repair

27. All defective components should be repaired or replaced.
28. Perform all the tests listed in either Table 8.3 (for 575-V motors) or Table 8.4 (for 4-kV motors). The results of the tests should meet the acceptance criteria before the motor can be reassembled.

Motor Rewind

If the windings of the motor have serious damage, the motor should be rewound.

APPENDIX: TYPICAL CAUSES OF WINDING FAILURES IN THREE-PHASE STATORS*

The life of a three-phase stator winding can be shortened dramatically when the motor is exposed to unfavorable operating conditions—electrical, mechanical or environmental. The winding failures illustrated below are typical of what can happen in such circumstances. They are shown here to help identify the causes of failure, so that, where possible, preventive measures may be taken.

---

*Reprinted with permission from *Failures in Three-Phase Stator Windings*, Electrical Apparatus Association, Inc., St. Louis, Mo.

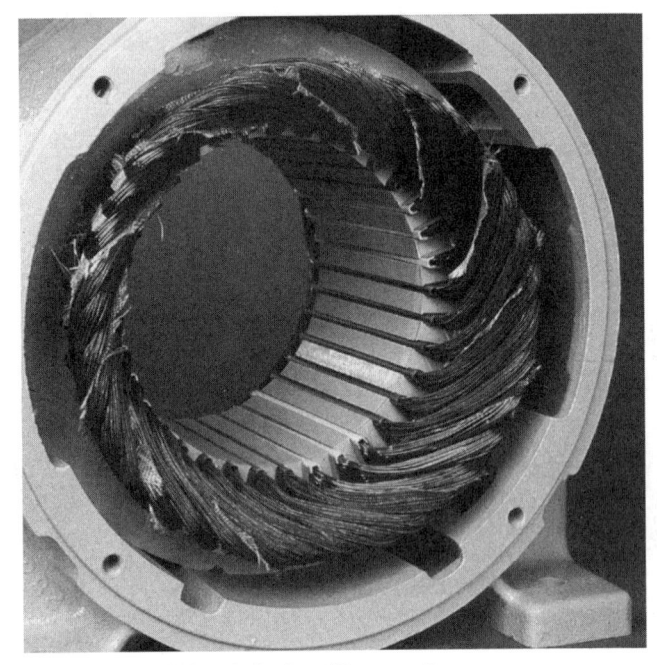

FIGURE 8.8 Winding single-phase (Y-connected).

A single-phase winding failure is the result of an open in one phase of the power supply to the motor (Figs. 8.8 and 8.9). The open is usually caused by a blown fuse, an open contactor, a broken power line, or bad connections.

Figures 8.10 to 8.15 illustrate insulation failures that typically are caused by contaminants, abrasion, vibration, or voltage surge.

Thermal deterioration of insulation in one phase of the stator winding can result from unequal voltage between phases (Fig. 8.16). Unequal voltages usually are caused by unbalanced loads on the power source, a poor connection at the motor terminal, or a high-resistance contact (weak spring). *Note:* A 1 percent voltage unbalance may result in a 6 to 10 percent current unbalance.

Thermal deterioration of the insulation in all phases of the stator winding typically is caused by load demands exceeding the rating of the motor. (Fig. 8.17). *Note:* Undervoltage and overvoltage (exceeding NEMA standards) will result in the same type of insulation deterioration.

Severe thermal deterioration of the insulation in all phases of the motor normally is caused by very high currents in the stator winding due to a locked-rotor condition (Fig. 8.18). It may also occur as a result of excessive starts or reversals.

Insulation failures like the one shown in Fig. 8.19 usually are caused by voltage surges. Voltage surges are often the result of switching power circuits, lightning strikes, capacitor discharges, and solid-state power devices.

REFERENCE

1. L. R. Higgins, *Maintenance Engineering*, 5th ed., McGraw-Hill, New York, 1995.

MAINTENANCE OF MOTORS 8.23

FIGURE 8.9 Winding single-phase (Δ-connected).

FIGURE 8.10 Winding shorted phase to phase.

FIGURE 8.11 Winding shorted turn to turn.

FIGURE 8.12 Winding with shorted coil.

MAINTENANCE OF MOTORS 8.25

FIGURE 8.13 Winding grounded at edge of slot.

FIGURE 8.14 Winding grounded in the slot.

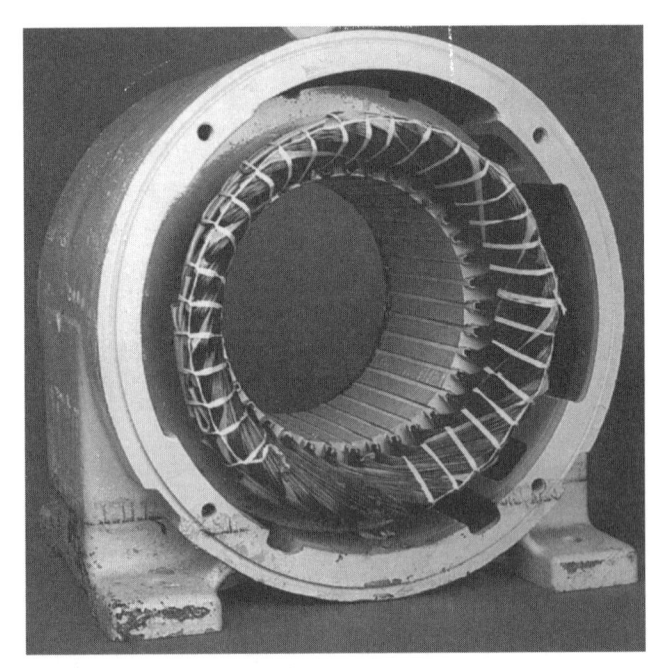

FIGURE 8.15 Shorted connection.

FIGURE 8.16 Phase damage due to unbalanced voltage.

MAINTENANCE OF MOTORS 8.27

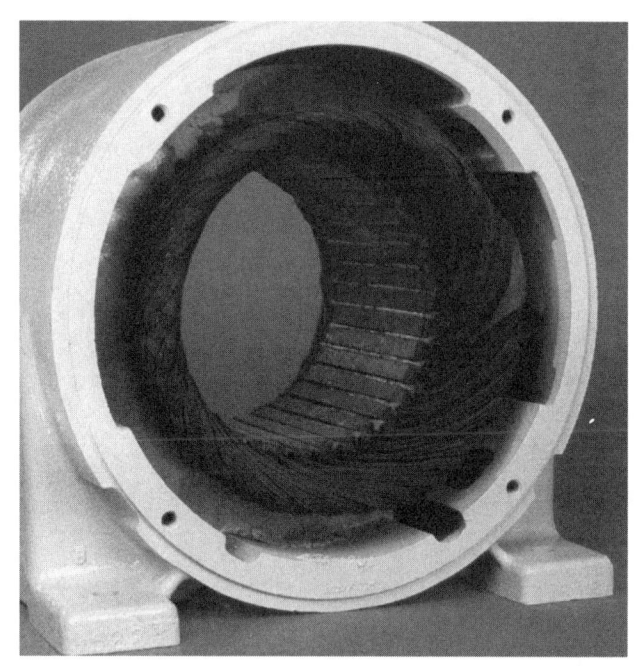

FIGURE 8.17 Winding damaged due to overload.

FIGURE 8.18 Damage caused by locked rotor.

FIGURE 8.19 Winding damaged by voltage surge.

CHAPTER 9
POWER ELECTRONICS, RECTIFIERS, AND PULSE-WIDTH MODULATION INVERTERS

INTRODUCTION TO POWER ELECTRONICS

The development of solid-state motor drive packages over the last 30 years has revolutionized the application of electric motors. Any power control problem can be solved using them. The solid-state drives allowed dc motors to run from ac power supplies or ac motors from dc power supplies. The frequency of ac power can be changed to any other frequency.

Furthermore, the reliability of solid-state drive systems has increased while their costs have decreased. The versatility of these solid-state controls and drives has provided many new applications for ac motors which were formerly covered by dc machines. The application of solid-state drives has also provided flexibility for dc motors.

The development and improvement of high-power solid-state devices have caused this major change.

POWER ELECTRONICS COMPONENTS

The most important types of semiconductor devices are

1. The diode
2. The two-wire thyristor (or PNPN diode)
3. The three-wire thyristor [or silicon controlled rectifier (SCR)]
4. The gate turnoff (GTO) thyristor
5. The DIAC
6. The TRIAC
7. The power transistor (PTR)
8. The insulated gate bipolar transistor (IGBT)

The Diode

A *diode* is a semiconductor device that conducts current in one direction only (Fig. 9.1). The current is conducted from its anode to its cathode, but not in the opposite direction.

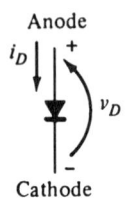

FIGURE 9.1 The symbol of a diode.

Figure 9.2 illustrates the voltage-current characteristic of a diode. A large current flows when a voltage is applied to the diode in the forward direction.

The current is limited to a very small value (microamperes or less) when a voltage is applied to the diode in the reverse direction. The diode will break and allow current in the reverse direction if the reverse voltage is large enough.

Diodes are rated by the maximum reverse voltage they can withstand before breaking down and by the amount of power they can safely dissipate. The power dissipated by a diode during forward direction is equal to $v_D i_D$. This power is limited to prevent overheating the diode. The *peak inverse voltage* (PIV) is the maximum reverse voltage of a diode. This value must be high enough to ensure that the diode will not conduct in the reverse direction.

Diodes are also rated by their *switching time*. It is the time required to go from the off state to the on state, and vice versa. Power diodes are very large, high-power devices that store large charge in their junctions. They switch states much more slowly than diodes found in electronic circuits. Essentially, all power diodes can be used in 50- to 60-Hz rectifiers because they can switch states fast enough for this application. However, pulse-width modulators (PWMs) require power diodes to switch states at rates higher than 10,000 Hz. Special diodes called *fast-recovery, high speed diodes* are used for these very fast switching applications.

The Two-Wire Thyristor or PNPN Diode

Thyristor is a name of a family of semiconductor devices which have up to four semiconductor layers. The two-wire thyristor is a member of this family. It is also known as the PNPN diode or *reverse-blocking diode-type thyristor* in the IEEE standard for graphic symbols (Fig. 9.3).

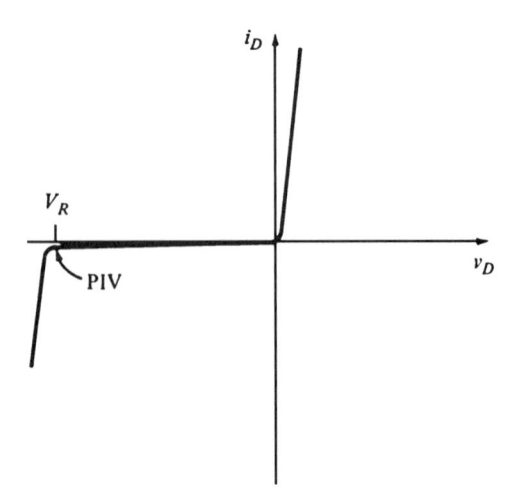

FIGURE 9.2 Voltage-current characteristic of a diode.

The PNPN is a rectifier or diode that has an unusual voltage-current characteristic in the forward-biased region (Fig. 9.4). Its characteristic curve consists of these regions:

1. The reverse-blocking region
2. The forward-blocking region
3. The conducting region

The PNPN diode behaves as follows:

1. It turns on when the applied voltage v_D exceeds V_{BO}.
2. It turns off when the current i_D drops below I_H.
3. It blocks all current flow in the reverse direction until the maximum reverse voltage is exceeded.

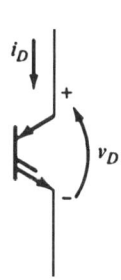

FIGURE 9.3 The symbol of a two-wire thyristor or PNPN diode.

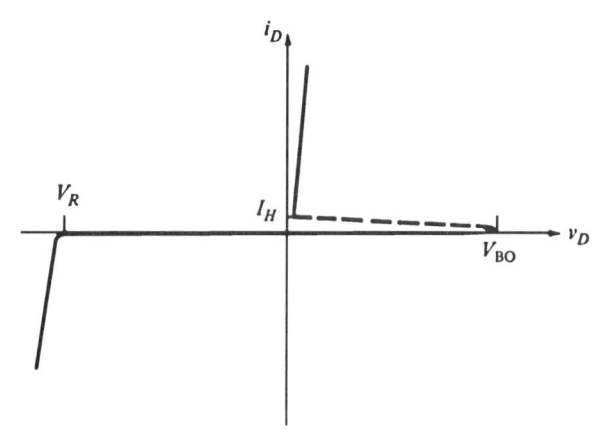

FIGURE 9.4 Voltage-current characteristic of a PNPN diode.

The Three-Wire Thyristor or SCR

This is the most important of the thyristor family. It is also known as the *silicon controlled rectifier* (SCR) (Fig. 9.5). Its voltage-current characteristic with the gate lead open is the same as that of a PNPN diode. However, the *breakover or turn-on voltage of an SCR can be adjusted* by a current flowing into its gate lead. Voltage V_{BO} drops when the gate current increases (Fig. 9.6). If the V_{BO} with no gate signal is larger than the highest voltage in the circuit, then a gate current is required to turn it on. Once it is on, it remains on until its current falls

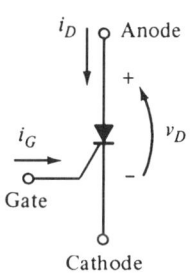

FIGURE 9.5 The symbol of a three-wire thyristor or SCR.

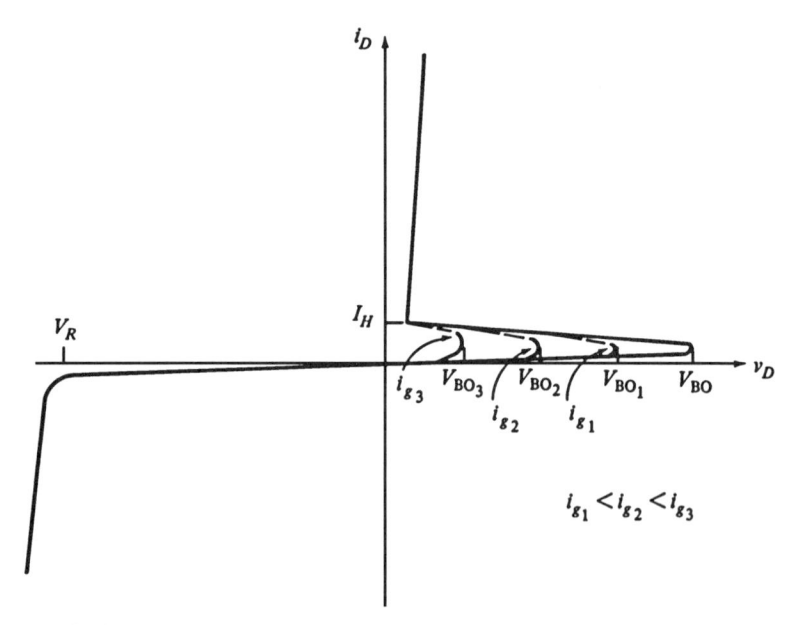

FIGURE 9.6 Voltage-current characteristics of an SCR.

below I_H. Therefore, once an SCR is triggered, its gate current can be removed without changing the state of the device.

The SCR is commonly used for switching or rectification applications. It is available in ratings ranging from a few amperes to a minimum of 3000 A. In summary, an SCR

1. Turns on when the voltage v_D applied to it exceeds V_{BO}
2. Has a breakover voltage V_{BO} whose level is controlled by the amount of gate current i_G present in the SCR
3. Turns off when the current i_D flowing through it drops below I_H
4. Blocks all current flow in the reverse direction until the maximum reverse voltage is exceeded

The Gate Turnoff Thyristor

A gate turnoff (GTO) thyristor is an SCR that can be turned off by a large enough negative pulse at its gate lead even if i_D exceeds i_H. These devices are becoming more popular because they eliminate the need for external components to turn off SCRs in dc circuits (Fig. 9.7a).

Figure 9.7b illustrates a typical gate current waveform for a high-power GTO thyristor. The gate current required to turn on a GTO thyristor is typically larger than that of an ordinary SCR. The gate current for a large high-power device is around 10 A. A large negative current pulse of 20- to 30-μs duration is required to turn off the device. The magnitude of this negative pulse must be one-fourth to one-sixth that of the current flowing through the device.

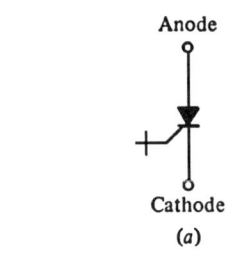

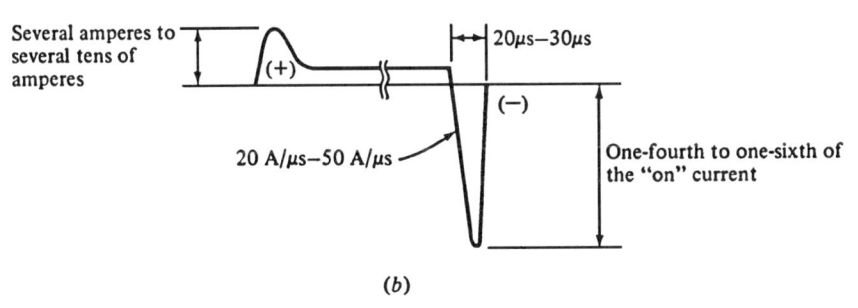

FIGURE 9.7 (a) The symbol of a gate turnoff thyristor. (b) The gate current waveform required to turn a GTO thyristor on and off.

The DIAC

The DIAC behaves as two PNPN diodes connected back to back. It is turned on when the applied voltage in either direction exceeds V_{BO}. Once it is turned on, it remains on until the current falls below I_H.

The TRIAC

The TRIAC behaves as two SCRs connected back to back with a common gate lead. The breakover voltage in a TRIAC decreases with increasing gate current, as with an SCR. However, a TRIAC responds to either positive or negative pulses at its gate. Once a TRIAC is turned on, it remains on until the current falls below I_H.

The Power Transistor

Figure 9.8a shows the symbol of a transistor. Figure 9.8b illustrates the collector-to-emitter voltage versus the collector current characteristic for a transistor. The collector current i_C is directly proportional to its base current i_B over a wide range of collector-to-emitter voltages (V_{CE}).

Power transistors are normally used to switch a current on or off. Figure 9.9 shows the $i_C = V_{CE}$ characteristic with the load line of the resistive load. Transistors are normally used as switches. They are completely on or completely off. A base current of zero will completely turn off the transistor.

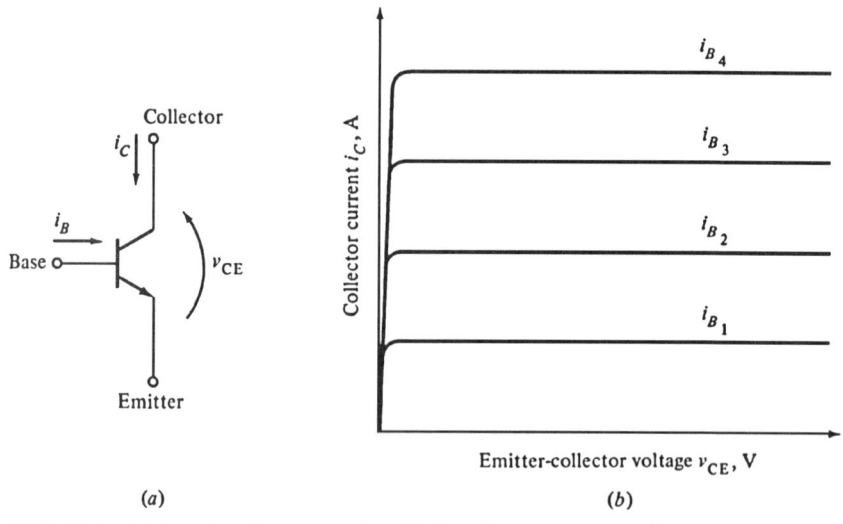

FIGURE 9.8 (a) The symbol of a power transistor. (b) The voltage-current characteristic of a power transistor.

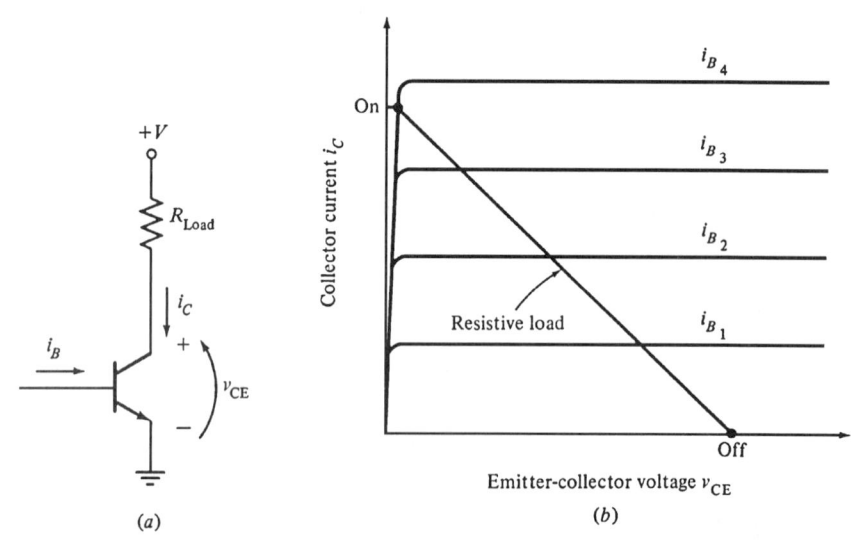

FIGURE 9.9 (a) A transistor with a resistive load. (b) The voltage-current characteristic of this transistor and the load.

If the base current is equal to i_{B_3}, the transistor will not be on or off. This is highly undesirable because a large collector current will flow, dissipating a lot of power in the transistor. If the base current is high enough, the transistor becomes completely saturated. It conducts without wasting a lot of power. A major drawback of large power transistors is their relatively slow switching from on to off, because a large base current is needed or removed to turn them on or off.

The Insulated Gate Bipolar Transistor (IGBT)

The IGBT is similar to the power transistor, except that it is controlled by the voltage applied to its gate rather than the current flowing into its base, as in power transistors. The current flowing in the gate of an IGBT is extremely small because the impedance of the control gate is very high. This device is equivalent to the combination of a metal-oxide semiconductor field effect transistor (MOSFET) and a power transistor (Fig. 9.10).

Since the current required to control an IGBT is very small, it can be switched much more quickly than a power transistor. The IGBTs are normally used in high-power, high-frequency applications.

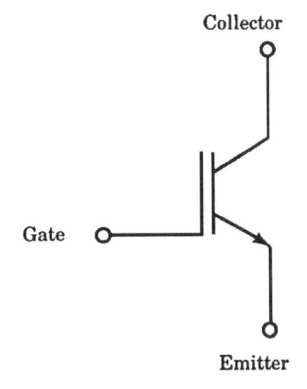

FIGURE 9.10 The symbol of an IGBT.

POWER AND SPEED COMPARISON OF POWER ELECTRONIC COMPONENTS

Figure 9.11 illustrates a comparison of the relative speeds and power-handling capabilities of these devices. The SCRs can handle higher power than any other devices. The GTO thyristors can handle almost the same power, but they are faster than SCRs. The power capability of power transistors is much less than that of both types of thyristors, but they can switch more than 10 times faster.

BASIC RECTIFIER CIRCUITS

A rectifier circuit converts ac power to dc power. The most common rectifier circuits are

1. The half-wave rectifier
2. The full-wave bridge rectifier
3. The three-phase half-wave rectifier
4. The three-phase full-wave rectifier

The *ripple factor* is a good measure of the smoothness of the dc voltage out of a rectifier circuit. The *percentage of ripple* in a dc power supply is

$$r = \frac{V_{ac,rms}}{V_{DC}} \times 100\%$$

where $V_{ac,rms}$ is the rms value of the ac components of the output voltage and V_{DC} is the dc component of the output voltage. Voltage V_{DC} is the *average* of the output voltage of the rectifier

$$V_{DC} = \frac{1}{T} \int v_o(t)\, dt$$

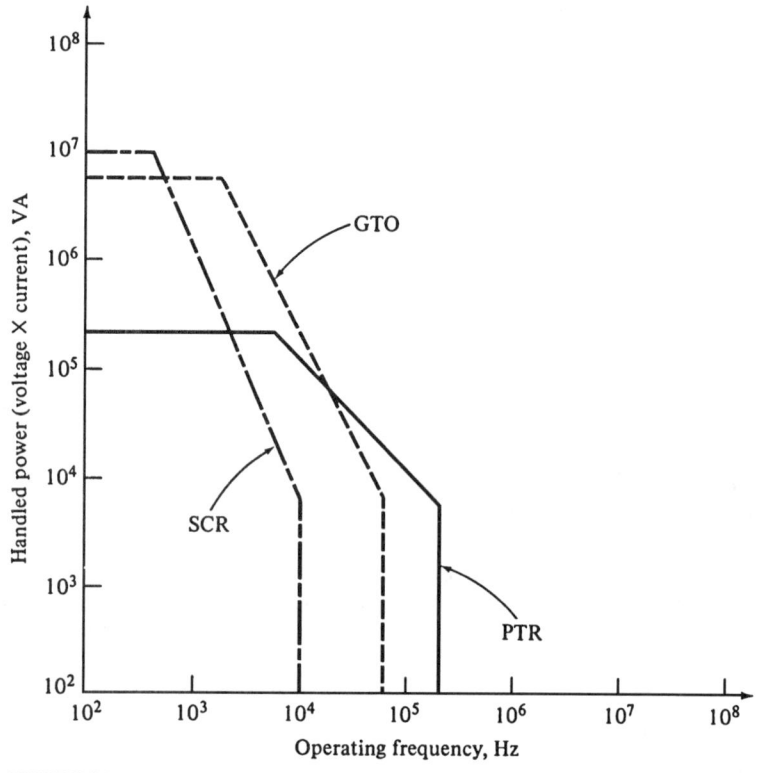

FIGURE 9.11 A comparison of the relative speeds and power-handling capabilities of SCRs, GTO thyristors, and power transistors.

The ripple r is given by

$$r = \sqrt{\left(\frac{V_{rms}}{V_{DC}}\right)^2 - 1} \times 100\%$$

where V_{rms} is the rms value of the total output voltage from the rectifier.

The Half-Wave Rectifier

Figure 9.12 illustrates a half-wave rectifier. The current flows through the diodes on the positive half cycle, and it is blocked on the negative half-cycle. This is a poor approximation of a constant dc waveform because it contains 60-Hz ac frequency components and all the harmonics. This rectifier has a ripple factor $r = 121$ percent; i.e., it has more ac voltage components in its output than dc voltage components.

The Full-Wave Rectifier

Figure 9.13a illustrates a full-wave bridge rectifier circuit. Diodes D_1 and D_3 conduct on the positive half-cycle, and D_2 and D_4 conduct on the negative half-cycle. This output voltage

POWER ELECTRONICS, RECTIFIERS, AND INVERTERS   9.9

(a)

(b)

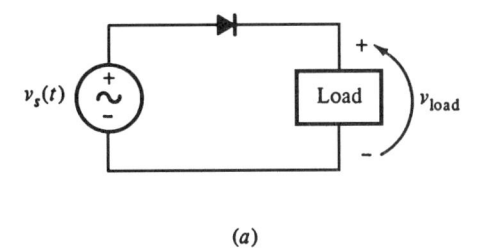

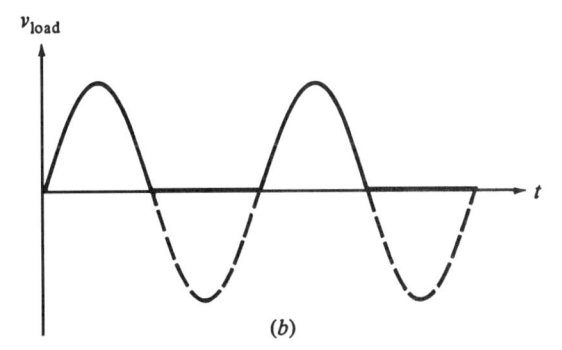

FIGURE 9.12 (a) A half-wave rectifier circuit. (b) The output voltage of the rectifier circuit.

is smoother than that of the previous rectifier, but it still contains ac frequency components at 120 Hz and its harmonics ($r = 48.2$ percent).

Figure 9.13b illustrates another possible full-wave rectifier circuit. Diode D_1 conducts on the positive half-cycle with the current returning through the center tap of the transformer, and D_2 conducts on the negative half-cycle with the current returning through the center tap of the transformer. The output is shown in Fig. 9.13c.

The Three-Phase Half-Wave Rectifier

Figure 9.14 illustrates a three-phase half-wave rectifier and its output voltage. At any instant, the diode with the largest voltage applied to it will conduct, and the other two diodes will be reversed-biased. The output voltage at any time is the highest of the three input voltages. The output voltage is smoother than that of the full-wave bridge rectifier circuit. It has voltage components at 180 Hz and its harmonic components ($r = 18.3$ percent).

The Three-Phase Full-Wave Rectifier

Figure 9.15 illustrates a three-phase full-wave rectifier. The first part of the circuit connects the highest of the three-phase voltages at any instant to the load. The second part consists of three diodes oriented with their cathodes connected to the supply voltages and anodes connected to the load. This arrangement connects the lowest of the three voltages to the load at any given time.

The three-phase full-wave rectifier connects the highest of the three voltages to one end of the load and the lowest voltage to the other end (Fig. 9.16). The three-phase full-wave rectifier provides smoother output than a three-phase half-wave rectifier ($r = 4.2$ percent).

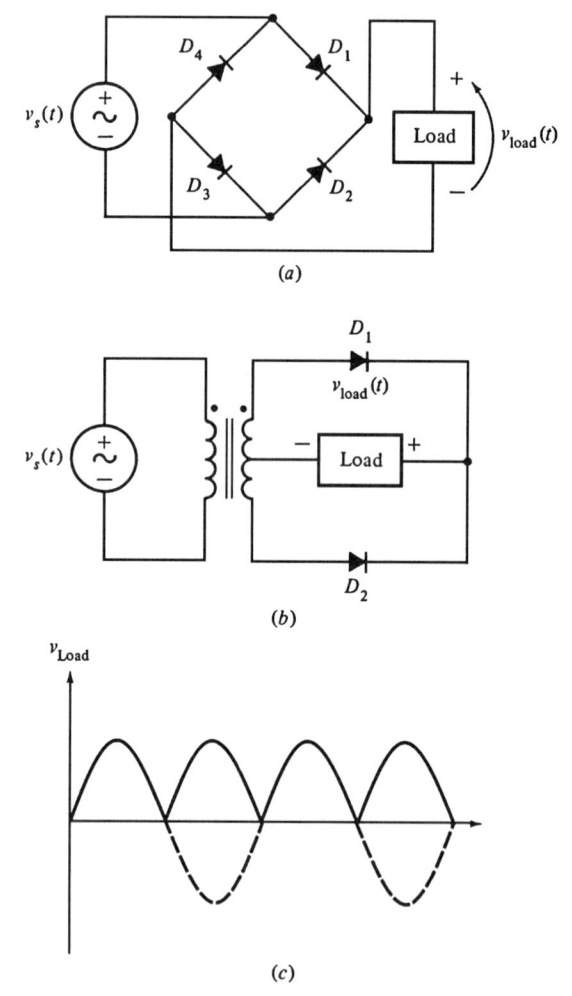

FIGURE 9.13 (a) A full-wave bridge rectifier circuit. (b) The output voltage of the rectifier circuit. (c) An alternate full-wave rectifier circuit using two diodes and a center-tapped transformer.

FILTERING RECTIFIER OUTPUT

Low-pass filters are used to smooth the output of any of these rectifiers by removing ac frequency components from the output. The common filter elements used to smoothen ac voltage changes are

1. Capacitors connected across the line.
2. Inductors connected in series with the line

Rectifier circuits use a single series inductor as a common filter or a *choke* (Fig. 9.17).

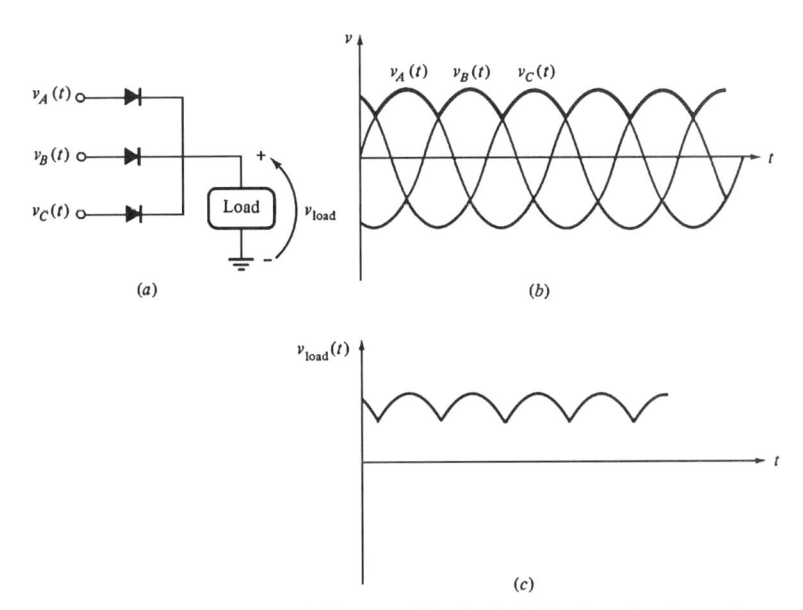

POWER ELECTRONICS, RECTIFIERS, AND INVERTERS   9.11

FIGURE 9.14 (*a*) A three-phase half-wave rectifier circuit. (*b*) The three-phase input voltages to the rectifier circuit. (*c*) The output voltage of the rectifier circuit.

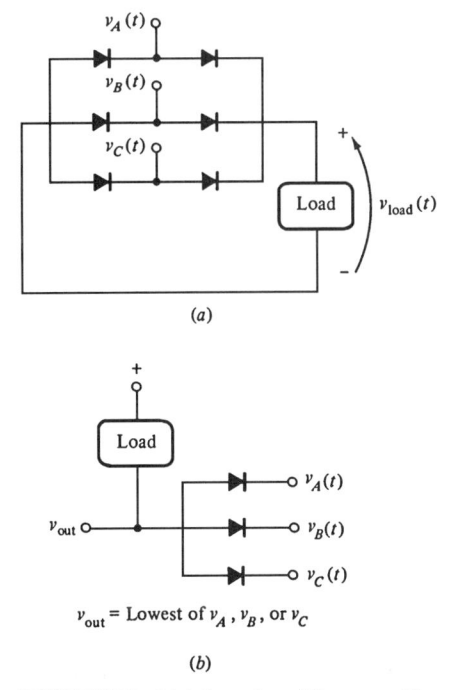

FIGURE 9.15 (*a*) A three-phase full-wave rectifier circuit. (*b*) This circuit places the *lowest* of its three input voltages at its output.

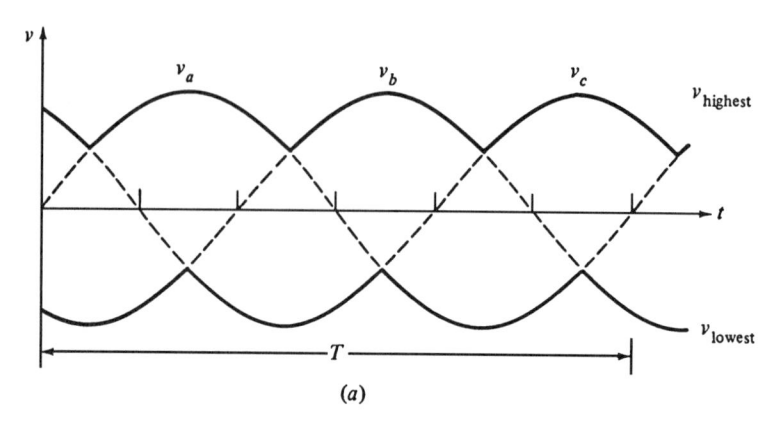

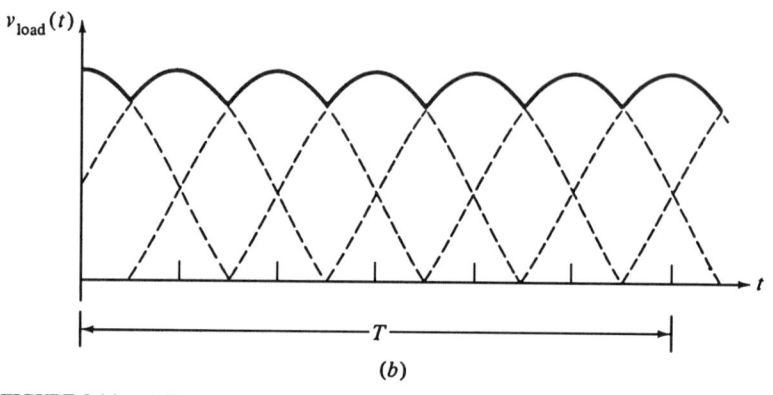

FIGURE 9.16 (*a*) The highest and lowest voltages in the three-phase full-wave rectifier. (*b*) The resulting output voltage.

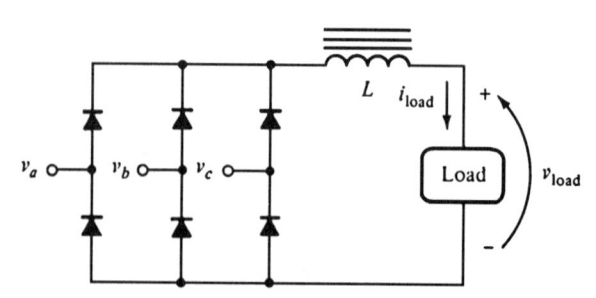

FIGURE 9.17 A three-phase full-wave bridge circuit with an inductive filter for reducing output ripple.

PULSE CIRCUITS

Some of the devices listed require a pulse of current to their gating circuits to operate. Analog and digital techniques are used to produce voltage and current pulses. Analog methods rely on PNPN diodes which have discrete nonconducting and conducting regions in their voltage-current characteristics.

A voltage and current pulse is generated by the transition from the conducting to the nonconducting region of the device. *Relaxation oscillator* is the name given to the circuit used to generate analog pulse.

Digital pulse generation circuits are commonly used in modern solid-state motor drives. They use a microcomputer which executes a program stored in *read-only memory* (ROM). The program uses different inputs to generate pulses at the proper time. Common inputs that the program considers are the desired speed of the motor, actual speed of the motor, rate of acceleration or deceleration, and specified voltage and current limits. Figure 9.18 illustrates a typical digital pulse generation circuit board from a pulse-width-modulated induction motor drive. The following simple analog circuits are examples of some basic types of pulse-producing circuits.

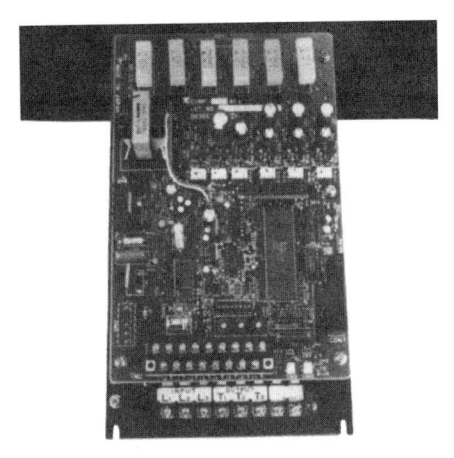

FIGURE 9.18 A typical digital pulse generation circuit board from a pulse-width-modulated (PWM) induction motor drive. (*Courtesy of MagneTek Drives and Systems.*)

A RELAXATION OSCILLATOR USING A PNPN DIODE

Figure 9.19 illustrates a relaxation oscillator or pulse-generating circuit built using a PNPN diode. The following conditions are assumed:

1. The power supply voltage V_{DC} must exceed V_{BO} for the PNPN diode.
2. V_{DC}/R_1 must be less than I_H for the PNPN diode.
3. R_1 must be much larger than R_2.

When the switch in the circuit is closed, capacitor C will charge with a time constant $\tau = R_1 C$. As the voltage increases, it will exceed V_{BO} and the PNPN diode will turn on.

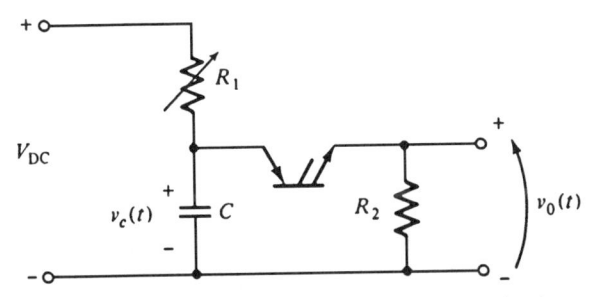

FIGURE 9.19 A relaxation oscillator (or pulse generator) using a PNPN diode.

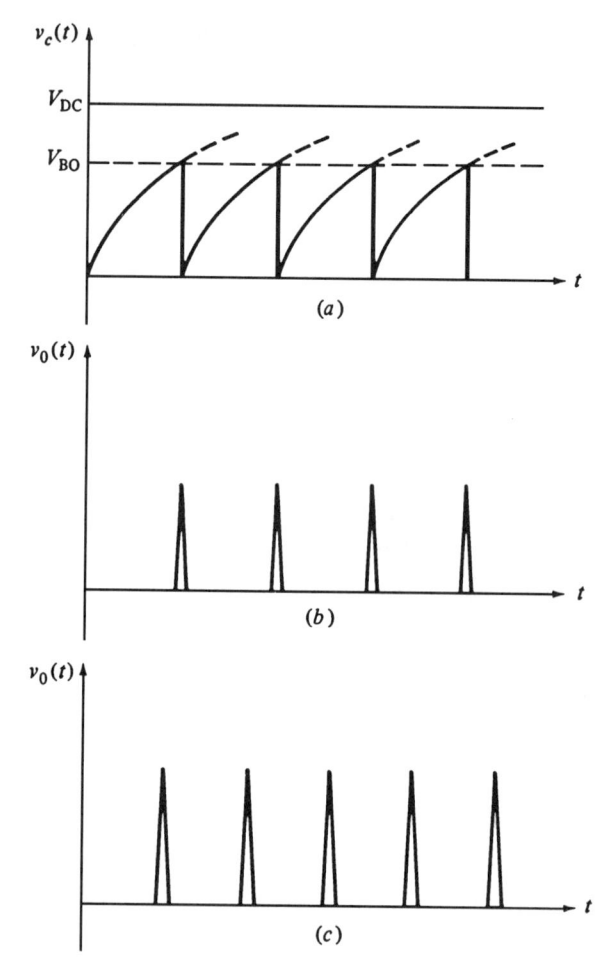

FIGURE 9.20 (*a*) The voltage across the capacitor in the relaxation oscillator. (*b*) The output voltage of the relaxation oscillator. (*c*) The output voltage of the oscillator after R_1 is decreased.

The capacitor will discharge through it. The PNPN diode will turn off. Figure 9.20*a* and *b* illustrate the voltage across the capacitor and the resulting output voltage and current.

The timing of the pulses can be changed by varying R_1. If R_1 is decreased, the capacitor will charge more quickly and the PNPN diode will be triggered sooner. The pulses will be closer together.

This circuit can be used to trigger an SCR by removing R_2 and connecting the SCR gate lead in its place (Fig. 9.21*a*). The circuit can be coupled to the SCR through a transformer (Fig. 9.21*b*). The pulse can be amplified by an extra transistor stage if more gate current is needed to drive the SCR (Fig. 9.21*c*). The circuit can also be built using a DIAC instead of the PNPN diode (Fig. 9.22).

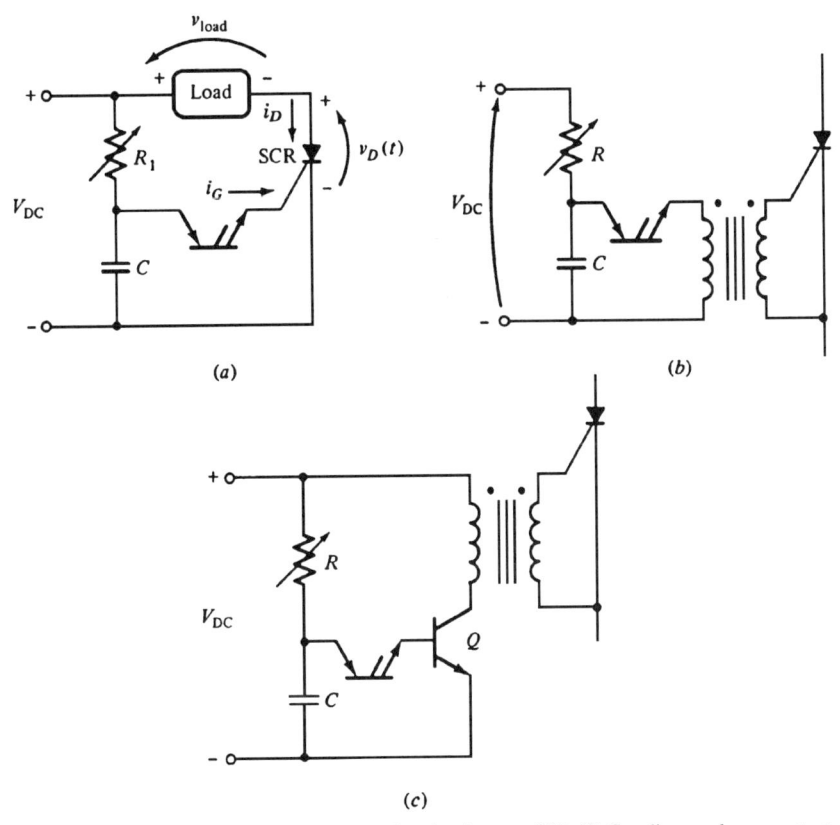

FIGURE 9.21 (*a*) Using a pulse generator to directly trigger an SCR. (*b*) Coupling a pulse generator to an SCR through a transformer. (*c*) Connecting a pulse generator to an SCR through a transistor amplifier to increase the strength of the pulse.

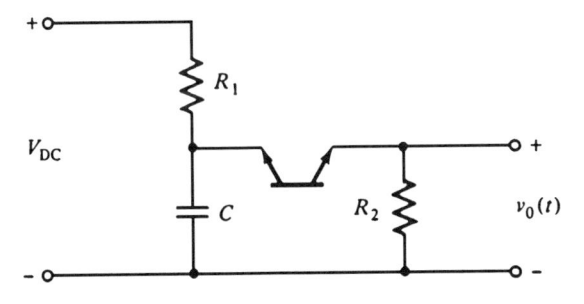

FIGURE 9.22 A relaxation oscillator using a DIAC instead of a PNPN diode.

PULSE SYNCHRONIZATION

The triggering pulse should be applied to the controlling SCRs at the same point in each ac cycle. This is normally done by synchronizing the pulse circuit to the ac power line supplying the SCRs. This is done by making the power supply to the triggering circuit the same as the power supply to the SCRs.

9.16 CHAPTER NINE

VOLTAGE VARIATION BY AC PHASE CONTROL

SCRs and TRIACs provide a convenient method for controlling the average voltage applied to a load by changing the phase angle of the source voltage.

AC Phase Control for a DC Load Driven from an AC Source

The concept of phase angle power control is illustrated in Fig. 9.23. The figure shows a voltage phase control circuit having a resistive dc load supplied by an ac source. The breakover voltage for the SCR for $i_G = 0$ A is greater than the highest voltage in the circuit. The PNPN diode has a very low breakover voltage (around 10 V).

Figure 9.24 illustrates the voltage V_1 at the terminals of the rectifier when switch S_1 is open. If switch S_1 is shut and S_2 is left open, the SCR will always be off because V_1 can never exceed V_{BO} for the SCR. Since the SCR is always an open circuit, the current through it and hence the voltage on the load will be zero. When S_2 is closed, the capacitor begins to charge. The capacitor continues to charge up to the breakover voltage of the PNPN diode, and the diode conducts. The current flows through the gate of the SCR, lowering V_{BO} and turning on the SCR. When the SCR turns on, currents flow through it and the load. The current continues to flow after the capacitor has discharged. The SCR turns off when its current falls below the holding current I_H (a few milliamperes). This occurs at the

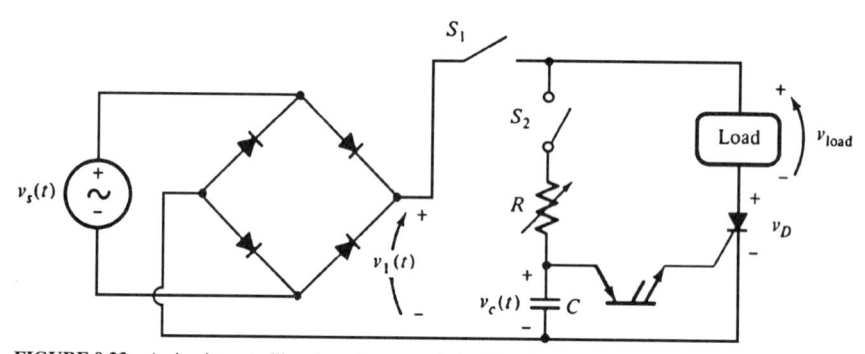

FIGURE 9.23 A circuit controlling the voltage to a dc load by phase angle control.

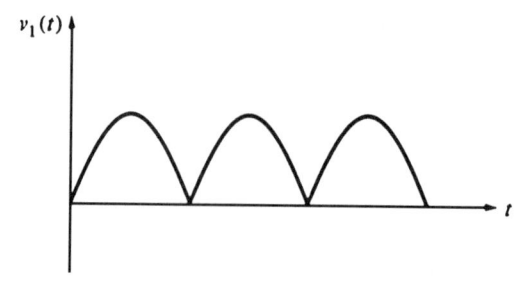

FIGURE 9.24 The voltage at the output of the bridge circuit with switch S_1 open.

POWER ELECTRONICS, RECTIFIERS, AND INVERTERS 9.17

$v_c(t)$

PNPN diode fires

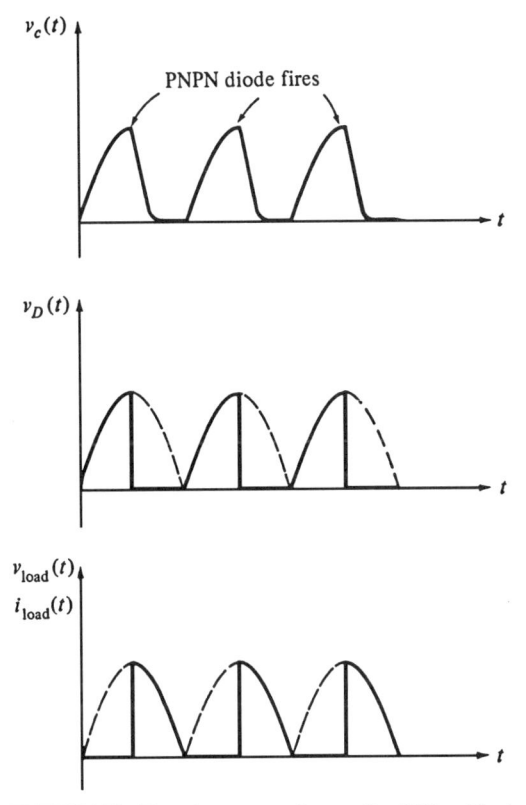

$v_D(t)$

$v_{load}(t)$
$i_{load}(t)$

FIGURE 9.25 The voltages across the capacitor, SCR, and load, and the current through the load, when switches S_1 and S_2 are closed.

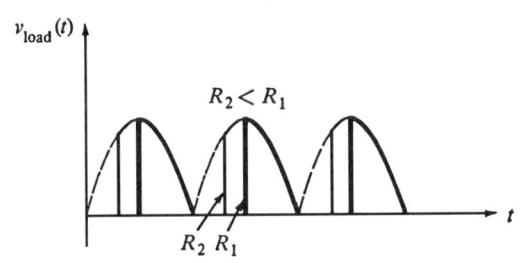

$v_{load}(t)$

$R_2 < R_1$

$R_2\ R_1$

FIGURE 9.26 The effect of decreasing R on the output voltage applied to the load in the circuit of Fig. 9.23.

extreme end of the half-cycle. Figure 9.25 illustrates the voltage and current waveforms for the circuit.

The power to the load can be changed by decreasing R. The capacitor will charge more quickly at the beginning of each half-cycle, and the SCR will fire soon. Since the SCR will be on for a longer period in the half-cycle, more power will be supplied to the load (Fig. 9.26). The resistor R controls the power flow to the load.

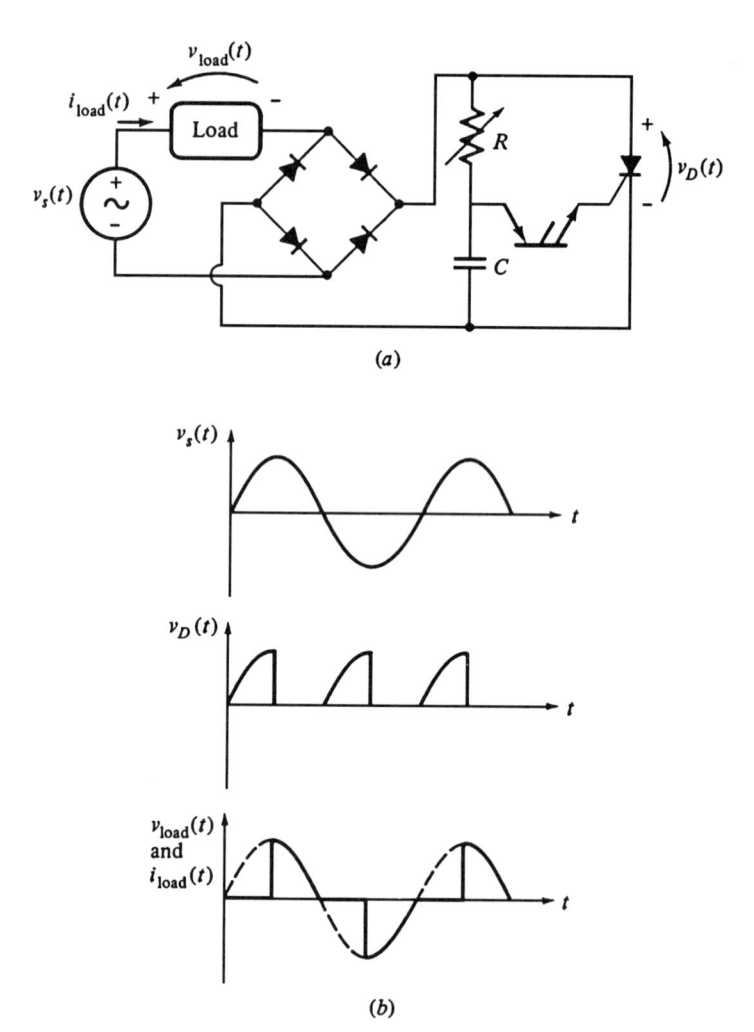

FIGURE 9.27 (a) A circuit controlling the voltage to an ac load by phase angle control. (b) Voltages on the source, the load, and the SCR in this controller.

AC Phase Control for an AC Load

An ac load can be controlled as shown in Fig. 9.27a. The load voltage and current are shown in Fig. 9.27b. A simpler ac power controller can be made by substituting the PNPN diode with a DIAC and the SCR by a TRIAC (Fig. 9.28).

THE EFFECT OF INDUCTIVE LOADS ON PHASE ANGLE CONTROL

Real machines are inductive loads normally. This introduces new complications to the operation of the controller. Since inductors store magnetic energy, *the current in an inductive*

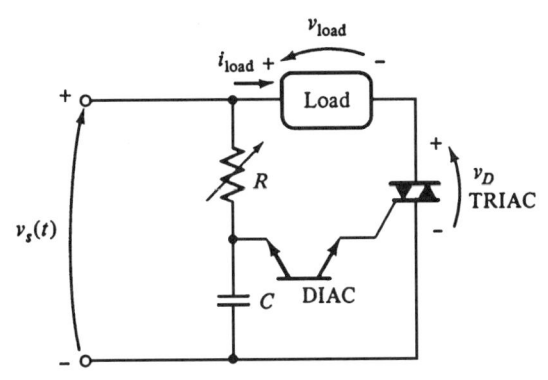

FIGURE 9.28 An ac phase angle controller using a DIAC and a TRIAC.

load cannot change instantaneously. This means that the current to the load will not rise immediately upon firing the SCR and will not stop flowing at exactly the end of the half-cycle. At the end of the half-cycle, the inductive voltage on the load will keep the SCR turned on for a short time, until the current flowing through the load and the SCR drops below I_H (Fig. 9.29).

INVERTERS

The static frequency conversion is the most rapidly growing area in modern electronics. It involves the conversion of ac power at one frequency to ac power at another frequency by using solid-state electronics. The traditional approaches to static ac frequency conversion employ the *cycloconverter* and the *rectifier-inverter.*

The cycloconverter is a device that directly converts ac power at one frequency to ac power at another frequency. The rectifier-inverter converts ac power to dc power and then converts dc power to ac power at a different frequency. A rectifier-inverter is divided into two components:

1. A *rectifier* to produce the dc power
2. An *inverter* to produce ac power from dc power

The Rectifier

The basic rectifier described earlier used diodes. These devices have a problem when they are used for motor control. Their output voltage is fixed for a given input voltage. This problem can be overcome by replacing the diodes with SCRs (Fig. 9.30). The average dc output voltage of this rectifier depends on when the SCRs are triggered during the positive half-cycle. If they are triggered at the beginning of the half-cycle, this circuit will be the same as the rectifier with diodes. The output voltage will be 0 V if the SCRs are never triggered. The dc output voltage will be between 0 V and the maximum for any other firing angle between 0° and 180°.

The output voltage of this circuit will have more harmonic content than a simple rectifier due to using SCRs. An inductor and a capacitor are placed at the output of the rectifier to assist in smoothing the dc output.

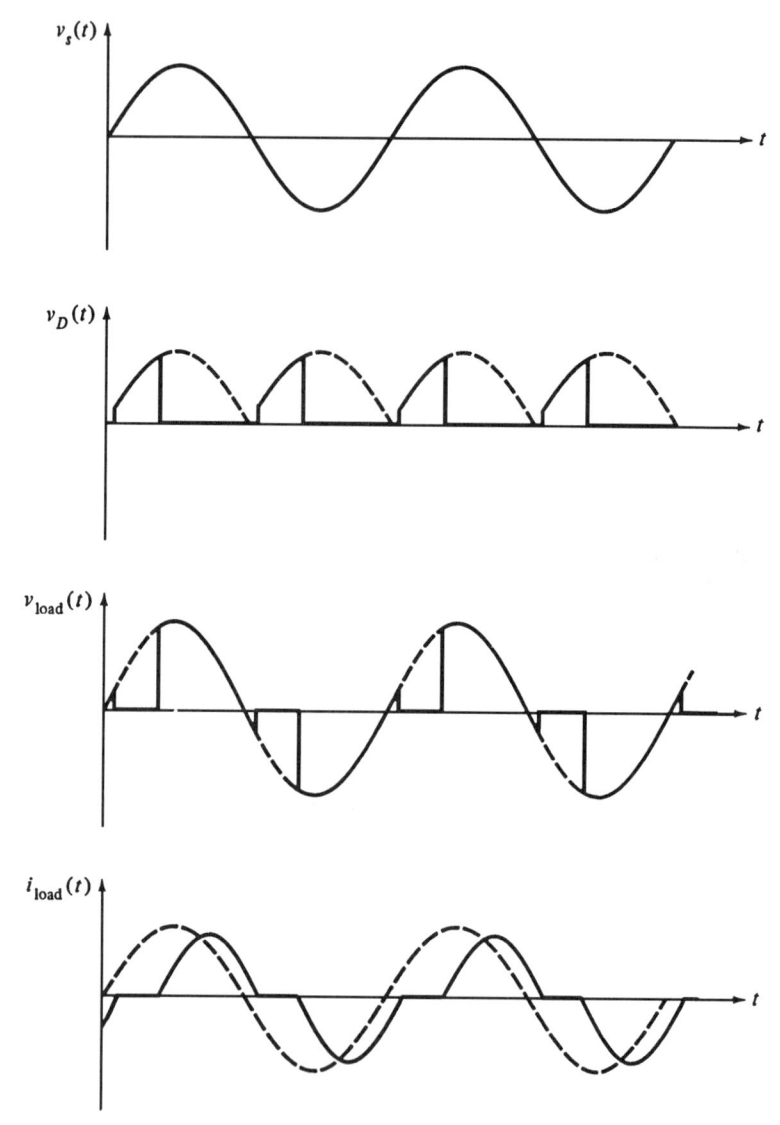

FIGURE 9.29 The effect of an inductive load on the current and voltage waveforms of the circuit shown in Fig. 9.27.

External Commutation Inverters

The two commutation techniques used for inverters are external commutation and self-commutation. *External commutation inverters* require energy from an external motor or power supply to turn off the SCRs (Fig. 9.31). A three-phase synchronous motor is connected to the inverter to provide the countervoltage needed to turn off one SCR when its companions are fired.

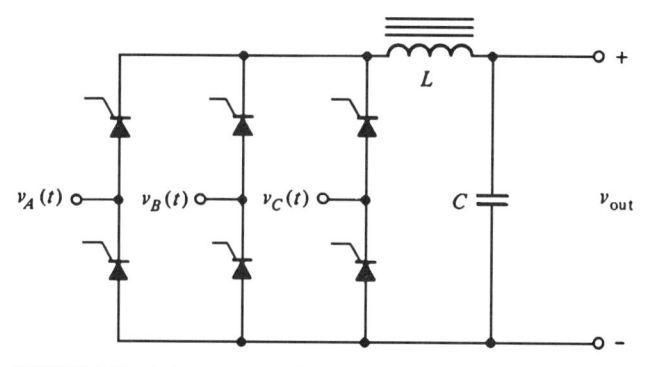

POWER ELECTRONICS, RECTIFIERS, AND INVERTERS 9.21

FIGURE 9.30 A three-phase rectifier circuit using SCRs to provide control of the dc output voltage level.

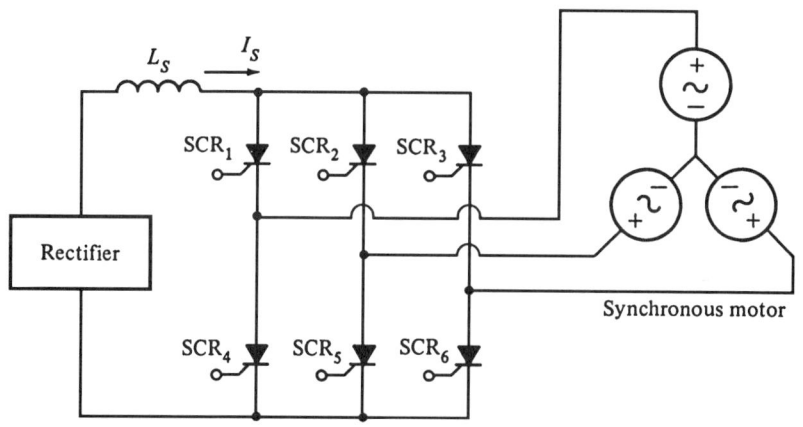

FIGURE 9.31 An external commutation inverter.

The SCRs are triggered in this order: SCR_1, SCR_6, SCR_2, SCR_4, SCR_3, SCR_5. When SCR_1 fires, the internal generated voltage in the synchronous motor provides the voltage needed to turn off SCR_3.

Self-Commutated Inverters

A load cannot always guarantee the proper countervoltage for commutation. A self-commutation inverter must be used. It is an inverter in which the active SCRs are turned off by energy stored in a capacitor when another SCR is switched on. Self-commutation inverters are designed also using GTOs or power transistors. In these cases, commutation capacitors are not required.

The types of self-commutation inverters are *current source inverters* (CSIs), *voltage source inverters* (VSIs), and pulse-width modulation (PWM) inverters. PWM inverters require faster switching components than CSIs and VSIs. Figure 9.32 shows a comparison between CSIs and VSIs.

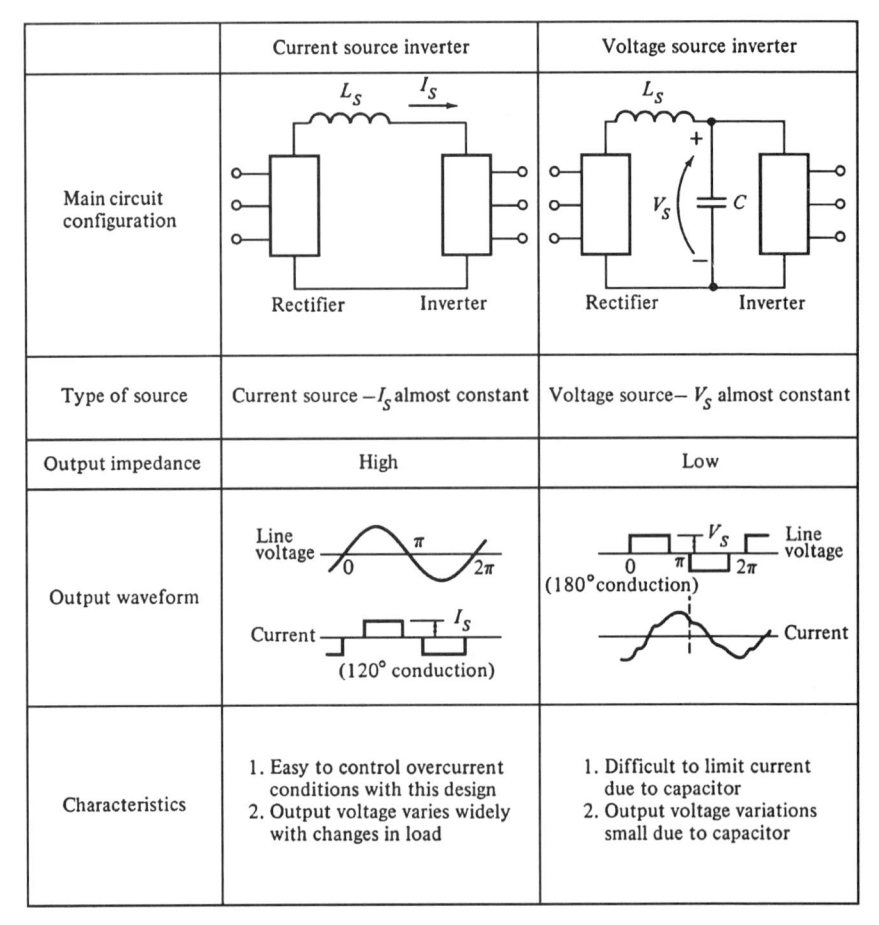

	Current source inverter	Voltage source inverter
Main circuit configuration	Rectifier — Inverter with L_S, I_S	Rectifier — Inverter with L_S, V_S, C
Type of source	Current source — I_S almost constant	Voltage source — V_S almost constant
Output impedance	High	Low
Output waveform	Line voltage (sinusoid); Current (120° conduction)	Line voltage V_S (180° conduction); Current (sinusoid)
Characteristics	1. Easy to control overcurrent conditions with this design 2. Output voltage varies widely with changes in load	1. Difficult to limit current due to capacitor 2. Output voltage variations small due to capacitor

FIGURE 9.32 Comparison of current source inverters and voltage source inverters.

The frequency of CSIs and VSIs is changed by changing the firing pulses on the gates of the SCRs. Both inverters can be used to drive ac motors at variable speeds.

Pulse-Width Modulation Inverters

Pulse-width modulation is the process of changing the width of pulses in a pulse train in direct proportion to a small control signal. The resulting pulses become wider when the control voltage is larger. A high-power waveform whose *average voltage* varies sinusoidally to drive ac motors can be generated by using a sinusoid at a desired frequency as the control voltage for a PWM circuit.

Figure 9.33 illustrates a single-phase PWM inverter circuit using IGBTs. The two comparators control the states of $IGBT_1$ to $IGBT_4$. A *comparator* compares the input voltage $v_{in}(t)$ to a reference signal. Based on the results of the test, it turns transistors on or off.

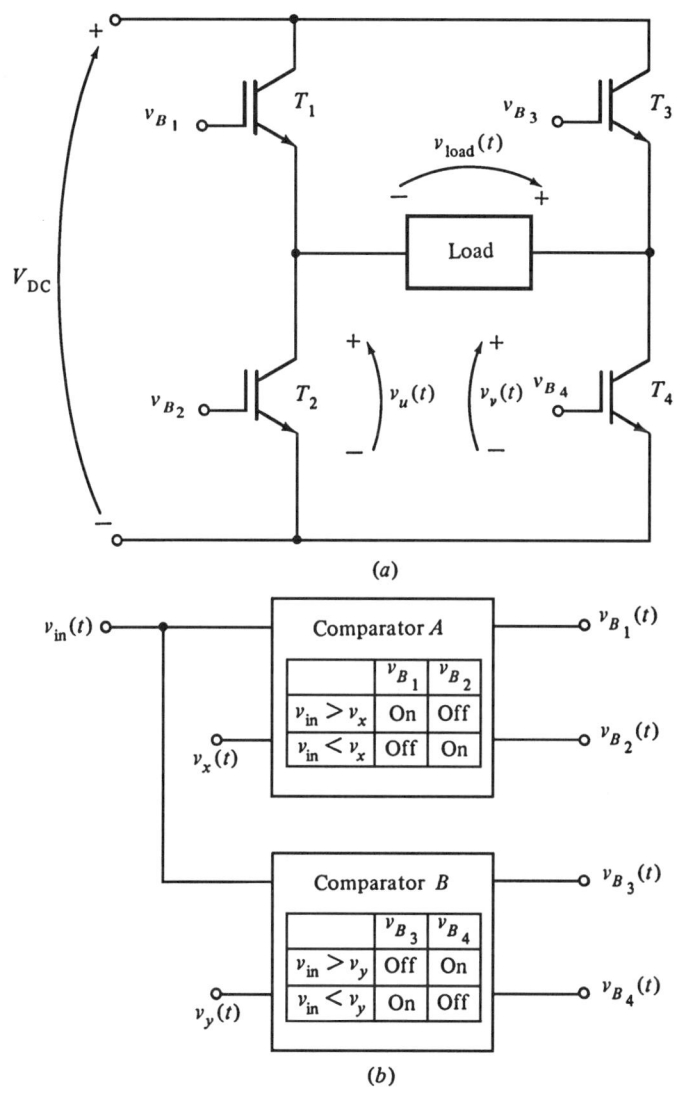

FIGURE 9.33 The basic concepts of pulse-width modulation. (*a*) A single-phase PWM circuit using IGBTs. (*b*) The comparators used to control the on and off states of the transistors.

Comparator A compares $v_{in}(t)$ to the reference voltage $v_x(t)$. Based on the results of the comparison, it controls the IGBT's T_1 and T_2.

Comparator B compares $v_{in}(t)$ to the reference voltage $v_y(t)$. It controls the IGBT's T_3 and T_4 based on the results of the test. If $v_{in}(t)$ is greater than $v_x(t)$, then comparator A will turn on T_1 and turn off T_2, and vice versa.

$v_x(t)$

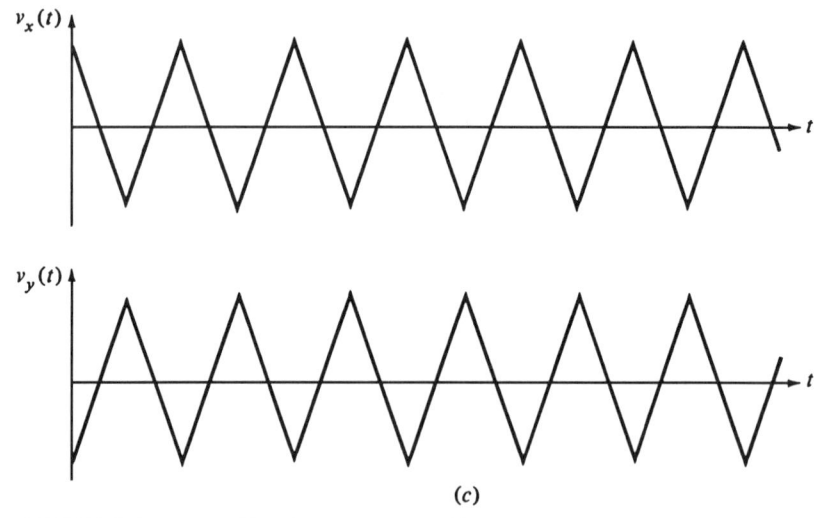

$v_y(t)$

(c)

FIGURE 9.33 (*Cont.*) (*c*) The reference voltages used in the comparators.

If the control voltage is 0 V, then voltages $v_u(t)$ and $v_v(t)$ are identical. The load voltage out of the circuit $v_{load}(t)$ is zero (Fig. 9.34). If a constant positive control voltage equal to one-half of the peak reference voltage is applied, the resulting output voltage will be a train of pulses with a 50 percent duty cycle (Fig. 9.35).

Finally, if a sinusoidal control voltage is applied (Fig. 9.36), the width of the resulting pulse train varies sinusoidally with the control voltage. The output is a high-power waveform whose average voltage is directly proportional to the average voltage of the control signal. The output waveform has the same *fundamental frequency* as the input control voltage. Evidently, there are harmonic components in the output voltage, but they do not usually have an effect in motor control applications. Additional heating may be generated in the motor due to harmonic components. However, this additional heating is compensated by using specially designed motors or by *derating* an ordinary motor.

A complete three-phase PWM inverter consists of three single-phase inverters with sinusoids shifted by 120° between phases. The variation of the input control voltage frequency provides the frequency control in a PWM inverter.

During a single cycle of the output voltage, a PWM inverter switches state numerous times. The frequencies of the reference voltages can be as high as 12 kHz. The components in a PWM inverter must change state up to 24,000 times every second. Components faster than CSIs or VSIs are required for this rapid switching. High-power and high-frequency components such as GTO thyristors, IGBTs, and/or power transistors are used in PWM inverters. A microcomputer mounted on a circuit board within the PWM motor controller provides the control voltage to the comparator circuits. The microcomputer can vary the control voltage to obtain different frequencies and voltage levels.

REFERENCE

1. S. J. Chapman, *Electric Machinery Fundamentals*, 2d ed., McGraw-Hill, New York, 1991.

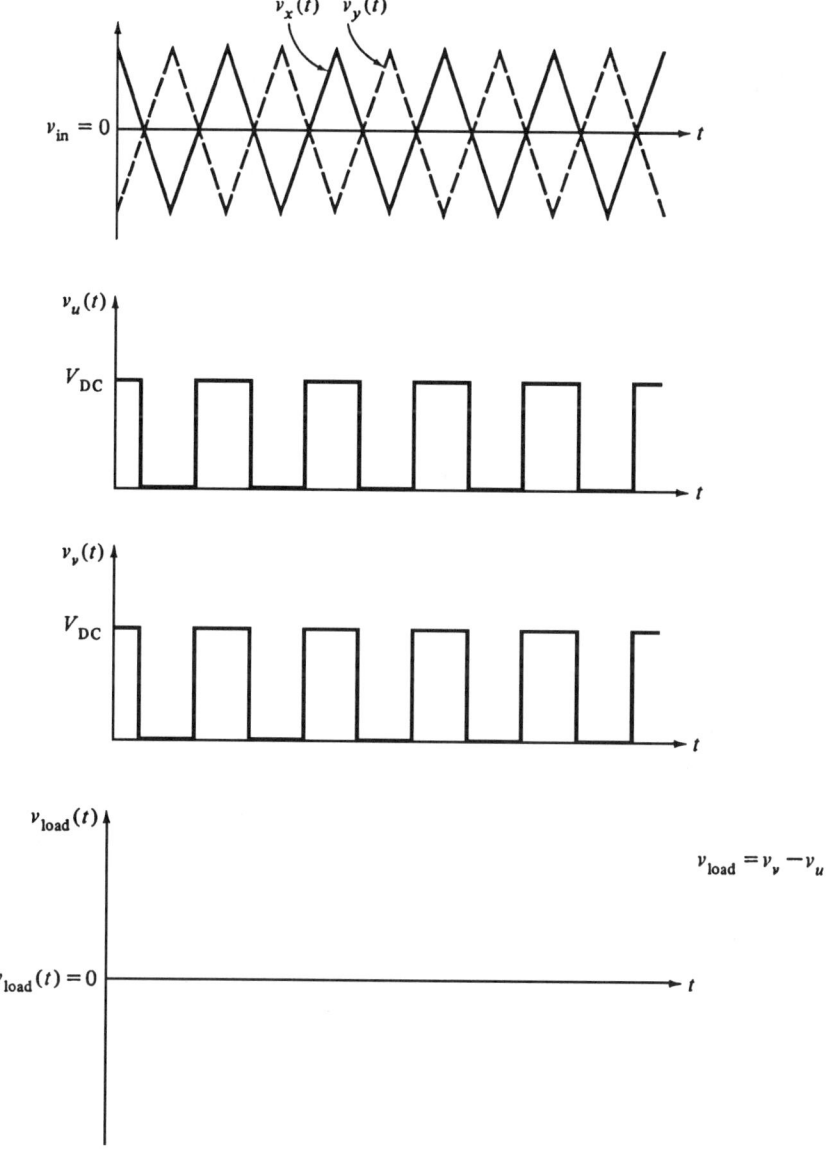

FIGURE 9.34 The output of the PWM circuit with an input voltage of 0 V. Note that $v_u(t) = v_v(t)$, so $v_{load}(t) = 0$.

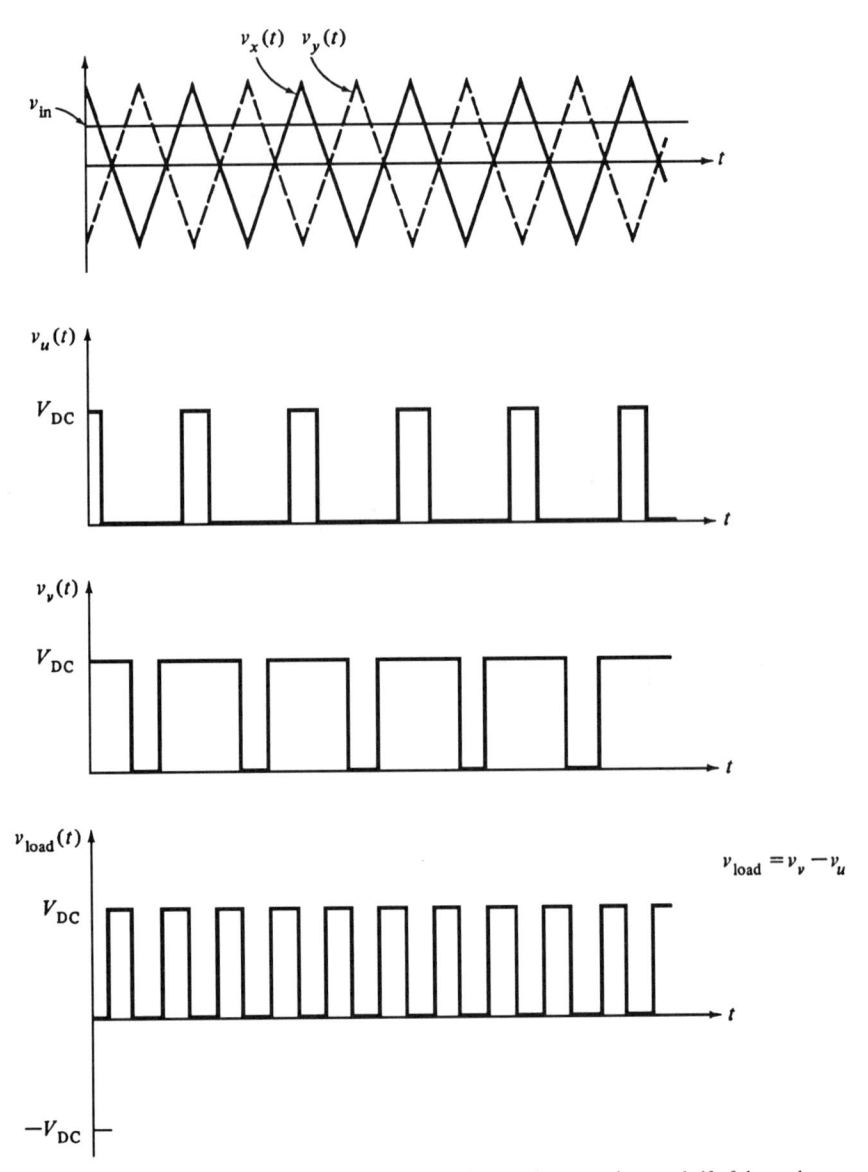

FIGURE 9.35 The output of the PWM circuit with an input voltage equal to one-half of the peak comparator voltage.

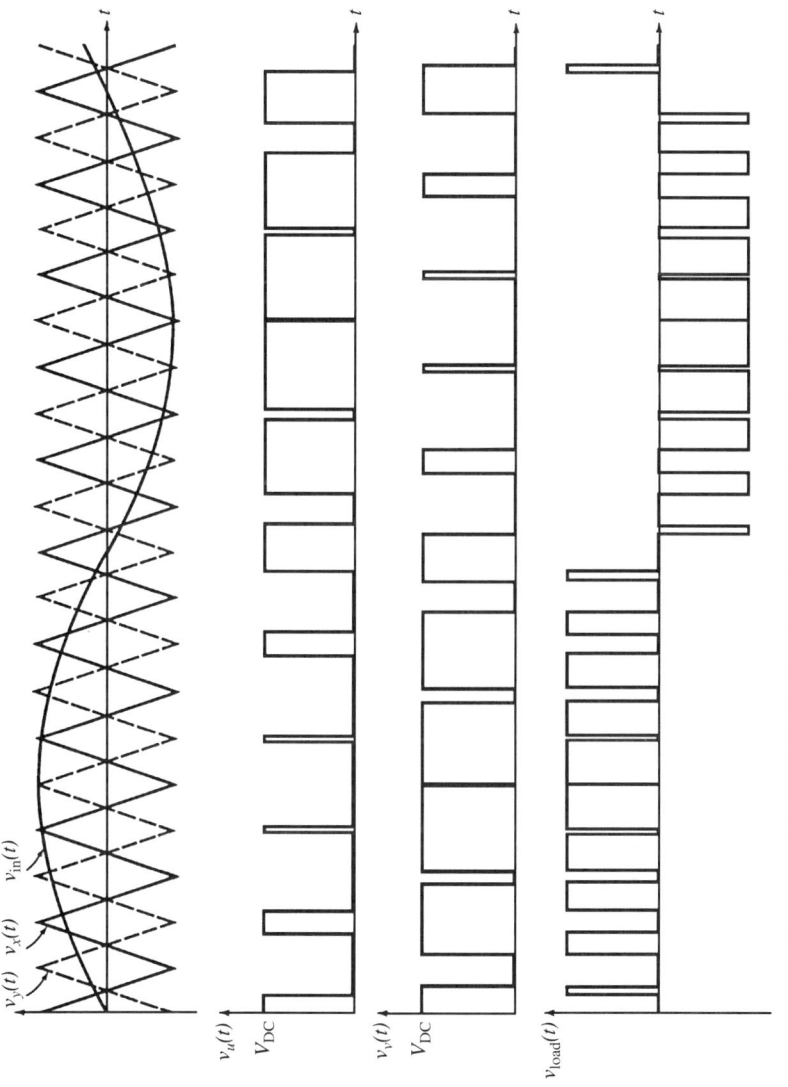

FIGURE 9.36 The output of the PWM circuit with a sinusoidal control voltage applied to its input.

CHAPTER 10
VARIABLE-SPEED DRIVES

BASIC PRINCIPLES OF AC VARIABLE-SPEED DRIVES

A *variable-speed drive* (VSD) is used to drive a motor at variable speed. The main parts of a VSD are

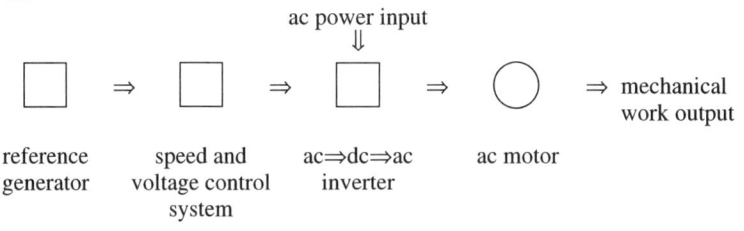

```
                          ac power input
                               ⇓
   ☐      ⇒     ☐      ⇒     ☐      ⇒     ◯      ⇒  mechanical
                                                        work output
reference     speed and      ac⇒dc⇒ac     ac motor
generator   voltage control   inverter
              system
```

The control system of the VSD adjusts the output voltage and frequency so that the ratio of voltage to frequency remains constant at all times. The two modes of operation are as follows:

Constant-Torque Region

In this region, the motor increases in speed from zero to the rated base speed while the torque remains at the rated value (Fig. 10.1). The motor produces its maximum (rated) power at the base speed.

Constant-Power (Extended Speed) Region

FIGURE 10.1 Constant-torque operation.

In this region, the motor operates beyond its base speed (Fig. 10.2). The frequency of the VSD is increased and the flux is decreased while the armature voltage is kept at its rated value. Since the motor was at its maximum power at base speed, the torque must be reduced when the speed is increased (power = torque × speed).

INVERTERS

The main component of a VSD is the inverter. It is a power converter that converts the fixed ac input voltage and frequency to a controlled variable voltage and frequency to operate a motor at the required speed (Fig. 10.3). An energy storage device separates the input from the output and allows each to operate independently. It is called a *link filter*. The incoming power can be structured to give a very high power factor and low harmonics without affecting the output (we will only deal with voltage controlled inverters).

10.1

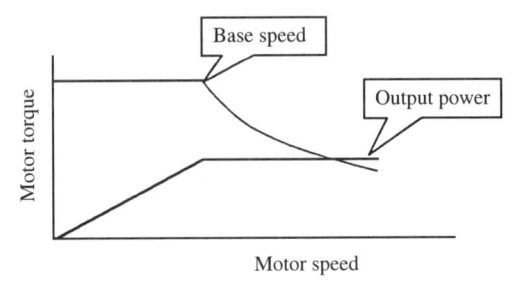

FIGURE 10.2 Constant-power operation.

Parts of an Inverter

A normal ac inverter has three parts:

1. An input converter to rectify ac power to dc power. It is normally called the *source bridge*.
2. An energy storage device which separates the input from the output and allows each to operate independently from the other. It is usually called a *link filter*.
3. A dc-to-ac inverter in the output stage. It is called an *inverter*. It generates the desired ac output voltage and frequency.

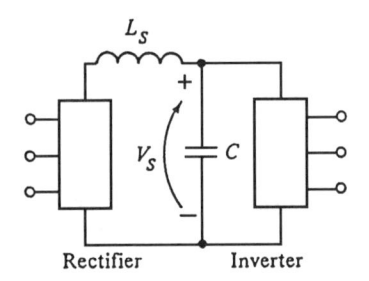

FIGURE 10.3 Voltage controlled inverter.

Pulse-Width-Modulated Inverters

PWM is referred to as *time ratio control*. From a constant dc input voltage, we get a variable output voltage and frequency by varying the percentage of time that the power control switch is closed. The output voltage will increase by increasing the percentage of time the switch is closed. The switch is either open or closed. There is no power dissipation across the switch in both states. Figure 10.4 illustrates a PWM circuit. Please note

$$E_L = E_{DF} = \frac{\text{time while switch is closed}}{\text{time while switch is closed} + \text{time while switch is open}} \times E_d$$

The main parts of a PWM circuit are

- The input *dc power*.
- The power switch can be any semiconductor that we command to turn on or off.
- The inductor stores the energy from the switch when it is closed, then releases it to the load when the switch is opened.
- The freewheeling diode gives a path to the current to flow when the switch is opened. The stored energy in the inductor is released to the load.

VARIABLE-SPEED DRIVES 10.3

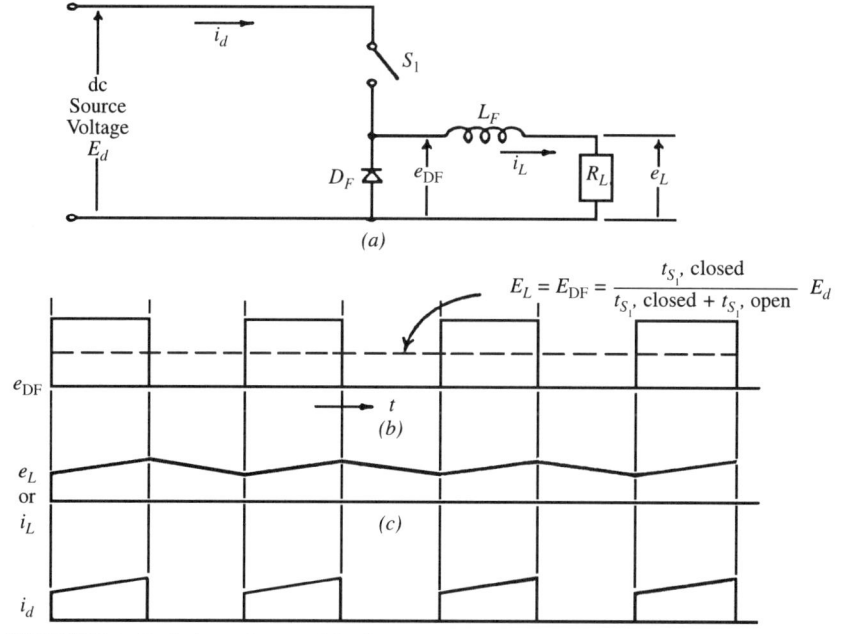

FIGURE 10.4 Basic time ratio control details.

Insulated Gate Bipolar Transistors. These bipolar power transistors are driven by an insulated gate metal-oxide transistor. A relatively simple 15-V gate driver signal is used to control the resulting high-current power transistor. The IGBT is a four-layer semiconductor similar to the SCR. Its main features are that

1. It has very fast switching on the order of 100 to 150 ns (1 ns = 10^{-9} s) and resulting high-voltage transients dv/dt of 5000 to 10,000 V/μs.
2. The IGBT chips are soldered in place and connected with discrete bond wires. They are very weak when it comes to thermal fatigue problems. The IGBT modules have significantly lower thermal fatigue capability than other semiconductors. The high dv/dt generated leads to problems with bearing currents and the insulation system.

Two-Level Pulse-Width-Modulated Inverter (PWM-2). This is a *voltage controlled* inverter. It uses a bank of electrolytic capacitors to store the intermediate energy. Its ratings are up to 600 V ac and 1800 A. It is usually used to control an induction motor up to 60 Hz. However, it can be used to go up to 120 or even 150 Hz in special applications. Figure 10.3 illustrates a nonregenerative PWM drive that uses IGBTs in the output inverter.

INPUT POWER CONVERTER (RECTIFIER)

Regeneration is not required in most cases. A simple diode bridge is used as an input power converter (Fig. 10.5).

10.4　　　　　　　　　　　　CHAPTER TEN

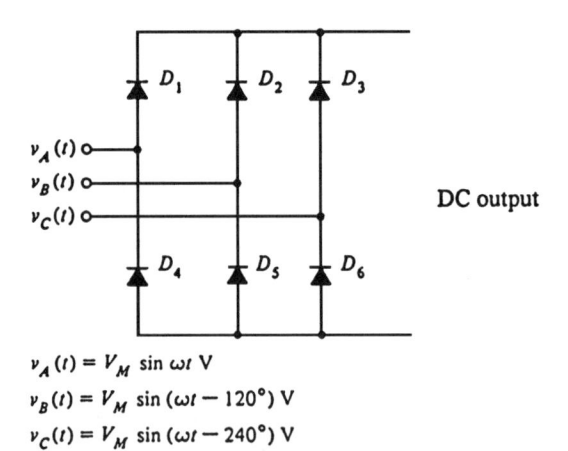

$v_A(t) = V_M \sin \omega t$ V
$v_B(t) = V_M \sin (\omega t - 120°)$ V
$v_C(t) = V_M \sin (\omega t - 240°)$ V

FIGURE 10.5　Diode bridge input stage (when regeneration is not required).

DC LINK ENERGY

This is a bank of electrolytic capacitors (high-capacitance). If normal capacitors were used, the diode bridge output current would be very discontinuous and the resulting ac input power factor would be very poor (as low as 50 percent) due to the very high levels of harmonics. An inductor in series with a capacitor (Fig. 10.3) is used to correct the problem.

OUTPUT IGBT INVERTER

Figure 10.6 illustrates a typical two-level PWM inverter circuit. It is similar to the SCR bridge, but it uses IGBTs for the switching devices. The energy storage capacitor is denoted by *C*. The motor connections are *a, b,* and *c*. The inverter operation is as follows:

Once the output frequency required to satisfy the speed regulator is given to the control system, it calculates the three-phase voltage commands (Fig. 10.7). A triangle voltage waveform (Fig. 10.8) is generated and synchronized with the desired IGBT switching frequency and phase. This is the PWM carrier waveform that sets the basic inverter switching frequency. The average width of the PWM waveforms generated approximates the sine wave reference. The inductances average and smooth the resulting waveform.

Figure 10.8 also shows the resulting PWM line-to-line output voltage compared to the original sine wave reference. When the transistors are on, the current charges the motor inductance. When they are off, the current freewheels through the corresponding diode. This causes the characteristic ripple shown in Fig. 10.9. The ripple causes motor heating in excess of that due to a sine wave current. Thus, motors must be thermally derated when used with two-level PWM inverters.

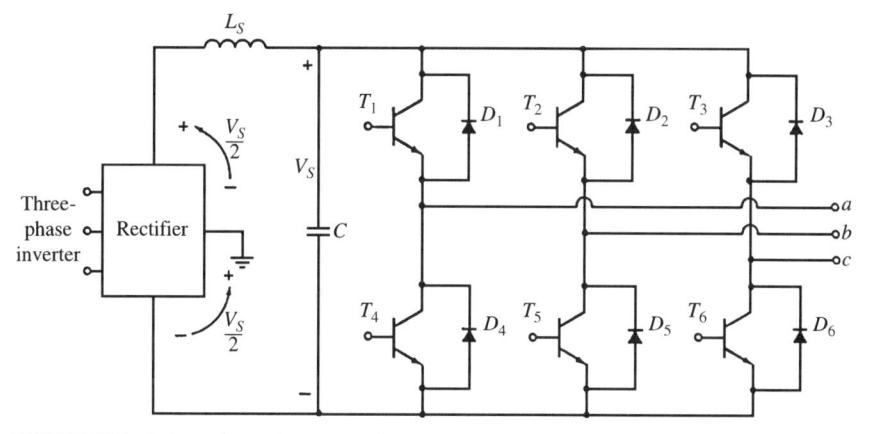

FIGURE 10.6 A three-phase voltage source inverter using power transistors.

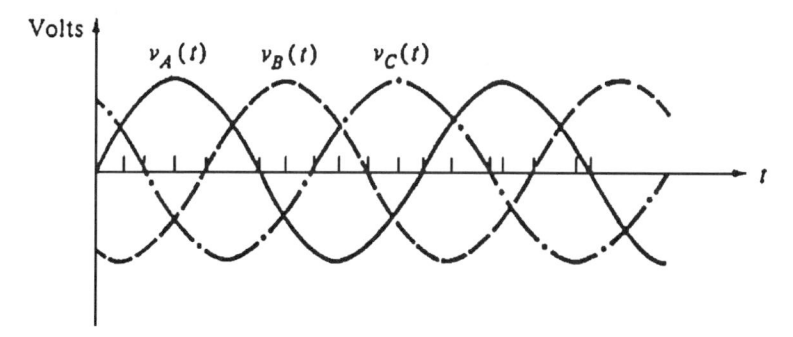

FIGURE 10.7 The three-phase voltage commands given to the VSD at the required output frequency.

INPUT SOURCES FOR REGENERATION OR DYNAMIC SLOWDOWN

If the inverter is made to regenerate (i.e., to try to pump back the mechanical energy in the load to the ac supply), the source diode bridge cannot reverse its current flow. Thus, the absorbed energy will cause the dc bus voltage to increase. If this higher voltage is not absorbed, the source bridge, load inverter, and capacitor bank could get damaged.

Dynamic Braking

The simplest way to absorb the load energy that is regenerated is by using a resistor across the dc link (dynamic braking resistor). This is simply a large power-dissipating resistor that

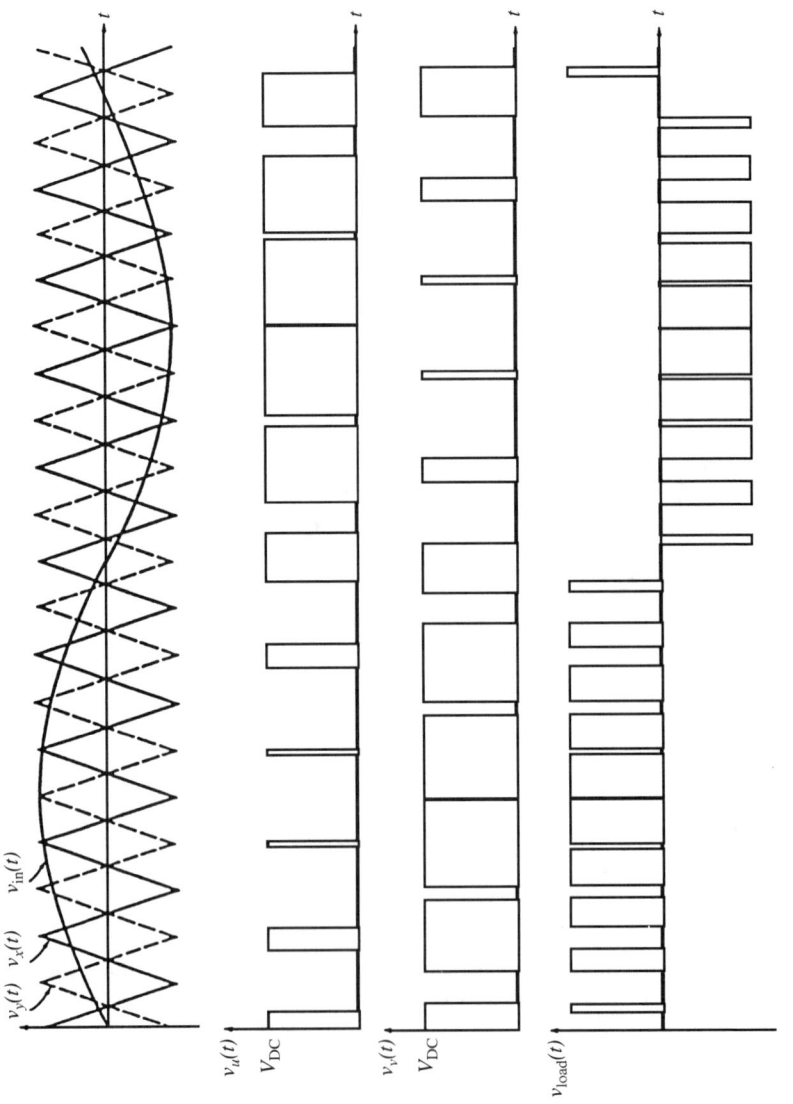

FIGURE 10.8 The output of the PWM circuit with a sinusoidal control voltage applied to its input.

[Graph showing current waveform oscillating between approximately +0.6 A and -0.6 A over time 0 to 6 seconds, with high-frequency harmonics superimposed on a sinusoidal shape]

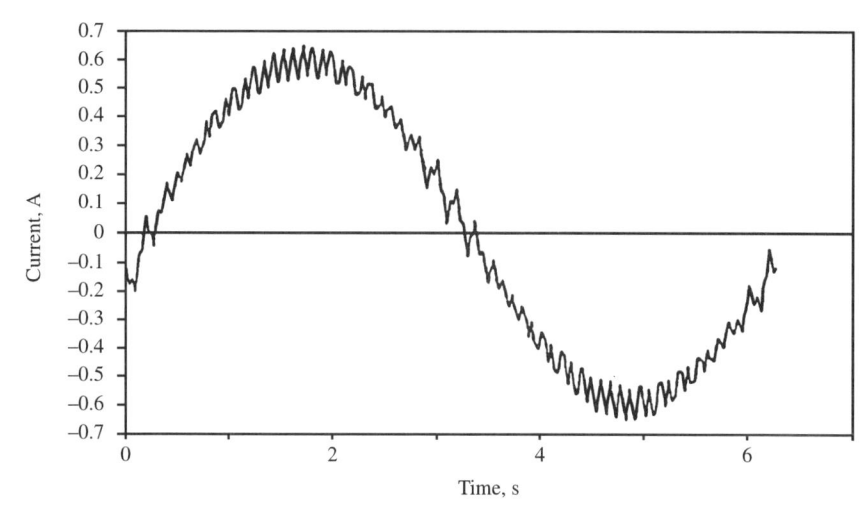

FIGURE 10.9 One phase of the resulting motor current (note the high-frequency harmonics present in this output waveform).

is switched across the dc bus. The dynamic braking resistance chopper is controlled so that when the link voltage rises above a preset limit, the resistor is switched in until the voltage drops below the preset limit. Therefore, the regulator controls automatically the amount of braking applied.

REGENERATION

The mechanical energy in the load can be regenerated back to the ac line (by lowering the synchronous speed below the actual mechanical speed of the load) by using a PWM bridge exactly like the one installed at the output. The second PWM bridge is installed at the input. It is identical to the one used on the output, except that it is turned around so that the ac signal is fed into the three-phase terminals and the dc signal is connected to the dc input as follows:

Three-phase ac input → PWM bridge → dc link including capacitors and inductors
→ PWM bridge → three-phase ac output

The PWM bridge at the input can be used to provide good control of the ac line power factor and harmonics. Figure 10.10 illustrates the ac input when the output is discontinuous (no inductors are used in the dc link).

Figure 10.11 illustrates the ac input line current when the PWM bridge is used at the input. This is a major improvement over the diode bridge. However, the current sine wave contains harmonics based on the IGBT carrier frequency (usually 1.4 to 5 kHz). These harmonics are, in reality, more troublesome than the discontinuous current obtained using the diode bridge.

The main advantage of using the PWM source bridge at the input is the ability to regulate the power factor, which is normally set to unity (in some cases, it can be set to lead to compensate for other equipment, depending on the kVA rating of the converter and the applied load). However, this option doubles the inverter cost.

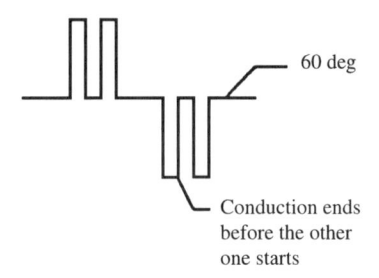

FIGURE 10.10 AC input current with discontinuous output.

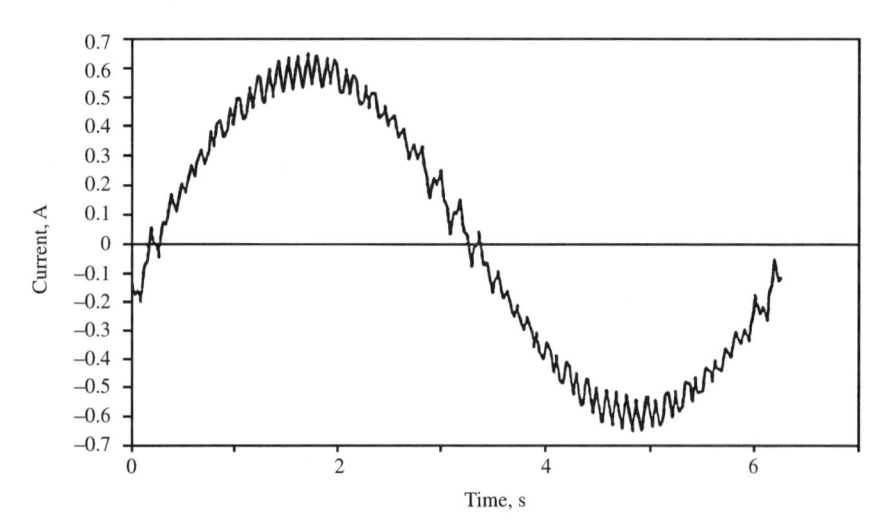

FIGURE 10.11 AC input harmonics caused by using a PWM bridge at the input.

PWM-2 CONSIDERATIONS

1. Motors must be derated due to the relatively high output harmonics; i.e., two-level PWM drive cannot be used with standard ac induction motors.
2. Due to the fast switching speed of the IGBTs (100 to 200 ns), high motor bearing currents are generated. They flow through the bearings from the rotor to the stator and frame.
3. The fast voltage transients and the capacitance between the stator windings and the rotor create motor bearing current. This current flows through the motor bearings from the rotor to the stator (frame). A special motor construction must be used to minimize the capacitance between the stator windings and the rotor.
4. The fast switching transients create voltage reflection problems at the motor due to the difference in impedance between the motor connection cables and the motor windings.

VARIABLE-SPEED DRIVES 10.9

These reflections result in an increase in the voltage transient levels on the terminals up to twice the nominal level. This problem is normally dealt with by using special motor insulation to withstand the higher voltages or by using a motor or an inverter filter.

TRANSIENTS, HARMONICS POWER FACTOR, AND FAILURES

Semiconductor Failure Rate

Although thyristors, diodes, and IGBTs are solid-state devices, they have wearout mechanisms just as insulation and other mechanical parts do. The wearout and failure rates of these devices can be calculated.

Figure 10.12 illustrates the general failure rate curve of SCRs, diodes, and IGBTs. The initial high failure rate is caused by manufacturing defects, application problems, and drive start-up stresses and lasts a few weeks. The high failure rate at the ends indicates the end of the life for the devices. In general, the lifetime of a device becomes shorter when it is operated harder and closer to its voltage rating.

Common Failure Modes

Differential Expansion (Mechanical Fatigue). This failure mode is mechanical fatigue or wearout caused by the difference in expansion rates as the temperature of the device changes. As the temperature of the device changes, different parts expand at different rates. These are the expansion coefficients for materials used in semiconductors:

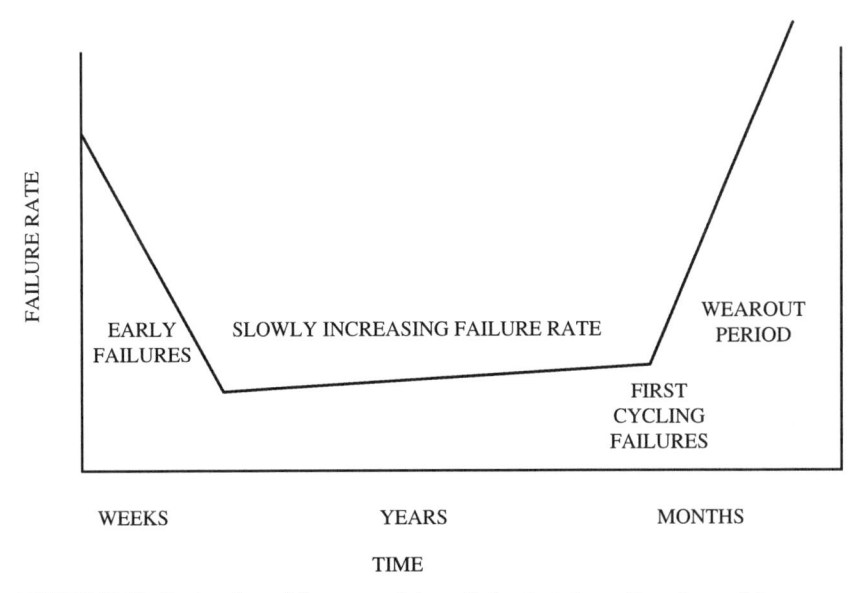

FIGURE 10.12 Semiconductor failure rate variation with time (note the nonlinear time scale).

10.10 CHAPTER TEN

Material	Expansion coefficient, in/(in·°C)
Silicon	4.2×10^{-6}
Copper	16.5×10^{-6}
Aluminum	8.5×10^{-6}
Iron	11.7×10^{-6}
Molybdenum	4.9×10^{-6}

Thus, the parts slide over each other, causing mechanical wearout. This failure is common to all semiconductors. It normally occurs at the end of life of these devices. Figure 10.13 shows the thermal cycling fatigue life. Note the dependency on size and temperature. Also note that soldered modules are much worse than compressed ones. Although this module is labeled IGBT, in reality it applies to all modules having soldered terminals (i.e., thyristors, diodes, and transistors). It is important to note the low number of thermal cycles required for failure if the junction temperatures are allowed to climb too high.

Fault Current Limit. This mode of failure is not applicable to IGBTs because they are not able to conduct currents in excess of their ratings. It is only applicable to thyristors and diodes GTOs. The junction temperature increases when the fault current increases. The maximum surge current that can be tolerated results in junction temperature excursion ΔT_j of 300°C. This temperature excursion can occur *once* in the lifetime of the equipment because it would have been damaged (maimed) by the high temperature. The number of current surges that can be tolerated increases rapidly if the peak current level (peak junction temperature) is reduced. The number of surges can be approximated by

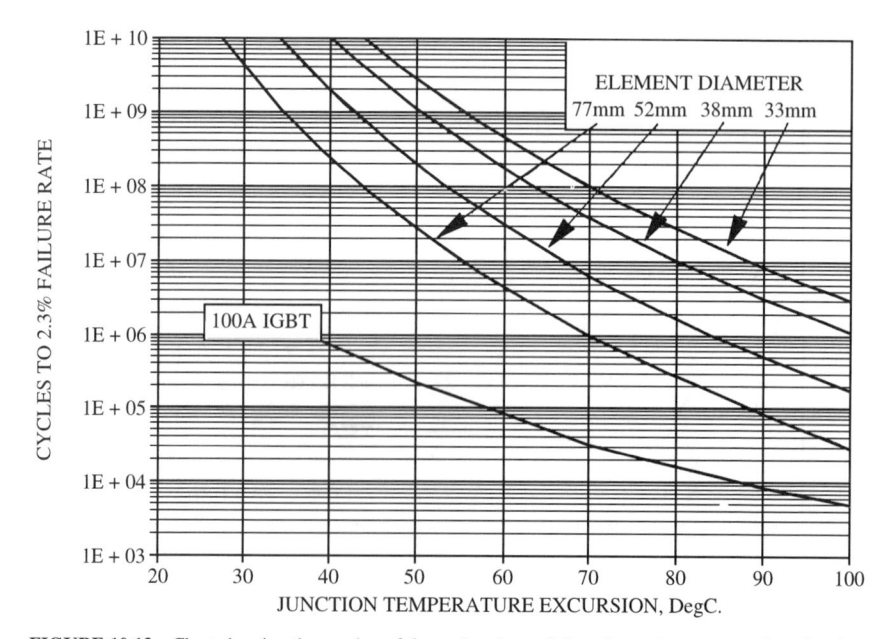

FIGURE 10.13 Chart showing the number of thermal cycles to failure for various press pack wafer sizes and soldered modules. (Note that the lower curve includes all modules where the wafers are soldered-thyristors, diodes, and IGBTs.)

$$N = \left(\frac{300}{\Delta T_j}\right)^9$$

This failure mode is normally caused by incorrect application. The designer of the system must ensure that the maximum fault current in the bridge cannot exceed these limits. This failure mode occurs normally during commissioning. However, it can occur at random intervals during the lifetime of the device.

Device Explosion Rating. This failure mechanism can occur in any of the power devices. Thyristors can break down in the reverse direction due to a fault. This is usually followed by a large surge current. The resulting arc at the edge of the device could be strong enough to blow open the ceramic housing. The explosion rating for a thyristor is normally 50 to 100 percent above the surge current rating.

This type of failure can have serious consequences because conductive plasma is vented from the failed cell into the bridge, resulting in extensive arcing and destruction throughout the whole bridge. These failures are normally caused by inadequate fault coordination (design deficiency). They usually occur during commissioning or at any time during the lifetime of the device.

Device Application. These failure modes normally occur in the middle region of Fig. 10.12 (slowly increasing failure rate) which extends over several years of operation. The failure rate depends on these application factors:

- Type of device application
- Voltage applied (as percentage of PIV)
- Junction temperature (at normal running load)

In general, the lifetimes of all semiconductors decrease when the applied voltage or temperature (as a percentage of the rating) increases.

THYRISTOR FAILURES AND TESTING

Recognizing Failed SCR or Diode

It is easy to recognize a failed SCR or diode because it normally (99.9 percent) becomes a short circuit. Any fuse in the circuit is usually blown due to the overcurrent. Shorted SCRs or diodes give a resistance around 10 Ω when measured by a volt-ohmmeter. The remaining 0.1 percent, the gate of an SCR, is open. The gate to cathode resistance of a good SCR is normally 15 to 30 Ω while that of an open-gate SCR is infinity.

Another failure mechanism occurs near the end of the useful life of the device. The leakage will increase due to degradation of all junctions. This results in operating problems such as numerous overcurrents. These symptoms should be taken as an indication that the devices should be replaced.

Testing of SCRs or Diodes

If the device is not experiencing problems such as intermittent overcurrent trips or blowing fuses, it should not be tested. The chance of damaging the devices by disconnecting and testing them is higher than that of finding a suspicious one. In general, *if the device is operating properly, do not test it.*

Comments about Failure Rates

1. The lifetime of all semiconductors can be calculated.
2. The thermal fatigue life decreases when
   - The temperature changes increase
   - The size of the device is larger
   - The device is soldered rather than press-packed
3. Damage to the device would occur at a fault current, resulting in 300°C change.
4. Fault coordination for each application should be done properly in the design phase to prevent explosion of the devices.
5. The failure rate depends on
   - Temperature
   - PIV
   - Bridge failure mode (catastrophic or not)
6. Failed devices are normally shorted (less than 10 Ω).
7. There is no need to routinely test the devices if the drive is operating properly.

AC DRIVE APPLICATION ISSUES

Introduction

The ac input characteristics are

1. Diode source current unbalance
2. AC power factor
3. AC input power change with AC input voltage

The ac output characteristics cover

1. IGBT switching transients
2. Cabling details for ac drives
3. Motor, cable, and power system grounding
4. Motor bearing currents

Diode Source Current Unbalance

Any *unbalance in the ac source voltage will cause very large ac unbalances to flow* in non-regenerative ac drives. This is due to the very low impedance of the dc link energy storage capacitor. In general, the ac current unbalance is about 10 to 20 times larger than the ac voltage unbalance (depending on whether there is a dc link inductor and what its size is). Since most ac power systems have an unbalance of 1 to 2 percent, these problems can occur:

- The rating of the ac cables must increase.
- The rating of the supply and any ac switchgear must also increase.

- The rating of the diode bridge may also require an increase due to the additional dissipation in the semiconductors. The consequences of the unbalance can be very significant because the voltage unbalance in typical power systems can be 2 to 3 percent. This problem will become more significant if the ac feeder has single-phase loads because these loads will increase the already existing voltage unbalances.

AC POWER FACTOR

The increase in harmonics in a power circuit (Fig. 10.9) will decrease the total power factor. It can drop to 60 percent for an inverter without a dc link reactor. Obviously, this is not acceptable. The problem is solved by adding a dc link reactor in series with the capacitor bank. If the reactor is chosen large enough, the diode bridge will always see an inductive load. The total power factor will increase to 0.95.

AC Input Power Changes with AC Input Voltage

The output inverter of the drive is a separate entity from the input source converter. They operate independently from each other. Assume that the drive is feeding a constant kilowatt load near its full rating. If the ac line voltage drops by 10 percent, the line alternating current increases by 10 percent. This will overload the drive.

The current increase due to the reduction of ac line voltage will be added to any current unbalance caused by a line voltage unbalance.

IGBT SWITCHING TRANSIENTS

Voltage transients are generated due to the switching in the PWM inverter. They propagate down the power cables to the motor. If the motor cables are not terminated properly, the switching waves will be reflected when they reach the motor. They will be transiently increased or decreased depending on the relative impedance between the line and the load. An alternative solution for this problem would be to slow down the rate of voltage rise.

If the connecting cable is long, the mismatch of impedances will generate a voltage reflection at the point where the line impedance changes (at the motor terminals). A transient rise in voltage of the wave will occur at the motor terminals due to the voltage reflection.

For a given switching rise time, the voltage rise at the terminals will increase with the length of the connecting cable. The wave front is reflected back to the inverter. If no mitigating actions are taken, the transient voltage at the motor terminals can double that of the inverter in the worst case. In most cases, *the reflection problem increases gradually with the length of the motor cable.* The rise in transient voltage is important for the following reasons.

Insulation Voltage Stress

The increase in voltage stress in the motor connecting cables and the motor insulation will shorten the lifetime of the insulation. If the peaks of any of the voltage transients exceed the insulation corona discharge level (partial discharge level), the insulation will degrade with each voltage pulse. It will eventually fail.

The new "inverter-rated" motors have triple-layer insulation. They have a 1600- to 2000-V partial discharge level. This allows them to withstand double the voltage peak transient from a 600-V inverter. Since most "standard" induction motors can withstand about 1200 V, its use in this application will shorten its life drastically. It can be as low as a few hours only.

Motor Winding Voltage Distribution

All high-frequency transient voltages tend to be unevenly distributed across the motor windings. The high frequencies develop greater voltages across the first windings rather than being evenly distributed across the whole length of the windings. The effect of high-frequency transients tends to be accentuated on the first few windings. This is where most motor insulation failures occur. The problem becomes worse when the frequency of the transients (i.e., the IGBT switching speed) is higher.

This problem does not occur with drives using older transistors or GTOs because the switching speeds are much lower (2 to 4 μs). When the drive uses IGBTs with switching speeds of 50 to 150 ns, the motor should be connected to the inverter by a cable shorter than 20 ft, or special precautions should be taken.

This problem can be solved by adding a filter to the output of the inverter to slow down the IGBT switching transients (Fig. 10.14). When the inverter switching speed is reduced, the length of the connecting motor cable can be increased.

Radiated Electromagnetic Interference (EMI)

The bandwidth (hertz) of any radiated EMI increases inversely proportionally to the rise time of the voltage transients. Since the IGBTs have a switching speed of 50 ns, EMI frequencies in the megahertz range are generated. These frequencies radiate very well.

This radiation is not limited to the motor cables only. They will also occur (to a lower level) on the ac source input to the inverter due to common-mode voltage problems. The radiation problem can be solved by installing a power line EMI filter.

The inverter switching problems can be solved by any of the following three solutions:

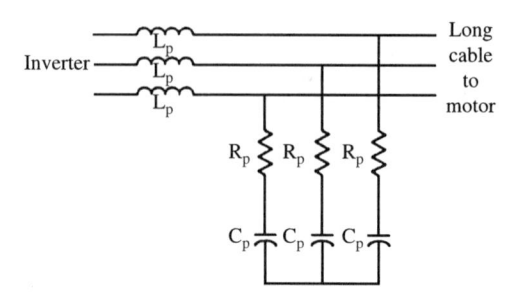

FIGURE 10.14 Inverter filter.

Cable Terminating (Matching) Impedance

An *RC* (resistance-capacitance) filter can be added to the motor terminals. Its impedance should match that of the connecting cable (Fig. 10.15). The reflected voltage will be canceled by removing the discontinuity of the impedance. This will reduce the motor and cable insulation stress. This solution has three disadvantages:

- Cost is high.
- It dissipates power.
- The filter must be added in a very inhospitable environment at the motor termination box. This solution prevents voltage reflection. Therefore, it reduces insulation stress. However, it does not change the switching speed of the inverter. Thus, the high voltage rise (dv/dt) will still be impressed on the motor insulation.

Inverter Output Filter

When a filter is added, like the one shown in Fig. 10.14, at the output of the inverter, the rise time of the voltage transients will be reduced to that of the GTOs due to the series inductors. The voltage reflection problems will be reduced significantly due to the decrease in voltage rise time. This will allow longer cables to the motor. The *RC* part of the filter is needed to dampen the switching transients. This filter is also costly, dissipates power, and requires large space in the inverter.

The most effective type of this filter is a large one that prevents switching voltages from propagating. In this case, there will not be any switching, insulation, or EMI problems. This type would be ideal for a drive used with an existing older motor having a suspect insulation. Since the filter's inductances have reduced the voltage rise times to the motor, the radiated EMI and the motor bearing currents are also reduced. *This is the best overall solution* (even with the increase in cost). In general, this is a standard part of the inverter, or it is available as an option in most commercial inverters.

Extra Insulation

This approach involves adding extra insulation to the magnet wire in the motor and connecting cable to handle the additional voltage caused by switching. A typical insulation for inverter duty motors has three layers. It can withstand the voltage transients of a 600-V inverter without corona discharge (standard motors cannot be used for these applications).

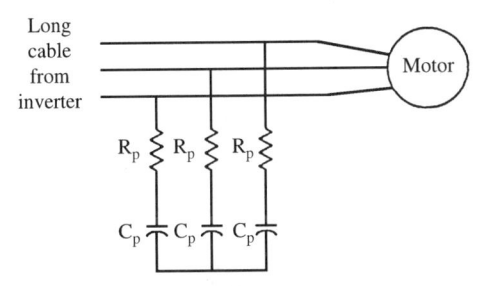

FIGURE 10.15 Motor filter.

The cable insulation problem must also be addressed. A 2000-V cable should be used (depending on the length of the connecting cable) for a 600-V IGBT inverter. This solution does not reduce the waveform rise time. Therefore, it does not address the motor bearing current or EMI radiation problems.

CABLING DETAILS FOR AC DRIVES

The cable connections from the inverter to the motor and ground are important details required to achieve a successful ac drive installation due to the fast switching rates. These details are applicable for the ac drives. However, they are *mandatory* for IGBT drives due to the higher switching rates. If this information is not properly followed, problems can occur in the motor and cable insulation, bearings, or EMI.

The most important criterion for power cabling is the *symmetry of the cable and grounding practices*. The symmetry is required to ensure cancellation of any stray fluxes. This will minimize the bearing currents and EMI radiation. The components within the inverter should also be symmetric.

CABLE DETAILS

The continuous corrugated aluminum cable [NEC type MC (metal clad), Fig. 10.16] is best suited for PWM. It costs about 30 percent more than the "standard" variety with three conductors, one ground, and interlocked aluminum armor. Note the symmetry of the power and the ground conductors (only one random ground conductor is installed in most standard 60-

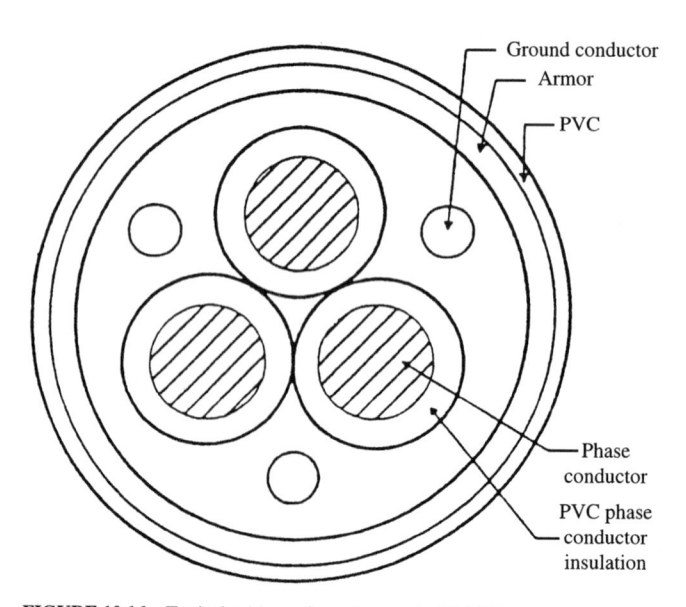

FIGURE 10.16 Typical cable configuration used with VSDs.

Hz power cables). The symmetry helps in canceling most of the stray fluxes. The symmetry in the cable also helps to reduce bearing currents and EMI (by providing better shielding).

Motor, Cable, and Power System Grounding

Proper termination of the cable is required to provide effective grounds and shielding for controlling the path of the currents. The proper termination must also be adequate for the life of the equipment (i.e., must not deteriorate due to corrosion). Good electrical and mechanical connections must be established at both ends of the cable between the continuous corrugated aluminum shield and the ground. The connection should provide low impedance for the high-frequency shield currents. The ground leads for the armor and cable should be properly terminated in the motor junction box. A machine tapered insert with internal threads should be used to terminate the armor. The contact between the shield and connector will be 360° (Figs. 10.17 and 10.18). The RC network from the center point of the inverter capacitor bank to ground provides good ac return for the cable shield currents and assists in limiting overcurrent transients during ground faults.

MOTOR BEARING CURRENTS

These currents flow through the motor bearings due to the transient switching voltages. Figure 10.19 illustrates the capacitances from all points on the stator windings to all points on the rotor. The total current flowing in the bearing is the sum of the current flowing through the capacitances due to the voltage transients ($I = C\ dV/dt$) and the current generated due to the common-mode voltage (Fig. 10.20). The latter current is generated due to the nonsymmetric switching pattern in the device. It is usually much larger than the first. The total current flowing through the bearing can be significant. It causes pitting on the running surfaces.

Figure 10.20 illustrates the equivalent circuit of the capacitance shown in Fig. 10.19. And R_b, C_b, and Z_b are the components of the bearing model. The common-mode voltage that is generated by the nonsymmetry in the switching pattern within the inverter is called the *zero sequence source*.

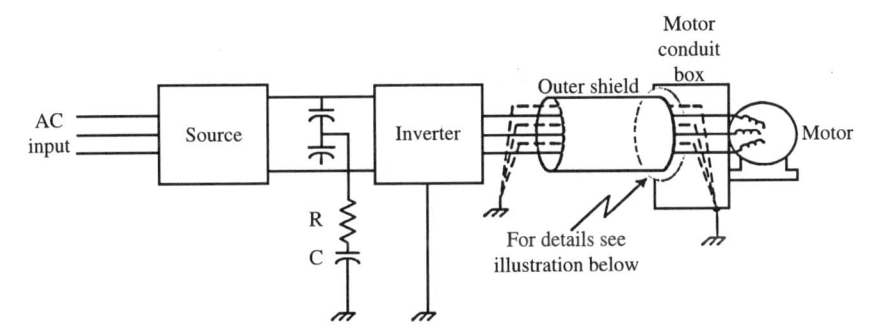

FIGURE 10.17 Inverter motor and cable grounding procedures.

10.18 CHAPTER TEN

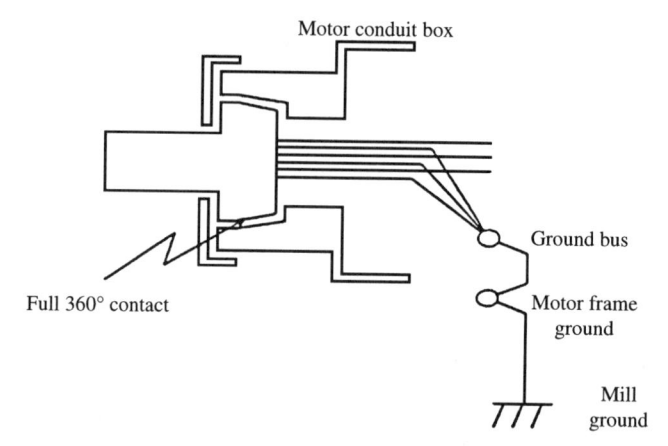

FIGURE 10.18 Details of the motor conduit box to cable interface.

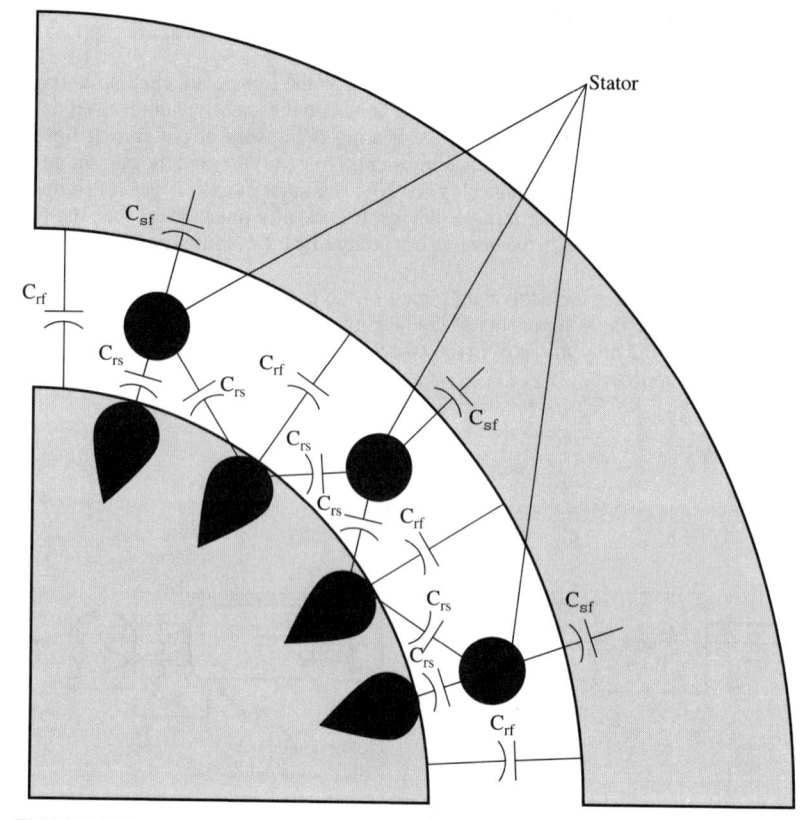

FIGURE 10.19 Motor capacitances from stator to rotor.

VARIABLE-SPEED DRIVES

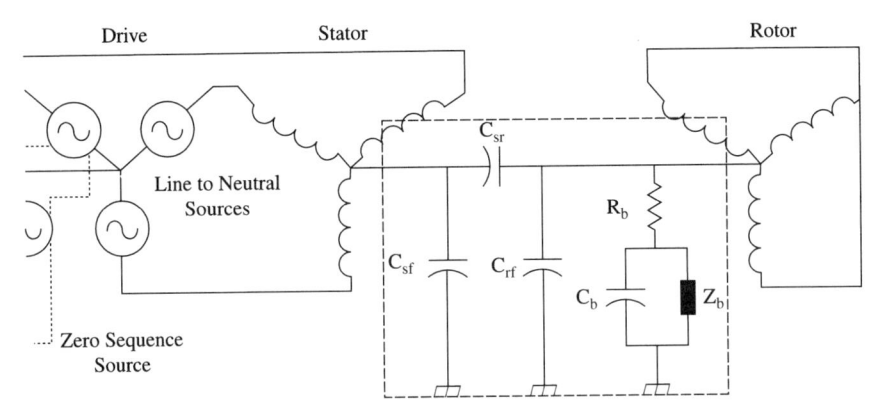

FIGURE 10.20 Motor and bearing equivalent circuit.

The bearing makes relatively good contact with the races during low-speed operation. Therefore, it is able to conduct substantial currents without incurring damage. At higher speeds, the bearing starts to operate on a thin film of oil (called hydroplaning). It is only a few micrometers thick and acts as both an insulator and a capacitor (C_b in Fig. 10.20). When the shaft voltage becomes very large and the oil-film capacitor breaks down, a high *electrostatic discharge current* (ESD) flows from the bearing to the race through Z_b. The ESD will eventually destroy the bearings.

The bearing currents must be eliminated or bypassed around the bearings. Otherwise, the bearing life will be very short. If insulation was placed on the motor bearings (in an attempt to stop the bearing current flow), the current will return to ground through the bearings of the attached machine (e.g., pump, compressor, etc.). These bearings, which are normally larger and more expensive than the motor bearings, will be destroyed.

The following techniques can be used to mitigate the problem with bearing current:

Add a Motor Shaft Grounding Brush

A conductive grounding brush can be attached to the motor frame. It rubs on a machined area of the motor shaft. In theory, most of the current should be shunted around the bearing directly to the motor frame.

These are the problems encountered with the solution:

1. The brushes are mounted inside the motor housing near one of the motor bearings. This leads to the following maintenance problems:

   - The bearing grease gets on the surface of the brush. It renders the brush nonconductive and therefore ineffective.
   - Inspection also becomes very difficult without disassembling the motor. Most of these field installations have been proved to be ineffective due to poor contact between the brush and ground.

2. The shaft grounding brush has inherently a relatively high inductance due to the wire connection. This inductance limits the amount of current that will be shunted to ground.

These grounding brushes have proved to be ineffective for this application.

Reduce the Stator to Rotor Capacitance Value

The bearing currents can be reduced by reducing the capacitive coupling from the stator windings to the rotor (C_{sr}) or the voltage switching transients. An electrostatic shield can be added between the stator and rotor. This is the best long-term solution, especially when it is combined with the inverter filter and a symmetric motor cable. The best solution is to use a large inverter filter that prevents harmonics from damaging the windings of the motor.

Use Conductive Grease in Motor Bearings

This solution has been found to be not very practical. This is so because the conductive particles in the grease cause high mechanical wear, resulting in a significant reduction in bearing life.

Motor Cable Wiring Practices

The bearing problem can be reduced by using the symmetric wiring described earlier because the common-mode voltage will be minimized.

The bearing currents will be reduced significantly if an inverter filter of the type shown in Fig. 10.14 is used because it results in drastic reduction in the voltage transients (dv/dt). It is recommended that a large filter be used because it can eliminate all the switching transients. The operation of the inverter will not result in voltage reflections, insulation degradation, or EMI problems. This is the best solution for a PWM inverter driving an older motor whose insulation may be suspect.

SUMMARY OF APPLICATION RULES FOR AC DRIVES

1. Unbalanced ac input voltages to a diode input source cause a current unbalance up to 20 times the voltage unbalance.
2. The power factor of a diode source bridge without a dc link reactor can be as low as 50 to 60 percent.
3. Older transistors and GTOs are much slower (50 to 20 μs) than IGBTs (100 to 150 ns).
4. *a.* Fast transient voltages can generate reflections that increase the voltage stress of the insulation system.
   b. The voltage reflections increase with the length of the motor cable and the switching speed.
5. *a.* The reflections can be stopped by a motor terminal (matching) filter. However, this filter retains the switching speed.
   b. The switching speeds are reduced by an inverter filter. *This is the preferred solution.*
   c. The motor and cable insulation voltage should be increased if neither filter is used.
6. EMI radiation is generated from the motor and ac line cables of all PWM inverters, especially the ones having IGBTs.

7. *a.* The motor connecting cables (including ground conductors) for PWM inverters having IGBTs must be symmetric.
   b. The best cable for these applications has continuous aluminum sheath, three conductors, and three grounds.
   c. The bearing currents and radiated EMI are reduced by the cable symmetry and continuous outer sheath.
8. The damage to the motor bearings is reduced by an inverter output filter or an electrostatic shield.

SELECTION CRITERIA OF VSDs

The following are the advantages of VSDs:

Variable Process Speed

Some processes operate at different speeds. For example, an ore grinding will operate at different speeds depending on the type, consistency, and size of the ore. VSDs are most suitable for these applications. Other applications requiring variable speed are conveyors and dynamometers.

Compressors and Pumps

Many applications require pumps and compressors to operate at part load. Conventional constant-speed motors and control valves have been used for these applications. This alternative is more expensive and causes flow-induced vibrations, cavitation, and erosion in the system. The motor operates at full load continuously regardless of the flow required. For 100-kW motor and 5 percent per kilowatthour utility rate, the operating cost of the motor will be $125.00 per day.

Typical pump characteristics (Fig. 10.21) confirm that the flow is proportional to the speed, while the torque varies with the square of the speed and the power with the cube of the speed. If the system requires 50 percent of the maximum flow rate, $113.00 will be wasted every day.

Motor Starting

Starting large motors can be very stressful on the windings and insulation systems due to high currents and torques. The power supply is also affected adversely due to voltage dips. A much smaller VSD (rated at a few percent of the motor) can be used to increase the speed of the motor gently, so it can be synchronized.

REGENERATION

During normal drive operation, the power is supplied to the VSD, then to the motor, and then to the load. Most of the energy is dissipated in the process, and the remainder is stored in load. A rotating load can be slowed down and energy removed from it and returned to

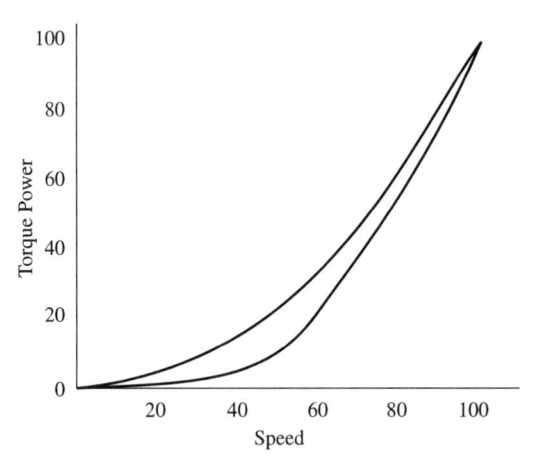

FIGURE 10.21 Typical pump power and torque versus speed.

the ac power line (regeneration). This is done by lowering the synchronous speed below the actual mechanical speed of the load (magnetic break). The amount of magnetic braking and power generation is proportional to the difference between the synchronous speed and actual mechanical speed. In the extreme case, the motor can be "plugged" (stopped almost instantly) by reversing the leads to any two phases; e.g., the synchronous speed of a two-pole motor is switched from 3600 to -3600 rpm instantly. These are the power flow diagrams during normal operation and regeneration:

Power flow in normal operation:

$$\text{AC utility power} \rightarrow \text{(VSD)} \rightarrow \text{motor} \rightarrow \text{load}$$

Power flow during regeneration:

$$\text{AC utility power} \leftarrow \text{(VSD)} \leftarrow \text{motor} \leftarrow \text{load}$$

If the cost of energy can be tolerated, the machine can coast down, dissipating its energy to the atmosphere. Mechanical brakes can be used to slow down the machine by converting the mechanical energy to heat. However, if the machine must be slowed down quickly or the cost of lost energy is high (e.g., frequent slowdown of large inertia), regeneration is required. These are some examples:

Dynamometer

These devices are used to test engines and transmissions. The dynamometers are used to load up the engine or transmission to determine its performance. Since regeneration is used to feed the mechanical energy in the load back to the power supply, the only power consumed is the losses in the system. In the past, a mechanical brake (water-cooled) was used for this function. The cost of lost energy in this case would be significant for dynamometers rated more than a few hundred horsepower.

Paper Machine Winder

This application has a very high load inertia (a few hundred times larger than the rotor inertia). The load must be started and stopped frequently as new rolls of paper are made. Brakes are needed to prevent the reels from coasting for a long time. In the past, water-cooled mechanical brakes were used to slow down the winder. Tremendous energy is lost in this process. The VSD can regenerate this energy back to the ac line. The only losses in this process are the ones that occur in the drive itself (2 to 3 percent of its rated output).

Dynamic Braking

In a dynamic brake, the mechanical energy in the load is dissipated in a water- or air-cooled resistor bank as shown:

Dynamic braking power flow:

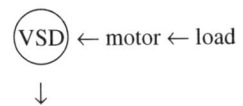

$$\text{(VSD)} \leftarrow \text{motor} \leftarrow \text{load}$$
$$\downarrow$$
$$\text{Dynamic braking resistor}$$

The resistor bank is installed across the dc bus inside the VSD. It is switched in when the link voltage rises above a preset limit. It remains on until the voltage drops below the preset limit. This braking method is used when the amount of energy that can be regenerated back to the ac line is not very large or the retrieval of energy is done very infrequently. This option costs about 10 percent of regeneration.

MAINTENANCE

All the components of the VSD have definite lifetimes that decrease significantly when their operating temperatures increase. These are some typical component lifetimes for a drive operating well within its design rating and ambient temperature between 0 and 40°C:

Electrolytic capacitors	5–7 years
Fans	3–7 years
IGBTs (assuming *no* overloads)	10–12 years
Power supplies	5–7 years
Motor bearings	2–7 years

The lifetime of all VSD components as well as bearings and insulation system drops significantly when the operating temperature increases. It is essential to maintain the ambient temperature within the acceptable range to achieve high reliability and longevity of this equipment.

COMMON FAILURE MODES

These are the most common failure modes of VSDs (most common listed first):

1. Motor bearings
2. Motor insulation
3. Drive electrolytic capacitors
4. AC line transients that damage the IGBTs and diodes

MOTOR APPLICATION GUIDELINES

Never use a "standard" induction motor with a VSD. All motors used with VSDs must be *inverter-rated.* They must be designed and manufactured for use with a VSD. These are the modifications from the standard induction motor design:

1. The quality and voltage capability of the winding insulation must be higher than that of a standard motor.
2. Thinner and higher-quality steel laminations are needed to reduce eddy current losses.
3. Improved motor cooling generates more heat due to nonsinusoidal PWM waveform.
4. Better-quality bearings are needed with shaft grounding brush or an electrostatic shield to reduce bearing current and increase bearing life. Consideration should be given to the motor cooling at low speed. Since the fan is mounted on the shaft, the motor cooling will be reduced at low speed. If the VSD is used with a pump or a compressor in which the load drops significantly with speed, overheating will not be a problem. However, if the VSD is used with a constant-torque load, a separate blower is required for cooling the motor. This will ensure that motor cooling is provided at low speed.

REFERENCE

1. A.C. Stevenson, *Power Converter Application Handbook*, Institute of Electrical and Electronics Engineers, New York, 1999.

CHAPTER 11
SYNCHRONOUS MACHINES

A synchronous machine operates at a fixed speed determined by the frequency of the power supply connected to it. The normal operating speed of the machine is known as the *synchronous speed*, given by

$$n_s = 120 \frac{f}{P} \quad \text{r/min}$$

where f = frequency of applied signal, Hz, and P = number of poles of the synchronous machine. In a synchronous machine, the relationship between the operating speed and the frequency of the power supply connected to it remains unchanged.

Synchronous machines have a wide range of output power applications. On the low end, the clock and timing motors and control alternators are in the milliwatt range. At the high end, the large alternators used in electric power generation have exceeded 1500-MW output power.

Synchronous machines are called alternators when used as generators. The smallest alternator is probably an ac tachometer. It is used as a speed sensor.

Synchronous motors have a wide range of applications. They are used in clocks and recording devices. The power rating of these motors varies from a few hundred watts to more than 100 MW. They operate at constant speed in a variety of applications including compressors, pumps, and drives for textile mills.

PHYSICAL DESCRIPTION

A synchronous machine has two electrical windings. They both provide excitation for the machine. The *armature winding* is where the main voltage is induced. The effect (excitation) that the current flowing in the armature winding has on the voltage developed in it is called the *armature reaction*. The *field winding* is the second winding in the machine. It enhances the magnetic excitation of the machine. In some machines, the field winding is replaced by permanent magnets. This configuration is called a permanent magnet synchronous machine.

The armature winding of a synchronous machine can be installed on the rotor or the stator. Most large synchronous machines have the armature winding on the stator and the field winding on the rotor.

Synchronous machines are classified according to the design of the rotor. There are salient-pole and nonsalient smooth rotors (also known as cylindrical rotors). In a salient-pole rotor, the windings are wound around an even number of poles that protrude out of the rotor. These poles, known also as projections, are made of magnetic materials called saliencies. In a cylindrical rotor, the poles do not protrude out of the rotor. Figures 11.1 and 11.2 illustrate a salient-pole and a cylindrical rotor, respectively.

11.2 CHAPTER ELEVEN

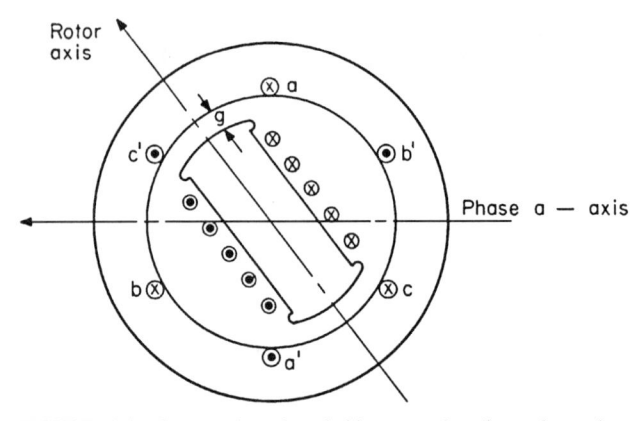

FIGURE 11.1 Cross section of a primitive two-pole salient-pole synchronous machine.

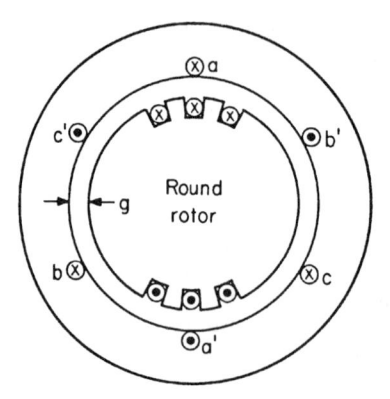

FIGURE 11.2 Cross section of a primitive two-pole cylindrical rotor synchronous machine.

POLE PITCH: ELECTRICAL DEGREES

The *pole pitch* λ is defined as

$$\lambda = \frac{360}{P}$$

where P is the number of poles. The pole pitch is an arc, given in degrees or radians. For example, if $P = 2$, then $\lambda = 180°$.

AIR GAP AND MAGNETIC CIRCUIT OF A SYNCHRONOUS MACHINE

Air Gap. The *radial distance* between the rotor and the stator is called the *air gap*. It is represented in Figs. 11.1 and 11.2 by the symbol g.

SYNCHRONOUS MACHINES   11.3

Stator. The stator is made of magnetic laminations stacked axially. Insulating material separates the laminations. It is made from either varnish applied in a liquid form or oxides formed in the heat-treating process. The laminations are held together by any of the following methods:

* Bolts
* Welding the outer circumference of the laminations
* Pressure bonding using the insulating material as the bonding agent

Radial bolts are used to fasten the stator stack to the housing. Figure 11.3 illustrates a typical stator lamination. The three sections in it are the slots, the teeth, and the stator yoke, or *back iron*. Figure 11.4 illustrates the three common geometries of stator slots in large synchronous machines.

The laminations are normally made of 3.5 percent silicon steel or carbon steel, such as ASA 1020. The thickness of a typical lamination in a large synchronous machine operating at 60 Hz is 0.37 to 0.635 mm (0.014 to 0.025 in). The stacking factor for large synchronous machines is 0.92 to 0.98.

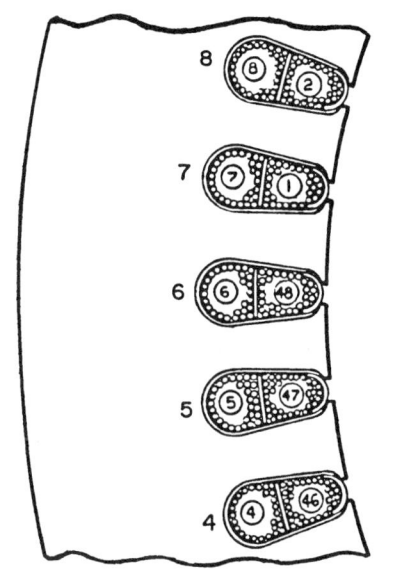

FIGURE 11.3 Portion of a typical synchronous machine stator lamination.

Rotor. The salient-pole rotor described above is used commonly in synchronous motors and slow-speed generators driven by hydraulic turbines. The cylindrical rotor is commonly used in high-speed generators driven by steam or gas turbines. The rotor of a synchronous machine has a damper, or amortisseur, winding in addition to the field winding. This winding is identical to the one in a squirrel-cage induction machine. It dampens the mechanical oscillations of the rotor in the synchronous machine by supplying a positive or negative induction torque. In some applications, it is used to start a synchronous machine as an induction motor.

The damper winding is made of copper or aluminum bars. The bars are shorted together electrically at each end by a shorting ring around their outer circumference. The shorting rings are made of the same material as the bars. In some cases, a lamination made of the same material as the bars and matching the steel laminations is used to short the bars. The bars are brazed to the end ring or end lamination to achieve good electrical connection.

The field winding is wound around a magnetic section of constant cross-sectional area under the pole face. In some designs, the field winding is preformed and installed over the inner pole section before mounting the pole-face section.

Rotors of relatively smaller machines are made of magnetic laminations. Large rotors are normally made of a single forging. This allows them to withstand the significant mechanical, electrical, and thermal stresses experienced during normal operation. The rotor laminations are made of the same material as the ones used in the stator. They are either silicon steel or carbon steel.

The slip rings are mounted at one end of the rotor shaft. The field winding is connected to the slip rings through radial bolts and up-shaft leads (known commonly as D bolts). The slip rings, radial bolts, and D bolts are insulated from the shaft. The external power supply is connected to the slip rings through copper-graphite or liquid-metal brushes.

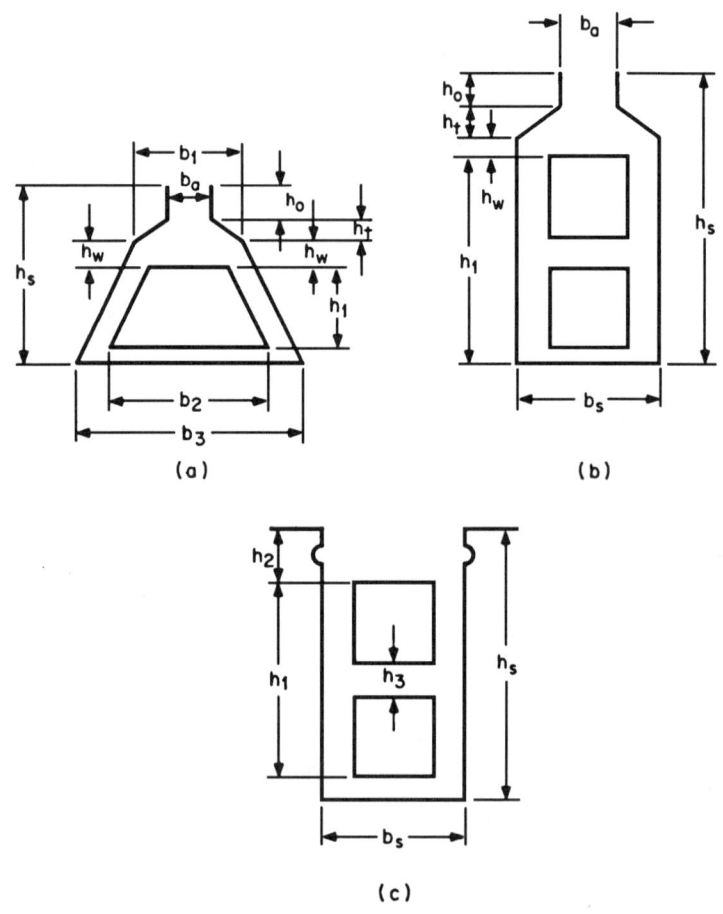

FIGURE 11.4 Slot geometries: (*a*) semiclosed slot with one-layer winding; (*b*) semiclosed slot with two-layer winding; (*c*) open slot.

SYNCHRONOUS MACHINE WINDINGS

The three types of windings used in synchronous machines are distributed, solenoid, and damper windings.

Distributed Windings. This type of winding is made of bundles of wire or insulated wire. It is inserted in slots around the rotor or stator air gap surface. In large machines, this winding is made of rectangular copper or aluminum bars. These bars are made from many elemental units (known as coils) insulated from each other with a cloth of nylon or Mylar. They are also coated with an insulated varnish and baked or thermally cured to form a rigid unit. The bars are inserted into slots similar to the ones shown in Fig. 11.4*c*. These bars are known as *preformed* coils.

In smaller-rated machines, these conductors are made of insulated wire known as magnet wire. The bundles of wire are inserted into slots in the core. They are coated with insulating varnish and baked or thermally cured for rigidity and high insulation resistance. These coils are known as random-wound, or mesh, coils.

Distributed windings are used as armature windings and as field windings in cylindrical machines.

Solenoid Windings. This type of winding is made of a multilayer of conductors separated by insulating strips. Hard electrical insulation separates the pole from this winding. This type of winding is used in salient-pole and dc machine fields, electromagnets, and power relays.

Damper Windings. The damper windings are installed at the outer surface of the rotor in a similar arrangement to the squirrel-cage bars in an induction machine.

FIELD EXCITATION

The field windings of a synchronous machine are supplied from a dc source. It originates from batteries, solar converters, dc generators, or a rectified ac source. This is called the *excitation* of the machine. Permanent magnets are also used to excite synchronous machines. The excitation current controls the terminal voltage, power factor, short-circuit current, torque, and transient response of the machine.

A dc or ac exciter is used for large synchronous machines. The exciter is normally mounted on the same shaft as the synchronous machine. The ac exciter is normally a conventional synchronous machine similar to the main machine. The three-phase ac output of the exciter is rectified and fed to the field winding of the main machine. Slip rings and brushes are also used to supply the dc field directly to synchronous machines. The excitation is normally controlled by varying the dc field by using a variable resistor.

Most excitation currents (except the ones supplied from batteries) contain ripple. This is highly undesirable because it increases harmonics* in the power output from the main machine. Thus, a filter is required to reduce the ripple in the excitation system of the machine to an acceptable value.

Rotating Rectifier Excitation

Small synchronous machines use "brushless" excitation known as *rotating rectifier* instead of slip rings and brushes. The reason for this is that synchronous machines using slip rings and brushes require high maintenance that includes replacement of the carbon brushes on power.

The excitation system of a rotating rectifier machine uses an exciter mounted on the shaft as the main synchronous machine. The field of the exciter is on the stator, and the armature is on the rotor. The ac power output from the exciter armature is rectified and fed to the field of the main machine. The excitation of the main machine is controlled by controlling the field in the stator of the exciter.

The main disadvantage of the rotating rectifier machine is the long response time required to change the voltage at the output from the main machine. Since the variation of

---

*Harmonics are ripples that distort the ac voltage and current in the power output. They are normally signals that are superimposed on the main (60- or 50-Hz) signal. They normally have a frequency equal to a multiple of the main frequency. For example, if the main frequency is 60 Hz, the harmonic signal will have a frequency of 5 × 60 or 8 × 60 Hz. Harmonics are highly undesirable because they increase the heat losses and electromagnetic interference from the machine.

the field current to the exciter must affect the exciter output before it can vary the voltage at the terminals of the main machine, delays occur in the variations of the voltage from the main machine. A typical response time of a rotating exciter is around 0.5 to 1.0 s. The response time required to change the voltage of a machine using slip rings and brushes is around 0.2 s.

The rotating rectifier machines cannot be used in large modern power plants due to the requirement of short response time of the excitation system. This is necessary to be able to vary the voltage quickly to stabilize the power out of the machine.

Series Excitation

The excitation of a synchronous machine may also be derived from the output power of the main machine. This type of excitation is accomplished in conjunction with a separate excitation system that originates from a separate source, such as a battery bank.

The series excitation is appropriate for isolated synchronous alternators which are used to start motors requiring high inrush current. This type of excitation is rarely used with large synchronous machines.

NO-LOAD AND SHORT-CIRCUIT VALUES

The *no-load* or *open-circuit* voltage is generated in the armature windings of a synchronous machine when the armature terminals are open-circuited and the rotor is rotating at synchronous speed while the field winding is energized. The induced open-circuit voltages in the three phases are given by

$$e_a = -E_m \cos \omega t$$

$$e_b = -E_m \cos (\omega t - 120°)$$

$$e_c = -E_m \cos (\omega t + 120°)$$

where $E_m = K\omega\phi$
K = constant that depends on generator size and design
ω = rotational speed of rotor
ϕ = flux created by field current

Figure 11.5 illustrates a typical open-circuit saturation curve of a synchronous machine. It represents the no-load characteristics of the machine. The air gap line represents the extension of the straight-line portion of the saturation curve. Figure 11.5 also illustrates a typical short-circuit saturation characteristic of a synchronous machine. It is obtained by shorting the terminals of the armature together while the rotor is rotating at synchronous speed and the field current is increased from zero to a small value. The short-circuit saturation curve is a plot of the variations of the current in the shorted armature windings versus the field current.

The zero power factor saturation characteristic is obtained by overexciting the machine while it is connected to a highly inductive load or an idle running synchronous motor. The terminal voltage of the machine being tested is varied while its armature current is held constant at the rated value by adjusting the excitation of the load and the machine being tested. This characteristic is important for the analysis of a synchronous motor.

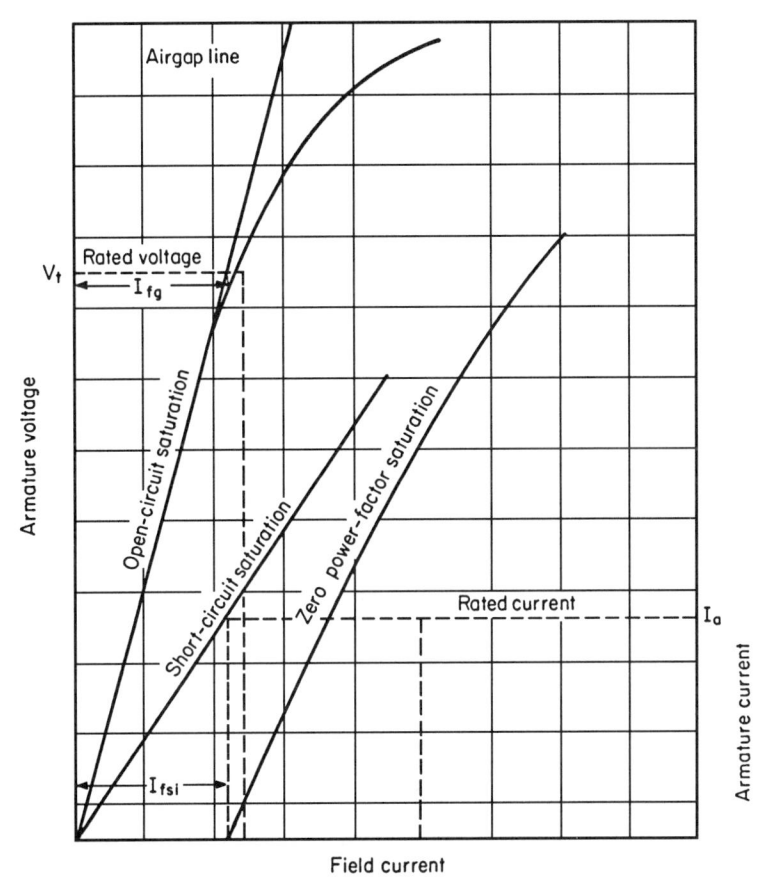

FIGURE 11.5 Open-circuit and short-circuit characteristics of a synchronous machine.

TORQUE TESTS

Some torque values are important for a synchronous machine. They include the following:

1. *Locked-rotor torque:* This is the torque developed when the rated voltage and frequency are applied to the motor while the rotor is prevented from turning (Fig. 11.6).
2. *Pull-out torque:* This is the maximum possible torque developed and sustained at synchronous speed for 1 min when the rated voltage and frequency are applied to the motor.
3. *Pull-in torque:* This is the maximum constant torque that the motor will develop as it approaches the synchronous speed when the voltage and frequency are applied.
4. *Pull-up torque:* This is the minimum torque developed during start-up at speeds below the synchronous speed when the rated voltage and frequency are applied to the motor.
5. *Speed-torque characteristic:* This is the variation of the torque developed by the motor versus the speed when the excitation is not applied to the motor and the rated voltage and frequency are applied to it.

11.8 CHAPTER ELEVEN

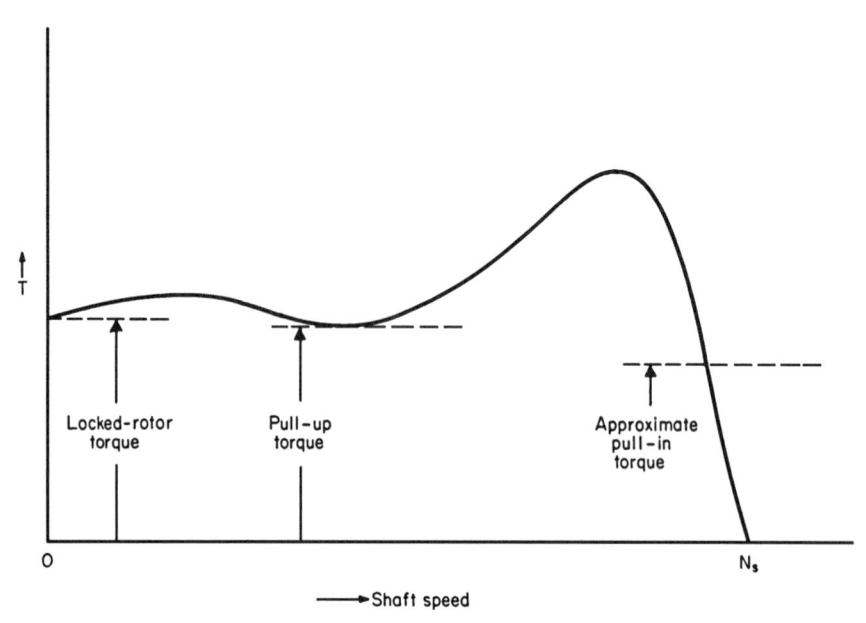

FIGURE 11.6 Speed-torque characteristic of a synchronous machine.

Speed-Torque Characteristic

The tests done to determine the speed-torque characteristic of a synchronous motor are identical to those of an induction motor. However, there are some differences between a synchronous motor and an induction motor. For example, the damper winding of a synchronous motor has a much higher resistance than a typical squirrel-cage winding used in an induction motor. The speed-torque characteristic of a synchronous motor depends on the material of the rotor bars and the condition of the field windings, open or closed.

Figure 11.7 illustrates the difference in speed-torque characteristic of a synchronous motor between an open and a shorted field circuit. It also shows the effect of the rotor bar material on the speed-torque characteristic of the motor. Most synchronous machines are designed to start with a shorted (or closed) field.

Pull-In Torque

The *pull-in torque* is the torque available to pull the rotor into synchronous speed. This characteristic is limited to synchronous machines only. It is difficult to calculate or measure the value of this parameter. It is expressed generally by the *nominal pull-in torque*, which is the torque developed by the motor at 95 percent of the synchronous speed (Fig. 11.6).

Pull-Out Torque

The *pull-out torque* is the maximum torque developed by a synchronous machine. It is also a characteristic limited to synchronous machines only. A synchronous machine

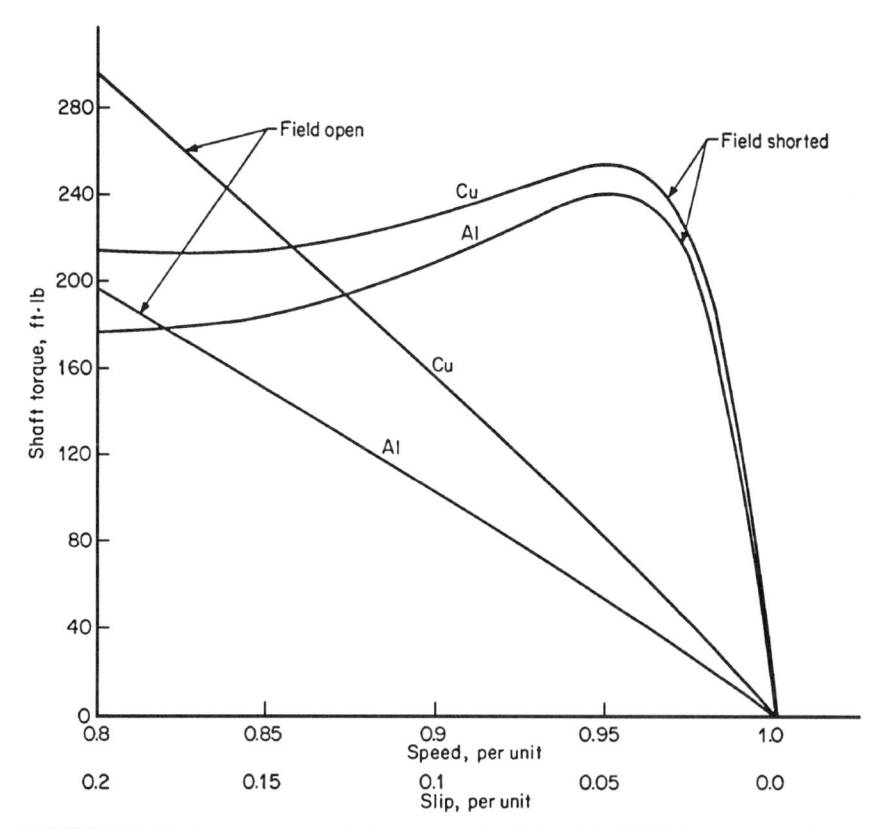

FIGURE 11.7 Synchronous motor speed-torque curves illustrating effect of field winding on motor torque. Effect of rotor bar material is also shown.

(motor or generator) will pull out of synchronous speed when the torque applied exceeds the pull-out torque.

EXCITATION OF A SYNCHRONOUS MACHINE

The excitation characteristics of a synchronous machine do not vary whether it is operated as a motor or as a generator. Figure 11.8 illustrates the excitation characteristics of a synchronous motor (the variations of armature current versus the field voltage) assuming a constant armature voltage. The power factor associated with a synchronous motor is determined based on "looking into the motor" from the terminals. In the underexcited region (i.e., to the left of the minima of the V curves in Fig. 11.8), the motor will be an inductive load (a lagging power factor load). In the overexcited region (i.e., to the right of the minima of the V curves), the motor will be a capacitive load (i.e., a leading power factor load). Thus, a synchronous motor can be used to deliver reactive power while consuming real power. The power factor associated with the two regions of the V curves shown in Fig. 11.8 (to the left and right of the minima of the curves) would be reversed if the machine were a synchronous generator. Figure 11.9 illustrates that a synchronous machine can operate in all four quadrants of the real power–reactive power plane.

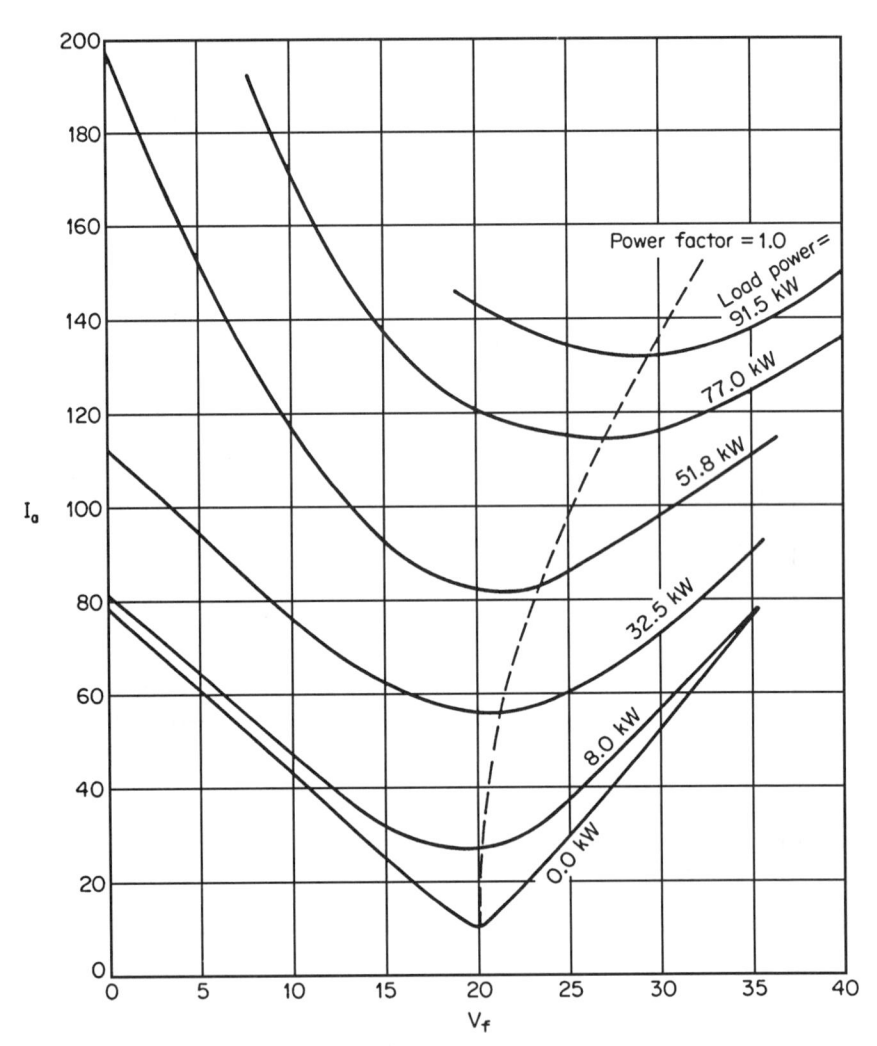

FIGURE 11.8 Synchronous motor V curves.

MACHINE LOSSES

Synchronous machine losses include the following:

Windage and Friction Loss

The friction that the gas in the air gap exerts on the rotor causes a mechanical loss known as *windage loss*. Since the efficiency of a generator has a significant impact on the cost of power generated, most units having a rating in excess of 20 MW are hydrogen-cooled to improve the efficiency. The following are the advantages of hydrogen:

SYNCHRONOUS MACHINES   11.11

- The density of hydrogen is 14 times lower than that of air. Thus, the windage losses of hydrogen-cooled generators are significantly lower than those of air-cooled ones.
- The heat-transfer properties of hydrogen are significantly better than those of air. This allows hydrogen-cooled generators to operate cooler than the ones cooled by air. Thus, hydrogen-cooled generators have many fewer problems than those cooled by air.
- Hydrogen is an inert gas. It does not support corrosion whereas air enhances the corrosion rate.
- Hydrogen extinguishes arc-initiated fires while oxygen is required to start them.

However, a large hydrogen leak from the generator can cause a fire or an explosion. The reason is that a mixture of hydrogen and oxygen containing hydrogen in a volumetric concentration of 4 to 76 percent can become explosive. Despite this hazard, most

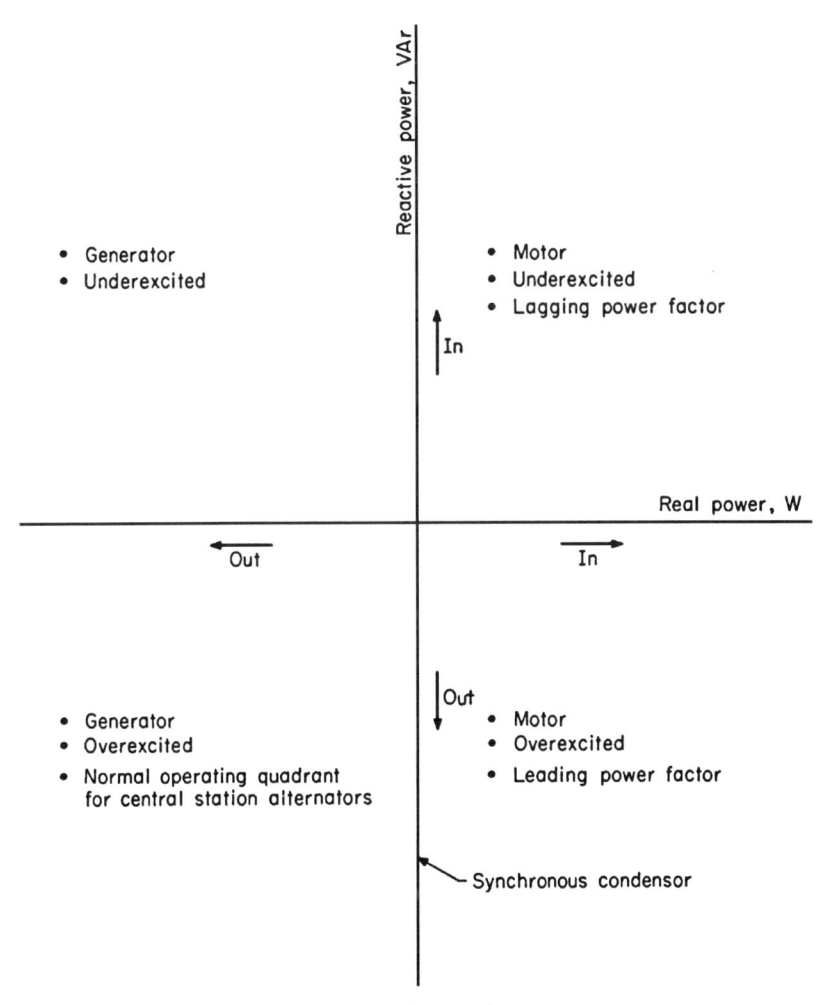

FIGURE 11.9 Four-quadrant synchronous machine operation.

hydrogen-cooled generators have operated safely. Only a few rare incidents have been reported. The hydrogen is normally maintained at 45 to 60 psi (300 to 400 kPa) inside the generator. It is purged using CO_2 to avoid explosion hazard.

Friction losses occur in the bearings and slip rings.

Core Losses

The core losses are the sum of hysteresis and eddy current losses. They can be reduced by decreasing the thickness of the laminations and using low-loss magnetic materials such as iron-nickel alloys, oriented silicon steel, or amorphous magnetic materials.

Stray-Load Loss

The stray-load loss is given by the increase in core losses with the current. It is caused by the induced losses from the leakage fluxes of the armature and variations of the flux distribution in the air gap. This loss includes the eddy current losses that occur in large armature conductors. This component of the stray-load loss is minimized by laminating the armature conductors using bundled conductors, strip conductors, etc. A typical value of this loss is around 1 percent of the power output.

Armature Conductor Loss

The *armature conductor loss* is defined as the sum of ohmic (or dc) loss and the *effective* (or ac) loss in the armature conductors. The effective loss is caused by the nonuniform flux distribution over the cross section of the conductor. This is known as the *skin effect*. It depends on the cross section of the conductor and the frequency of the armature current. The skin effect can increase the armature copper loss significantly if the conductor is large. The armature conductors are normally laminated or segmented to reduce this loss.

Excitation Loss

The excitation loss includes the loss of the field conductor and the automatic voltage regulator which controls the voltage at the terminals of the machine. The heat losses from the field conductor are normally included in the armature heat losses.

REFERENCE

1. A. Syed Nasar, *Handbook of Electric Machines*, McGraw-Hill, New York, 1987.

CHAPTER 12
SYNCHRONOUS GENERATORS

Synchronous generators or *alternators* are synchronous machines that convert mechanical energy to ac electric energy.

SYNCHRONOUS GENERATOR CONSTRUCTION

A direct current is applied to the rotor winding of a synchronous generator to produce the rotor magnetic field. A prime mover causes the generator rotor to rotate the magnetic field in the machine. A three-phase set of voltages is induced in the stator windings by the rotating magnetic field.

The rotor is a large electromagnet. Its magnetic poles can be salient (protruding or sticking out from the surface of the rotor), as shown in Fig. 12.1 or nonsalient (flush with the surface of the rotor), as shown in Fig. 12.2. Two- and four-pole rotors have normally nonsalient poles, while rotors with more than four poles have salient-pole rotors.

Small generator rotors are constructed of thin laminations to reduce eddy current losses, while large rotors are not constructed from laminations due to the high mechanical stresses encountered during operation.

The field circuit of the rotor is supplied by a direct current. The common methods used to supply the dc power are

1. By means of *slip rings* and *brushes*
2. By a special dc power source mounted directly on the shaft of the rotor

Slip rings are metal rings that encircle the rotor shaft but are insulated from it. Each of the two slip rings on the shaft is connected to one end of the dc rotor winding, and a number of brushes ride on each slip ring.

The positive end of the dc voltage source is connected to one slip ring, and the negative to the second. This ensures that the same dc voltage is applied to the field windings regardless of the angular position or speed of the rotor.

Slip rings and brushes require high maintenance because the brushes must be checked for wear regularly. Also, the voltage drop across the brushes can be the cause of large power losses when the field currents are high.

Despite these problems, all small generators use slip rings and brushes because all other methods used for supplying dc field power are more expensive.

Large generators use *brushless exciters* for supplying dc field power to the rotor. They consist of a small ac generator having its field circuit mounted on the stator and its armature circuit mounted on the rotor shaft.

12.2 CHAPTER TWELVE

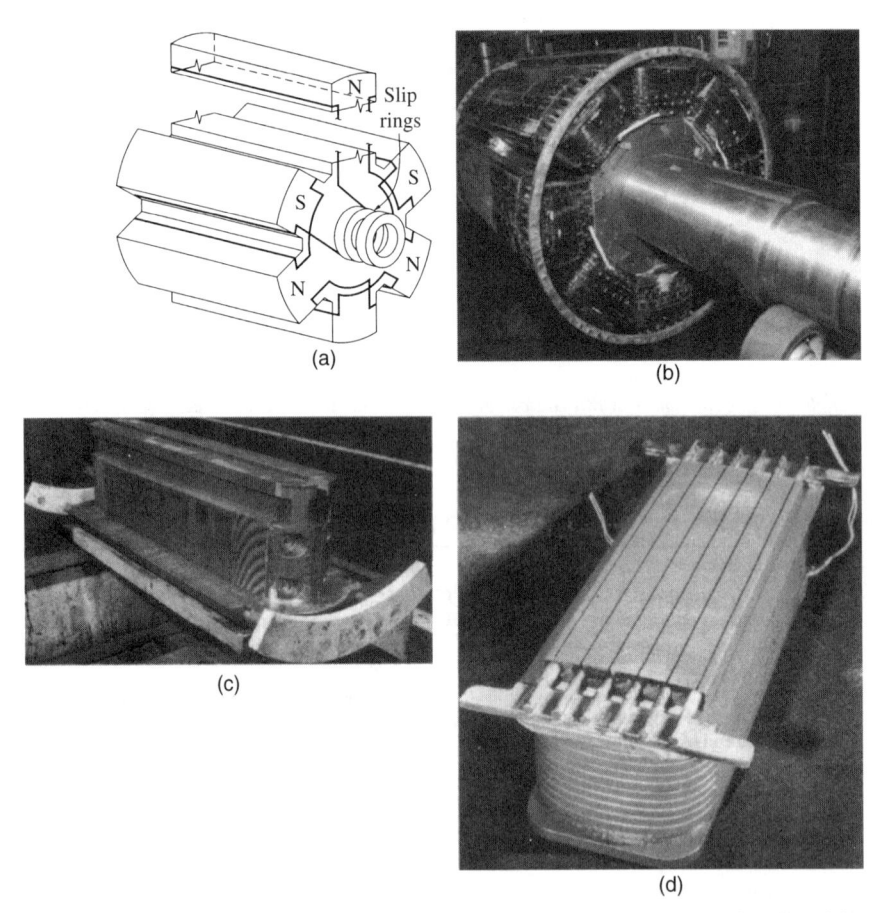

FIGURE 12.1 (*a*) A salient six-pole rotor for a synchronous machine. (*b*) Photograph of a salient eight-pole synchronous machine rotor showing the windings on the individual rotor poles. (*Courtesy of General Electric Company.*) (*c*) Photograph of a single salient pole from a rotor with the field windings not yet in place. (*Courtesy of General Electric Company.*) (*d*) A single salient pole shown after the field windings are installed but before it is mounted on the rotor. (*Courtesy of Westinghouse Electric Company.*)

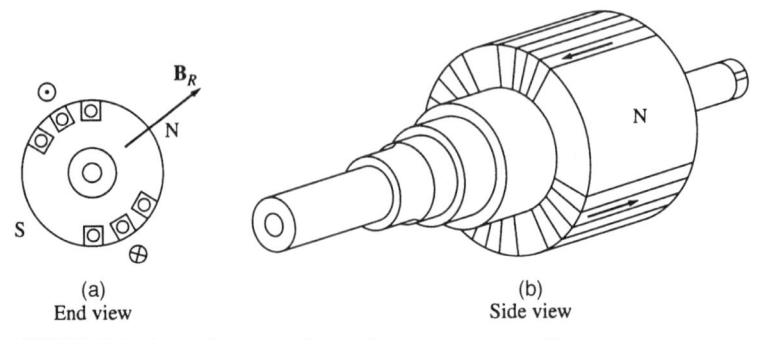

FIGURE 12.2 A nonsalient two-pole rotor for a synchronous machine.

The exciter generator output (three-phase ac) is converted to dc power by a three-phase rectifier circuit also mounted on the rotor. The dc power is fed to the main field circuit. The field current for the main generator can be controlled by the small dc field power of the exciter generator, which is located on the stator (Figs. 12.3 and 12.4). A brushless excitation system requires much less maintenance than slip rings and brushes because there is no mechanical contact between the rotor and the stator.

The generator excitation system can be made *completely* independent of any external power sources by using a small pilot exciter. It consists of a small ac generator with *permanent magnets* mounted on the rotor shaft and a three-phase winding on the stator.

The pilot exciter produces the power required by the field circuit of the exciter, which is used to control the field circuit of the main generator. When a pilot exciter is used, the generator can operate without any external electric power (Fig. 12.5).

Most synchronous generators that have brushless exciters also use slip rings and brushes as an auxiliary source of field dc power in emergencies. Figure 12.6 illustrates a cutaway of a complete large synchronous generator having a salient-pole rotor with eight poles, and a brushless exciter.

THE SPEED OF ROTATION OF A SYNCHRONOUS GENERATOR

The electrical frequency of synchronous generators is synchronized (locked in) with the mechanical rate of rotation. The rate of rotation of the magnetic fields (mechanical speed) is related to the stator electrical frequency by:

$$f_e = \frac{n_m P}{120}$$

where f_e = electrical frequency, Hz
n_m = mechanical speed of magnetic field, r/min (= speed of rotor for synchronous machines)
P = number of poles

For example, a two-pole generator rotor must rotate at 3600 r/min to generate electricity at 60 Hz.

THE INTERNAL GENERATED VOLTAGE OF A SYNCHRONOUS GENERATOR

The magnitude of the voltage induced in a given stator phase is given by

$$E_A = K\phi\omega$$

where K is a constant that depends on the generator construction, ϕ is the flux in the machine, and ω is the frequency or speed of rotation. Figure 12.7a illustrates the relationship between the flux in the machine and the field current I_F. Since the internal generated voltage E_A is directly proportional to the flux, the relationship between E_A and I_F is similar to the one between ϕ and I_F (Fig. 12.7b). The graph is known as the *magnetization curve* or *open-circuit characteristic* of the machine.

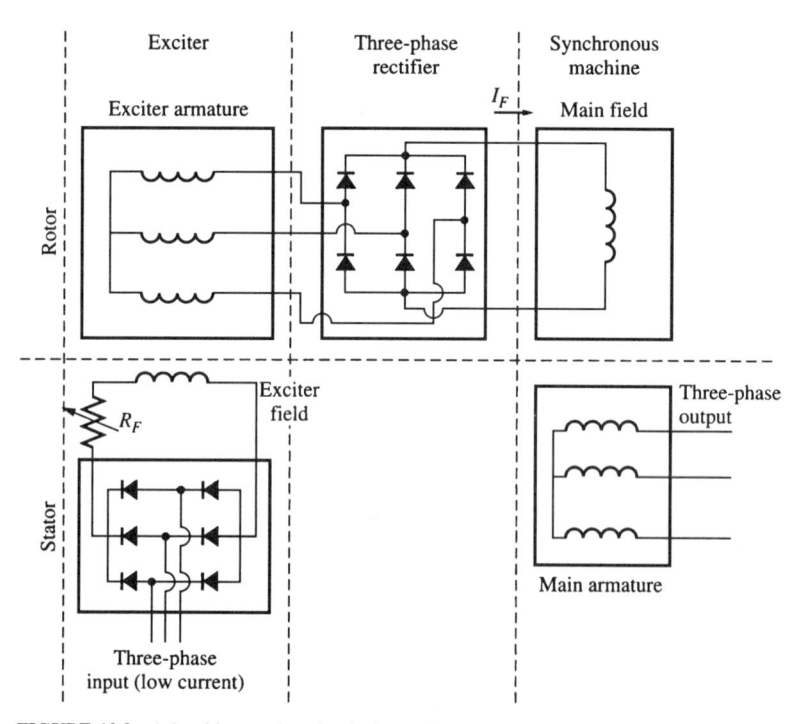

FIGURE 12.3 A brushless exciter circuit. A small three-phase current is rectified and used to supply the field circuit of the exciter, which is located on the stator. The output of the armature circuit of the exciter (on the rotor) is then rectified and used to supply the field current of the main machine.

FIGURE 12.4 Photograph of a synchronous machine rotor with a brushless exciter mounted on the same shaft. Notice the rectifying electronics visible next to the armature of the exciter.

SYNCHRONOUS GENERATORS 12.5

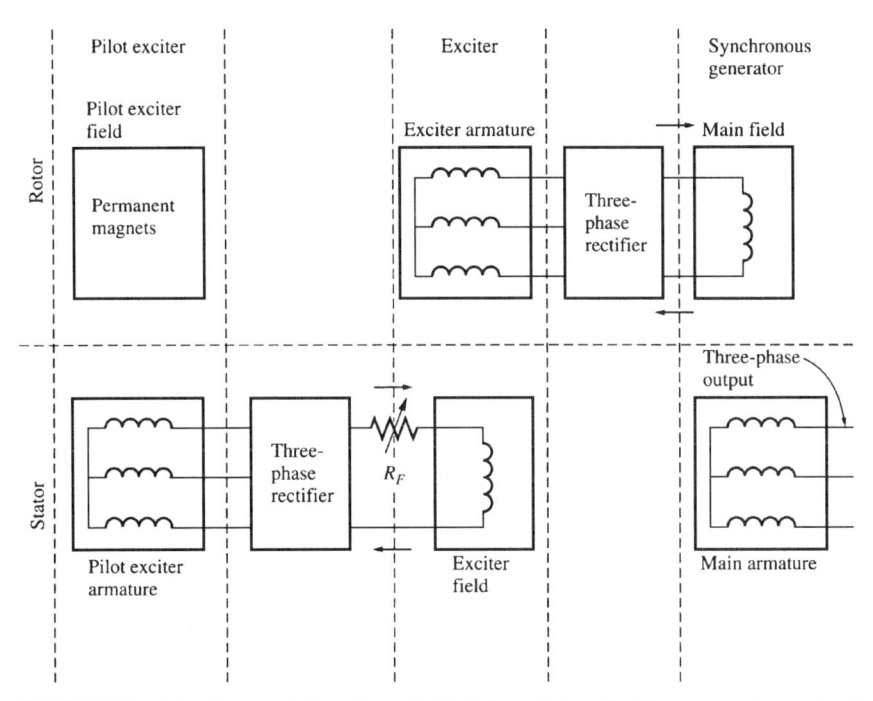

FIGURE 12.5 A brushless excitation scheme that includes a pilot exciter. The permanent magnets of the pilot exciter produce the field current of the exciter, which in turn produces the field current of the main machine.

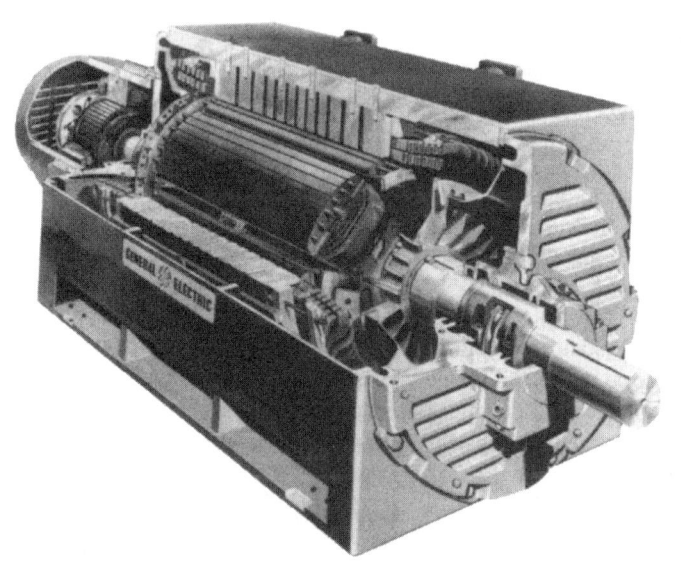

FIGURE 12.6 A cutaway diagram of a large synchronous machine. Note the salient-pole construction and the on-shaft exciter. (*Courtesy of General Electric Company.*)

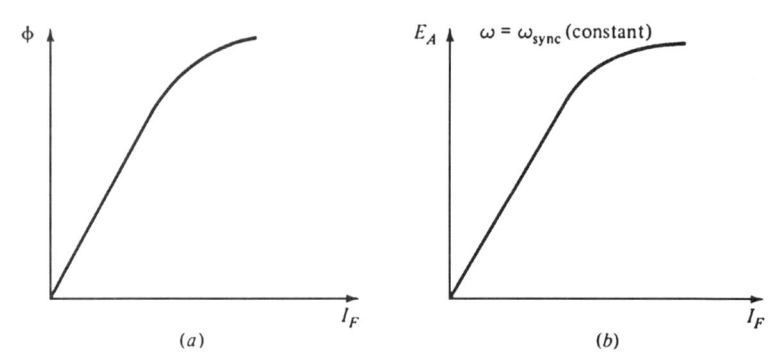

FIGURE 12.7 (a) Plot of flux versus field current for a synchronous generator. (b) The magnetization curve for the synchronous generator.

THE EQUIVALENT CIRCUIT OF A SYNCHRONOUS GENERATOR

Voltage $\mathbf{E}_A$ is the internal generated voltage induced in one phase of a synchronous generator. However, this is not the usual voltage that appears at the terminals of the generator. In reality, the internal voltage $\mathbf{E}_A$ is the same as the output voltage $\mathbf{V}_\phi$ of a phase only when there is no armature current flowing in the stator.

These factors cause the difference between $\mathbf{E}_A$ and $\mathbf{V}_\phi$:

1. The *armature reaction,* which is the distortion of the air gap magnetic field by the current flowing in the stator
2. The self-inductance of the armature (stator) windings
3. The resistance of the armature windings

The armature reaction has the largest impact on the difference between $\mathbf{E}_A$ and $\mathbf{V}_\phi$.

The voltage $\mathbf{E}_A$ is induced when the rotor is spinning. If the generator's terminals are attached to a load, a current flows.

The three-phase current flowing in the stator will produce its own magnetic field in the machine. This *stator* magnetic field distorts the magnetic field produced by the rotor, resulting in a change of the phase voltage. This effect is known as the *armature reaction* because the current in the armature (stator) affects the magnetic field which produced it in the first place.

Figure 12.8a illustrates a two-pole rotor spinning inside a three-phase stator when there is no load connected to the machine. An internal generated voltage $\mathbf{E}_A$ is produced by the rotor magnetic field $\mathbf{B}_R$ whose direction coincides with the peak value of $\mathbf{E}_A$. The voltage will be positive out of the top conductors and negative into the bottom conductors of the stator.

When the generator is not connected to a load, there is no current flow in the armature. The phase voltage $\mathbf{V}_\phi$ will be equal to $\mathbf{E}_A$. When the generator is connected to a lagging load, the peak current will occur at an angle behind the peak voltage (Fig. 12.8b).

The current flowing in the stator windings produces a magnetic field called $\mathbf{B}_S$ whose direction is given by the right-hand rule (Fig. 12.8c). A voltage is produced in the stator $\mathbf{E}_{\text{stat}}$ by the stator magnetic field $\mathbf{B}_S$. The total voltage in a phase is the sum of the internal voltage $\mathbf{E}_A$ and the armature reaction voltage $\mathbf{E}_{\text{stat}}$:

$$\mathbf{V}_\phi = \mathbf{E}_A + \mathbf{E}_{\text{stat}}$$

SYNCHRONOUS GENERATORS 12.7

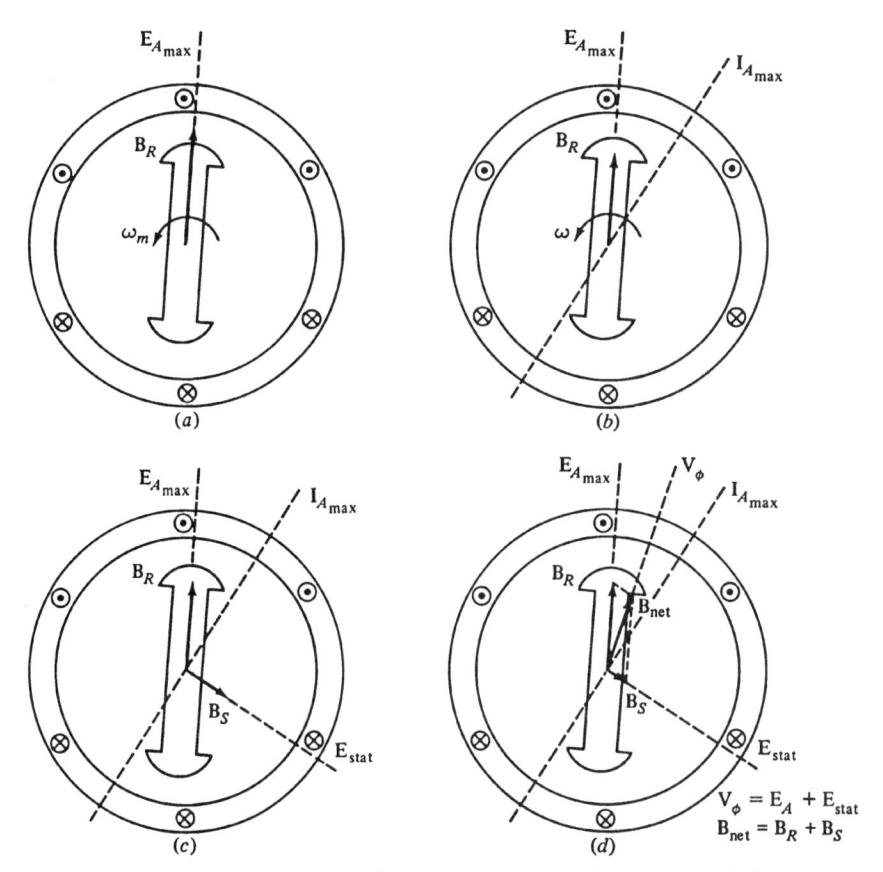

FIGURE 12.8 The development of a model for armature reaction: (*a*) A rotating magnetic field produces the internal generated voltage $\mathbf{E}_A$. (*b*) The resulting voltage produces a lagging current flow when connected to a lagging load. (*c*) The stator current produces its own magnetic field $\mathbf{B}_S$, which produces its own voltage $\mathbf{E}_{stat}$ in the stator windings of the machine. (*d*) The field $\mathbf{B}_S$ adds to $\mathbf{B}_R$, distorting it into $\mathbf{B}_{net}$. The voltage $\mathbf{E}_{stat}$ adds to $\mathbf{E}_A$, producing $\mathbf{V}_\phi$ at the output of the phase.

The net magnetic field $\mathbf{B}_{net}$ is the sum of the rotor and stator magnetic fields:

$$\mathbf{B}_{net} = \mathbf{B}_R + \mathbf{B}_S$$

The angle of the resulting magnetic field $\mathbf{B}_{net}$ coincides with the one of the net voltage $\mathbf{V}_\phi$ (Fig. 12.8*d*). The angle of voltage $\mathbf{E}_{stat}$ is 90° behind the one of the maximum current $\mathbf{I}_A$. Also, the voltage $\mathbf{E}_{stat}$ is directly proportional to $\mathbf{I}_A$. If X is the proportionality constant, the *armature reaction* voltage can be expressed as

$$\mathbf{E}_{stat} = -jX\mathbf{I}_A$$

The voltage of a phase is

$$\mathbf{V}_\phi = \mathbf{E}_A - jX\mathbf{I}_A$$

Figure 12.9 shows that the armature reaction voltage can be modeled as an inductor placed in series with the internal generated voltage.

When the effects of the stator windings self-inductance L_A (and its corresponding reactance X_A) and resistance R_A are added, the relationship becomes:

$$\mathbf{V}_\phi = \mathbf{E}_A - jX\mathbf{I}_A - jX_A\mathbf{I}_A - R_A I_A$$

When the effects of the armature reaction and self-inductance are combined (the reactances are added), the *synchronous reactance* of the generator is

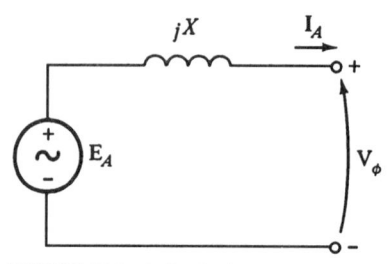

FIGURE 12.9 A simple circuit (see text).

$$X_S = X + X_A$$

The final equation becomes

$$\mathbf{V}_\phi = \mathbf{E}_A - jX_S\mathbf{I}_A - R_A I_A$$

Figure 12.10 illustrates the equivalent circuit of a three-phase synchronous generator. The rotor field circuit is supplied by dc power which is modeled by the coil's inductance

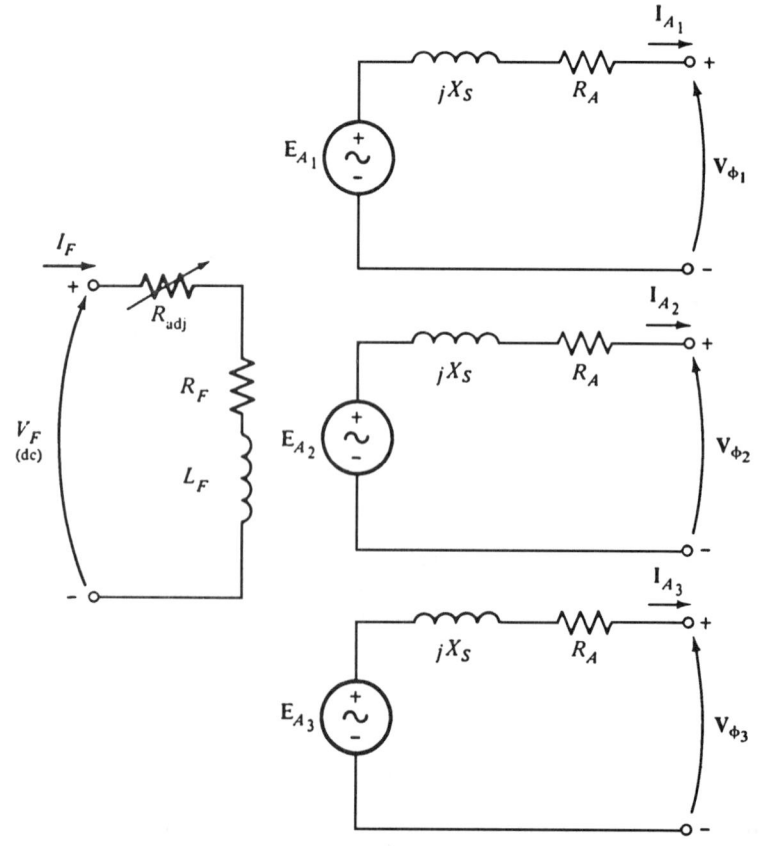

FIGURE 12.10 The full equivalent circuit of a three-phase synchronous generator.

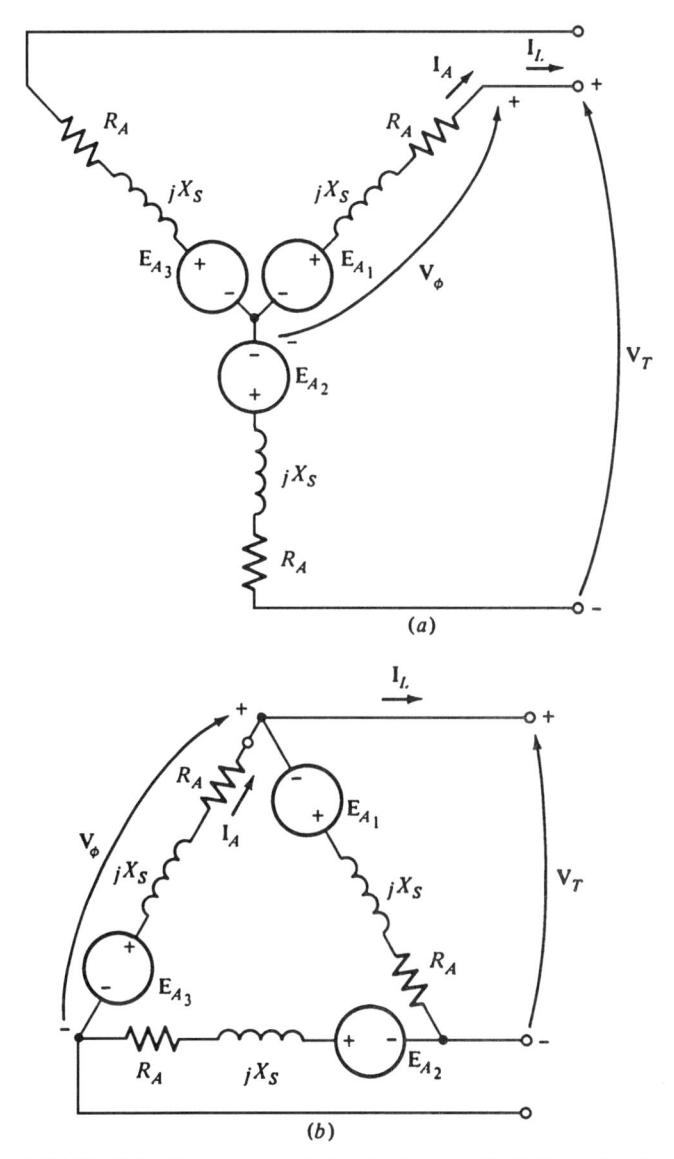

FIGURE 12.11 The generator equivalent circuit connected (*a*) in Y and (*b*) in Δ.

and resistance in series. The adjustable resistance R_{adj} controls the field current. The internal generated voltage for each of the phases is shown in series with the synchronous reactance X_S and the stator winding resistance R_A. The three phases are identical except that the voltages and currents are 120° apart in angle.

Figure 12.11 illustrates that the phases can be either Y- or Δ-connected. When they are Y-connected, the terminal voltage V_T is related to the phase voltage V_ϕ by

$$V_T = \sqrt{3}\, V_\phi$$

When they are Δ-connected, then

$$V_T = V_\phi$$

Since the three phases are identical except that their phase angles are different, the *per-phase equivalent circuit* is used (Fig. 12.12).

THE PHASOR DIAGRAM OF A SYNCHRONOUS GENERATOR

Phasors are used to describe the relationships between ac voltages. Figure 12.13 illustrates these relationships when the generator is supplying a purely resistive load (at unity power factor). The total voltage E_A differs from the terminal voltage V_ϕ by the resistive and inductive voltage drops. All voltages and currents are referenced to V_ϕ, which is assumed arbitrarily to be at angle $0°$.

Figure 12.14 illustrates the phasor diagrams of generators operating at lagging and leading power factors. Notice that for a given phase voltage and armature current, lagging loads require larger internal generated voltage E_A than leading loads. Therefore, a larger field current is required for lagging loads to get the same terminal voltage, because

$$E_A = K\phi\omega$$

where ω must remain constant to maintain constant frequency. Thus, for a given field current and magnitude of load current, the terminal voltage for lagging loads is lower than the one for leading loads.

In real synchronous generators, the winding resistance is much smaller than the synchronous reactance. Therefore, R_A is often neglected in qualitative studies of voltage variations.

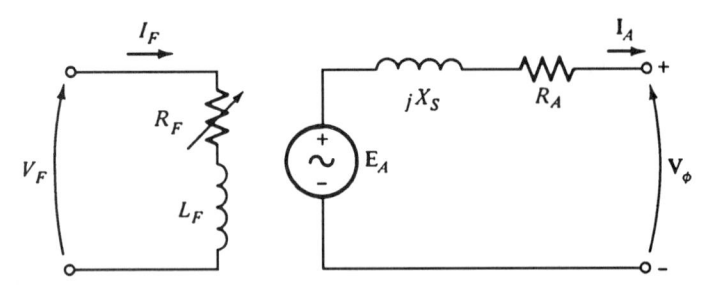

FIGURE 12.12 The per-phase equivalent circuit of a synchronous generator. The internal field circuit resistance and the external variable resistance have been combined into a single resistor R_F.

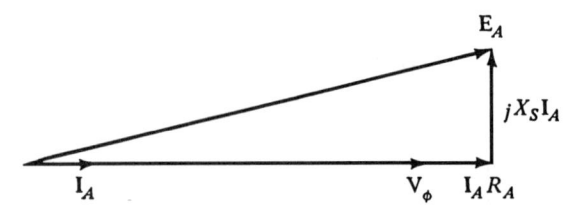

FIGURE 12.13 The phasor diagram of a synchronous generator at unity power factor.

SYNCHRONOUS GENERATORS   12.11

(a)

(b)

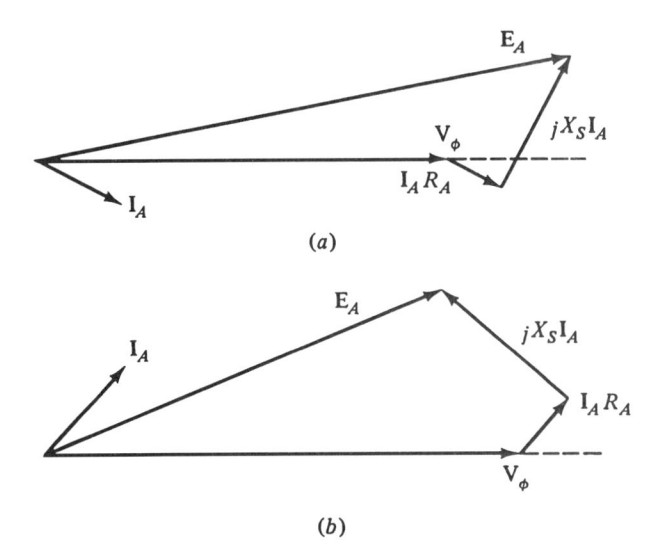

FIGURE 12.14 The phasor diagram of a synchronous generator at (a) lagging and (b) leading power factor.

POWER AND TORQUE IN SYNCHRONOUS GENERATORS

A synchronous generator is a machine that converts mechanical power to three-phase electric power. The mechanical power is given usually by a turbine. However, the rotational speed must remain constant to maintain a steady frequency.

Figure 12.15 illustrates the power flow in a synchronous generator. The input mechanical power is $P_{in} = \tau_{app}\omega_m$, while the power converted from mechanical to electric energy is

$$P_{conv} = \tau_{ind}\omega_m$$

$$P_{conv} = 3E_A I_A \cos \gamma$$

where γ is the angle between $\mathbf{E}_A$ and $\mathbf{I}_A$. The real electric output power of the machine is

$$P_{out} = \sqrt{3} V_T I_L \cos \theta$$

or in phase quantities as

$$P_{out} = 3V_\phi I_A \cos \theta$$

The reactive power is

$$Q_{out} = \sqrt{3} V_T I_L \sin \theta$$

or in phase quantities as

$$Q_{out} = 3V_\phi I_A \sin \theta$$

A very useful expression for the output power can be derived if the armature resistance R_A is ignored (since $X_S \geq R_A$).

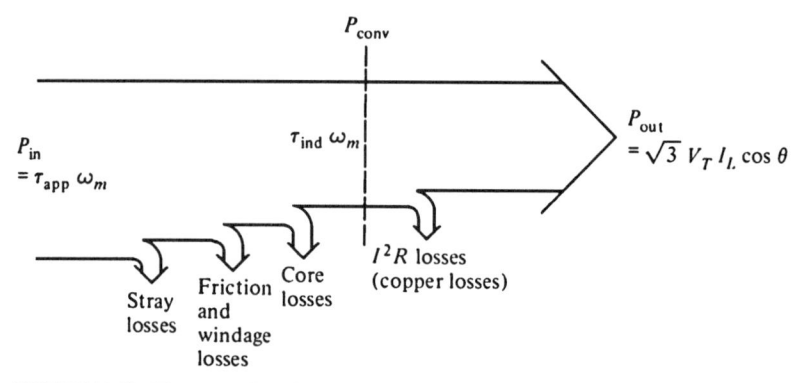

FIGURE 12.15 The power flow diagram of a synchronous generator.

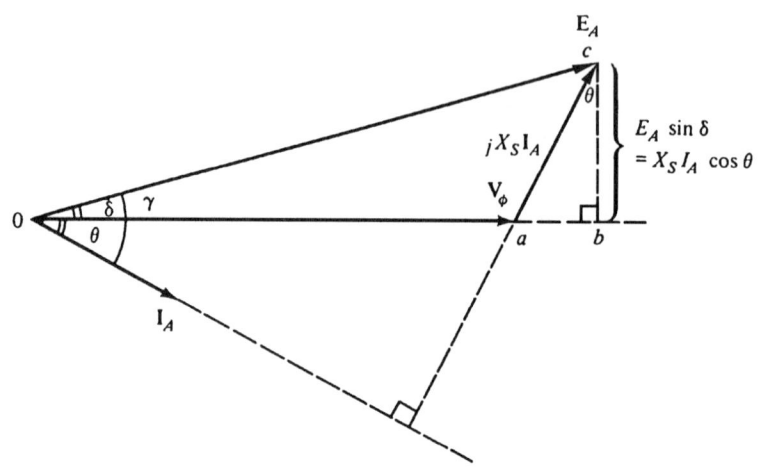

FIGURE 12.16 Simplified phasor diagram with armature resistance ignored.

Figure 12.16 illustrates a simplified phasor diagram of a synchronous generator when the stator resistance is ignored. The vertical segment bc can be expressed as either $E_A \sin \delta$ or $X_S I_A \cos \theta$. Therefore,

$$I_A \cos \theta = \frac{E_A \sin \delta}{X_S}$$

and substituting into the output power equation gives

$$P = \frac{3V_\phi E_A \sin \delta}{X_S}$$

There are no electrical losses in this generator because the resistances are assumed to be zero, and $P_{conv} = P_{out}$.

The output power equation shows that the power produced depends on the angle δ (torque angle) between V_ϕ and E_A. Normally, real generators have a full-load torque angle of 15 to 20°. The induced torque in the generator can be expressed as

$$\tau_{ind} = k\mathbf{B}_R \times \mathbf{B}_S$$

or as

$$\tau_{ind} = k\mathbf{B}_R \times \mathbf{B}_{net}$$

The magnitude of the expressed torque is

$$\tau_{ind} = kB_R B_{net} \sin \delta$$

where δ (the torque angle) is the angle between the rotor and net magnetic fields.

An alternative expression for the induced torque in terms of electrical quantities is

$$\boxed{\tau_{ind} = \frac{3V_\phi E_A \sin \delta}{\omega_m X_S}}$$

THE SYNCHRONOUS GENERATOR OPERATING ALONE

When a synchronous generator is operating under load, its behavior varies greatly depending on the power factor of the load and if the generator is operating alone or in parallel with other synchronous generators.

Throughout the upcoming sections, the effect of R_A is ignored, and the speed of the generators and the rotor flux will be assumed constant.

The Effect of Load Changes on a Synchronous Generator Operating Alone

Figure 12.17 illustrates a generator supplying a load. What are the effects of load increase on the generator? When the load increases, the real and/or reactive power drawn from the generator increases. The load increase increases the load current drawn from the generator.

The flux ϕ is constant because the field resistor did not change, and the field current is constant. Since the prime mover governing system maintains the mechanical speed ω constant, *the magnitude of the internal generated voltage* $E_A = K\phi\omega$ *is constant.* Since E_A is constant, which parameter is varying with the changing load?

If the generator is operating at a lagging power factor and an additional load is added at the *same power factor,* then the magnitude of I_A increases, but the angle θ between I_A and

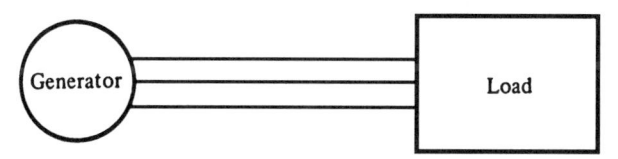

FIGURE 12.17 A single generator supplying a load.

V_ϕ remains constant. Therefore, the armature reaction voltage $jX_S I_A$ has increased while keeping the same angle. Since

$$E_A = V_\phi + jX_S I_A$$

$jX_S I_A$ must increase while the magnitude of E_A remains constant (Fig. 12.18a). Therefore, when the load increases, the voltage V_ϕ decreases sharply.

Figure 12.18b illustrates the effect when the generator is loaded with a unity power factor. It can be seen that V_ϕ decreases slightly. Figure 12.18c illustrates the effect when the generator is loaded with leading power factor loads. It can be seen that V_ϕ increases.

The *voltage regulation* is a convenient way to compare the behavior of two generators. The generator voltage regulation (VR) is given by

$$\boxed{VR = \frac{V_{nl} - V_{fl}}{V_{fl}} \times 100\%}$$

where V_{nl} and V_{fl} are the no-load and full-load voltages, respectively, of the generator. When a synchronous generator is operating at a lagging power factor, it has a large positive voltage regulation. When a synchronous generator is operating at a unity power factor, it has a small positive voltage regulation, and a synchronous generator operating at a leading power factor has a negative voltage regulation.

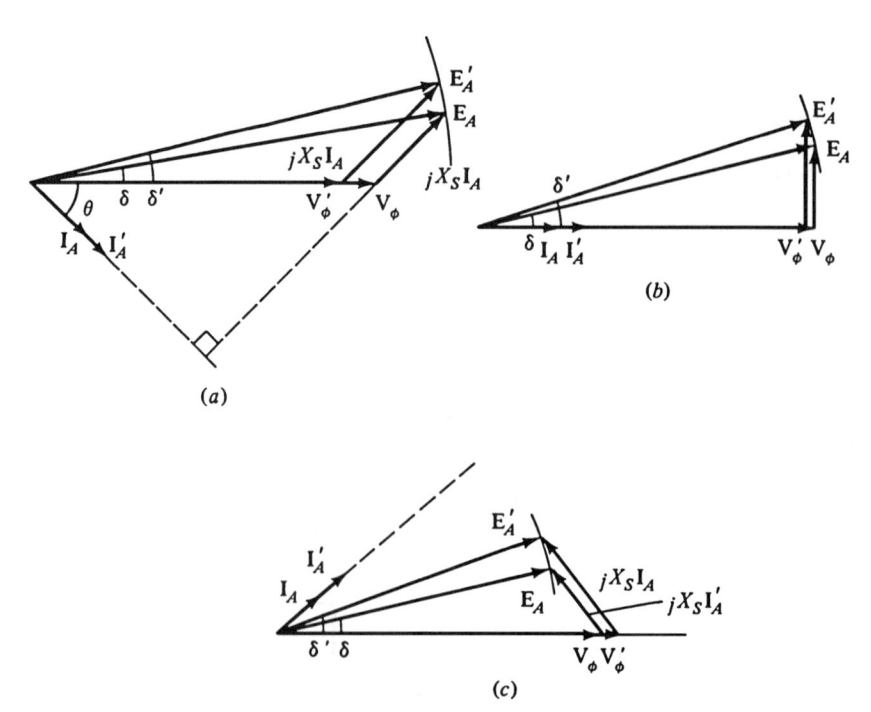

FIGURE 12.18 The effect of an increase in generator loads at constant power factor upon its terminal voltage: (a) Lagging power factor; (b) unity power factor; (c) leading power factor.

During normal operation, it is desirable to maintain the voltage supplied to the load constant even when the load varies. The terminal voltage variations can be corrected by varying the magnitude of $\mathbf{E}_A$ to compensate for changes in the load. Since $E_A = K\phi\omega$ and ω remains constant, E_A can be controlled by varying the flux in the generator. For example, when a lagging load is added to the generator, the terminal voltage will fall. The field resistor R_F is decreased to restore the terminal voltage to its previous level. When R_F decreases, the field current I_F increases. This causes the flux to increase, which results in increasing E_A and, therefore, the phase and terminal voltage.

This process is reversed to decrease the terminal voltage.

PARALLEL OPERATION OF AC GENERATORS

In most generator applications, there is more than one generator operating in parallel to supply power to various loads. The North American grid is an extreme example of a situation where thousands of generators share the load on the system.

The major advantages for operating synchronous generators in parallel are as follows:

1. The reliability of the power system increases when many generators are operating in parallel, because the failure of any one of them does not cause a total power loss to the loads.
2. When many generators operate in parallel, one or more of them can be taken out when failures occur in power plants or for preventive maintenance.
3. If one generator is used, it cannot operate near full load (because the loads are changing), then it will be inefficient. When several machines are operating in parallel, it is possible to operate only a fraction of them. The ones that are operating will be more efficient because they are near full load.

The Conditions Required for Paralleling

Figure 12.19 illustrates a synchronous generator G_1 supplying power to a load with another generator G_2 that is about to be paralleled with G_1 by closing the switch S_1. If the switch is closed at some arbitrary moment, the generators could be severely damaged and the load may lose power. If the voltages are different in the conductors being tied together, there will be *very* large current flow when the switch is closed.

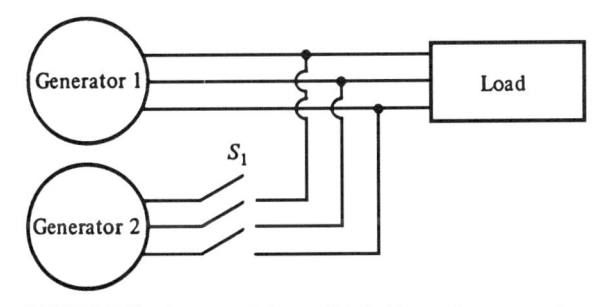

FIGURE 12.19 A generator being paralleled with a running power system.

This problem can be avoided by ensuring that each of the three phases has *the same voltage magnitude and phase angle* as the conductor to which it is connected. To ensure this match, these *paralleling conditions* must be met:

1. The two generators must have the same rms line voltages.
2. The *phase sequence* must be the same in the two generators.
3. The two *a* phases must have the same phase angles.
4. The frequency of the *oncoming generator* must be slightly higher than the frequency of the running system.

If the sequence in which the phase voltages peak in the two generators is different (Fig. 12.20a), then two pairs of voltages are 120° out of phase, and only one pair of voltages (the *a* phases) is in phase. If the generators are connected in this manner, large currents will flow in phases *b* and *c*, causing damage to both machines.

The phase sequence problem can be corrected by swapping the connections on any two of the three phases on one of the generators.

If the frequencies of the power supplied by the two generators are not almost equal when they are connected together, large power transients will occur until the generators stabilize at a common frequency. The frequencies of the two generators must differ by a small amount so that the phase angles of the oncoming generator will change slowly relative to the phase angles of the running system. The angles between the voltages can be observed, and switch S_1 can be closed when the systems are exactly in phase.

The General Procedure for Paralleling Generators

If generator G_2 is to be connected to the running system (Fig. 12.20), the following steps should be taken to accomplish paralleling:

1. The terminal voltage of the oncoming generator should be adjusted by changing the field current until it is equal to the line voltage of the running system.
2. The phase sequences of the oncoming generator and of the running system should be the same. The phase sequence can be checked by using the following methods:
   a. A small induction motor can be connected alternately to the terminals of each of the two generators. If the motor rotates in the same direction each time, then the phase sequences of both generators are the same. If the phase sequences are different, the motors will rotate in opposite directions. In this case, two of the conductors on the incoming generator must be reversed.
   b. Figure 12.20b illustrates three lightbulbs connected across the terminals of the switch connecting the generator to the system. When the phase changes between the two systems, the lightbulbs become bright when the phase difference is large and dim when the phase difference is small. When *the systems have the same phase sequence, all three bulbs become bright and dim simultaneously.* If the systems have opposite phase sequence, the bulbs get bright in succession.

The frequency of the oncoming generator should be slightly higher than the frequency of the running system. A frequency meter is used until the frequencies are close; then changes in phase between the the generator and the system are observed.

The frequency of the oncoming generator is adjusted to a slightly higher frequency to ensure that when it is connected, it will come on-line supplying power as a generator, instead of consuming it as a motor.

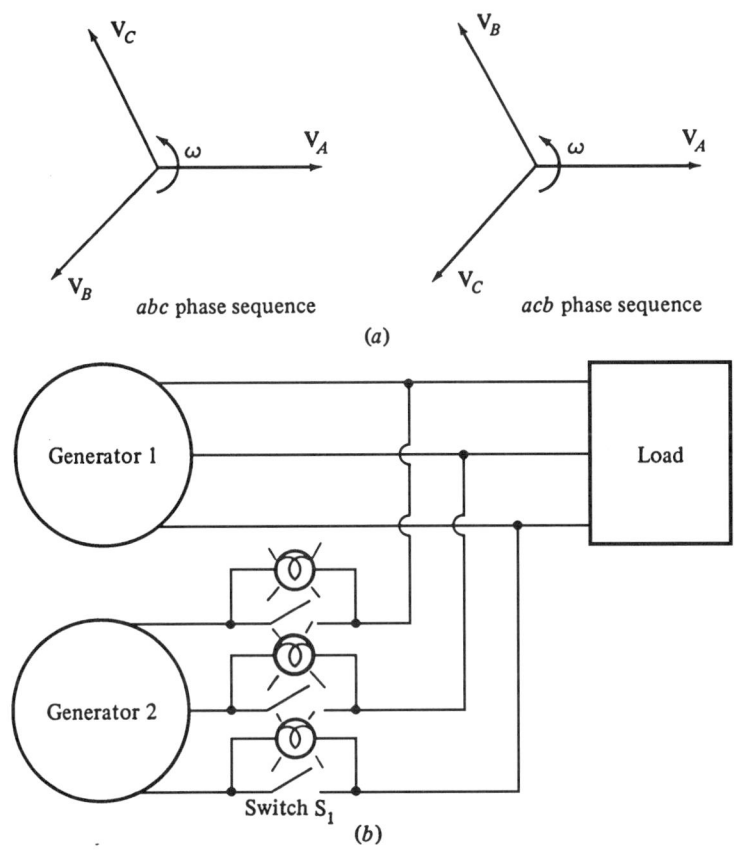

FIGURE 12.20 (a) The two possible phase sequences of a three-phase system. (b) The three-lightbulb method for checking phase sequence.

Once the frequencies are almost equal, the voltages in the two systems will change phase relative to each other very slowly. This change in phase is observed, and the switch connecting the two systems together is closed when the phase angles are equal (Fig. 12.21). A confirmation that the two systems are in phase can be achieved by watching the three lightbulbs. The systems are in phase when the three lightbulbs all go out (because the voltage difference across them is zero).

This simple scheme is useful, but it is not very accurate. A synchroscope is more accurate. It is a meter that measures the difference in phase angle between the a phases of the two systems (Fig. 12.22). The phase difference between the two a phases is shown by the dial. When the systems are in phase (0° phase difference), the dial is at the top. When they are 180° out of phase, the dial is at the bottom.

The phase angle on the meter changes slowly because the frequencies of the two systems are slightly different. Since the oncoming generator frequency is slightly higher than the system frequency, the synchroscope needle rotates clockwise because the phase angle advances. If the oncoming generator frequency is lower than the system frequency, the

12.18 CHAPTER TWELVE

(a)

(b)

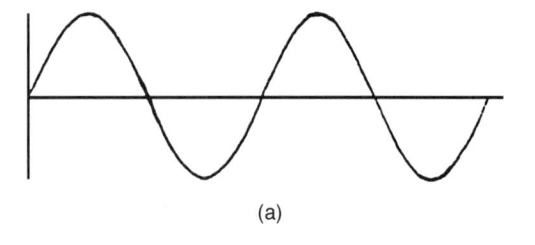

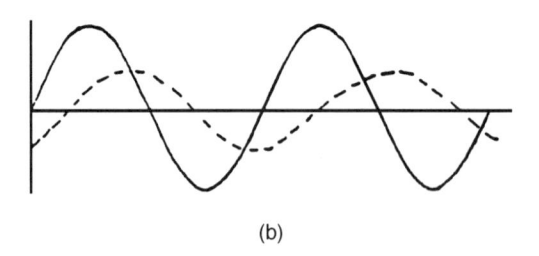

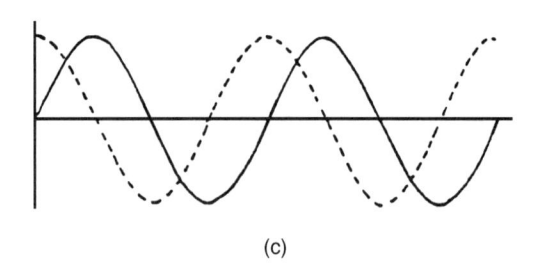

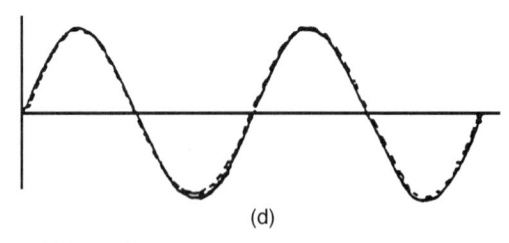

(c)

(d)

FIGURE 12.21 Steps taken to synchronize an incoming ac generator to the supply system. (*a*) Existing system voltage wave. (One phase only shown.) (*b*) Machine voltage wave shown dotted. Out of phase and frequency. Being built up to equal the system max. volts by adjustment of field rheostat. (*c*) Machine voltage now equal to system. Voltage waves out of phase but frequency being increased by increasing speed of prime mover. (*d*) Machine voltage now equal to system, in phase and with equal frequency. Synchroscope shows 12 o'clock. Switch can now be closed.

needle rotates counterclockwise. When the needle of the synchroscope stops in the vertical position, the voltages are in phase and the switch can be closed to connect the systems.

However, the synchroscope provides the relationship for only one phase. It does not provide information about the phase sequence.

The whole process of paralleling large generators to the line is done by a computer. For small generators, the operator performs the paralleling steps.

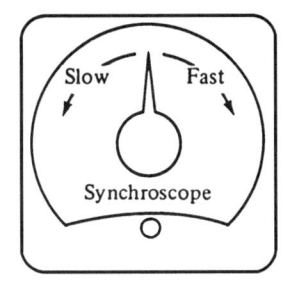

FIGURE 12.22 A synchroscope.

Frequency-Power and Voltage-Reactive Power Characteristics of a Synchronous Generator

The mechanical source of power for the generator is a *prime mover* such as diesel engines or steam, gas, water, and wind turbines. All prime movers behave in a similar fashion. As the power drawn from them increases, the rotational speed decreases. In general, this decrease in speed is nonlinear. However, the governor makes this decrease in speed linear with increasing power demand.

Thus, the governing system has a slight speed drooping characteristic with increasing load. The speed droop (SD) of a prime mover is defined by

$$SD = \frac{n_{nl} - n_{fl}}{n_{fl}} \times 100\%$$

where n_{nl} is the no-load speed of the prime mover and n_{fl} is the full-load speed of the prime mover. The speed droop of most generators is usually 2 to 4 percent. In addition, most governors have a set-point adjustment to allow the no-load speed of the turbine to be varied. A typical speed-power curve is shown in Fig. 12.23.

Since the electrical frequency is related to the shaft speed and the number of poles by

$$f_e = \frac{n_m P}{120}$$

the power output is related to the electrical frequency. Figure 12.23b illustrates a frequency-power graph. The power output is related to the frequency by:

$$P = S_P(f_{nl} - f_{sys})$$

where P = power output of generator
f_{nl} = no-load frequency of generator
f_{sys} = operating frequency of system
S_P = slope of curve, kW/Hz or MW/Hz

The reactive power Q has a similar relationship with the terminal voltage V_T. As previously described, the terminal voltage drops when a lagging load is added to a synchronous generator. The terminal voltage increases when a leading load is added to a synchronous generator. Figure 12.24 illustrates a plot of terminal voltage versus reactive power.

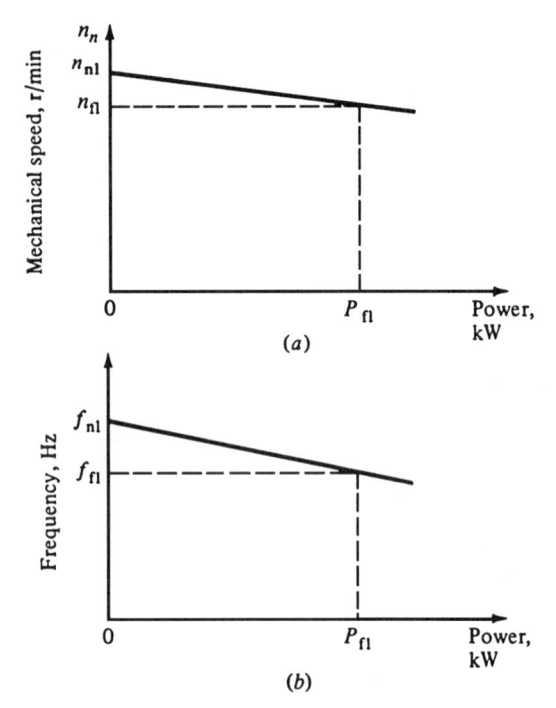

FIGURE 12.23 (*a*) The speed-power curve for a typical prime mover. (*b*) The resulting frequency-power curve for the generator.

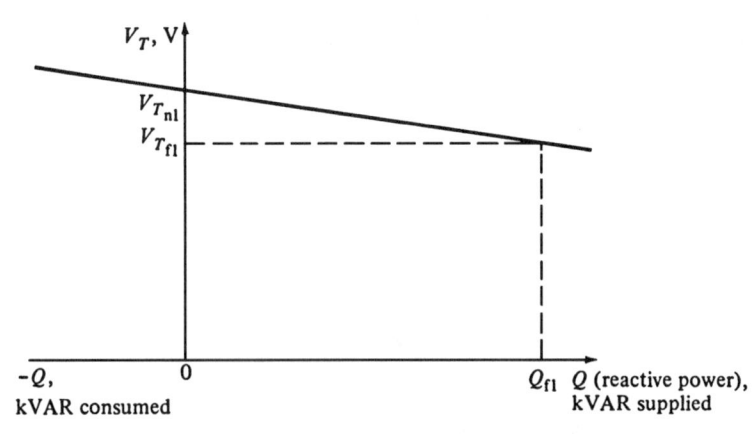

FIGURE 12.24 The terminal voltage V_T-reactive power Q curve for a synchronous generator.

This plot has a drooping characteristic that is not generally linear, but most generator voltage regulators have a feature to make this characteristic linear.

When the no-load terminal voltage set point on the voltage regulator is changed, the curve can slide up and down.

The frequency-power and terminal voltage-reactive power characteristics play important roles in parallel operation of synchronous generators. When a single generator is operating

SYNCHRONOUS GENERATORS 12.21

alone, the real power P and reactive power Q are equal to the amounts demanded by the loads.
The generator's controls cannot control the real and reactive power supplied. Therefore, for a given real power, the generator's operating frequency f_e is controlled by the governor set points, and for a given reactive power, the generator's terminal voltage V_T is controlled by the field current.

OPERATION OF GENERATORS IN PARALLEL WITH LARGE POWER SYSTEMS

The power system is usually very large so that *nothing* the operator of a synchronous generator connected to it does will have any effect on the power system. An example is the North American power grid, which is very large so that any action taken by one generator cannot make an observable change in the overall grid frequency.

This principle is idealized by the concept of an infinite bus, which is a very large power system such that its voltage and frequency do not change regardless of the amounts of real and reactive power supplied to or drawn from it. Figure 12.25 illustrates the power-frequency and reactive power-terminal voltage characteristics of such a system.

The behavior of a generator connected to an infinite bus is easier to explain when the automatic field current regulator is not considered. Thus, the following discussion will ignore the slight differences caused by the field regulator (Fig. 12.26). When a generator is connected in parallel with another generator or a large system, *the frequency and terminal voltage of all the generators must be the same because their output conductors are tied together.* Therefore, a common vertical axis can be used to plot the real power-frequency and reactive power-voltage characteristics back to back.

If a generator has been paralleled with the infinite bus, it will be essentially "floating" on line. It supplies a small amount of real power and little or no reactive power (Fig. 12.27). If the generator that has been paralleled to line has a slightly lower frequency than the running system (Fig. 12.28), the no-load frequency of the generator will be less than the operating frequency. In this case, the power supplied by the generator is negative (it consumes electric energy because it is running as a motor).

The oncoming generator frequency should be adjusted to be slightly higher than the frequency of the running system to ensure that the generator comes on line supplying power instead of consuming it.

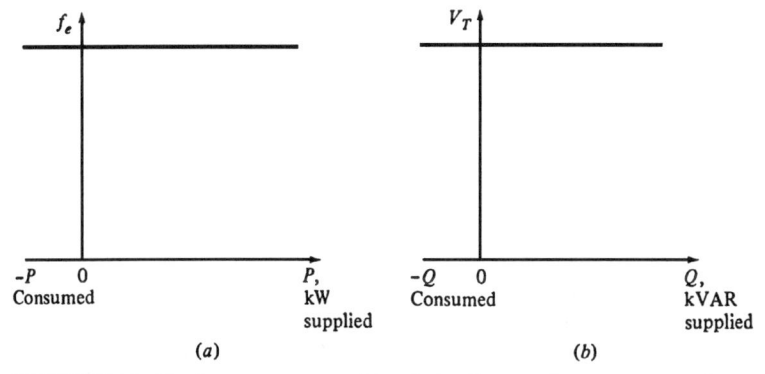

FIGURE 12.25 The frequency-power and terminal voltage-reactive power curves for an infinite bus.

12.22 CHAPTER TWELVE

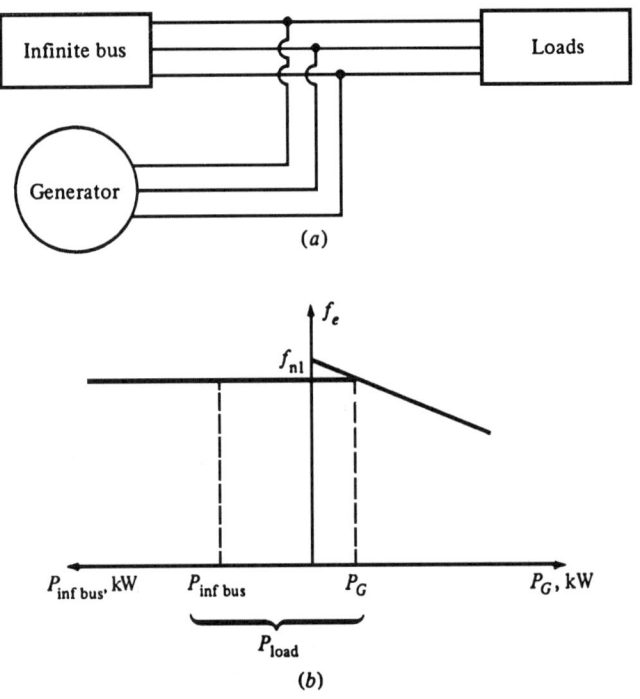

FIGURE 12.26 (*a*) A synchronous generator operating in parallel with an infinite bus. (*b*) The frequency-power diagram (or *house diagram*) for a synchronous generator in parallel with an infinite bus.

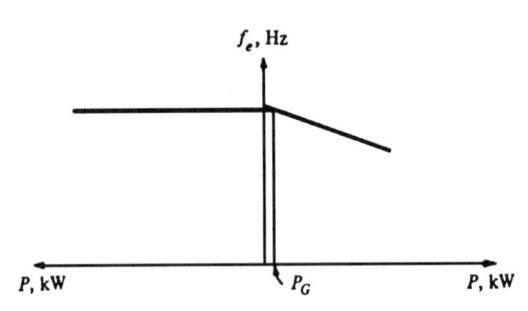

FIGURE 12.27 The frequency-power diagram at the moment just after paralleling.

In reality, most generators have reverse-power trip connected to them. They must be paralleled when their frequency is higher than that of the running system. If such a generator starts to "motor" (consume power), it will be automatically disconnected from the line.

Once the generator is connected, the governor set point is increased to shift the no-load frequency of the generator upward. Since the frequency of the system remains constant (the frequency of the infinite bus cannot change), the generator output power increases. The house diagram and the phasor diagram are illustrated in Fig. 12.29*a* and *b*. Notice in

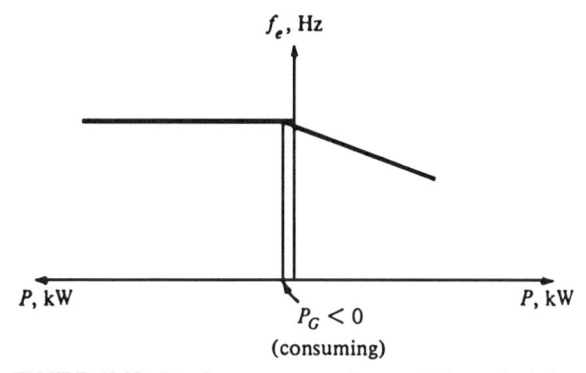

FIGURE 12.28 The frequency-power diagram if the no-load frequency of the generator were slightly *less* than the system frequency before paralleling.

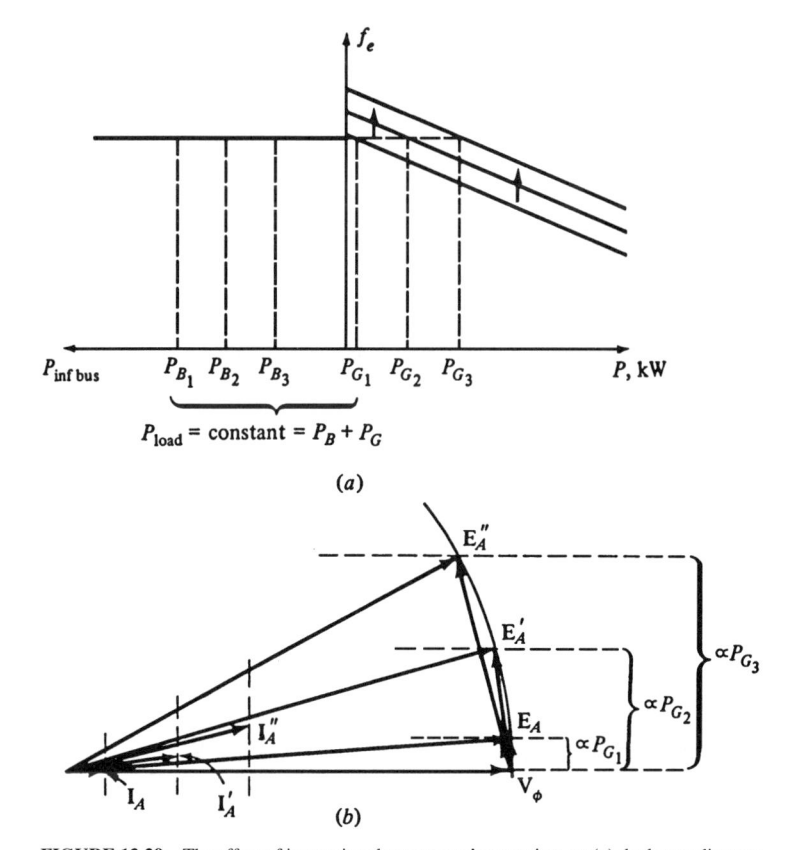

FIGURE 12.29 The effect of increasing the governor's set points on (*a*) the house diagram and (*b*) the phasor diagram.

the phasor diagram that the magnitude of $\mathbf{E}_A$ ($= K\phi\omega$) remains constant because I_F and ω remained unchanged, while $E_A \sin \delta$ (which is proportional to the output power as long as V_T remains constant) has increased.

When the governor set point is increased, the no-load frequency and the output power of the generator increase. As the power increases, the magnitude of E_A remains constant while $E_A \sin \delta$ is increased further.

If the output power of the generator is increased until it exceeds the power consumed by the load, the additional power generated flows back into the system (infinite bus). By definition, the infinite bus can consume or supply any amount of power while the frequency remains constant. Therefore, the additional power is consumed.

Figure 12.29b illustrates the phasor diagram of the generator when the real power has been adjusted to the desired value. Notice that at this time, the generator has a slightly leading power factor. It is acting as a capacitor, requiring reactive power. The field current can be adjusted so the generator can supply reactive power. However, there are some constraints on the operation of the generator under these circumstances. The first constraint on the generator is that when I_F is changing, *the power must remain constant*. The power given to the generator is $P_{in} = \tau_{app} \omega_m$.

For a given governor setting, the prime mover of the generator has a fixed torque-speed characteristic. When the governor set point is changed, the curve moves. Since the generator is tied to the system (infinite bus), its speed *cannot* change. Therefore, since the governor set point and the generator's speed have not changed, the power supplied by the generator must remain constant. Since the power supplied does not change when the field current is changing, then $I_A \cos \theta$ and $E_A \sin \delta$ (the distance proportional to the power in the phasor diagram) cannot change.

The flux ϕ increases when the field current is increased. Therefore, E_A ($= K\phi\omega$) must increase. If $\mathbf{E}_A$ increases while $E_A \sin \delta$ remains constant, then phasor $\mathbf{E}_A$ must slide along the constant-power line shown in Fig. 12.30. Since $\mathbf{V}_\phi$ is constant, the angle of $jX_S\mathbf{I}_A$ changes as shown. Therefore, the angle and magnitude of $\mathbf{I}_A$ change.

Notice that the distance proportional to Q ($I_A \sin \theta$) increases. This means that *increasing the field current in a synchronous generator operating in parallel with a power system (infinite bus) increases the reactive power output of the generator*.

In summary, when a generator is operating in parallel with a power system (infinite bus), the following are true:

1. The power system connected to the generator controls the frequency and the terminal voltage.
2. The real power supplied by the generator to the system is controlled by the governor set point.
3. The reactive power supplied by the generator to the system is controlled by the field current.

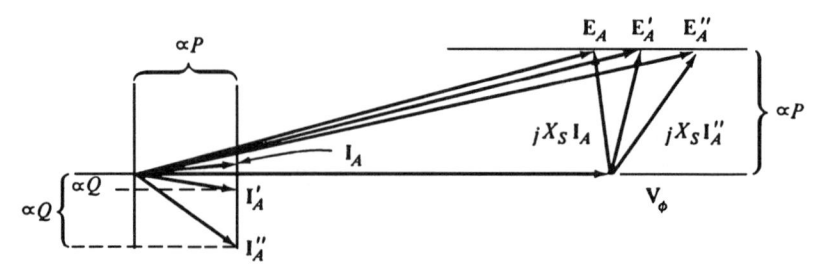

FIGURE 12.30 The effect of increasing the generator's field current on the phasor diagram of the machine.

SYNCHRONOUS GENERATOR RATINGS

There are limits to the output power of a synchronous generator. These limits are known as *ratings* of the generator. Their purpose is to protect the generator from damage caused by improper operation. The synchronous generator ratings are *voltage, frequency, speed, apparent power (kilovoltamperes), power factor, field current,* and *service factor.*

The Voltage, Speed, and Frequency Ratings

The common system frequencies used today are 50 Hz (in Europe, Asia, etc.) and 60 Hz (in the Americas). Once the frequency and the number of poles are known, there is only one possible rotational speed.

One of the most important ratings for the generator is the voltage at which it operates. Since the generator's voltage depends on the flux, the higher the design voltage, the higher the flux. However, the flux cannot increase indefinitely because the field current has a maximum value.

The main consideration in determining the rated voltage of the generator is the breakdown value of the winding insulation. The voltage at which the generator operates must not approach the breakdown value.

A generator rated for a given frequency (say, 60 Hz) can be operated at 50 Hz as long as some conditions are met. Since there is a maximum flux achievable in a given generator, and since $E_A = K\phi\omega$, the maximum allowable E_A must change when the speed is changed. For example, a generator rated for 60 Hz can be operated at 50 Hz if the voltage is derated to 50/60, or 83.3 percent, of its design value. The opposite effect will occur when a generator rated for 50 Hz is operated at 60 Hz.

Apparent Power and Power Factor Ratings

The factors that determine the power limits of electric machines are the shaft torque and the heating of the windings. In general, the shaft can handle larger power than the machine is rated for. Therefore, the steady-state power limits are determined by the heating in the windings of the machine. The windings that must be protected in a synchronous generator are the armature windings and the field windings.

The maximum allowable current in the armature determines the maximum apparent power for the generator. Since the apparent power S is given by

$$S = 3V_\phi I_A$$

if the rated voltage is known, the maximum allowable current in the armature determines the rated apparent power of the generator.

The power factor of the armature current does not affect the heating of the armature windings. The stator copper loss (SCL) heating effect is

$$P_{SCL} = 3I_A^2 R_A$$

These effects are independent of the angle between the I_A and V_ϕ. These generators are not rated in megawatts (MW), but in megavoltamperes (MVA).

The field winding copper losses are

$$P_{RCL} = I_F^2 R_F$$

Therefore, the maximum allowable heating determines the maximum field current for the machine. Since $E_A = K\phi\omega$, this also determines the maximum acceptable E_A. Since there is a maximum value for I_F and E_A, there is a minimum acceptable power factor of the generator when it is operating at the rated MVA.

Figure 12.31 illustrates the phasor diagram of a synchronous generator with the rated voltage and armature current. The current angle can vary, as shown. Since $\mathbf{E}_A$ is the sum of $\mathbf{V}_\phi$ and $jX_S\mathbf{I}_A$, there are some current angles for which the required E_A exceeds $E_{A,\max}$. If the generator is operated at these power factors and the rated armature current, the field windings will burn.

The angle of $\mathbf{I}_A$ that results in the maximum allowable $\mathbf{E}_A$ while $\mathbf{V}_\phi$ is at the rated value determines the generator rated power factor. The generator can be operated at lower power factor (more lagging) than the rated value, but only by reducing the MVA output of the generator.

SYNCHRONOUS GENERATOR CAPABILITY CURVES

The generator *capability diagram* expresses the stator and rotor heat limits and any external limits on the generator. The capability diagram illustrates the complex power $S = P = jQ$. It is derived from the generator's phasor diagram, assuming that $\mathbf{V}_\phi$ is constant at the generator's rated voltage.

Figure 12.32 illustrates the phasor diagram of a synchronous generator operating at its rated voltage and lagging power factor. The orthogonal axis is drawn with units of volts. The length of the vertical segment AB is $X_S I_A \cos\theta$, and horizontal segment OA is $X_S I_A \sin\theta$. The generator's real power output is

$$P = 3V_\phi I_A \cos\theta$$

The reactive power output is

$$Q = 3V_\phi I_A \sin\theta$$

The apparent power output is

$$S = 3V_\phi I_A$$

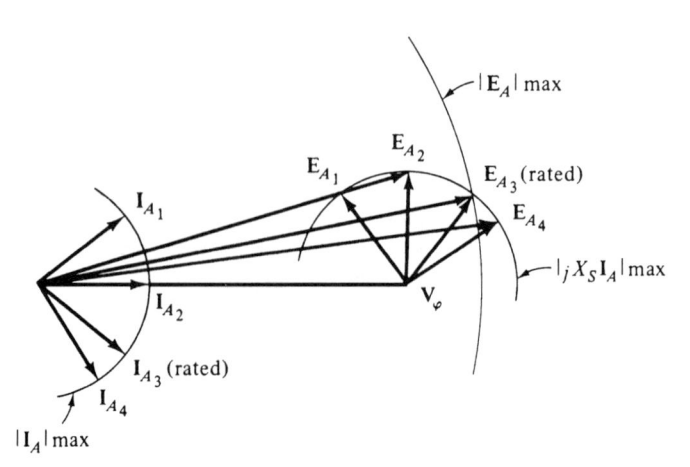

FIGURE 12.31 How the rotor field current limit sets the rated power factor of a generator.

SYNCHRONOUS GENERATORS

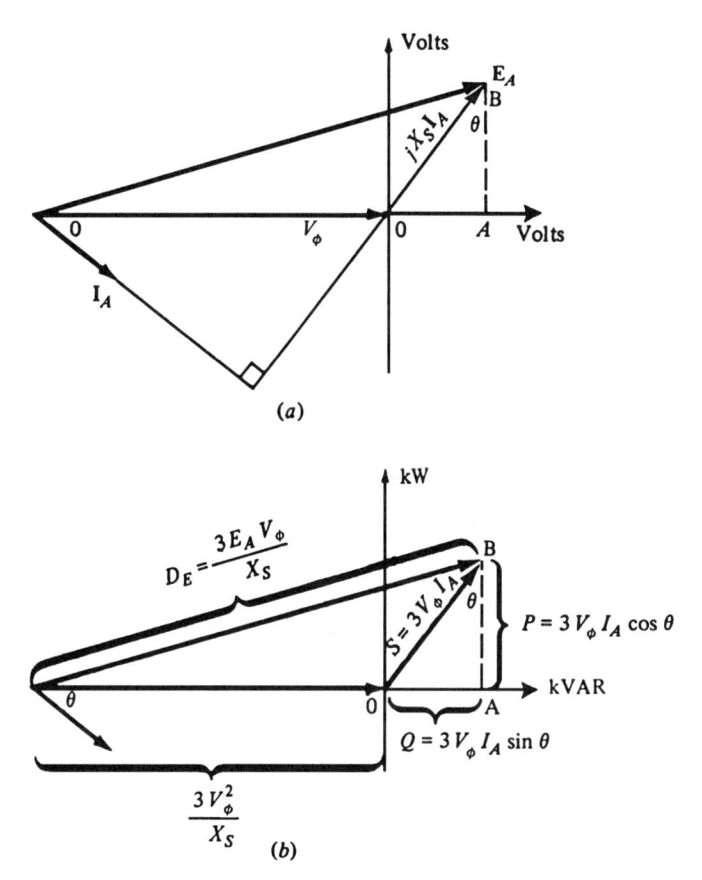

FIGURE 12.32 Derivation of a synchronous generator capability curve: (*a*) The generator phase diagram. (*b*) The corresponding power units.

Figure 12.32*b* illustrates how the axes can be recalibrated in terms of real and reactive power. The conversion factor used to change the scale of the axis from volts to voltamperes is $3V_\phi/X_S$.

$$P = 3V_\phi I_\phi \cos\theta = \frac{3V_\phi}{X_S}(X_S I_A \cos\theta)$$

$$Q = 3V_\phi I_\phi \sin\theta = \frac{3V_\phi}{X_S}(X_S I_A \sin\theta)$$

On the voltage axes, the origin of the phasor diagram is located at $-V_\phi$. Therefore, the origin on the power diagram is located at

$$Q = \frac{3V_\phi}{X_S}(-V_\phi)$$

$$= -\frac{3V_\phi^2}{X_S}$$

On the power diagram, the length corresponding to E_A is

$$D_E = \frac{3E_A V_\phi}{X_S}$$

The length that corresponds to $X_S I_A$ on the power diagram is $3V_\phi I_A$.

Figure 12.33 illustrates the final capability curve of a synchronous generator. It illustrates a plot of real power P versus reactive power Q. The lines representing constant armature current I_A are shown as lines of constant apparent power $S = 3V_\phi I_A$, which are represented by concentric circles around the origin. The lines representing constant field current correspond to lines of constant E_A. These are illustrated by circles of magnitude $3E_A V_\phi / X_S$ centered at

$$Q = -\frac{3V_\phi^2}{X_S}$$

The armature current limit is illustrated by the circle corresponding to the rated I_A or MVA. The field current limit is illustrated by the circle corresponding to the rated I_F or E_A.

Any point located within both circles is a safe operating point for the generator. Additional constraints such as the maximum prime mover power can also be shown on the diagram (Fig. 12.34).

SHORT-TIME OPERATION AND SERVICE FACTOR

The heating of the armature and field windings of a synchronous generator is the most important limit in steady-state operation. The power level at which the heating limit usually occurs is much lower than the maximum power that the generator is mechanically and

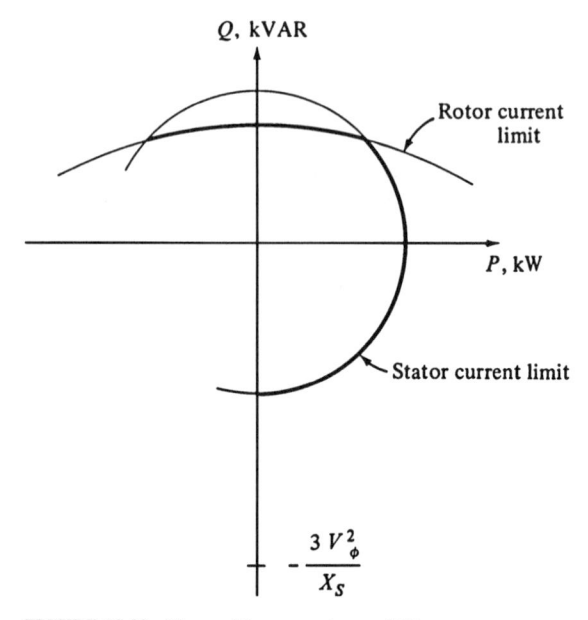

FIGURE 12.33 The resulting generator capability curve.

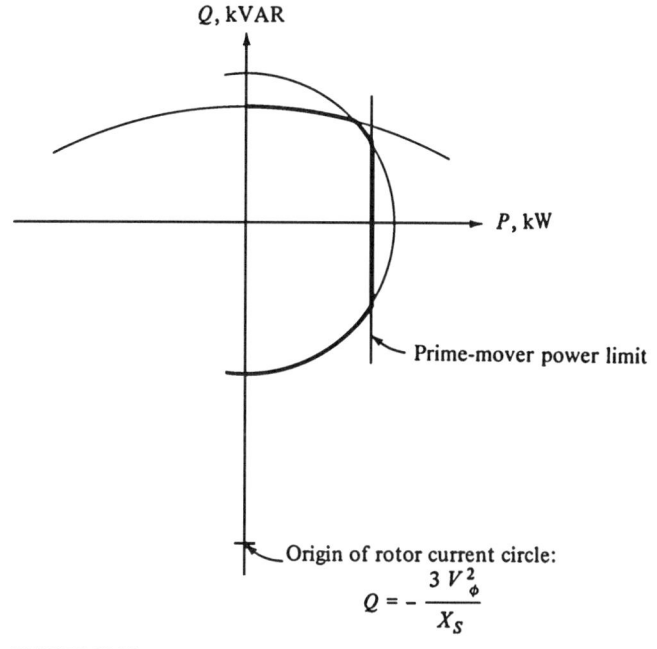

FIGURE 12.34 A capability diagram showing the prime mover power limit.

magnetically able to supply. In general, a typical synchronous generator can supply up to 300 percent of its rated power until its windings burn up.

This ability to supply more power than the rated amount is used for momentary power surges which occur during motor starting and other load transients. A synchronous generator can supply more power than the rated value for longer periods of time as long as the windings do not heat up excessively before the load is removed. For example, a generator rated for 1 MW is able to supply 1.5 MW for 1 min without causing serious damage to the windings. This generator can operate for longer periods at lower power levels.

The *insulation class* of the windings determines the maximum temperature rise in the generator. The standard insulation classes are A, B, F, and H. In general, these classes correspond to temperature rises above ambient of 60, 80, 105, and 125°C, respectively. The power supplied by a generator increases with the insulation class without overheating the windings.

In motors and generators, overheating the windings is a *serious problem*. In general, when the temperature of the windings increases by 10°C above the rated value, the average lifetime of the machine is reduced by one-half. Since the increase in the temperature of the windings above the rated value drastically reduces the lifetime of the machine, a synchronous generator should not be overloaded unless it is absolutely necessary.

The *service factor* is the ratio of the actual maximum power of the machine to its nameplate rating. A 1.15 service factor of a generator indicates that it can operate indefinitely at 115 percent of the rated load without harm. The service factor of a motor or a generator provides a margin for error in case the rated loads were improperly estimated.

REFERENCE

1. S. J. Chapman, *Electric Machinery Fundamentals*, 2d ed., McGraw-Hill, New York, 1991.

CHAPTER 13
GENERATOR COMPONENTS, AUXILIARIES, AND EXCITATION

Figure 13.1 illustrates a sectional view of a large generator. Hydrogen is used to cool most generators having a rating larger than 50 MW.

THE ROTOR

The rotor is made from a single steel forging. The steel is vacuum-degassed to minimize the possibility of hydrogen-initiated cracking. Reheating and quenching also harden the forging. Stress-relieving heat treatment is done following rough machining. Ultrasonic examination is performed at various stages of the rotor. Figure 13.2 illustrates the winding slots in the rotor. Figure 13.3 illustrates a rotor cross section and the gas flow.

The generator countertorque increases to 4 to 5 times the full-load torque when a short circuit occurs at the generator terminals. The rotor and turbine-end coupling must be able to withstand this peak torque.

Rotor Winding

Each winding turn is assembled separately in half-turns or in more pieces. The joints are at the centers of the end turns or at the corners. They are brazed together after assembling each turn, to form a series-connected coil. The coils are made of high-conductivity copper with a small amount of silver to improve the creep properties. The gas exits through radially aligned slots.

Slot liners of molded glass fiber insulate the coils. These separators of glass fiber are used between each turn. They insulate against almost 10 V between adjacent turns (Fig. 13.4). The end rings and end disks are separated from the end windings by thick layers of insulation. Insulation blocks are placed in the spaces between the end windings to ensure the coils do not distort. The winding slots are cut in diametrically opposite pairs. They are equally pitched over two-thirds of the rotor periphery, leaving the pole faces without winding slots. This results in a difference between the stiffness in the two perpendicular axes. This difference leads to vibration at twice the speed. Equalizing slots are cut in the pole faces (Fig. 13.5) to prevent this problem from occurring. The slots are wider and shallower than the winding slots. They are filled with steel blocks to restore the magnetic properties. The blocks contain holes to allow the ventilating gas to flow.

The average winding temperature should not exceed 115°C. The hydrogen enters the rotor from both ends under the end windings and emerges radially from the wedges. Figure 13.6 illustrates the fans used to drive the hydrogen through the stator.

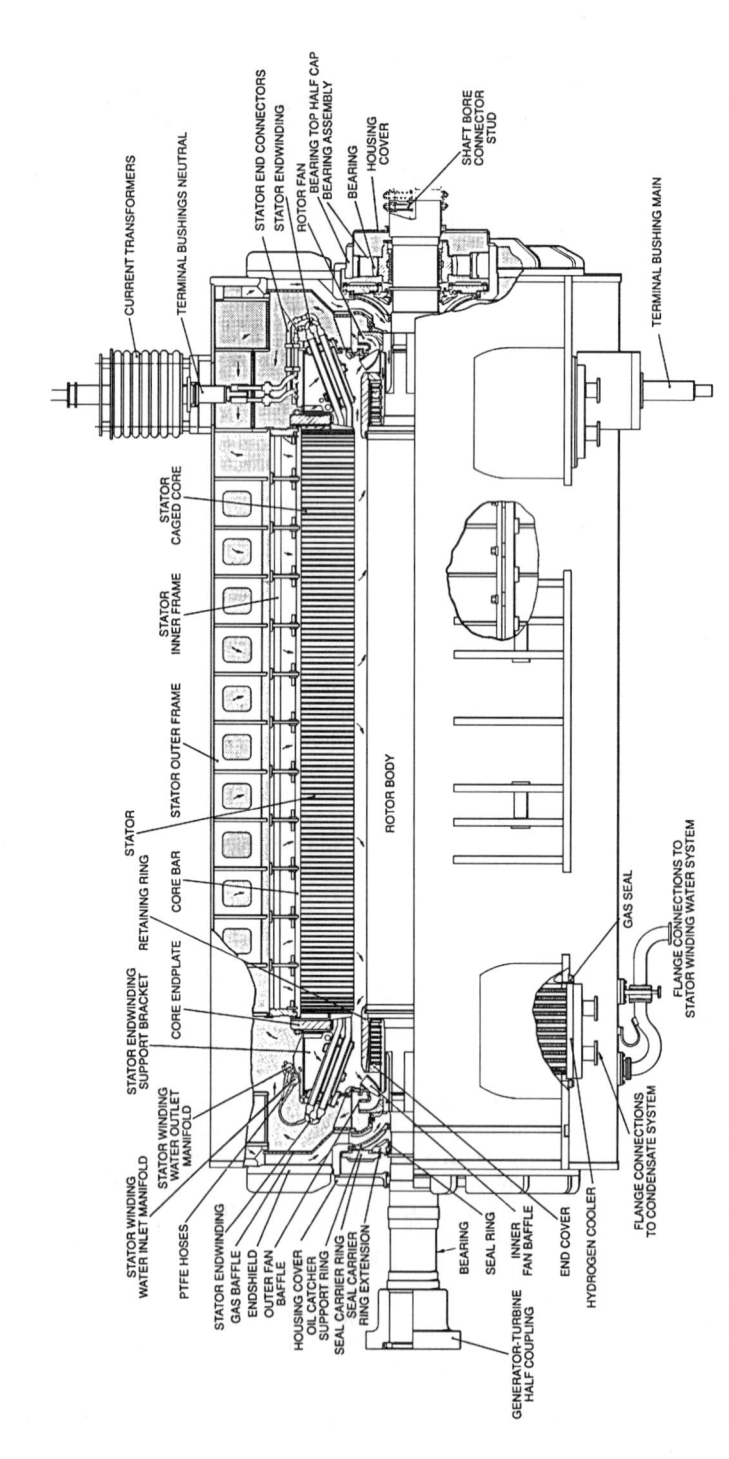

FIGURE 13.1 Sectional view of a 660-MW generator.

GENERATOR COMPONENTS, AUXILIARIES, AND EXCITATION 13.3

FIGURE 13.2 Cutting winding slots in a rotor.

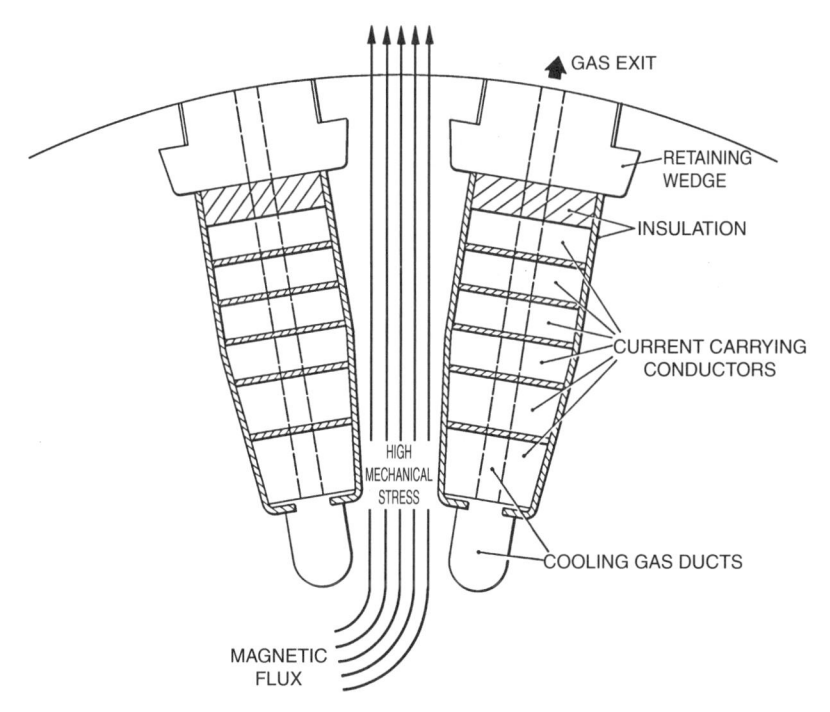

FIGURE 13.3 A section of a rotor.

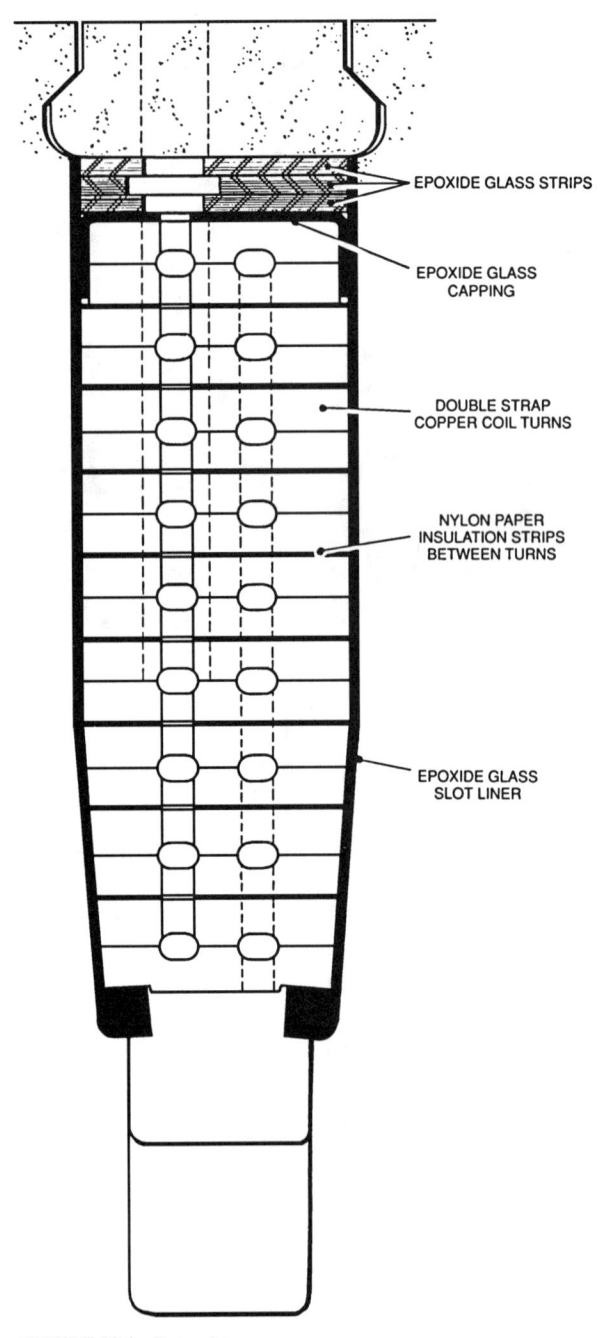

FIGURE 13.4 Rotor slot.

GENERATOR COMPONENTS, AUXILIARIES, AND EXCITATION   13.5

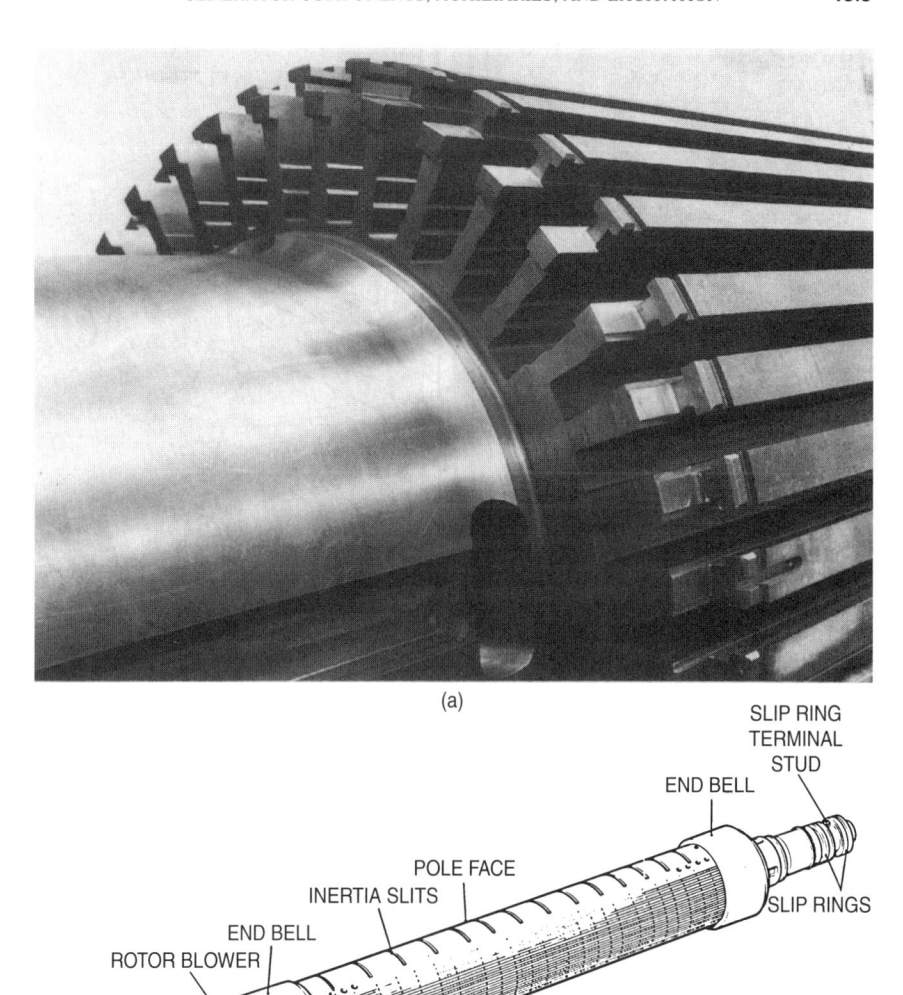

(a)

(b)

FIGURE 13.5 Stiffness compensation.

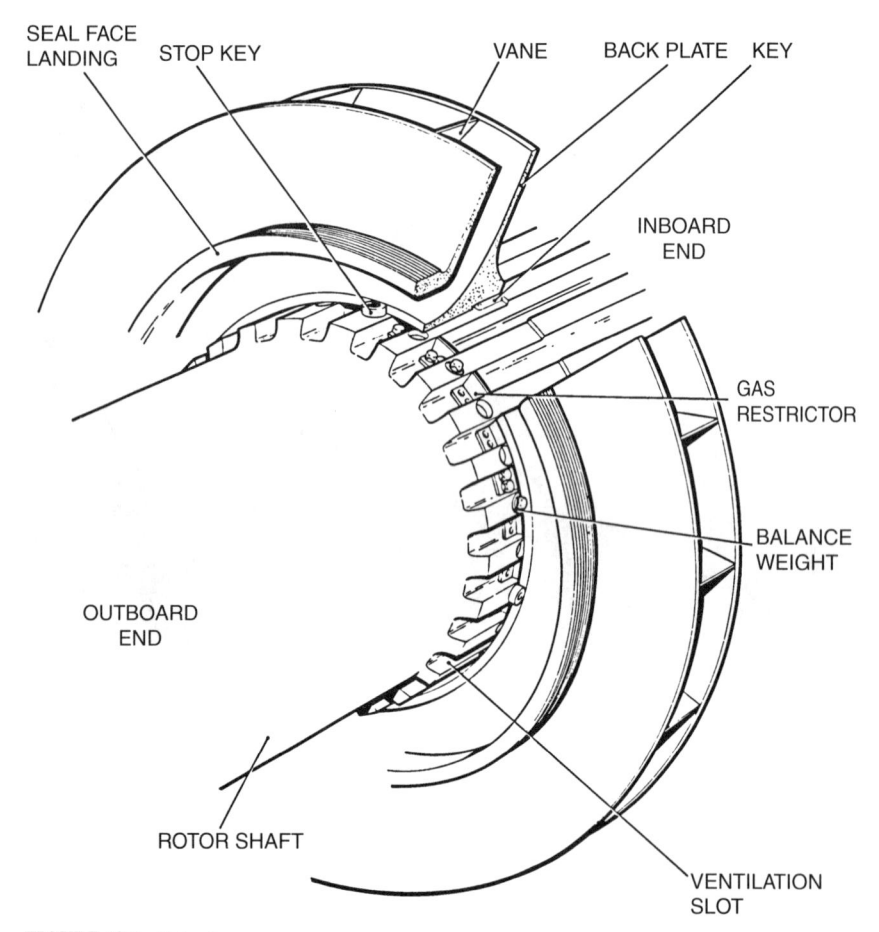

FIGURE 13.6 Rotor fan.

Flexible leads made of thin copper strips are connected to the ends of the winding. These leads are placed in two shallow slots in the shaft. Wedges retain them. The leads are connected to radial copper studs, which are connected to D-shaped copper bars placed in the shaft bore. Hydrogen seals are provided on the radial studs. The D leads are connected to the slip rings by radial connection bolts (Fig. 13.7).

Rotor End Rings

The end rings (Fig. 13.8) are used to restrain the rotor end windings from flying out under centrifugal forces. These rings have traditionally been made from nonmagnetic austenitic steel, typically 18% Mn, 4% Cr. A ring is machined from a single forging. It is shrunk-fit at the end of the rotor body. The material of the end rings was proved to be liable to stress corrosion cracking. A protective finish is given to all the surfaces except the shrink fit to ensure that hydrogen, water vapor, etc., do not contact the metal. The rings should be removed during long maintenance outage (every 8 to 10 years) and inspected for detailed surface cracking using a fluorescent dye. Ultrasonic scanning is not sufficient due to the coarse grain structure. A recent

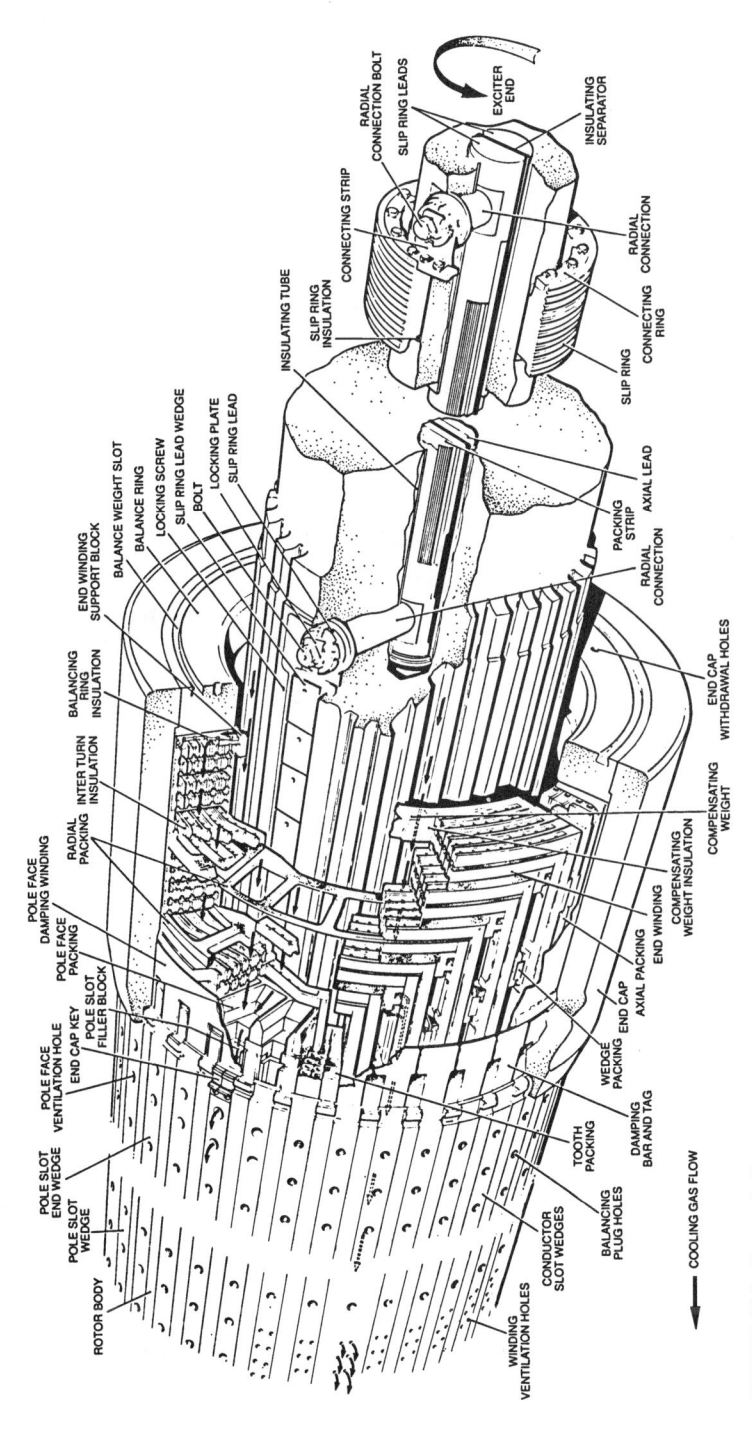

FIGURE 13.7 Rotor winding.

13.7

FIGURE 13.8 Rotor end ring.

development has proved that austenitic steel containing 18% Mn and 18% Cr is immune to stress corrosion cracking. New machines use this alloy. It is also used for replacement rings. This eliminates the need for periodic inspection. It is important to mention that a fracture of an end ring can result in serious damage to the machine and at least a few months' outage. It is highly recommended to replace the traditional material with the new material.

The rings must be heated to 300°C to expand sufficiently for the shrink surface. Induction heating is preferred to direct heating to prevent possible damage to the rings. The end ring is insulated from the end winding with a molded-in glass-based liner or a loose cylinder sleeve. Hydrogen enters the rotor in the clearance between the end winding and the shaft. The outboard end of the ring is not permitted to contact the shaft, to prevent the shaft flexure from promoting fatigue and fretting damage at the interfaces. A balancing ring is also included in the end disk for balancing the rotor.

Wedges and Dampers

Wedges are used to retain the winding slot contents. Wedges are designed to withstand stresses from the windings while allowing the hydrogen to pass through holes. They must also be nonmagnetic to minimize the flux leakage around the circumference of the rotor. They are normally made of aluminum. One continuous wedge is used for each slot.

During system faults or during unbalanced electrical loading, negative phase sequence currents and fluxes occur, leading to induced currents in the surface of the rotor. These currents will flow in the wedges which act as a "damper winding" similar to the bars in the rotor of an induction motor. The end rings act as shorting rings in the motor. Arcing and localized pitting may occur between the end rings and the wedges.

Slip Rings, Brush Gear, and Shaft Grounding

The D leads in the bore are connected through radial copper connectors (which normally have backup hydrogen seals) and flexible connections to the slip rings (Fig. 13.9). The excitation current is around 5000 A dc for a 660-MW generator. The surface area of the slip rings must be large to run cool while transferring the current. Figure 13.10 illustrates the brush gear including brushes and holders of a removable bracket. The holders can be replaced on power. Constant-pressure springs are used to maintain brush pressure. A brush life should be at least 6 months. A separate compartment houses the brush gear. A shaft-mounted fan provides separate ventilation so that brush dust is not spread on other excitation components. Small amounts of hydrogen may pass through the connection seals and may accumulate in the brush gear compartments during extended outages. The fan dilutes them safely during start-up before excitation current is applied. The brush gear can be easily inspected through windows in the cover. Figure 13.11 illustrates brushless rotor connections.

A large generator produces normally an on-load voltage of 10 to 50 V between its shaft ends due to magnetic dissymmetry. This voltage drives an axial current through the rotor body. The current returns through bearings and journals. It causes damage to their surfaces. Insulation barriers are installed to prevent such current from circulating. The insulation is installed at all locations where the shaft could contact earthed metal, e.g., bearings, seals, oil scrapers, oil pipes, and gear-driven pumps.

Some designs have two layers with a "floating" metallic component between them. The integrity of insulation is confirmed by a simple resistance measurement between the floating component and earth.

If the insulation remains clean and intact, a difference in voltage will exist between the shaft at the exciter end and ground. This provides another method to confirm the integrity of the insulation. The shaft voltage is monitored by shaft-riding brush. An alarm is initiated when the shaft voltage drops below a predetermined value.

It is important to maintain the shaft at the turbine end of the generator at ground level. A pair of shaft riding brushes ground the shaft through a resistor. Since carbon brushes develop a high-resistance glaze when operated for extended periods without current flow, a special circuit introduces a *wetting current* into and out of the shaft through the brushes. This circuit also detects loss of contact between the brush and the shaft.

Fans

Fans drive the hydrogen through the stator and the coolers. Two identical fans are mounted at each end of the shaft. Centrifugal or axial-type fans are used (Fig. 13.12).

Rotor and Alignment Threading

The stator bore is about 25 cm larger than the rotor diameter. The rotor is inserted into the stator by supporting the inserted end of the rotor on a thick steel skid plate which slides into the stator, while the outboard end is supported by a crane.

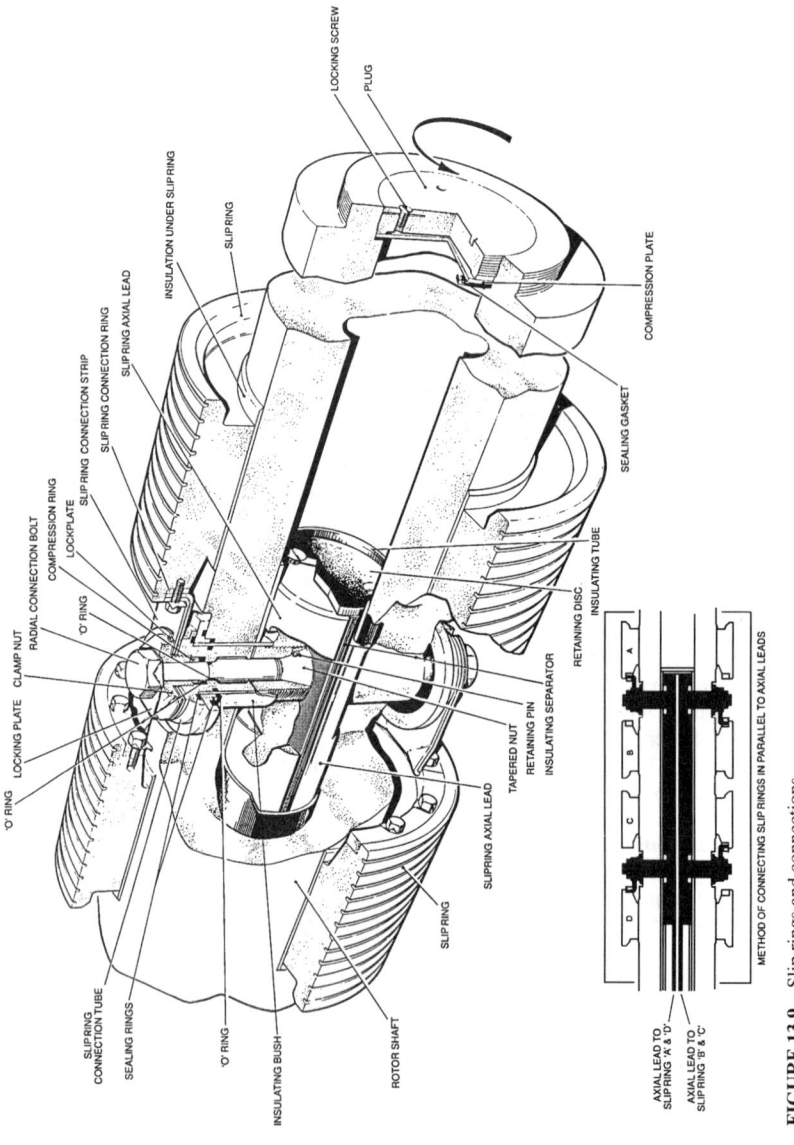

FIGURE 13.9 Slip rings and connections.

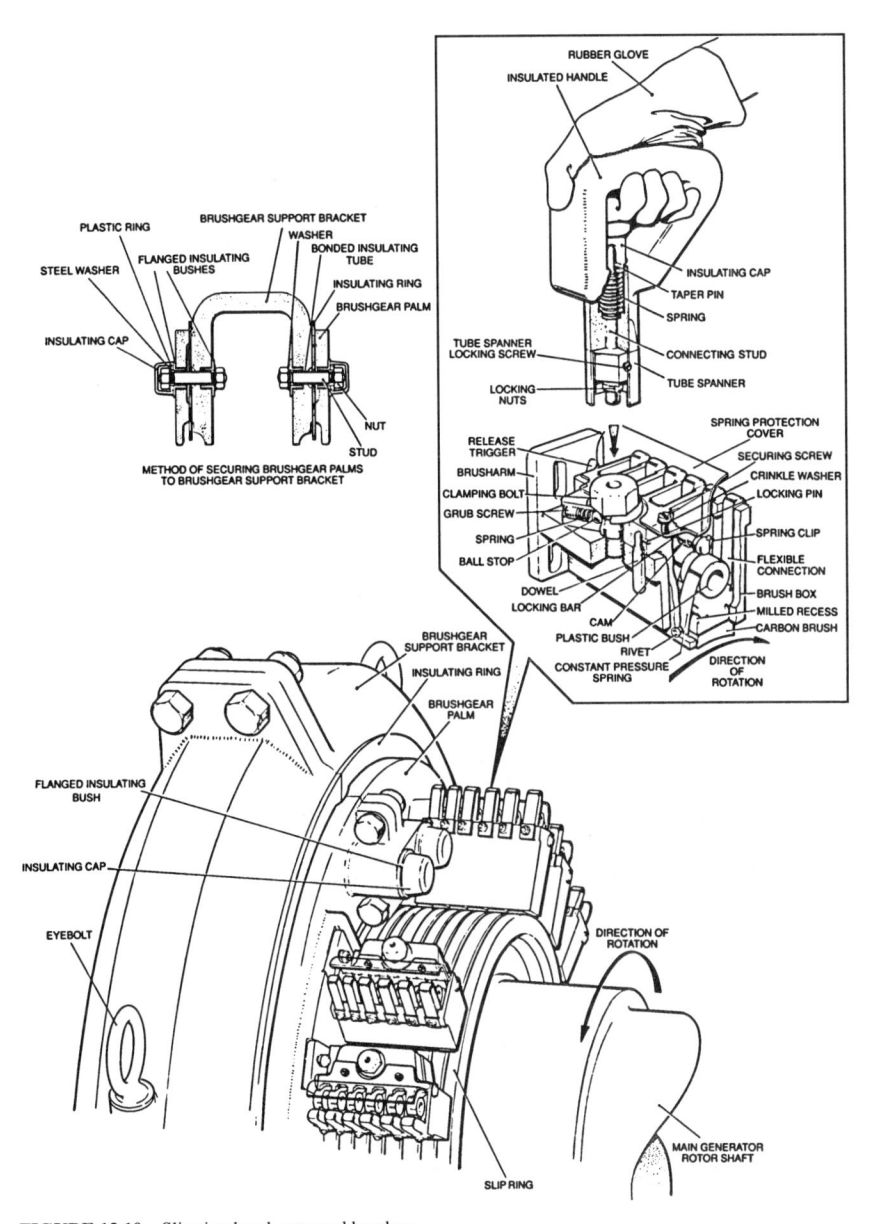

GENERATOR COMPONENTS, AUXILIARIES, AND EXCITATION   13.11

FIGURE 13.10 Slip ring brush gear and brushes.

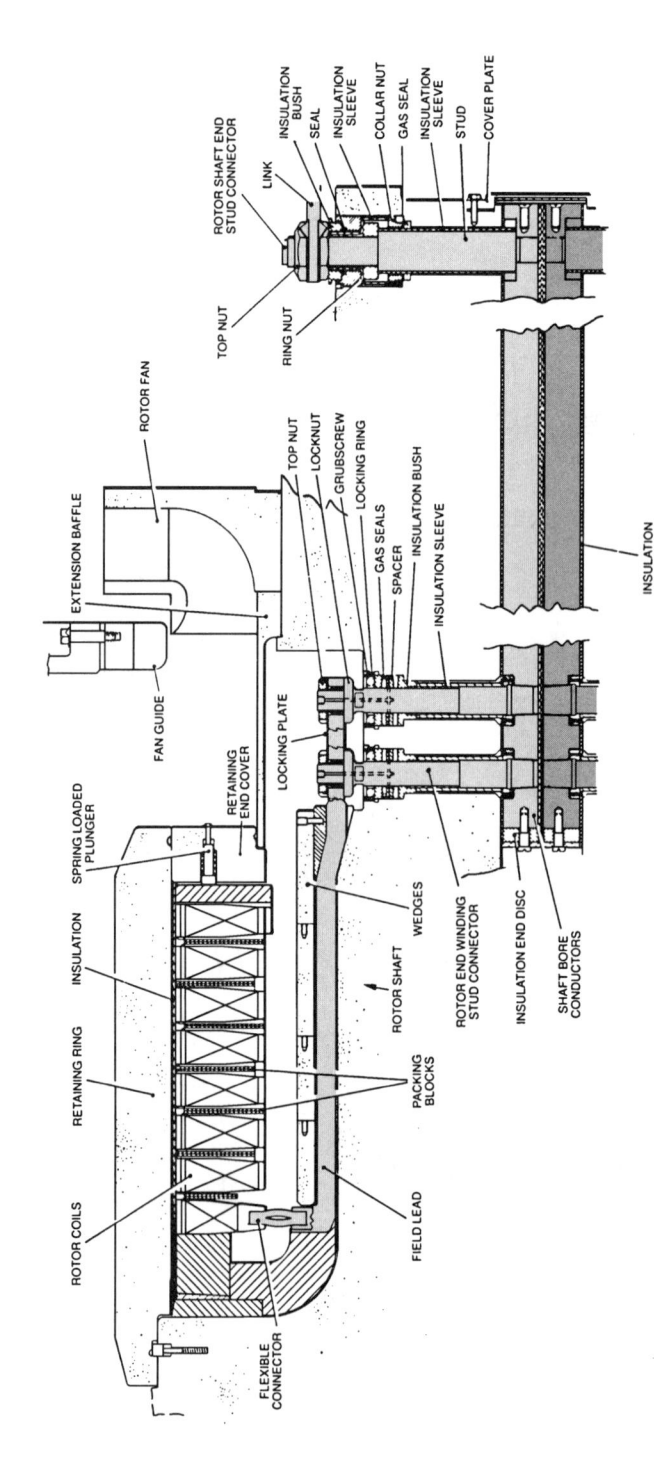

FIGURE 13.11 Brushless rotor connections.

13.12

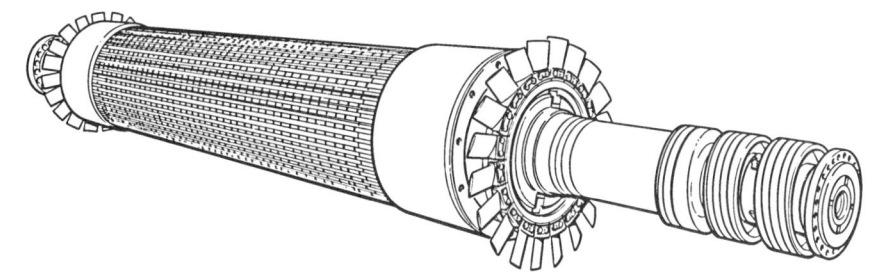

FIGURE 13.12 Axial-flow fans on rotor.

Vibration

Generator rotors rated at 500 to 600 MW have two main critical speeds (natural resonance in bending). Simple two-plane balancing techniques are not adequate to obtain the high degree of balance required and to ensure low vibration levels during run-up and rundown. Therefore, balancing facilities are provided along the rotor in the form of taped holes in cylindrical surfaces. The manufacturer balances the rotor at operating speed. The winding is then heated, and the rotor is operated at 20 percent overspeed. This allows the rotor to be subjected to stresses higher than the ones experienced in service. Trim balancing is then conducted, if required. There is a relationship between vibration amplitude and temperature in some rotors. For example, uneven ventilation can create a few degrees difference in temperature between two adjacent poles. This effect can be partially offset by balancing to optimize the conditions at operating temperature (Fig. 13.13).

Uneven equalization of stiffness will cause vibration having a frequency of twice the operating speed. It is important to distinguish between the vibration caused by unbalance (occurring at 1 times the operating speed) and equalization of stiffness. A large crack in the rotor will have relatively larger effect on the double frequency vibration component. Vibration signals during rundown are analyzed and compared with the ones obtained in previous rundowns. Oil whirl in bearings can cause vibration at one-half the speed. The amplitude and phase of vibration are recorded at the bearings of the generator and exciter, using accelerometers mounted on the bearing supports and by proximity probes, which detect shaft movements.

Bearings and Seals

The generator bearings are spherically seated to facilitate alignment. They are pressure-lubricated, have jacking-oil taps, and are insulated from the pedestals. Seals (Fig. 13.14) are provided in the end shields to prevent hydrogen from escaping along the shaft. Most seals have a nonrotating white-metal ring bearing against a collar on the shaft. Oil is fed to an annular groove in the ring. It flows radially inward across the face into a collection space and radially outward into an atmospheric air compartment. The seal ring must be maintained against the rotating collar. Therefore, it must be able to move axially to accommodate the thermal expansion of the shaft. Figure 13.15 illustrates a seal that resembles small journal bearings (radial seal). The oil is applied centrally. It flows axially inward to face the hydrogen pressure. It also flows axially outboard into an atmospheric compartment. The seal does not have to move axially, because the shaft can move freely inside it. This is a major advantage over the seal design illustrated in Fig. 13.14. Most generators use radial seals.

13.14 CHAPTER THIRTEEN

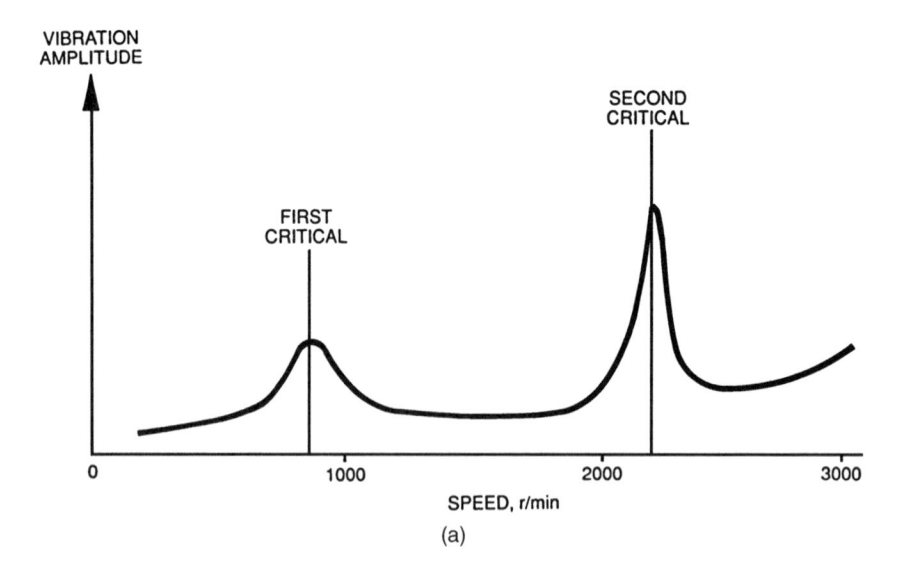

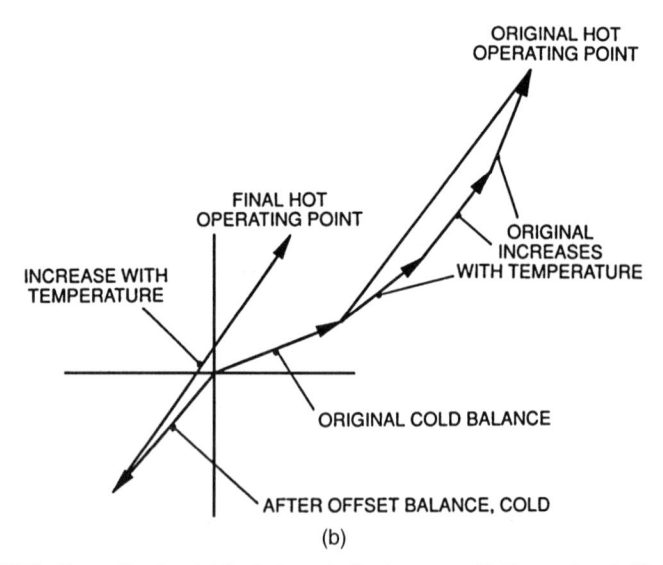

FIGURE 13.13 Rotor vibration. (*a*) Typical speed-vibration curve. (*b*) Vector plot of offset balancing. (Vectors represent amplitude and phase angle of shaft displacement or sinusoidal velocity.)

Size and Weight

The rotor of a 660-MW generator is up to 16.5 m long and weighs up to 75 tons. The rotor must never be supported on its end rings. The weight must be supported by the body surface. The rotor must also be protected from water contamination, while in transient or storage. A weatherproof container with an effective moisture absorbent must be used. If the rotor is left inside an open stator, dry air must be circulated.

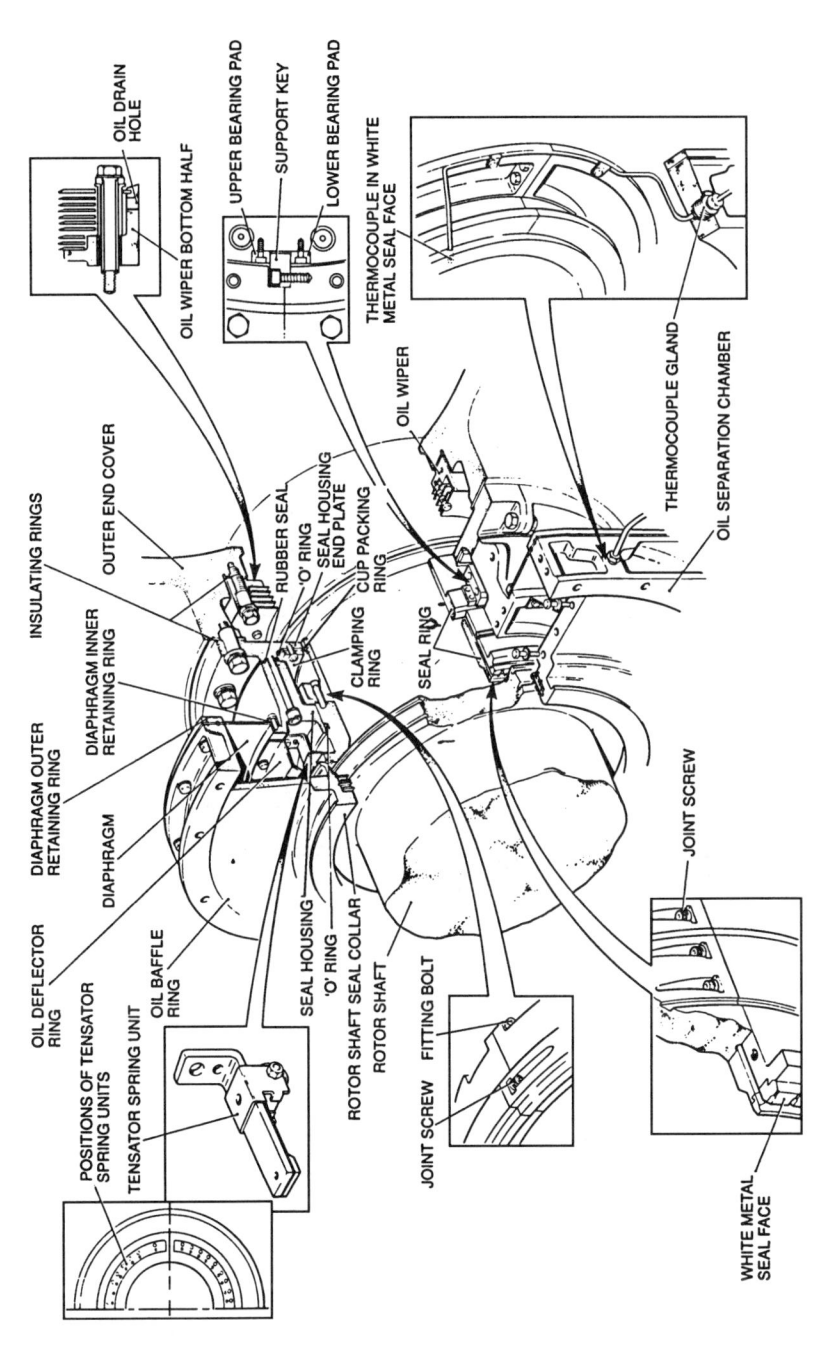

FIGURE 13.14 Thrust-type shaft seal.

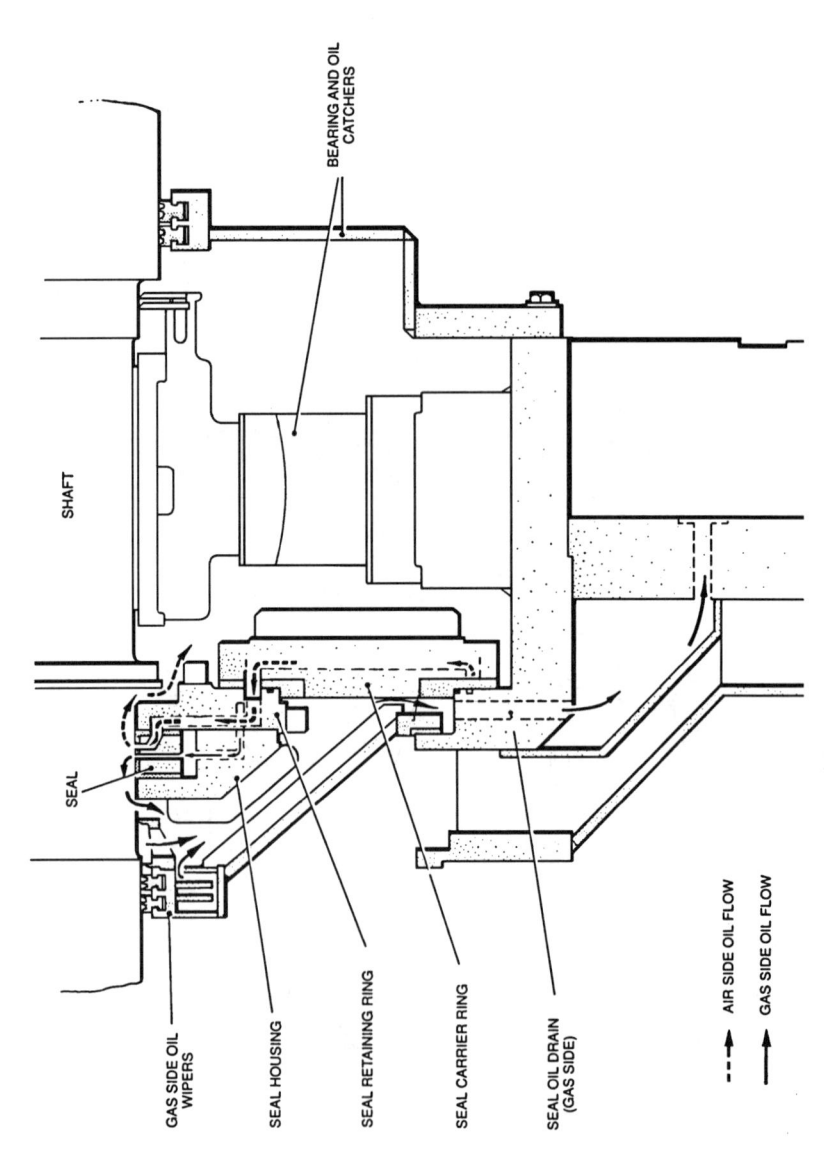

FIGURE 13.15 Double-flow ring seal (radial seal).

TURBINE-GENERATOR COMPONENTS: THE STATOR

Stator Core

The core laminations are normally 0.35 or 0.5 mm thick. They are coated with thin layers of baked-on insulating varnish. Core flux tests are done on the complete core with a flux density of 90 to 100 percent of the rated value. If there is contact between two adjacent plates, local hot spots will develop. The stator bore is scanned using an infrared camera to identify areas of higher-than-normal temperature during such a test. A bonding agent is used in some designs to ensure that individual plates and particularly the teeth do not vibrate independently. Packing material is used to correct any waviness in core buildup.

Some designs use grain-oriented sheets of steel. They have deliberately different magnetic properties in the two perpendicular axes (Fig. 13.16). The low-loss orientation is arranged for the flux in the circumferential direction. This allows higher flux density in the back of the core compared with nonoriented steel, for the same specific loss. The core plates of grain-oriented steel are specially annealed after punching.

The net axial length of magnetic steel that the flux can use is less than the measured stacked length by a factor of 0.9 to 0.95. This is known as the *stacking factor*. This is caused by the varnish layers and airspaces between the laminations due to uneven plate thickness and imperfect consolidation.

Hysteresis and eddy current losses in the core constitute a significant portion of the total losses. In some designs, the heat produced by these losses is removed by hydrogen circulating radially through the ducts and axially through holes (Fig. 13.17).

Thermocouples are installed in the hottest areas of the core. If a hot spot develops in service, it will not normally be detected by existing thermocouples. A flux test is the way to detect a hot spot. If accidental contacts occur at the tooth tips or there is damage to the slot surfaces, circulating currents could occur. The magnitude of the current depends on the contact resistances between the back of the core plates and the core frame bars on which the plates are assembled. In most designs, all these bars (except for the one which grounds the core) are insulated from the frame to reduce the possibility of circulating currents.

Core Frame

Figure 13.18 illustrates the core frame. The core end plate assembly is normally made from a thick disk of nonmagnetic steel. Conducting screens of copper or aluminum about 10 mm thick cover the outer surfaces of the core end plates (Fig. 13.19). They are called the end plate flux shield. The leakage flux creates circulating currents in these screens. These currents prevent the penetration of an unacceptable amount of flux into the core end plate or the ends of the core.

Stator Winding

In large two-pole generators, the winding of each phase is arranged in two identical parallel circuits, located diametrically opposite each other (Fig. 13.20). If the conductor is made of an assembly of separate strips, the leakage flux (the lines of induction that do not engage the rotor) density of each strip increases linearly with distance from the bottom strip (Fig. 13.21). This alternating leakage flux induces an alternating voltage along the lengths of the strips that varies with the square of the distance of the strip from the bottom of the slot. If

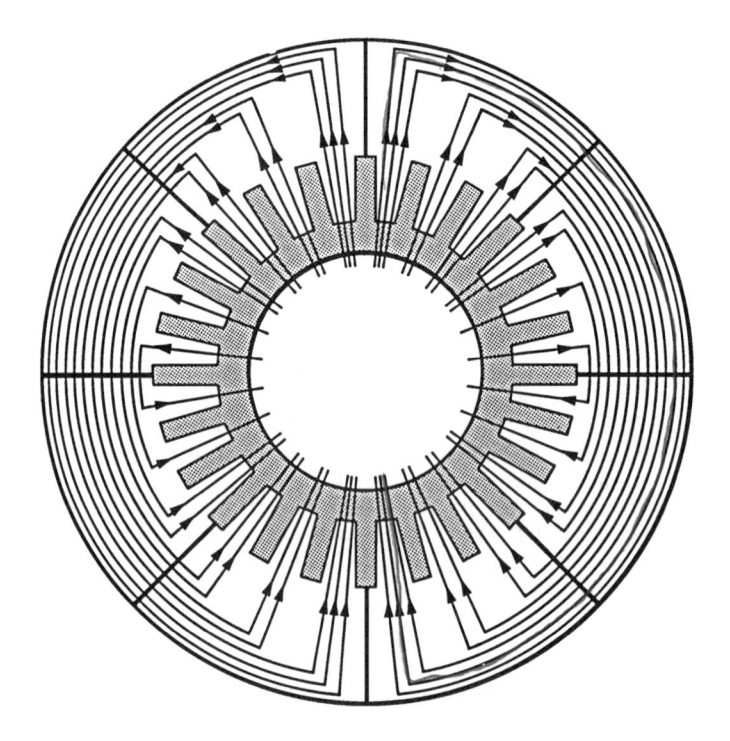

DIAGRAMMATIC REPRESENTATION OF MAGNETIC FLUX
IN STATOR CORE, OPEN-CIRCUIT CONDITIONS

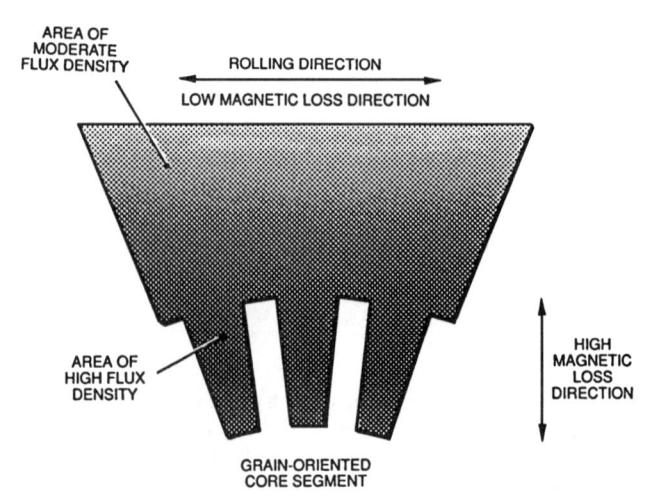

FIGURE 13.16 Flux in stator core.

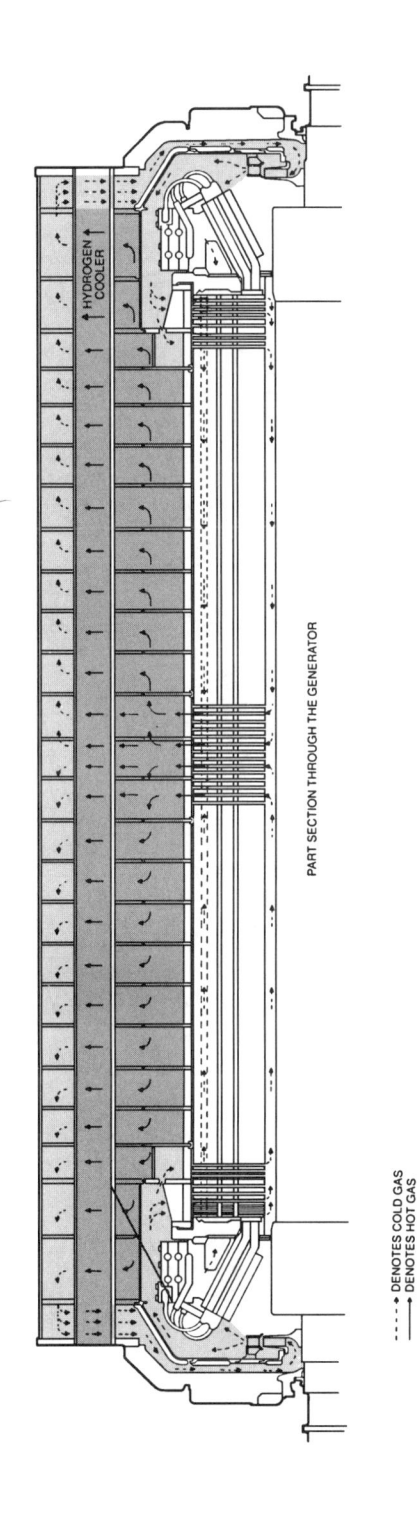

FIGURE 13.17 Stator ventilation.

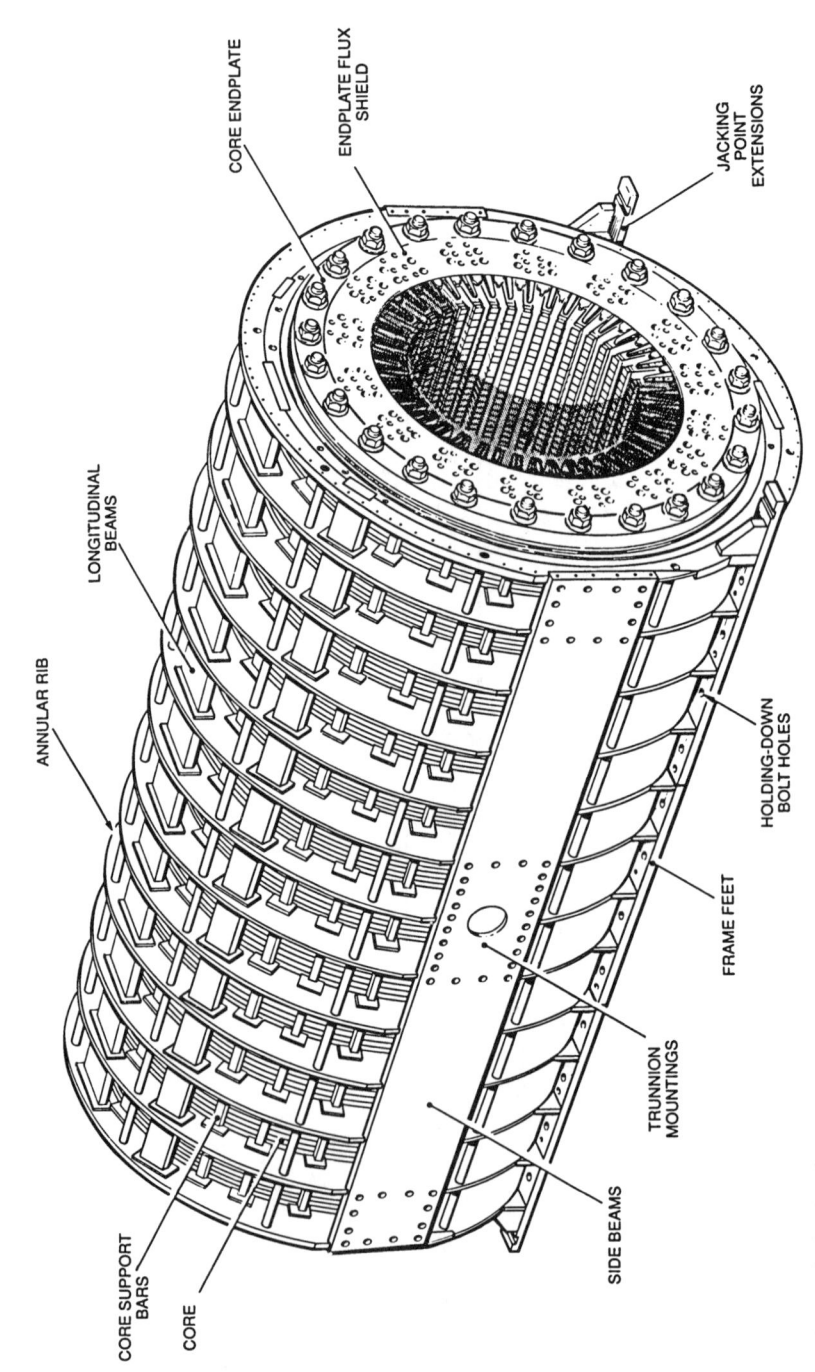

FIGURE 13.18 Core frame.

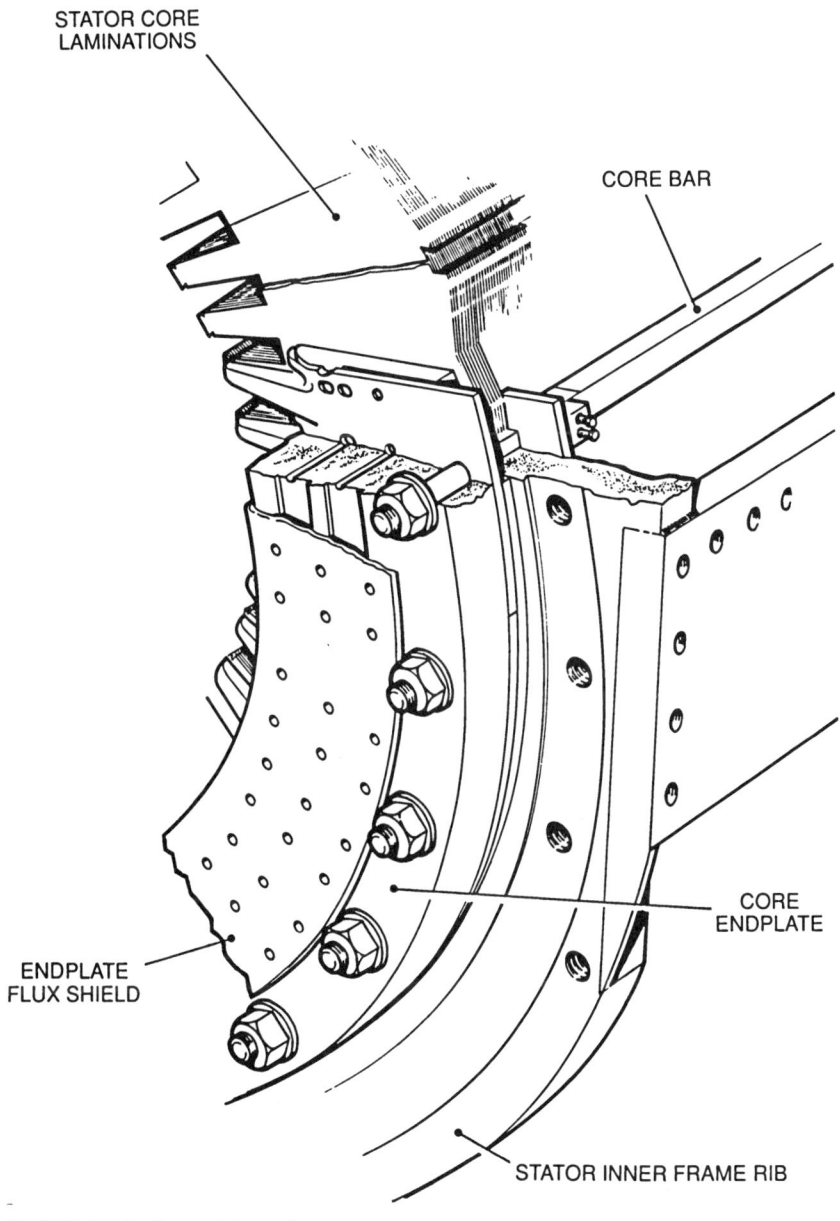

FIGURE 13.19 Core end plate and screen.

13.22 CHAPTER THIRTEEN

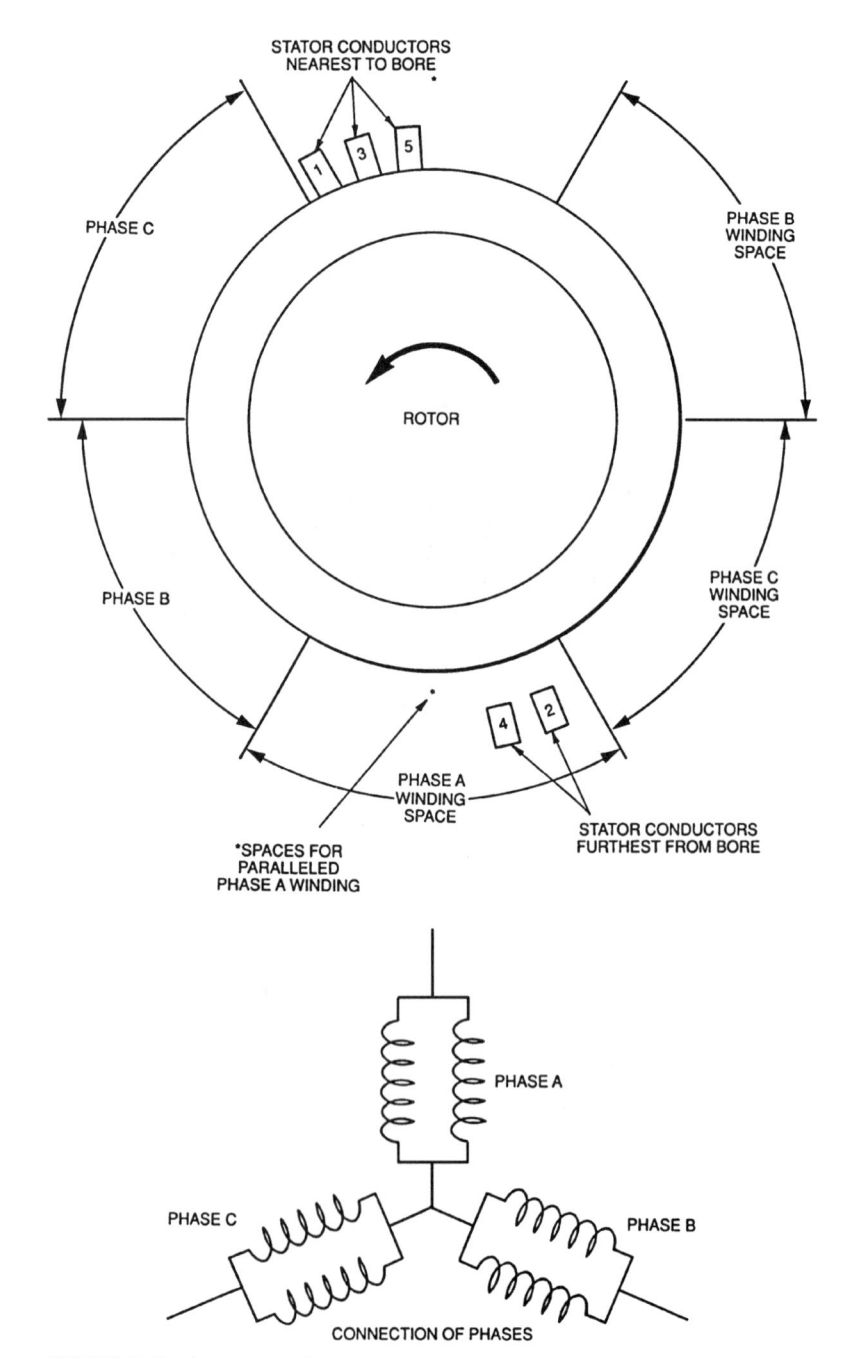

FIGURE 13.20 Arrangement of stator conductors.

FIGURE 13.21 Variations of eddy currents in stator conductors.

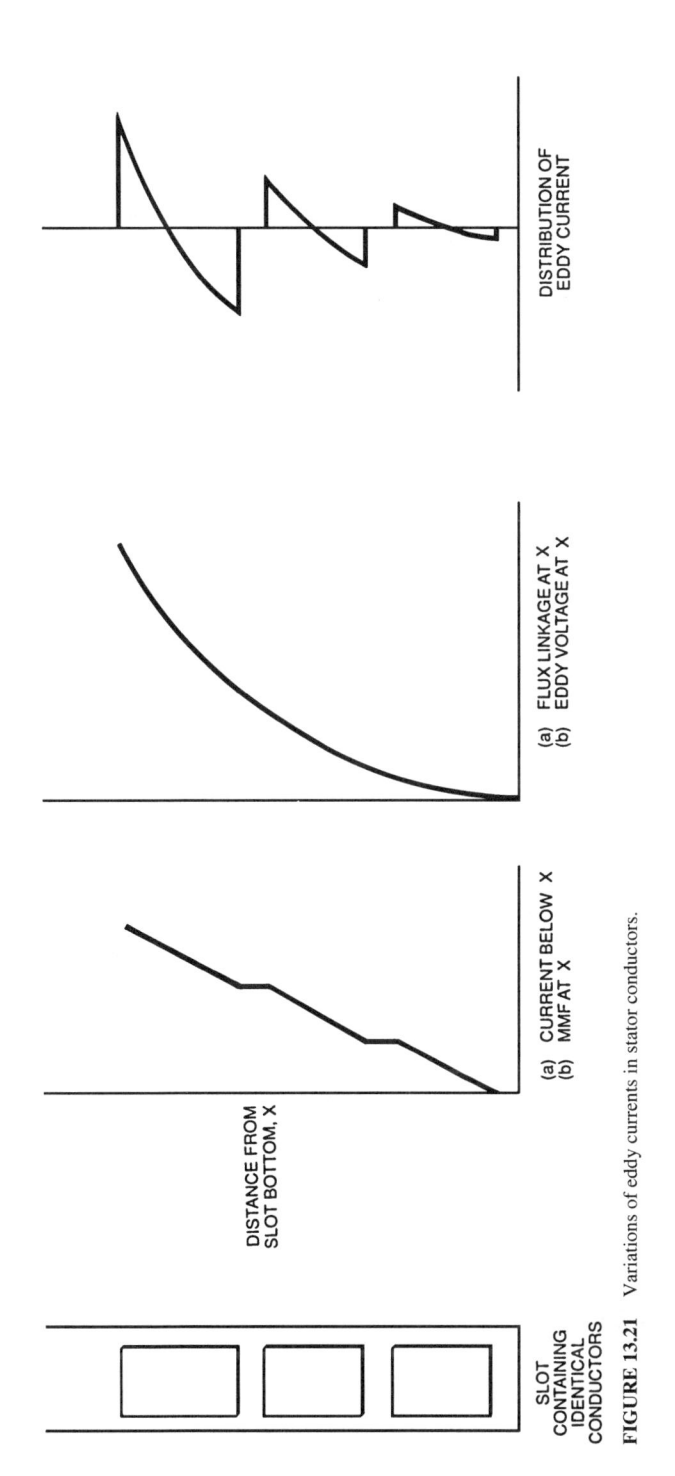

13.24　　　　　　　　　　　　　　　CHAPTER THIRTEEN

a solid conductor is used, or if the strips are parallel to each other and connected together at the core ends, currents will circulate around the bar due to the unequal voltages. This will cause unacceptable eddy current losses and heating. This effect is minimized by dividing the conductor into lightly insulated strips. These strips are arranged in two or four stacks in the bar width. They are transposed along the length of the bar by the Roebel method (Fig. 13.22). Each strip occupies every position in the stack for an equal axial distance. This arrangement equalizes the eddy current voltages, and the eddy currents will not circulate between the strips. Demineralized water circulates in the rectangular section tubes to remove the heat from the strips (Fig. 13.23).

The conductors are made of hard-drawn copper having a high conductivity. Each strip has a thin coating of glass-fiber insulation. The insulation wound along the length of the bar consists of a tape of mica powder loaded with a synthetic resin, with a glass-fiber backing. Electrical tests are performed to confirm the integrity of the insulation. A semiconducting material is used to treat the slot length of each bar to ensure that bar-to-slot electrical discharges do not occur. The surface discharge at the ends of the slots is limited by applying a high-resistance stress grading finish.

The bars experience large forces because they carry large currents and they are placed in a high flux density. These forces are directed radially outward toward the bottom (closed end) of the slot. They alternate at 120 Hz. The closing wedges are not therefore needed to restrain these bars against these forces. However, the bars should not vibrate. The wedges are designed to exert a radial force by tapered packers or by a corrugated glass spring member. Some designs have a sideways restraint by a corrugated glass spring packer in the slot side. Insulation materials consisting of packers, separators, and drive strips are also used in the slot (Fig. 13.23).

The stator winding electrical loss consists of I^2R heating (R is measured using the dc resistance of the winding phases at operating temperature) and *stray* losses which include:

FIGURE 13.22　Roebel transpositions.

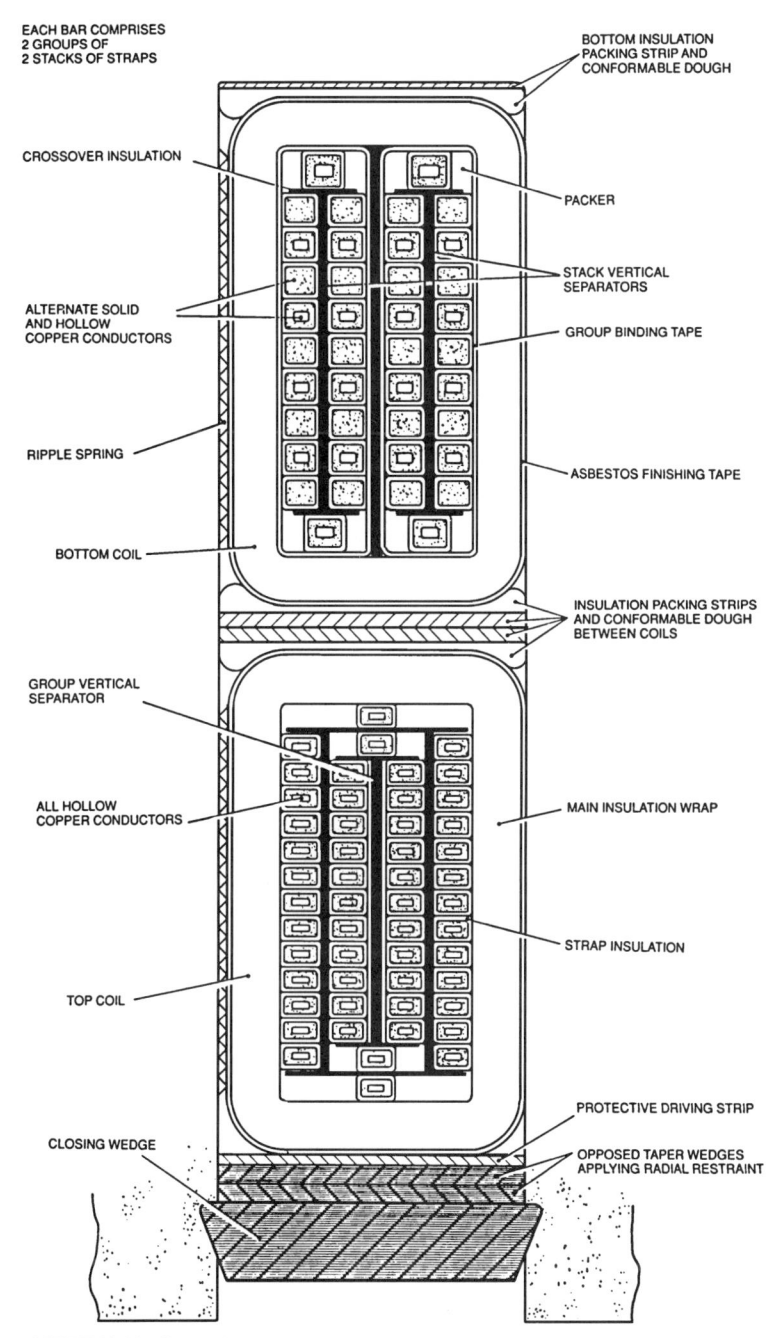

GENERATOR COMPONENTS, AUXILIARIES, AND EXCITATION   13.25

FIGURE 13.23 Stator slot.

1. AC resistance that is larger than dc resistance (skin effect).
2. Eddy currents (explained earlier).
3. Currents induced in core end plates, screens, and end teeth.
4. Harmonic currents induced in the rotor and end ring surfaces.
5. Currents induced in the frame, casings, end shields, fan baffles, etc. Appropriate cooling methods are needed for these losses in order to avoid localized hot spots.

End Winding Support

Bands of conductors are arranged side by side in the end windings. They all carry the same currents (some are in phase with each other and others are not). Large electromagnetic forces are produced in the end windings during normal operation, and especially during fault conditions when large current peaks occur. The end turns must be strongly braced to withstand these peak forces and minimize the 120-Hz vibrations.

A large magnetic flux is produced in the end regions by the mmf in the end windings of the stator and rotor. Metallic components cannot be used to fasten the end winding because

1. They would have eddy currents induced in them. This will cause additional loss and possibly hot spots.
2. Metallic components also vibrate and tend to become loose, or wear away their surrounding medium.

Therefore, nonmetallic components such as molded glass fiber are normally used. Large support brackets are bolted to the core end plate. They provide a support for a large glass-fiber conical support ring (Fig. 13.24). The vibration of the end windings must be limited because it can create fatigue cracking in the winding copper. This can have particularly serious consequences if it occurs in a water-carrying conductor because hydrogen will leak into the water system. Resonance near 120 Hz must be avoided because the core ovalizing and the winding exciting force occur at this frequency. Vibration increase in the end windings due to slackening of the support is monitored by accelerometers. The amplitude of vibration depends highly on the current. Any looseness developed after a period of operation is corrected by tightening the bolts, inserting or tightening wedges, and/or pumping a thermosetting resin into rubber bags located between conductor bars.

Electrical Connections and Terminals

The high end (line end) conductor bars and the low voltage end (neutral end) of a phase band are electrically connected to tubular connectors. These connectors run circumferentially behind the end windings at the exciter end to the outgoing terminals. The connectors have internal water cooling. However, they must be insulated from the line voltage. Figure 13.25 illustrates a terminal bushing. It is a paper-insulated item, cooled internally from the stator winding water system. The insulation is capable of withstanding the hydrogen pressure in the casing without having any leakage.

Stator Winding Cooling Components

Demineralized water is used for cooling the stator windings. It must be pure enough to be electrically nonconducting. The water is degassed and treated continuously in an ion exchanger. These are the target values:

FIGURE 13.24 View of a 660-MW generator stator end windings.

Conductivity: 100 μS/m
Dissolved oxygen: 200 μg/L maximum (in some systems >2000 is acceptable)
Total copper: 150 μg/L maximum
pH value: 9 maximum

These levels have proved to have no aggressive attack on the winding copper after many years of service. Water enters one or more manifolds made of copper or stainless steel pipes. The manifolds run circumferentially around the core end plate. Flexible PTFE hoses connect

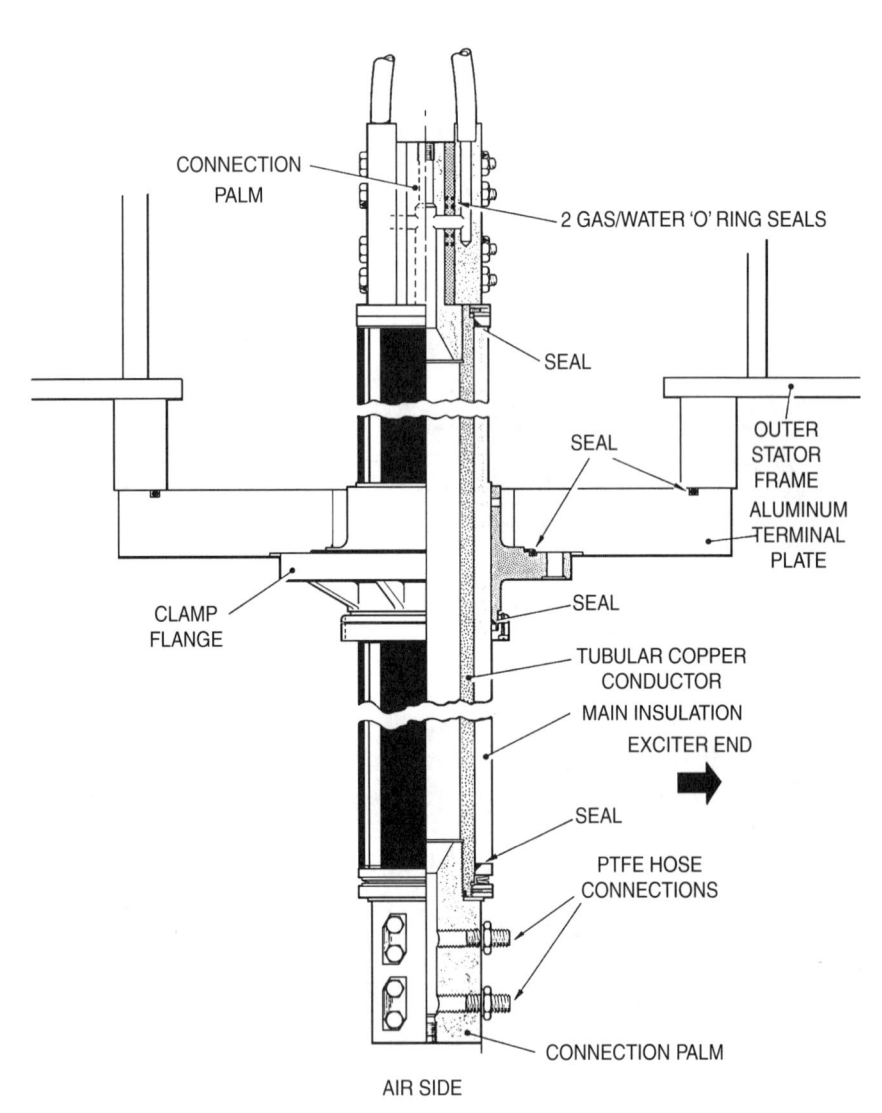

FIGURE 13.25 Generator terminals.

the manifolds to all water inlet ports on the stator conductor joints. In a two-pass design, water flows through both bars in parallel. It is then transferred to the two connected bars at the other end. The water returns through similar hoses to the outlet manifold (Fig. 13.26).

The hydrogen is maintained at a higher pressure than the water. If a leak develops, hydrogen enters the water. The winding insulation will be damaged if the water enters the hydrogen system. The water temperature increases by less than 30°C. The inlet temperature of the water

GENERATOR COMPONENTS, AUXILIARIES, AND EXCITATION   13.29

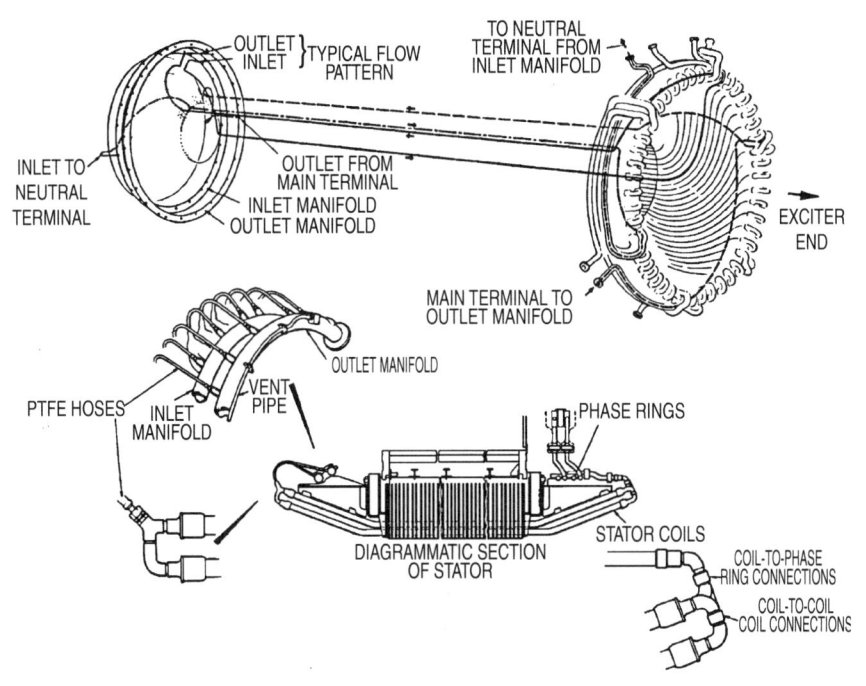

FIGURE 13.26 Stator winding water cooling system components.

is 40°C. There is a significant margin before boiling occurs at 115 to 120°C (at the working pressure). The water temperature of each bar is monitored by thermocouples in the slots or in the water outlets. This allows detection of reduced water flow.

Hydrogen Cooling Components

Hydrogen is brought into the casing by an axially oriented distribution pipe at the top. Carbon dioxide is used to scavenge the hydrogen (air cannot be used for this function because an explosive mixture of hydrogen and oxygen will form when the volumetric concentration is between 4 and 76 percent). The carbon dioxide is admitted through a similar pipe at the bottom. The rotor fans drive hydrogen over the end windings and through the cores of the stator and rotor. During normal operation, the hydrogen temperature increases by about 25°C during the few seconds required to complete the circuit. Two or four coolers are mounted inside the casing. They consist of banks of finned or wire-wound tubes. The water flows into the tubes while hydrogen flows over them (Fig. 13.27).

The headers of the coolers are accessible. The tubes can be cleaned without degassing the casing. The supports of the tubes and the cooler frame are designed to avoid resonance near the principal exciting frequencies of 60 and 120 Hz. It is important to prevent moisture condensation on the stator end windings (electrical breakdown can occur). The dew point of the hydrogen emerging from the coolers is monitored by hygrometers. This dew point must be at least 20°C lower than the temperature of the cooled hydrogen emerging from the coolers. During normal operation, the stator winding temperature is above 40°C. Thus if condensation occurred, it will be on the hydrogen coolers first. During start-up, the

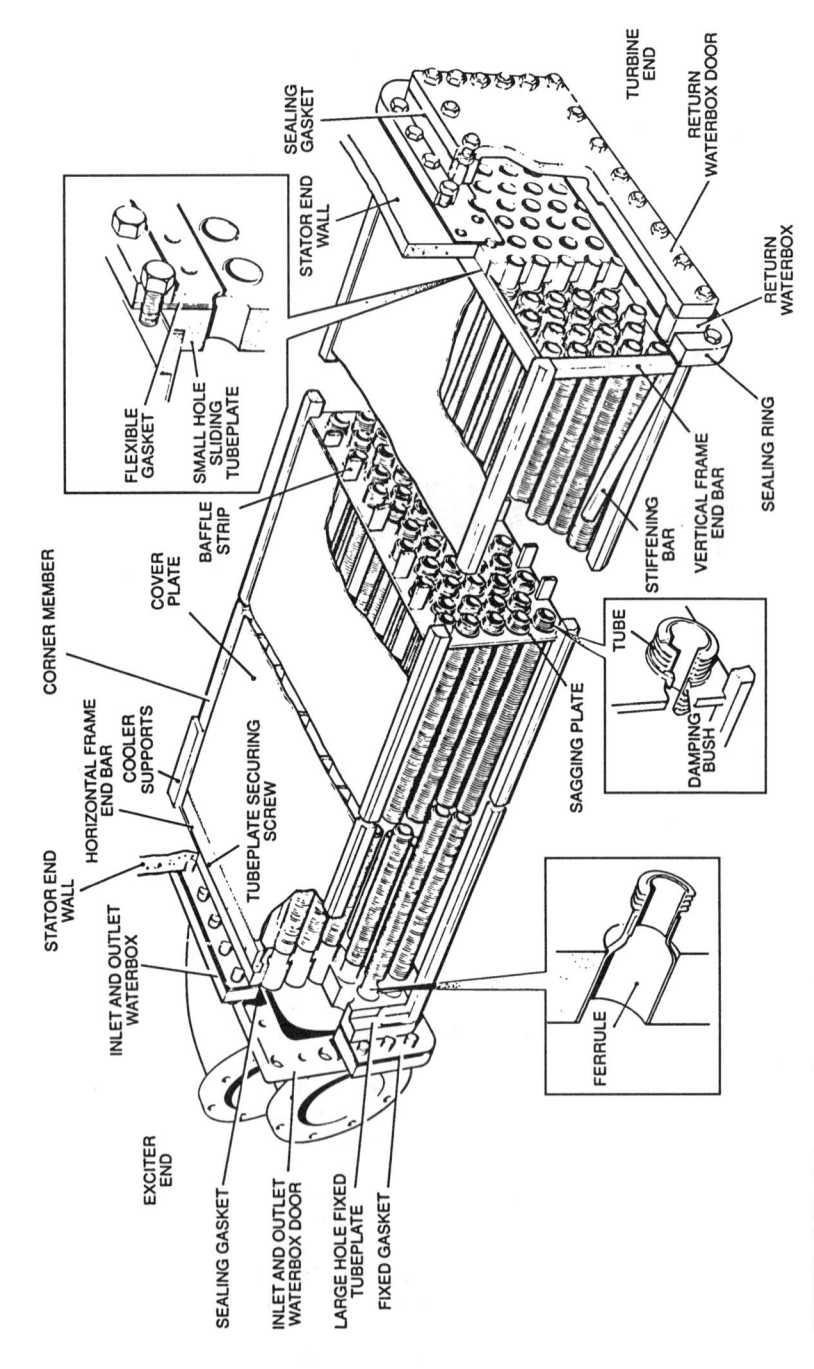

FIGURE 13.27 Hydrogen cooler.

cooling water of the stator windings is cold. It is preheated electrically, or circulated for a period of time to increase the winding temperature before exciting the generator. This prevents the possibility of having condensation on the windings.

Stator Casing

The stator core and core frame are mounted inside the casing. The casing must withstand the load and fault torques. It must also provide a pressure-tight enclosure for the hydrogen. Annular rings and axial members are mounted inside the casings to strengthen them and allow the hydrogen to flow (Figs. 13.28 and 13.29).

The end shields are made of thick circular steel plates. They are reinforced by ribs to withstand the casing pressure with minimal axial deflection. The stationary components of the shaft seal are housed in the end shields. The outboard bearing is also housed inside the end shields in some designs. The sealing of the end shield and casing joints must be leak-free against hydrogen pressure.

A hydrostatic pressure test is conducted on the whole casing. The casing must also be leak-tight when the hydrogen pressure drops from 4 to 0.035 bar in 24 h. Any leaks of oil or water are drained from the bottom of the casing to liquid leakage detectors. These detectors initiate an alarm. A temperature sensor is installed at the CO_2 inlet. It initiates an alarm if the incoming CO_2 has not been heated sufficiently. Cool gas can create unacceptably high localized thermal stresses. Electric heaters are mounted in the bottom half of the casing. They prevent condensation during outages.

COOLING SYSTEMS

The efficiency of a large generator is about 98.5 percent. In some designs, the losses are transferred to the boiler feedwater system.

Hydrogen Cooling

Hydrogen has several advantages over air for heat removal from the generator:

1. The density of hydrogen is one-fourteenth that of air. The windage losses (caused by churning the gas around the rotor) are much less with hydrogen.
2. The heat removal capability of hydrogen (at operating pressure) is about 10 times higher than the one for air.
3. The degradation by oxidation processes cannot occur because hydrogen is free from oxygen.
4. Hydrogen does not support fire, which can start by arcing.

The main disadvantage of hydrogen is that it forms an explosive mixture when it combines with air within a volumetric concentration in the range of 4 to 76 percent. Sophisticated sealing arrangements are required to ensure leak-tight casing.

Hydrogen Cooling Systems

It is essential to prevent air and hydrogen mixture inside the generator. Carbon dioxide is used as a buffer gas between air and hydrogen. The process is called *scavenging* or *gassing up* and

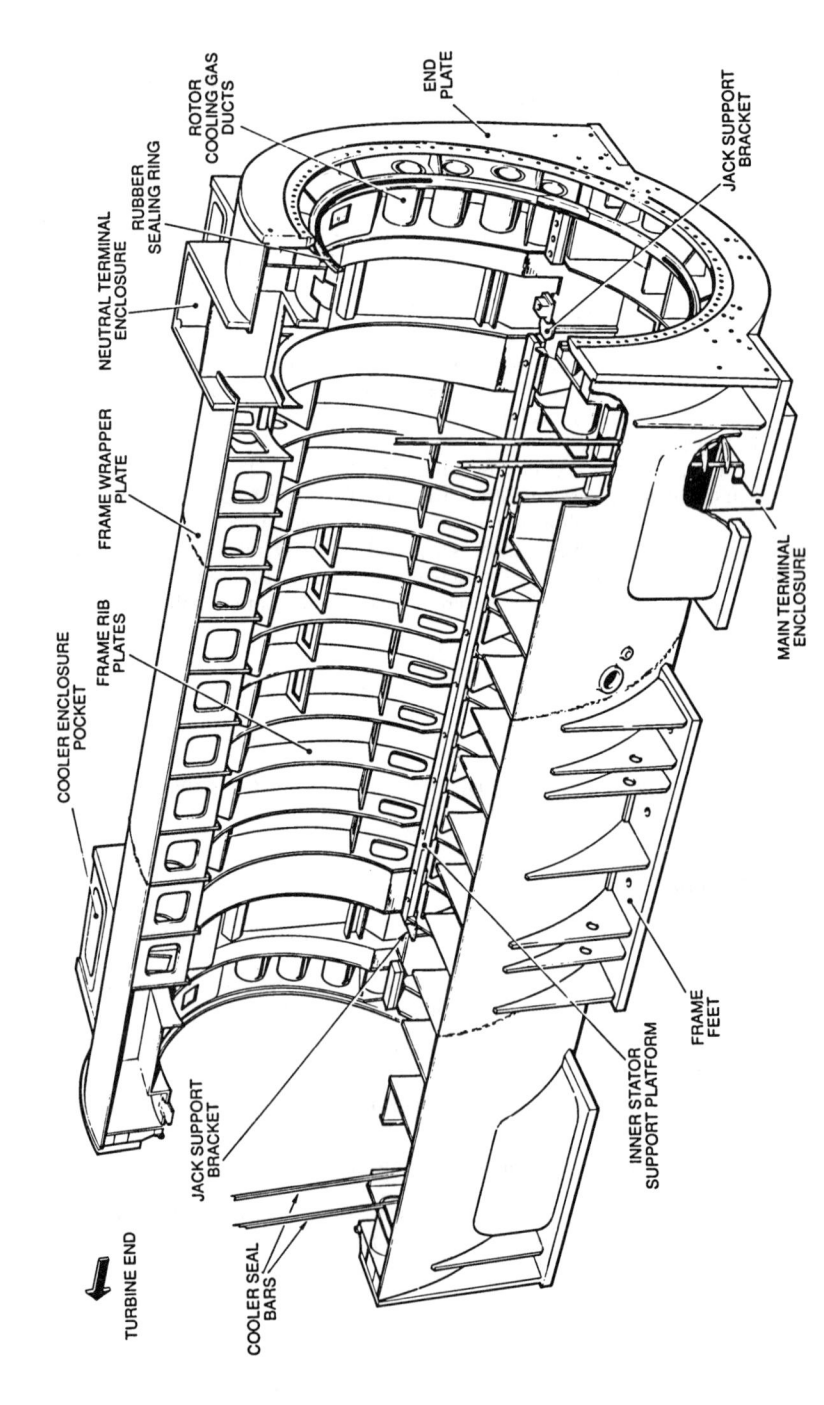

FIGURE 13.28 Outer stator casing.

FIGURE 13.29 Core frame being inserted into casing.

degassing. Carbon dioxide is normally stored as a liquid. It is expanded to a low pressure above atmospheric. It is also heated to prevent it from freezing due to the expansion process. CO_2 is fed into the bottom of the casing through a long perforated pipe. It displaces the air from the top via the hydrogen inlet distribution pipe to atmosphere outside the station. The proportion of CO_2 in the gas passing to atmosphere is being monitored by a gas analyzer. When the CO_2 concentration becomes sufficiently high, the flow of CO_2 is interrupted (Fig. 13.30).

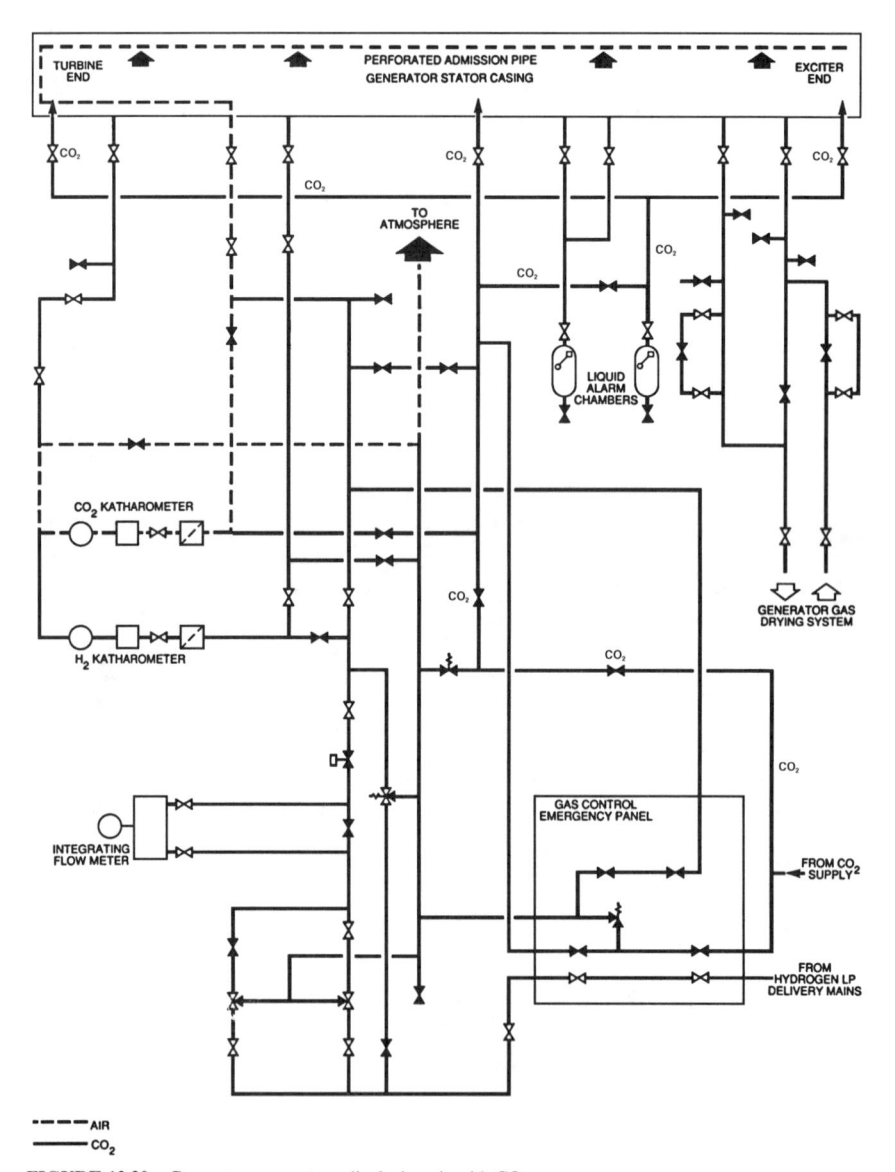

FIGURE 13.30 Generator gas system, displacing air with CO_2.

High-purity hydrogen is fed to the casing from a central storage tank or electrolytic process. The hydrogen reaches the gas control panel at about 10 bar. Its pressure is reduced before it flows through the top admission pipe into the casing. Since hydrogen is much lighter than CO_2, it displaces the CO_2 from the bottom of the casings through the CO_2 pipe to atmosphere. The reverse of this procedure is followed to remove hydrogen from the generator for long outages.

Separate procedures are used to scavenge tanks to prevent dangerous mixtures. The reverse of the listed procedure is done using CO_2 and dry compressed air to remove hydrogen from the generator. The hydrogen purity is normally high because air cannot enter a pressurized system.

A sample of casing hydrogen is circulated continuously through a Katharometer-type purity monitor (the sample is driven by the differential pressure developed across the rotor fans). The monitor initiates an alarm if the purity falls below 97 percent. Pure gases from the piped supplies are used to calibrate the purity monitor (and the gas analyzer).

Hydrogen is admitted by a pressure-sensitive valve when the casing pressure drops. A relief valve releases hydrogen to the atmosphere if the pressure becomes excessive. The hydrogen makeup is normally monitored.

Several thermocouples are used to monitor the hydrogen temperature. Hydrogen is flowing at 30 m^3/s typically (in a 500-MW generator). The heat absorbed by the hydrogen is about 5 MW. The increase in temperature of the hydrogen is about 30°C. The cooled gas should not be at a higher temperature than 40°C. Thus, the hydrogen entering the coolers should not be hotter than 70°C.

The water pressure in the stator windings and hydrogen coolers is lower than the hydrogen pressure. Therefore, water cannot leak into the hydrogen system from the stator windings or hydrogen cooler. However, water can be released from the oil used for shaft sealing. The water concentration will increase if the oil is untreated turbine lubricating oil which has picked up water from the glands of the steam turbines. The moisture concentration in the hydrogen should be kept low to prevent condensation on the windings. The differential pressure across the rotor fans is used to send a hydrogen flow through a drier. When the rotor is not turning, a motor-driven blower maintains a flow through the rotor (Fig. 13.31).

Hygrometers are used to monitor the humidity of the hydrogen. The maximum permissible dew point is more than 20°C below the cold gas temperature (measured at casing pressure).

In the event of a serious seal failure, hydrogen will escape rapidly. If it encounters an ignition source such as the shaft rubbing, it will burn intensely. In this case, the hydrogen in the casing should be vented to atmosphere. CO_2 should be admitted into the casing.

SHAFT SEALS AND SEAL OIL SYSTEMS

The seals are located in the end shields. They seal the hydrogen in the machine where the rotor shaft emerges from the casing and shields. The main types of seals are thrust and journal seals.

Thrust-Type Seal

Figure 13.14 illustrates a thrust-type seal. The seal ring acts as a thrust face acting on a shaft collar. Oil is supplied to a central circumferential groove in the white-metal face of the seal ring. The oil pressure is higher than the casing hydrogen. Most of the oil flows outward due to centrifugal forces over the thrust face. It then drains into a well. A small oil flow moves inward against centrifugal forces. This flow is driven by the difference between the oil and the hydrogen into a drainage compartment, which is at hydrogen pressure. Entrained air and water can be released from this oil, resulting in contamination of the hydrogen. Therefore, it is important to minimize this oil flow.

The housing of the seal ring must be able to move axially about 30 mm to accommodate the thermal expansion of all the coupled rotors. The housing is designed to move inside a stationary member. It uses rubber sealing rings to contain the oil and exert axial pressure at the seal face.

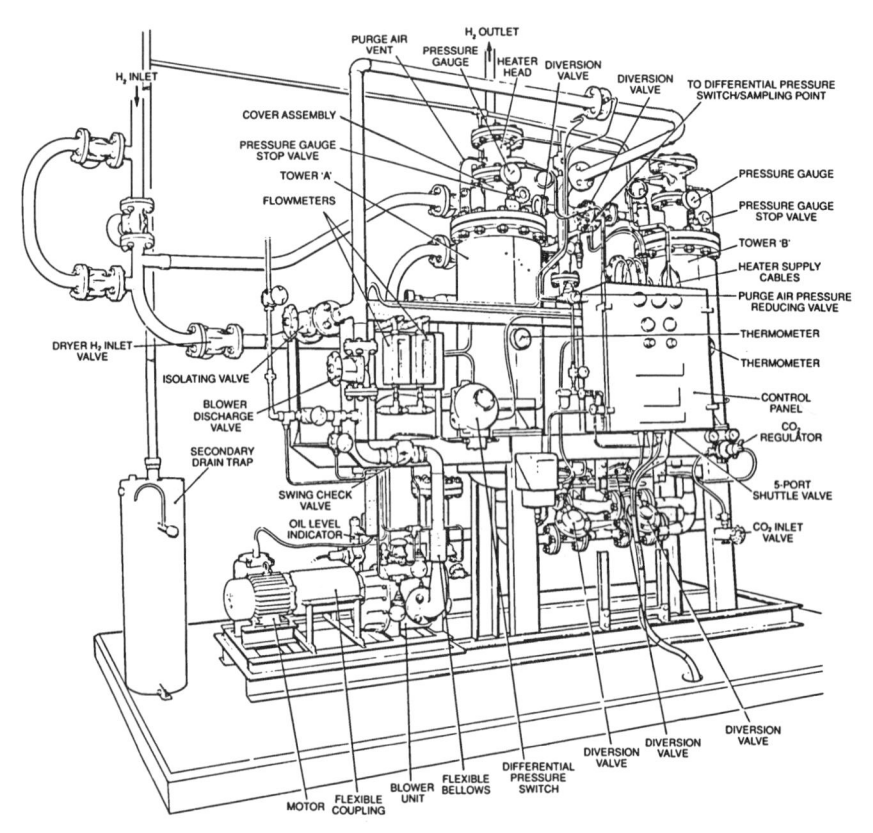

FIGURE 13.31 Gas drier and blower.

Some seal designs have an additional chamber between the fixed and sliding components. It is fed with oil at varying pressures to control the overall pressure at the seal face. Other seal designs have additional pressure provided by springs.

Journal-Type Seal

This seal design is similar to a journal bearing floating on the shaft. This design allows the shaft to move axially through the seal. Thus, it does not need to accommodate the thermal expansion of the shaft. Again, oil is supplied to an annular groove in the white-metal ring. It flows in the clearances between the shaft and the bore of the seal. The flow is outward to a drain and inward to the space pressurized by hydrogen. The inward flow rate is much larger than the inward flow rate in the thrust-type seal because it is not inhibited by centrifugal forces. Thus, this flow is capable of contaminating the hydrogen significantly. The oil fed to the seals is subjected to vacuum treatment to reduce the contamination level of the hydrogen. The treatment involves the removal of air and water from the oil. Despite this disadvantage, the journal-type seal is considered better able to handle the axial movement of the shaft.

A more sophisticated design of the journal-type seal involves two separate oil supplies (Fig. 13.15). They are for the inward and outward flows. This design eliminates the need for vacuum treatment. The oil supplied is different from the turbine lubricating oil supply, which is the main source of entrained water.

GENERATOR COMPONENTS, AUXILIARIES, AND EXCITATION 13.37

Seal Oil System

In conventional design (Fig. 13.32), the shaft-driven lubricating oil pump supplies the oil for the main seal. The oil pressure is controlled by a diaphragm valve, which maintains a constant differential pressure above the hydrogen pressure at the seals. A water-cooled heat exchanger is used to cool the oil. The oil is sent through a fine filter to

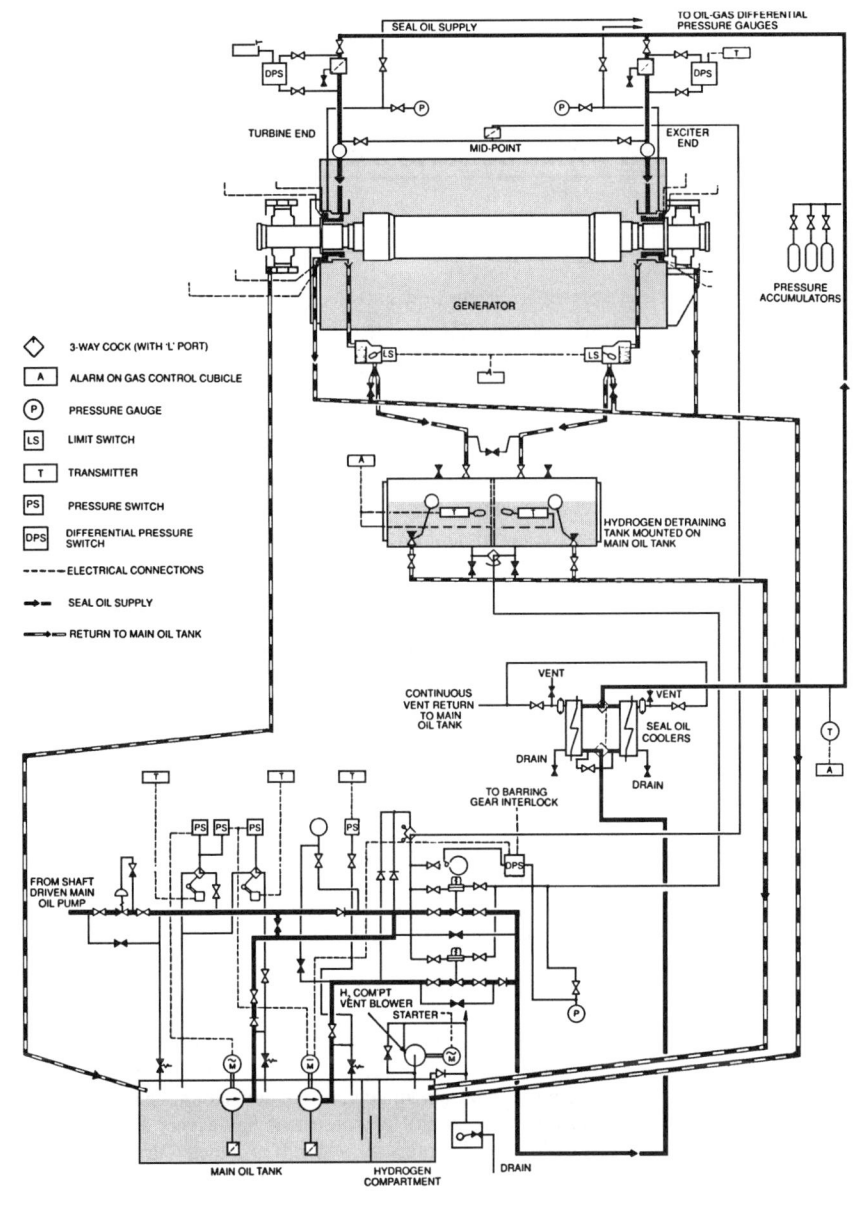

FIGURE 13.32 Seal oil system.

prevent metallic particles from reaching the tight clearances in the seal. When the unit is shut down, motor-driven pumps are used to supply the seals with oil. They are used as emergency backup. They are initiated by the dropping pressure of the seal oil. They are normally vertical submerged pumps mounted on top of the lubricating oil tank. A dc pump is also provided in case of emergencies (loss of ac power). This pump is expected to operate for a few hours only, while the hydrogen is scavenged. There is a possibility that hydrogen will enter the drain tank. Low-level alarms are normally installed. A blower is used to exhaust the gas above the oil in the tank to atmosphere. The blower reduces the pressure in the bearing housings by creating a vacuum in the tank to reduce the egress of oil vapor at the bearings.

STATOR WINDING WATER COOLING SYSTEMS

Figure 13.33 illustrates the demineralized water system used to remove the heat from the stator bars. The main criteria are as follows:

1. Very low conductivity prevents current flow and electrical flashover.
2. High-integrity insulation is used to transfer water into the conductors.
3. Low water velocity prevents erosion. Corrosion must also be prevented. Erosion or corrosion could result in a buildup of conducting material, leading to an electrical flashover.
4. The maximum water pressure must be lower than the hydrogen pressure. If leakage occurs, the hydrogen enters the water circuit. If water is allowed to enter the hydrogen, the winding insulation could be damaged.
5. The maximum water temperature should be well below saturation (boiling occurs at 115°C at system pressure) to ensure adequate heat removal capability. The normal inlet and outlet temperatures are around 40 and 67°C, respectively.

A portion of the water is circulated through a demineralizer (Fig. 13.34). All the metals in contact with the water are nonferrous or stainless steel. If the metals have even a small amount of ferrous materials, magnetite will form. They will be held by electromagnetic forces. The water flows in flexible translucent hoses made of PTFE (polytetrafluorethylene) into and out of the conductors (Fig. 13.35). In the double-pass design, the water supply enters a circular manifold. The manifold is supported from the stator core end plate. PTFE hoses are used to connect the manifold to the bars and between the top and bottom bars. The water flows in parallel channels in these bars. At the exciter end, the water is transferred through another PTFE hose to the outlet manifold. The inlet and outlet manifolds are located alongside each other. The terminal bushings and phase connections are cooled by a small flow. A higher pressure is required for the double-pass design compared with the single-pass design. However, only one-half of the number of hoses is needed. This reduces the chance of leakage. In the single-pass arrangement, the manifolds are at opposite ends. The water flows through the bars in parallel.

The water temperature rises rapidly if the flow is reduced. The differential pressure across an orifice plate or the stator windings is used to detect a reduction in flow. In this case, the standby pump should be started immediately, or the unit should be tripped.

An initial test is done on the water circuit to confirm it has a very low leak rate. However, a small quantity of hydrogen still enters the water. It is detected in a settling tank installed on the outlet side of the generator. Most of the gas is largely detrained in a header

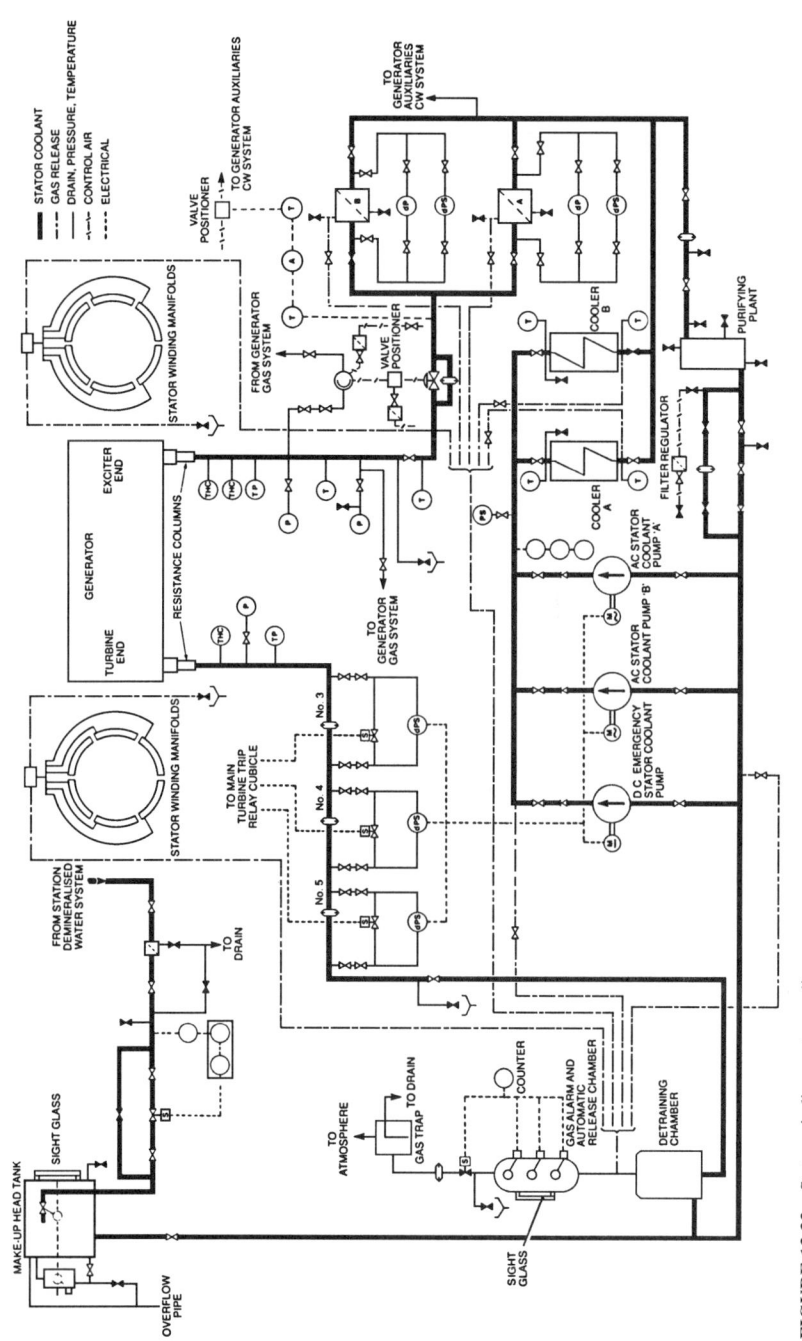

FIGURE 13.33 Stator winding water cooling system.

13.40 CHAPTER THIRTEEN

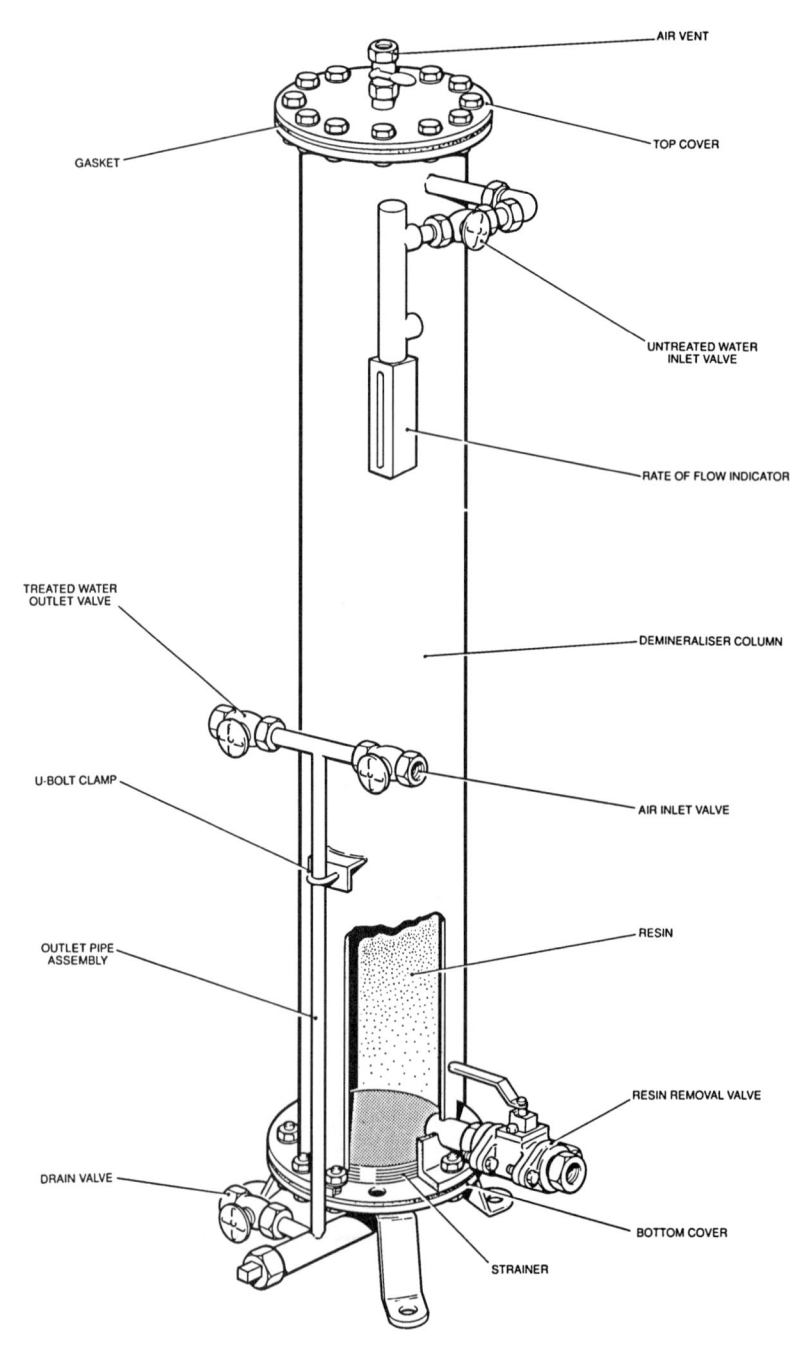

FIGURE 13.34 Demineralizer.

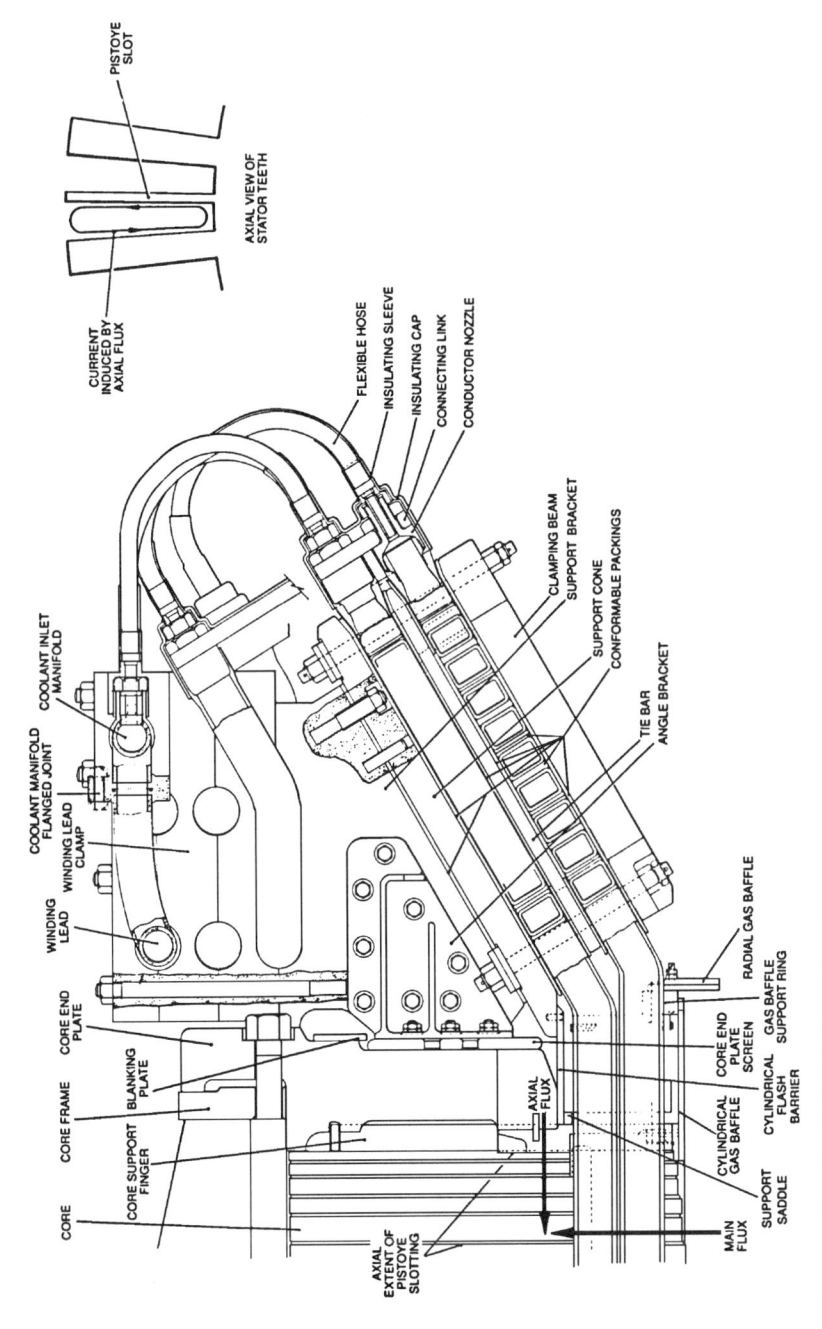

FIGURE 13.35 Water flow in the conductors.

tank. The gas is collected in a chamber equipped with timed-release valves. An alarm is initiated if the release rate exceeds a predetermined level (Fig. 13.36).

Thermocouples are installed in each of the winding slots. They detect low-flow conditions. Modern machines have a thermocouple in each outlet hose. They provide a direct indication of low flow. As noted earlier, condensation should not occur on the windings. Some generators have an electric heating element. Other designs have an automatic cooler bypassing system. It prevents cold water from circulating in the windings during start-up and low-load conditions.

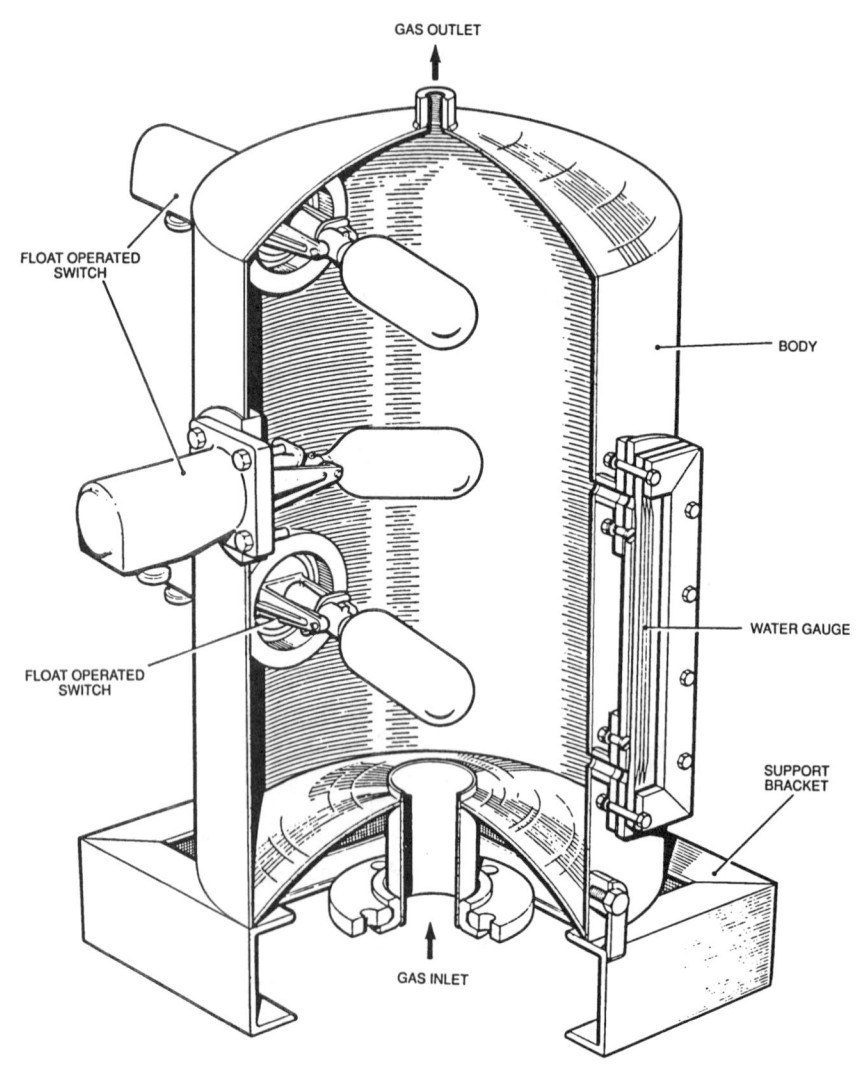

FIGURE 13.36 Gas in water detection chamber.

OTHER COOLING SYSTEMS

The hydrogen is cooled by passing it through a water-cooled heat exchanger mounted in the casing. The heat exchanger has nonferrous tubes. The heat exchangers have a double-pass water circulation to the inlet and outlet water connections at the same end. Demineralized water is used for these coolers.

Lake water is not used for these coolers due to the danger of corrosion. The hydrogen is maintained at a higher pressure than the cooling water in the heat exchanger. In the event of a leak, hydrogen will leak into the water (water ingress into the hydrogen can have serious consequences). In modern design, the water circuit has hydrogen detectors (Fig. 13.37). The hydrogen coolers have some redundancy. It is possible to operate with one hydrogen cooler isolated. The loss of cooling water is detected by an increase in the cooling temperature. The rapid increase in hydrogen temperature will cause the unit to trip. The rotating exciters and slip ring/brush gear or rotating rectifier chambers have air cooling systems. A closed air circuit with a water-cooled heat exchanger is used for the rotating exciter. Open air ventilation is normally used for the slip rings.

EXCITATION

AC Excitation Systems

Figure 13.38 illustrates a typical ac excitation scheme. It shows the shaft-mounted main and pilot exciters together with their brush gear. Permanent-magnet pilot exciters are used to minimize dependency on external power supplies. The pilot exciter provides the excitation power for the automatic voltage regulator (AVR) control equipment. A 660-MW plant has a salient-pole pilot exciter with ratings near 100 kW. The main and pilot exciters are cooled by air. Shaft-mounted fans are used to provide the cooling. The performance is monitored by measuring the temperature at the inlet and outlet of the cooling system.

Exciter Transient Performance

The ceiling requirements for exciters are considerably higher than for rated full-load conditions. The transient performance of an exciter is given by

$$\text{Exciter response ratio} = \frac{\text{average rate of increase in excitation open-circuit voltage (V/s)}}{\text{nominal excitation voltage}}$$

In a typical exciter, the output voltage needs to be increased from 100 to 200 percent within 3.5 s. Figure 13.39 illustrates the average rate of increase of excitation open-circuit voltage.

The Pilot Exciter

The permanent-magnet generator (PMG) pilot exciters used for 660-MW units are salient-pole design (Fig. 13.40). This design provides a constant voltage supply to the thyristor converter

13.44 CHAPTER THIRTEEN

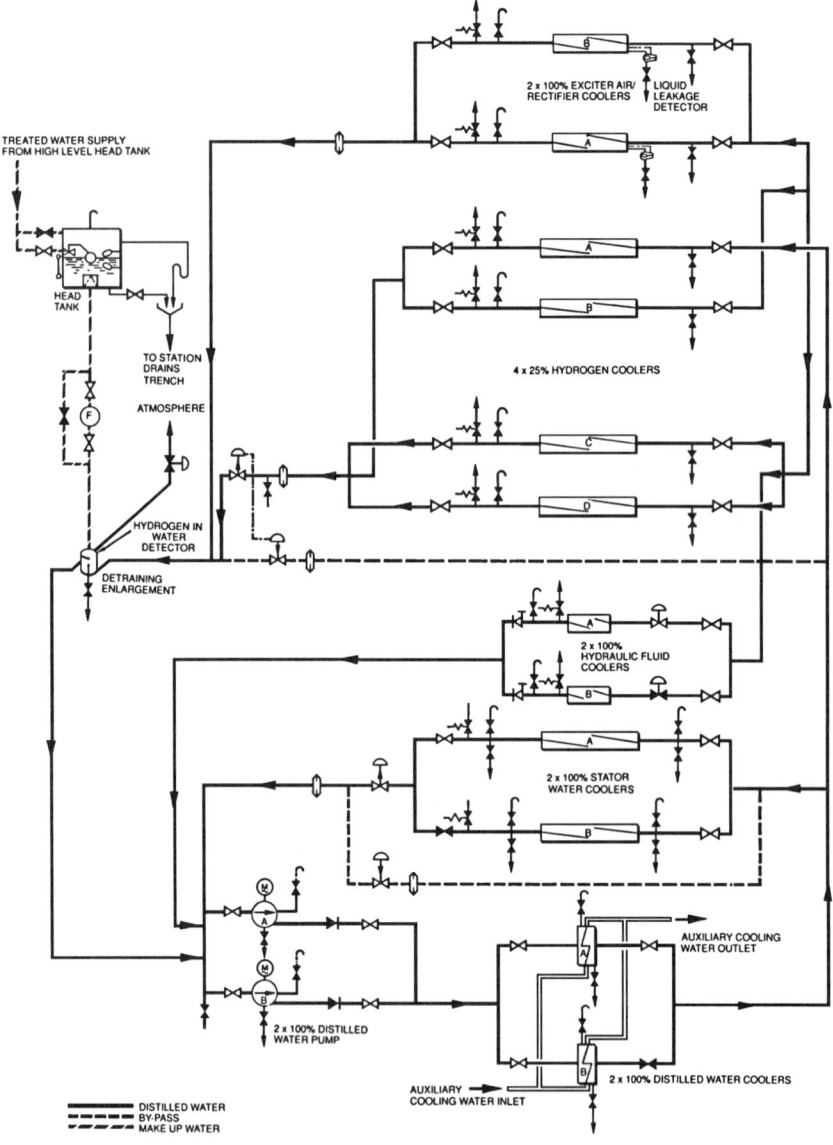

FIGURE 13.37 Distilled water cooling system.

and AVR control circuits. A high-energy material like Alcomax is used for the permanent-magnet poles. The poles are bolted to a steel hub and held in place by pole shoes. A nonmagnetic steel is used for the bolts to prevent the formation of a magnetic shunt. The pole shoes are skewed in some designs to improve the waveform of the output voltage and reduce electrical noise. The stator windings are arranged in a two-layer design. The stator conductors are insulated with polyester enamel. The coil insulation is a class F epoxy glass material.

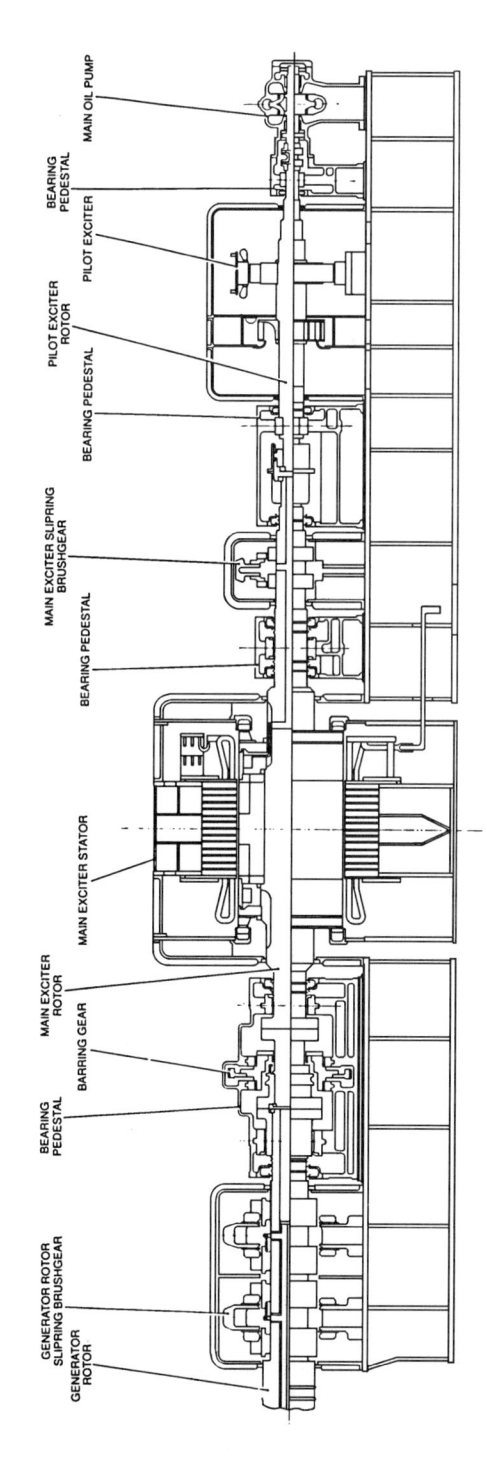

FIGURE 13.38 Section through main and pilot exciters.

13.45

13.46 CHAPTER THIRTEEN

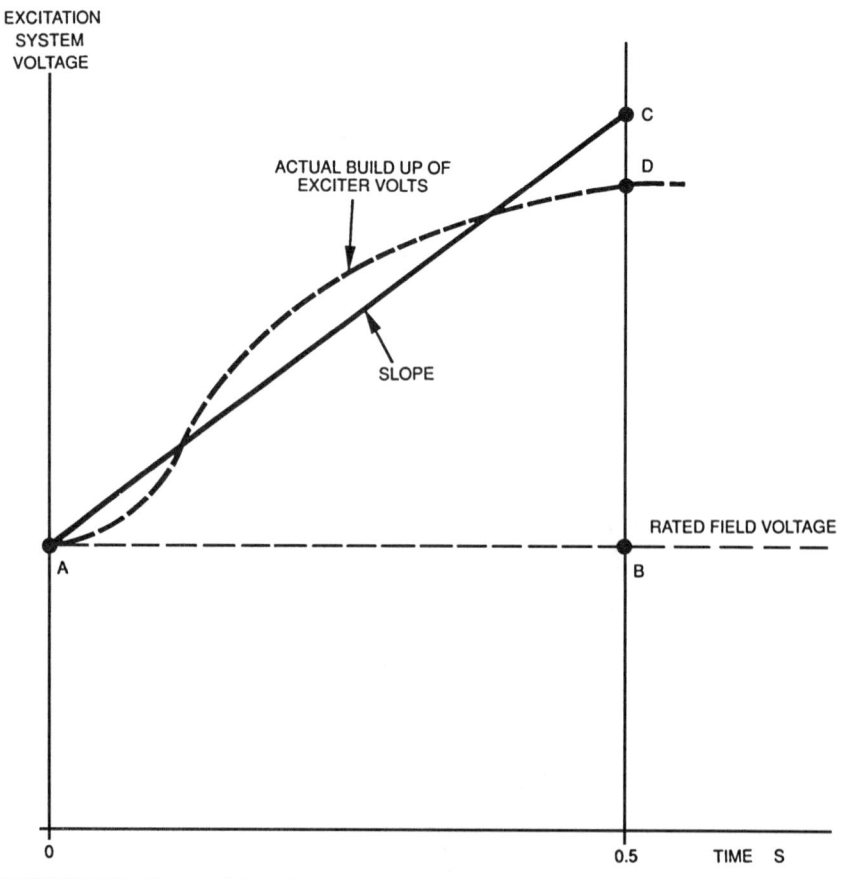

FIGURE 13.39 Concept of the exciter response ratio.

The Main Exciter

The main ac exciter has normally four or six poles (Fig. 13.41).

Exciter Performance Testing

The manufacturer of the exciters is required to perform these tests: open- and short-circuits, overspeed balancing, and high-voltage.

Pilot Exciter Protections

The pilot exciter delivers its full current during field forcing. Modern AVRs have a time/current limiter. It allows the pilot exciter to deliver maximum current during a determined interval. The current is brought back to a normal value following this interval. The main exciter, like the pilot exciter, has a considerably higher margin than required. It has a

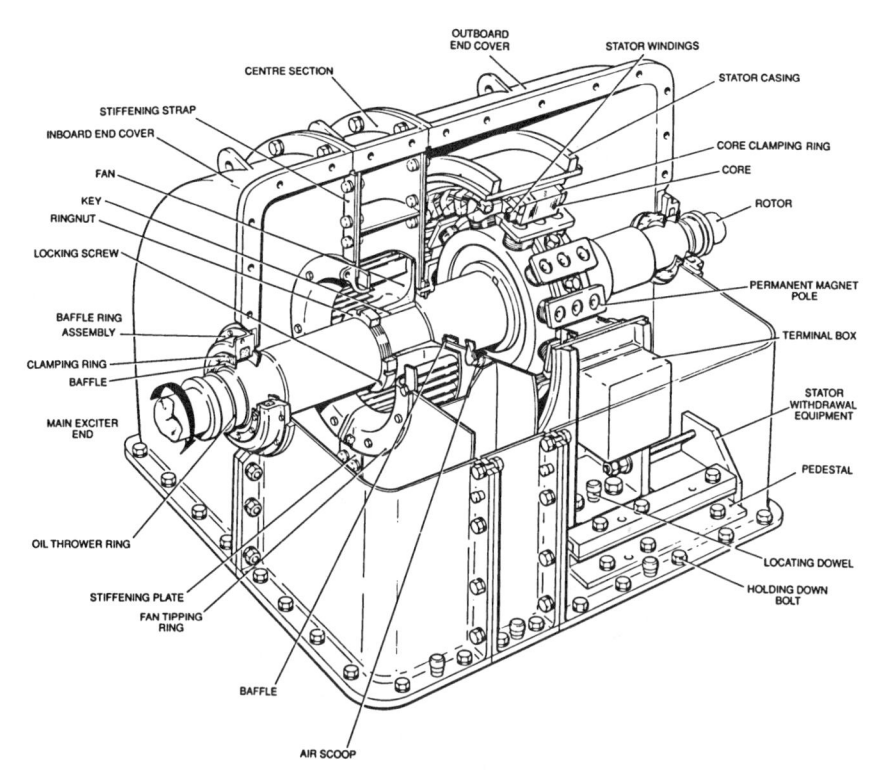

FIGURE 13.40 Salient-pole, permanent-magnet generator.

3.3-kV winding insulation despite having a working voltage of 500 V. The voltage ceiling of the main exciter is 1000 V. Its rated current is much lower than the maximum current.

Brushless Excitation Systems

Most modern gas turbines use brushless excitation systems. The rotating diodes are arranged as a three-phase bridge. The bridge arm consists of two diodes in series. If one of them fails (due to a short circuit), the second diode will continue to operate. Thus, the bridge continues to operate normally. If both diodes fail in the same arm, the fault is detected by a monitoring circuit which trips the machine. Essential measurements such as ground fault indication, field current, and voltage are taken by telemetry or instrument slip rings.

The Rotating Armature Main Exciter

Brushless machines require less maintenance than conventional ones. They also do not have sliding or rubbing electrical contacts that cause sparking and carbon dust. The main exciter is a three-phase rotating armature ac generator. The dc field is in the stator, and the ac winding is on the rotor. A typical rotating armature main exciter is illustrated in Fig. 13.42.

The exciter armature is made of low-loss steel laminations. The laminations are shrunk onto a shaft forged from annealed carbon steel. Cooling air enters axial slots along the rotor

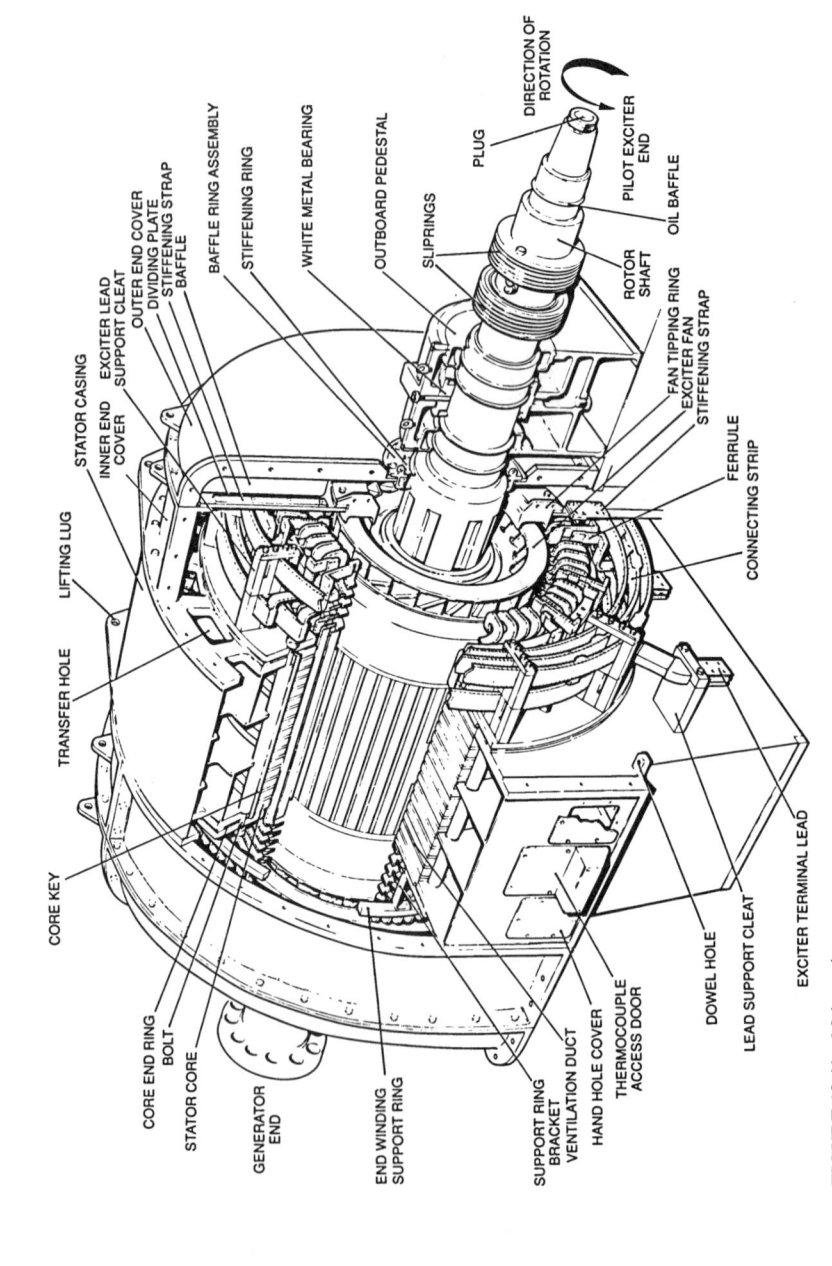

FIGURE 13.41 Main exciter.

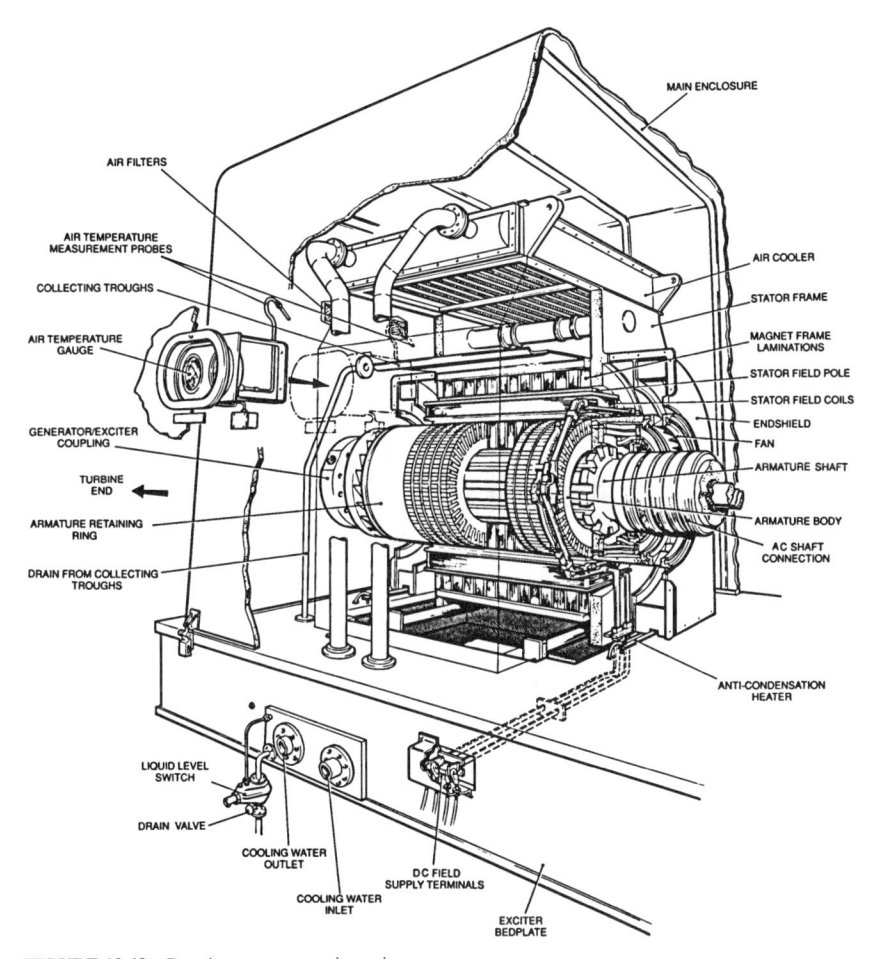

FIGURE 13.42 Rotating armature main exciter.

body. The rotor conductors are made of braided strips in parallel. They are radially transposed to reduce eddy current losses.

The rotating rectifier of a 660-MW generator is illustrated in Fig. 13.43. It is mounted on the outboard of the main ac exciter. The three-phase ac power is supplied from the main exciter to the silicon diode rectifier by axial conductors taken along the surface of the shaft. A steel retaining ring contains the components of the rectifier against centrifugal forces. The retaining ring is shrunk on the outside of the hub.

THE VOLTAGE REGULATOR

Background

Early voltage regulators used mechanical components. They had a large deadband and long response time and required regulator maintenance. Modern automatic voltage regulators

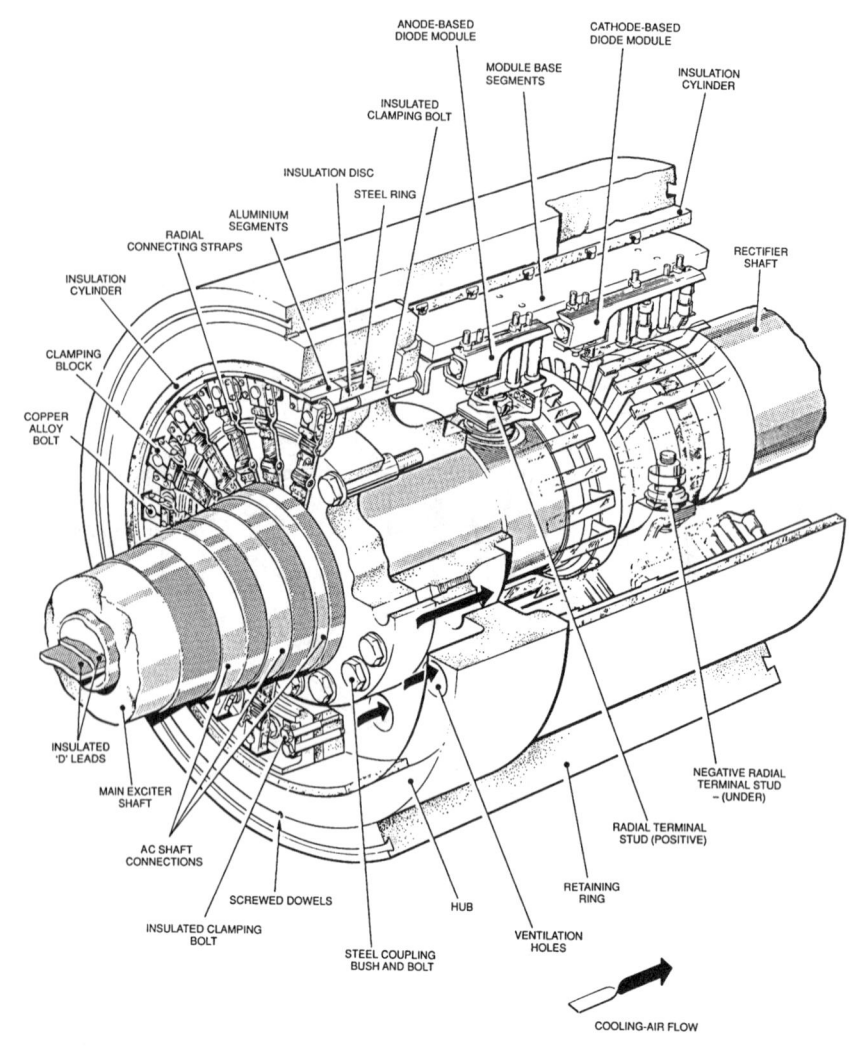

FIGURE 13.43 Rotating rectifier.

use integrated circuits or digital microprocessor techniques. Figure 13.44 illustrates a modern dual-channel arrangement.

System Description

The main function of the AVR is to maintain constant generator terminal voltage while the load conditions are changing. A dual-channel AVR with manual backup is normally used. The reliability of this design is high because the loss of one channel does not affect the operational performance. The faulty channel can be repaired during operation.

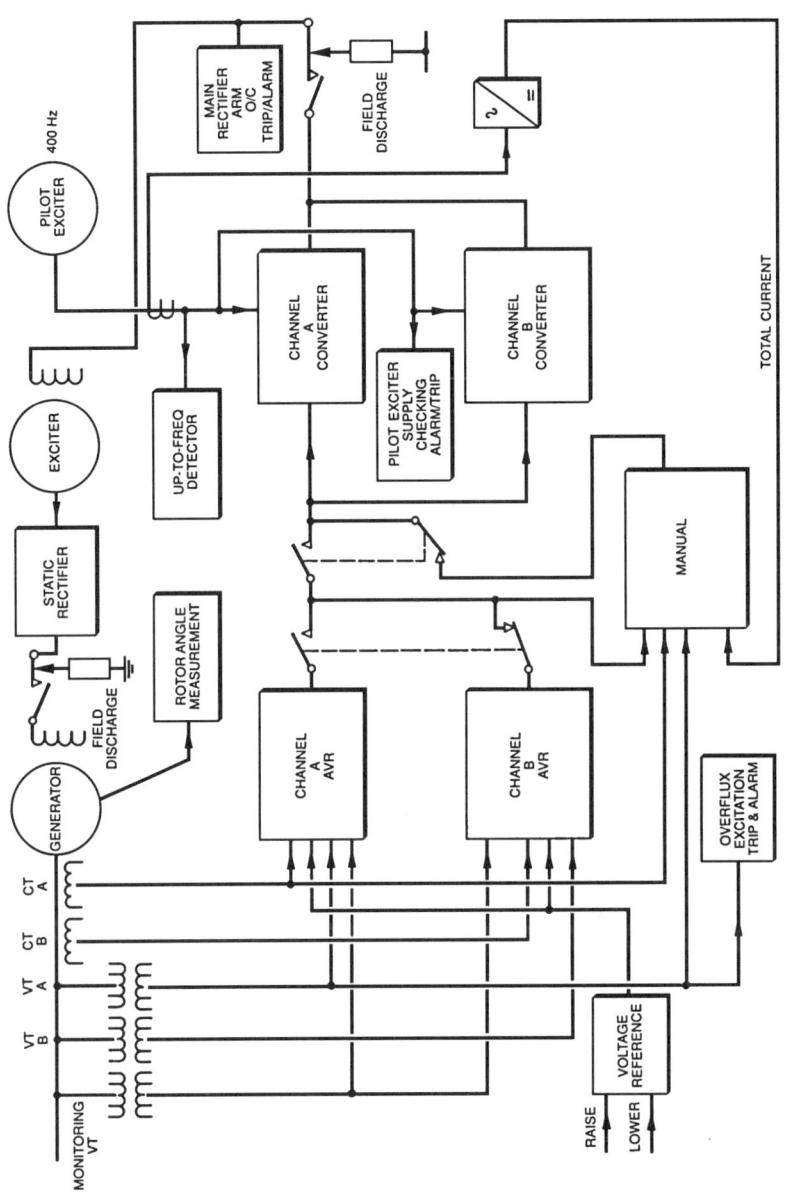

FIGURE 13.44 Dual-channel AVR.

The Regulator

The AVR is a closed-loop controller. It compares a signal proportional to the terminal voltage of the generator with a steady voltage reference. The difference (error) is used to control the exciter output.

When the load changes, the error increases. The channel A AVR applies a proportional-integral-derivative (PID) algorithm to the error and provides a corrective signal. This signal is amplified by the channel A converter. It is then sent to vary a field resistance. The excitation current will change, and the terminal voltage will change accordingly. It is critical to have a fast, stable response from the AVR. Special signal conditioning networks are introduced in the PID control to prevent instability. Accurate tuning (selection of PID coefficients) of the voltage response is achieved by having adjustable time constants.

The AVR receives the generator terminal voltage signal through its own interposing voltage transformer. The voltage signal is rectified and filtered before being compared with the reference voltage.

Auto Follow-up Circuit

In a dual-channel AVR, both channels can be active simultaneously. Each channel provides one-half of the excitation requirements. An alternative design allows one channel to be active while the other follows passively. If a channel trips, the other picks up the full excitation requirement in a "bumpless" manner. A follow-up circuit is used to achieve this function. It tracks the primary (or active) channel and drives the output of the standby channel to match the output of the primary channel.

Manual Follow-up

This is a manual follow-up system similar to the auto follow-up system. When the AVR fails, the manual control takes over in a bumpless manner.

AVR Protection

The AVR plays a critical role in the overall protection scheme of the generator because it controls the suppression of the field after faults. The generator should also be protected against AVR component failure which could jeopardize its operation. An overvoltage relay monitors the terminal voltage of the generator. If the voltage exceeds a safe level, the field current is reduced in minimum time. This relay is only active when the generator is not synchronized.

An overfluxing relay is also active only during unsynchronized operation. If the safe voltage/frequency ratio is exceeded, the generator transformer could be overfluxed. A special relay detects this condition and initiates an alarm. The AVR controls reduce excitation to a safe level. If this condition persists, the excitation is tripped. A component failure within the AVR results in over- or underexcitation. The active channel output is compared with the minimum and maximum field currents. When a limit is exceeded for a few seconds, the channel is tripped.

The Digital AVR

The use of microprocessors in AVRs has many advantages. The reliability will increase due to the reduction of the number of components. Most of the control logic in solid-state AVRs is done by electromechanical relays. These relays will be replaced by a specified microprocessor software. The cost of microprocessor-based AVRs is lower than that of conventional solid-state AVRs. This is due to the replacement of the customized printed circuit boards by standard memory circuits. However, the main advantage of microprocessor-based AVRs lies in the wide range of sophisticated control features. One type of controller, called the adaptive regulator, is capable of adjusting its structure to accommodate the changing plant conditions.

Excitation Control

Modern excitation equipment includes a number of limiter circuits. These limiters operate as parallel controllers. Their signals replace the generator voltage, which is the controlled variable when the input signals exceed predefined limits.

Rotor Current Limiter

All exciters have the capability of supplying a field current significantly higher than the one required during normal operation. This field forcing capability or margin is needed during a fault to increase the reactive power. However, the duration of the increase in current must be limited to prevent overheating of the rotor, which would lead to degradation of the insulation system. During a system fault, the AVR boosts excitation. This situation lasts normally milliseconds before the circuit breaker clears the fault. However, the backup protection is allowed to last up to 5 s (or more). After this delay, the rotor current limit circuit sends a signal that overrides the one from the AVR, causing a reduction in excitation current.

Overfluxing Limit

Modern AVRs have overfluxing limiter circuits in addition to the overfluxing protection circuit. The overfluxing limiter circuit is a closed-loop controller. It monitors the voltage/frequency ratio when the generator is not synchronized. When a predefined ratio is exceeded, the limiter reduces excitation.

THE POWER SYSTEM STABILIZER

When a generator is synchronized with the grid, it is magnetically coupled to hundreds of other generators. This coupling is not rigid like a mechanical coupling. It is a flexible coupling similar to a connection with elastic bands. During normal operation, the generator oscillates slightly with respect to the grid. These oscillations are similar to vibrations of a mass attached to a rigid surface by a spring. These electromechanical oscillations normally have a frequency of 0.2 to 2.0 Hz. This frequency depends on the load and location of the generator with respect to other large generators. Each machine can have different modes of oscillation. The frequency of these oscillations can be 0.3 Hz or 1 to 2 Hz normally. Therefore, the electric power produced by the generator is not matching the mechanical power produced by the turbine at every instant. However, the average mechanical power produced matches the electric power generated by the unit.

In some cases, groups of generators at one end of a transmission line oscillate with respect to those at the other end. For example, in a four-unit generating plant, the four generators tend to be coherent. They tend to oscillate as a group. An oscillation of 10 to 50 MW (above and below the 600-MW rating) is expected. These oscillations are called *power system oscilla-*

tions. They depend on the load. They must be prevented. Otherwise, they can severely limit the MW transfer across the transmission system.

Following a system fault, an accelerating torque will be applied to the generator as a result of changes in the electrical transmission system. The generator must produce a breaking torque in this situation to counter the accelerating torque. The *damper winding* will produce a countertorque (breaking or damping).

Note that the damper winding are bars normally made of copper or brass. They are inserted in the pole face slots and connected at the ends. They form closed circuits as in squirrel-cage winding. During normal operation, the generator is operating at synchronous speed. The damper winding also moves at the same speed. Thus, it is inactive. During a transient, the generator speed changes. The damper winding is now moving at a different speed from the synchronous speed. The currents induced in the damper winding generate an opposing torque to the relative motion. This action helps return the rotor to its normal speed.

The losses such as windage and bearing friction are speed-dependent. They also produce a countertorque that will help to reduce the overspeed. (Windage losses increase with the cube of the speed. The journal bearing losses increase with the square of the speed. The axial thrust bearing loss increases with the speed.)

The *power system stabilizer* (PSS) is added when there is insufficient countertorque (damping). All units over 10 MW must be reviewed for need of PSS. Most of them require a PSS. The PSS measures the shaft speed and real power generated. It determines the difference between the mechanical power and the electric power. It produces a signal based on this difference that changes the speed of the machine (it produces a component of generator torque in phase with the speed changes). *The objective of the PSS is to keep the power leaving the machine constant.* It changes the excitation current at the same frequency as the electromechanical oscillations. This action changes the generator voltage with respect to the voltage in the grid. Power will flow from the grid to the unit to provide the countertorque required when the speed of the shaft exceeds the synchronous speed. It is important to mention that an improperly tuned PSS can lead to disastrous consequences. This is due to the voltage variations that it creates during operation. If the voltage variations are incorrect, excessive torque changes can occur, leading to significant damage.

In summary, these are the salient features about the PSS:

1. The PSS acts as a shock absorber to dampen the power swings.
2. The AVR cannot handle power swings because it is monitoring the voltage only.
3. A fault on the line can excite a unit severely if the damping is poor.
4. The PSS monitors the change in power and change in speed.
5. During steady-state operation, the PSS will not interfere.
6. When a fault occurs, the voltage drops. The field current must be increased to push as much active power out to increase the synchronizing torque. However, this increase in synchronizing torque (active power out) lasts for a few seconds only, for these reasons:
    a. Active power flow between the generator and the load is proportional to

$$\frac{V_{generator} V_{load}}{(\text{Reactance between generator and load})(\sin \alpha)}$$

where α is the angle between the phasors of $\mathbf{V}_{generator}$ and $\mathbf{V}_{load}$ and the reactance includes the step-up transformer only. If excitation is increased, $\mathbf{V}_{generator}$ increases, but α changes within a few seconds so that the active power flow remains unchanged.
    b. The active power flow is determined by the turbine.

It is also important to mention that the reactive power flow is proportional to

$$\frac{V_{generator} - V_{load}}{\text{Reactance between generator and load}}$$

Any change in $V_{generator}$ will result in significant change in the reactive power sent to the grid. Therefore, when the excitation changes, $V_{generator}$ will change, resulting in significant transfer of reactive power.

CHARACTERISTICS OF GENERATOR EXCITER POWER (GEP) SYSTEMS

The characteristics of GEP are established by extensive system investigations. All plant operating modes must be examined to identify the conditions of marginal stability. In general, the periods of low system demand (at night) are the most critical. The generator operates at a leading power factor during these times. In pumped-storage plants (where the turbine is used to pump the water upstream during the night) this is done because the price of electricity is very low at night. The same water is allowed down through the turbine during the day because the price of electricity is higher, so the situation is more critical. This is because of the large rotor angle (angle between the rotor flux and stator flux—this angle normally increases with the load when the generator operates) in comparison with the remaining machines on the system during the night. The generator is operating as a motor in this situation.

A number of simulations are done on the unit (including AVR and PSS). The PSS settings are adjusted for optimum performance of excitation under all critical operating conditions. These settings are then used during plant commissioning. This is done to reduce on-site testing, which can be expensive.

Excitation System Analysis

The generator excitation system has the primary responsibility for power system dynamic and transient stability. *Dynamic stability* refers to the performance of the system following small load changes. This can result in sustained oscillations around 0.5 Hz when large power is transferred over long distances. These oscillations must be rapidly attenuated. Otherwise, the transmission system will be severely limited. *Transient stability* refers to the ability of a generator or group of generators to maintain synchronous operation following system faults.

Following a fault, a boost of synchronous torque is required to maintain the generator in synchronism. (*Note:* The synchronous torque is the torque used to maintain the generator in synchronism. It is created by active power sent to the grid. This is done by increasing the field current.) In this situation, the AVR bucks (resists, opposes) and/or boosts the field current to develop the additional synchronizing torque. Therefore, the AVR must be properly tuned to play an essential role in maintaining stable system operation under all operating conditions.

GENERATOR OPERATION

Running up to Speed

Before running up to speed, air will have been scavenged from the generator casing. Hydrogen will fill the casing to almost rated pressure. The hydrogen pressure increases with temperature. Rated pressure is achieved on steady load. The stator windings should remain warmer than the hydrogen to prevent condensation.

It is recommended to go through the first and second critical speeds (around 900 and 2200 r/min) of the rotor quickly to avoid high vibrations (Fig. 13.13). As the rated speed is approached, excitation is applied automatically by the voltage regulator (or manually) by closing the switches of the exciter and the main field. The voltage/frequency control device prevents the voltage from exceeding the rated voltage/frequency. This is done to prevent overfluxing of the generator transformer. The rated voltage should be established at rated speed with the machine on open-circuit.

Open-Circuit Conditions and Synchronizing

Generators are generally operated near their rated voltage. If the grid requires a different voltage, the transformer tap changers will accommodate this request. A voltage range of ±5 percent is normally specified. The open-circuit characteristic is normally determined

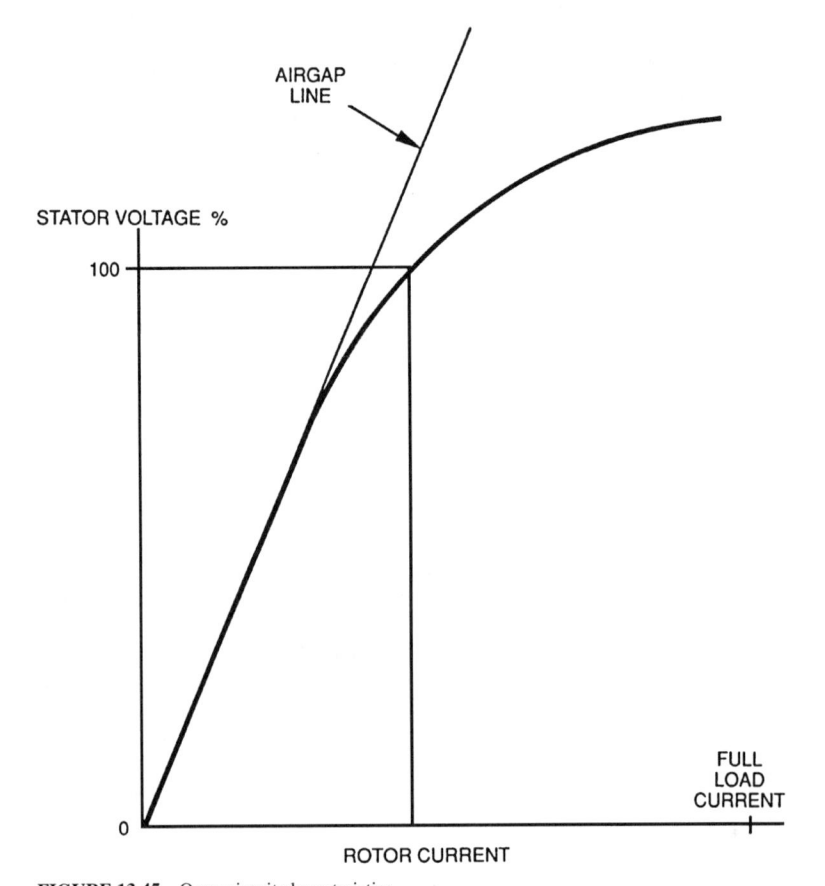

FIGURE 13.45 Open-circuit characteristics.

by the manufacturer. Several measurements of rotor currents and stator voltage are taken and plotted (Fig. 13.45). The relationship is linear (the air gap line) up to about 75 percent of rated voltage.

Note: The magnetomotive force (mmf) is applied across the reluctance of the air gap and the reluctance of the core. The reluctance of the air gap dominates because it is much larger than that of the core. When saturation is reached, the reluctance of the iron starts to change. This occurs at the knee of the curve (Fig. 13.45).

During a long outage, the open-circuit characteristic should be checked by measuring the parameters at a few points along the curve. Improper synchronization can have serious consequences. If the magnitude or angular position of the voltage phasors were significantly different when the circuit breaker was closed, large current would circulate from the system through the stator windings due to the voltage difference. This causes high forces in the windings.

If there were a significant difference in frequency, a large torque would be imposed on the rotor due to the sudden pulling into synchronism. A backup device confirms adequate synchronization conditions before allowing the circuit breaker to close.

The Application of a Load

If the generator voltage phasors (magnitude and angular position) match exactly, there will be no current flow and no electrical torque. An imbalance in phasors must be created in order to generate a load. The steam turbine governing valves are opened gradually. The rotor starts to accelerate due to the additional torque. It moves forward relative to its no-load position while still remaining in synchronism with the grid. The difference in voltage phasor created by this angular change generates current in the stator windings. An electrical torque is generated which balances the increased mechanical torque.

Capability Chart

Figure 13.46 illustrates the capability chart of a generator. It is a MW-MVAR diagram. A constant MW limit is drawn at the rated power output of the turbine. The rated stator current locus cuts the rated MW line at the rated MVA and power factor point. The rated rotor current imposes a limit on MW and the lagging power factor. The capability chart shows the limits of generated MW and MVAR.

Neutral Grounding

The neutral ends of the three stator winding phases are connected together outside the casing. The star point is connected to ground through a neutral grounding device. It is designed to limit the fault current upon a ground fault in the stator winding. The neutral grounding device consists of a single-phase transformer. Its primary is connected between the generator star point and ground. Its secondary is connected to a resistor. This arrangement is chosen because the apparent impedance of the resistor appears on the primary side as a^2Z, where $a = N_p/N_s$, and Z is the impedance of the resistor. This creates a very high impedance that limits the fault current to 15 A.

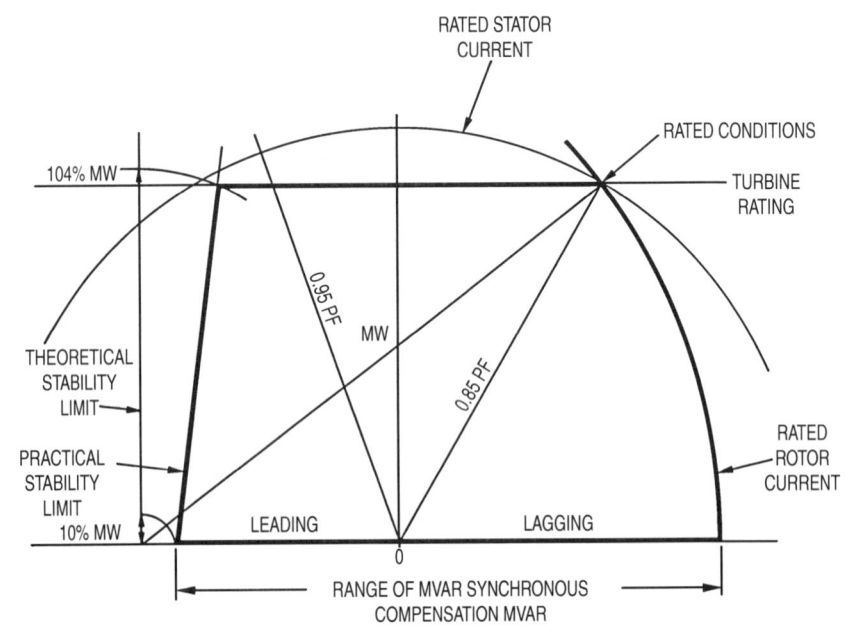

FIGURE 13.46 Capability chart.

Rotor Torque

During electrical faults, the stator currents are many times larger than the rated value. The associated electromagnetic torques have similar magnitudes. The shaft and coupling must be designed to withstand stipulated fault conditions without failure. However, the coupling bolts exhibit distortion in some cases after a severe electrical fault.

REFERENCE

1. British Electricity International, *Modern Power* Station Practice, vol. C, *Turbines, Generators and Associated Plant*, 3d ed., Pergamon Press, Oxford, England, 1991.

CHAPTER 14
GENERATOR MAIN CONNECTIONS

INTRODUCTION

The main connections bus bar (Fig. 14.1) connects the generator to its associated generator transformer. An aluminum tube (conductor) surrounded by a concentric enclosure of the same material is used for each phase of the generator. Since each enclosure is isolated from its neighbor, the term *isolated phase bus bar* (IPBB) is used.

The conductor of each phase is supported at the center of the enclosure by equispaced insulators. These insulators are rigidly fixed around the circumference of the enclosure. They allow limited radial movement of the conductor. The rating of the busbar installation is based on the temperature rise above a specified ambient at maximum load and its ability to withstand ground (earth) and three-phase short-circuit faults anywhere in the generator system without damage.

ISOLATED PHASE BUS BAR CIRCULATORY CURRENTS

The currents flowing in the conductors produce large magnetic fields around them. These fields induce currents in the enclosures and generate forces between all the conductors and the enclosures. The extreme ends of each phase enclosure are bonded together to reduce these forces. This allows a balanced current flow in the enclosures of the three phases (Fig. 14.2). This arrangement is called *electrically continuous IPBB with short-circuit.*

The enclosures are supported by insulated foot mountings to insulate them from the grounds. They are also grounded at one point. The circulating currents in the enclosures are controlled by isolating them from all the systems in the plant that are connected to the conductors. Rubber bellows are used to connect the enclosures to the systems in the plant. These bellows provide electrical isolation of the enclosures from the plant systems as well as physical protection of the conductors. The large magnetic field produced by the conductor currents can induce circulating currents and heat in the adjacent steelwork. This problem should be considered during the design phase because excessive heat is an unnecessary loss and can cause expansion that can be hazardous to personnel. All generating units having a rating lower than 660 MW use the naturally cooled main connections system (Fig. 14.3). Units having higher rating rely normally on forced-cooling designs.

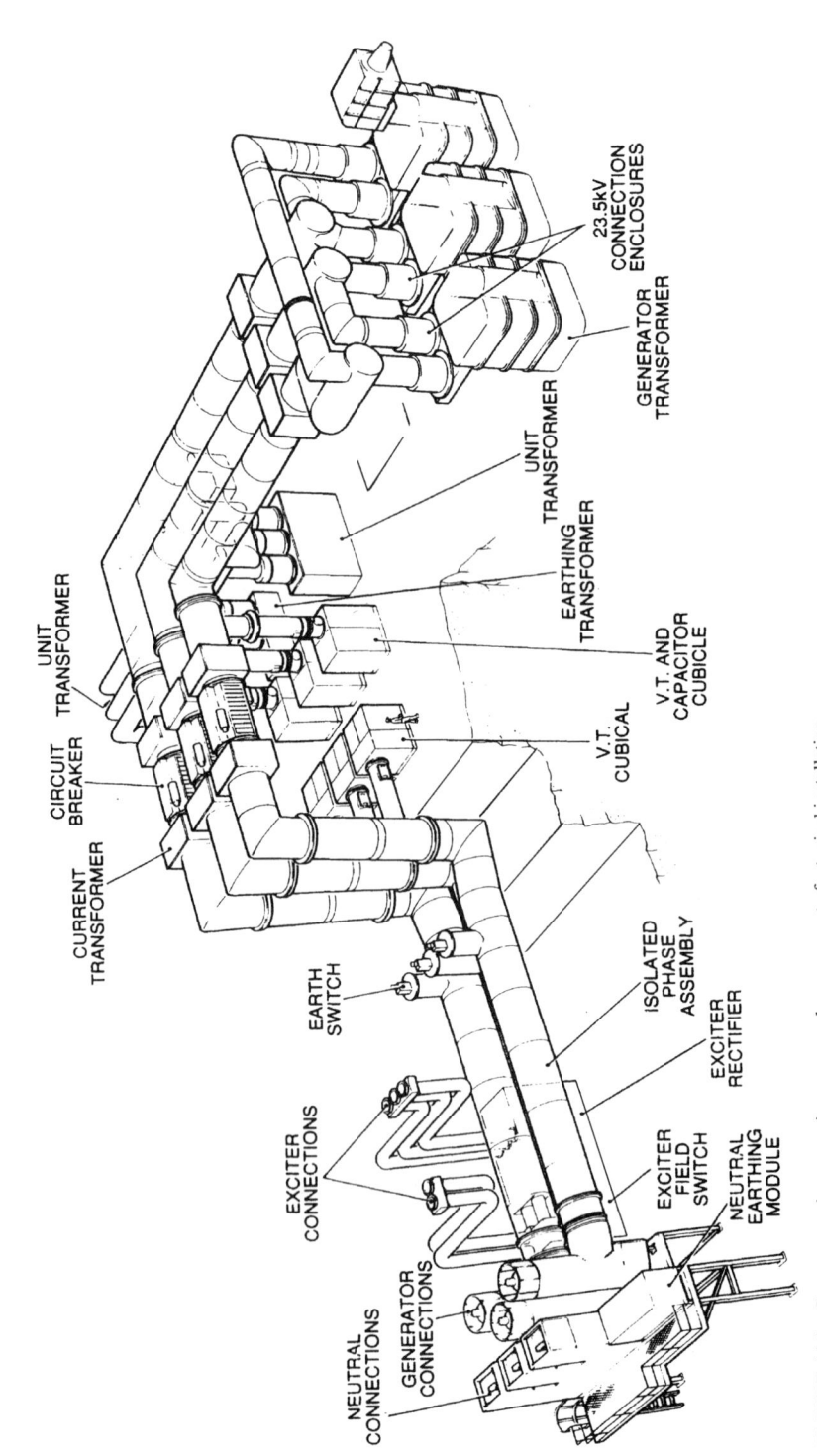

FIGURE 14.1 Generator main connections—general arrangement of a typical installation.

GENERATOR MAIN CONNECTIONS

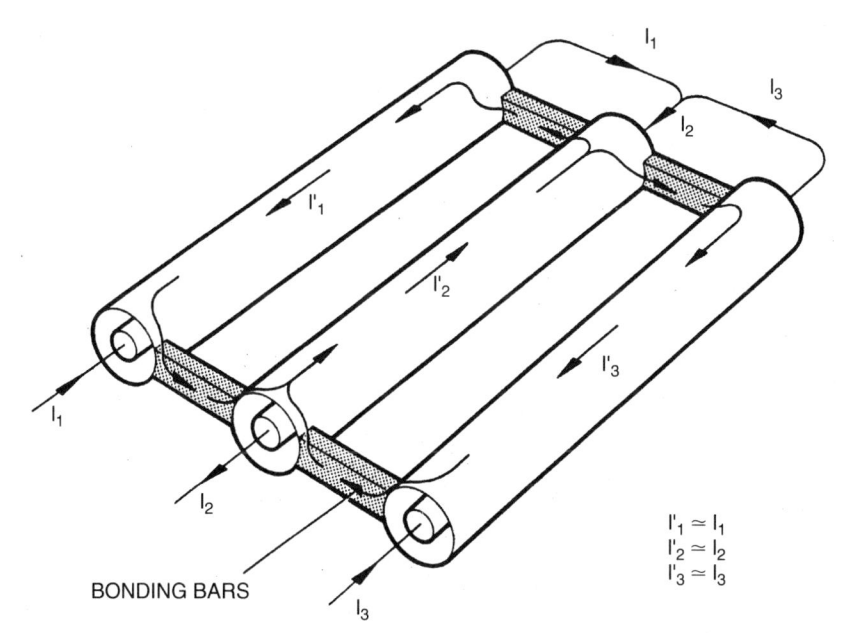

FIGURE 14.2 Phase isolated bus bar—continuous sheath.

SYSTEM DESCRIPTION

Figure 14.4 illustrates a typical generator main connections system. The windings of the generator are star-connected. This star point is formed at the neutral end of the winding outside the generator. The generator transformer is normally wound as a delta at the low-voltage (LV) end and star at the high-voltage (HV) end. It consists of a three-phase transformer or three single-phase transformers. The HV terminals of the unit auxiliary transformers and the three-phase grounding (earth) transformers (if installed) are supplied from the main bus bar. This connection is known as a tee-off from the main bus bar.

The voltage on the main connections system is monitored for various reasons including synchronization, tariff metering, instrumentation, automatic voltage regulation, power system stabilization, and protection systems. The current is also monitored for various reasons including protection systems. Voltage and current transformers are installed as shown in Fig. 14.4 to take these measurements.

REFERENCE

1. British Electricity International, *Modern Power Station Practice—Electrical Systems and Equipment, Volume D*, 3d ed., Pergamon Press, Oxford, United Kingdom, 1992.

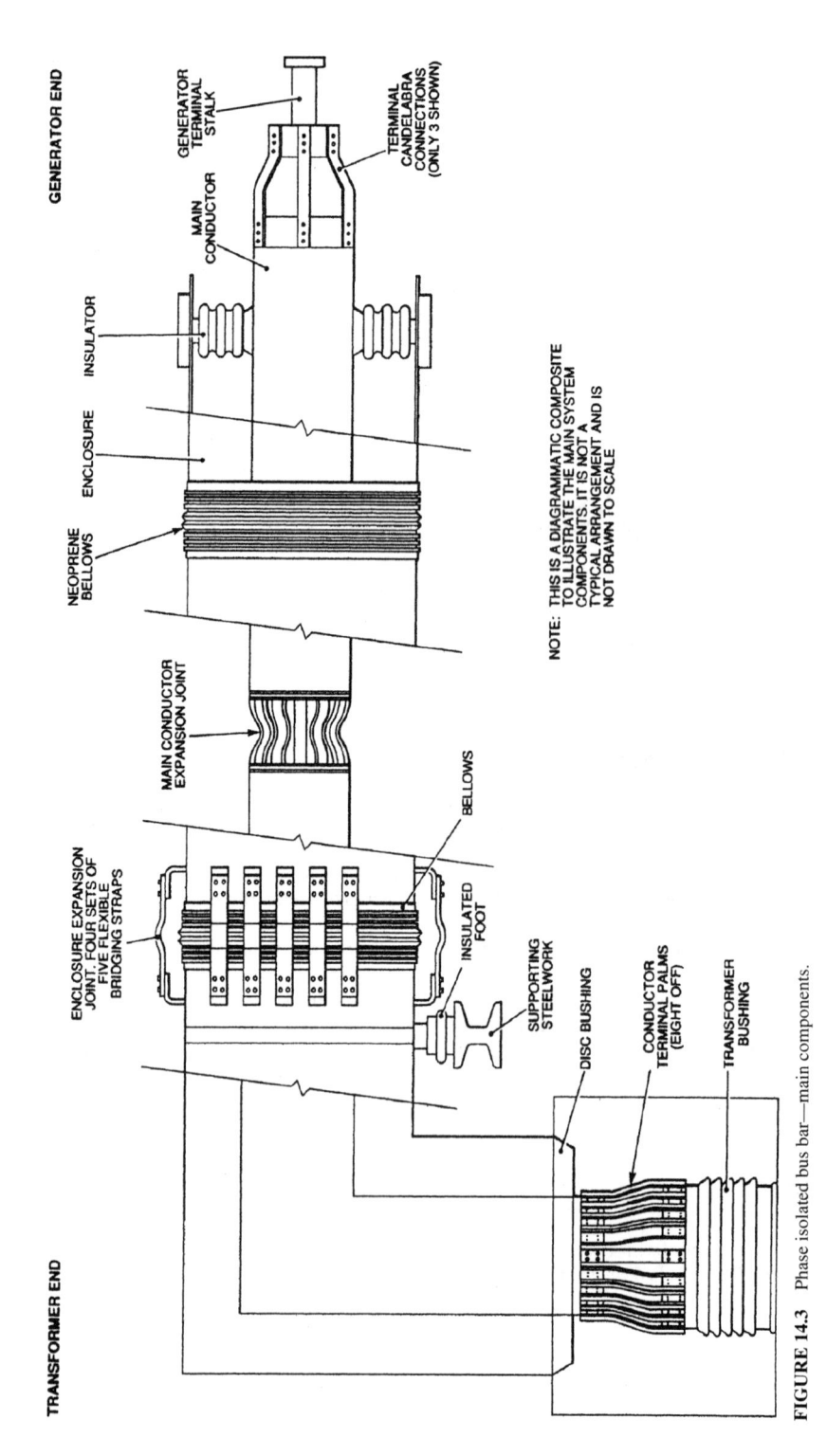

FIGURE 14.3 Phase isolated bus bar—main components.

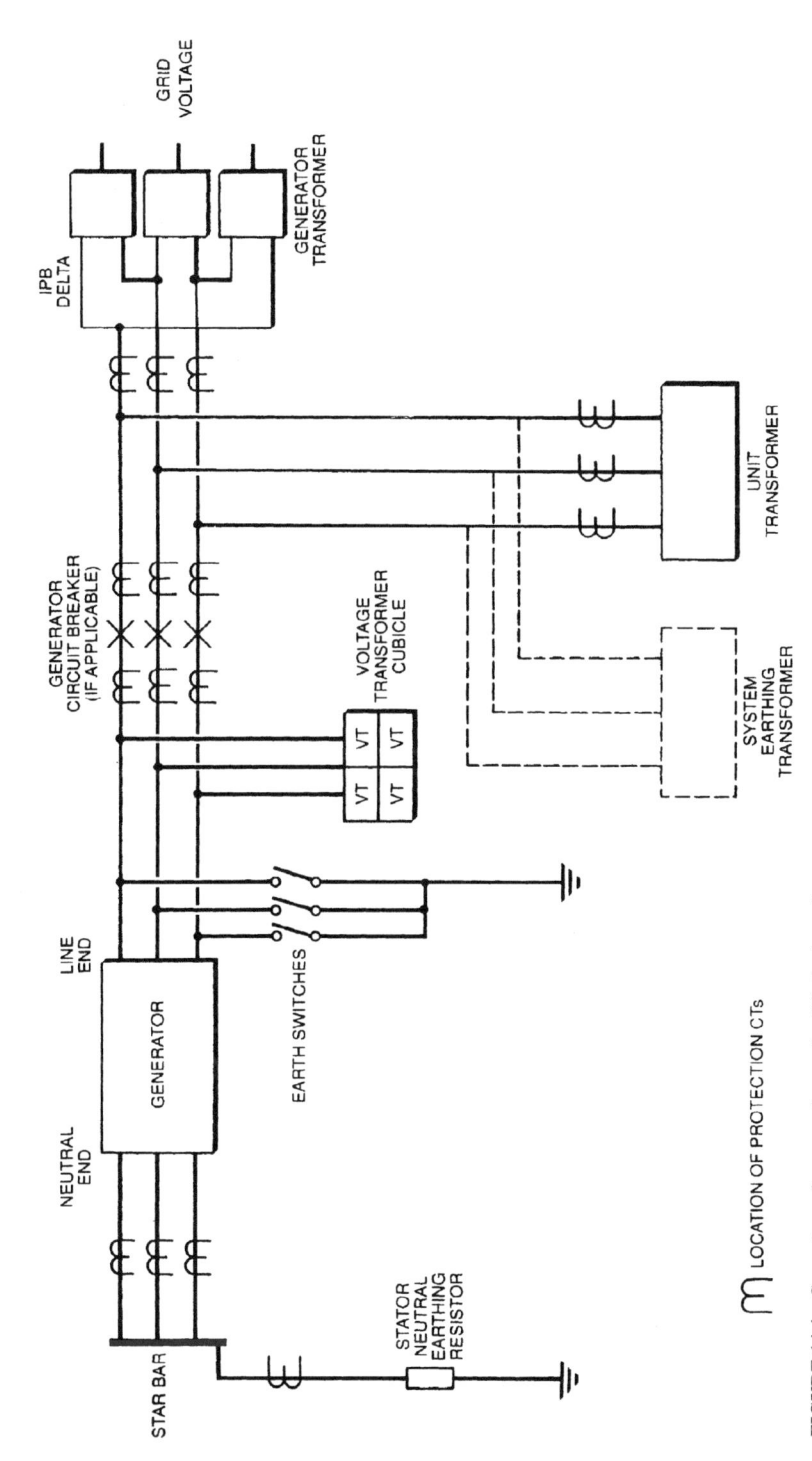

FIGURE 14.4 Generator main connections—simplified schematic.

CHAPTER 15
PERFORMANCE AND OPERATION OF GENERATORS

GENERATOR SYSTEMS

The following sections describe the excitation, cooling, and lubricating systems associated with the generator.

Excitation

The excitation system controls the output voltage and the reactive power (VAR) delivered by the generator. Brushless excitation relies on ac machines with diode bridges to convert the ac power to dc. The diode bridges can be stationary or rotating on the shaft. Self-excitation is also used by supplying power through a transformer connected to the terminals of the generator. The output of the transformer supplies either a diode or a thyristor bridge.

Hydrogen Cooling

The following are the ranges of air and hydrogen cooling in generators:

- Generators having a rating lower than 20 MW are air-cooled.
- Generators having a rating between 20 and 200 MW are either air- or hydrogen-cooled. However, the majority of generators having a rating over 100 MW are hydrogen-cooled.
- Generators having a rating over 200 MW are hydrogen-cooled.

Hydrogen has proved to be an excellent cooling medium for generators. Air-cooled generators having a rating higher than 100 MW had frequent forced outages and shorter lifetime due to poor cooling.

The hydrogen pressure inside the generator is 4 to 5 bar. It cools the stator and the rotor. The purity of the hydrogen should be maintained within specifications. Carbon dioxide and nitrogen have been used as a purging gas in generators. Note that carbon dioxide is not inert. It should not be stored in the generator for long periods.

The hydrogen cooling capability decreases with the pressure. If the hydrogen pressure cannot be maintained at the rated value, due to an operational problem, the rotor current and the capability of the machine should be reduced accordingly. The stator core temperature should be continuously monitored. The recommended temperature should not be exceeded.

Cooling of the Stator Conductors

The stator conductors of large generators are cooled by demineralized water to maximize the rating of the machine. The purity of the water should be maintained within strict limits. Its conductivity should be lower than 2 μS/cm. It should also be kept free of oxygen to prevent the formation of oxides and the subsequent blockage of the water channels in the conductors. The coolant circuit should be free of ferrous materials to prevent the formation of magnetite. The water filters should be kept in a good condition.

Since the purpose of the water is to cool the stator bars, it is useful for the operator to understand the effects of the following conditions:

- Total or partial impairment of the coolant pumps
- Blockage of the water channels on the conductors due to either debris or the formation of oxide layer
- Reduction in coolant flow due to vapor (air or hydrogen) locks
- Hydrogen leakage into the water system

The manufacturer provides the limits for any of these conditions.

The temperature of the conductor bars will increase when the flow is reduced. If boiling occurs, the pressure will fluctuate. The boiling temperature is normally around 120°C because the water pressure is above atmospheric (the water is normally supplied from a raised header tank).

Hydrogen Seals

The hydrogen seals are supplied with oil for lubrication. A hydrodynamically generated oil film is established between the white metal of the seal and the mating rotor collar. Its thickness varies between 0.013 and 0.038 mm when the unit operates at normal speed. The film thickness decreases significantly when the rotational speed drops below 400 rpm. Thus, the faces of the seal will be poorly lubricated when the unit is on turning gear (barring phase) or when the speed is below 400 rpm. Therefore, it is necessary to maintain the flatness and the surface finish of the seal faces within specifications to prevent seal failure.

The hydrogen seals have very small clearances. Any debris present in the oil could affect the performance of the seal if it is larger than the clearances. The situation is worsened when the unit is on turning gear (under barring conditions) due to thinning of the oil film. Thus, the seal oil filters should be able to remove particles down to at least 1 μm to improve the integrity of the seal.

The seal should be able to move axially about 4 to 5 cm to accommodate the expansion and contraction of the shaft. Elastomers in the form of lip seal or O-ring are used in the hydrogen seal to provide oil and gas sealing as it moves axially. An increase in the white metal temperature or excessive leakage of oil or hydrogen is indicative of wear in the hydrogen seal. The contact face of the seal collar should not exceed 120°C. Since there is 4.5 to 6.0 mm of metal between the contact face of the seal and the thermocouple, the maximum operating temperature of the thermocouple should be limited to 100°C due to the 20 to 25°C temperature drop across the metal. This temperature drop will decrease due to wear. Thus, an allowance should be made if there is good reason to believe that wear has occurred.

The following are the three common operational problems that occur frequently on hydrogen seals:

- Hydrogen leaks
- Oil leaks into the generator
- High white metal temperatures

General advice to deal with these problems is difficult. However, the following actions have improved the condition of the seal in some applications:

- With high metal temperatures, increase the collar/seal gap or improve the cooling of the seal.
- With gas leakage, reduce the collar/seal gap.
- With oil leaks, reduce the differential pressure of the oil.

CONDITION MONITORING

Various condition monitors are installed on generators and protective systems to identify fault conditions early and to take corrective action to prevent a major failure.

Temperature Monitoring—Thermocouples

Thermocouples are installed in the generator system at the following locations:

- The inlet and the outlet of the hydrogen flow to the heat exchanger
- The inlet and the outlet of the stator water flow to the heat exchanger
- The stator core temperatures
- Hydrogen seal face temperatures

The thermocouple readings provide critical information about the generator. For example, an increase in the stator water temperature at the outlet of the heat exchanger is indicative of a possible blockage in the water channels of the stator bars. Note that the change in the reading of the thermocouple occurs around 30 min after the fault has occurred. This is so because of the delay associated with the conduction of heat from the hot spot to the thermocouple.

Direct measurement of the rotor conductor bar (winding) temperature is not possible. However, the average rotor bar temperature can be deduced from an ohmmeter connected to the rotor current and the voltage across the slip rings.

Hydrogen Gas Analysis

Gaseous products and liquid aerosol are produced at the surface of an organic insulation when its temperature exceeds 200°C. These products become entrained with the hydrogen gas inside the generator when there is an insulation fault. There are two types of instruments used to detect these products online. They rely on analyzing either the gases or the particulates generated when the fault occurs.

A fault that originates in the stator core is called a *core fault*. It normally results in the melting of a large number of core laminations. There is no protection scheme to isolate the generator when this fault occurs. It is only detected when a ground fault occurs due to failure of the insulation between the stator winding and the core.

An *incipient core fault* is a defect that causes shorts between the stator laminations. It can be caused by a scratch or metallic debris. It may be discovered during flux tests or if it is in the vicinity of a thermocouple. It does not normally cause an outage because its damage is limited. However, it can grow during operation to cause a major core fault. Its growth will most likely be caused by pole slipping. This is a condition that occurs due to either instability or a decay in the rotor flux under decreased excitation.

Studies have shown that the interlamination insulation will break down under these conditions, resulting in growth of the damaged area. It is of utmost importance to keep the generator area clean when it is opened up for maintenance. Any foreign material left in the air gap can damage the laminations and cause a failure.

The ion-chamber particle detector, also known as a *core monitor,* is used to detect stator winding faults caused by thermal degradation of the insulation. These faults were detected in the past when the insulation failed, causing an interwinding or a ground fault. The hydrogen coolant enters a chamber in this equipment lined with either thorium oxide or americium pellets. The lining ionizes some of the hydrogen molecules by emitting alpha particles. The ionized molecules then pass between electrodes maintained at a specified voltage difference. The free ions are collected at the electrodes, causing a small electric current to be produced in a circuit.

When aerosol enters the chamber, the current generated by the ions decreases. Thus, a reduction in the current indicates insulation degradation. Oil mist entrained with the hydrogen also causes a deflection in the current reading and an alarm.

Figure 15.1 illustrates the response of a generator core monitor (GCM) to a core fault. Figure 15.2 shows the procedure that should be followed upon receiving an alarm from a GCM.

Hydrogen Dew Point Monitoring and Control

It is important to control the concentration of moisture in hydrogen to prevent arcing in the generator and aqueous stress-corrosion cracking in the end rings. The objective of the moisture control is to prevent condensation inside the generator. Table 15.1 provides a summary of the dew point requirements for hydrogen-cooled generators under various conditions. The hydrogen driers will normally be able to maintain the required dew point during normal operation unless the concentration of moisture in oil exceeds the specified limit. Also note that the dew point increases with the pressure, as shown in Fig. 15.3. For example, a dew point of $-20°C$ at atmospheric pressure corresponds to a dew point of $-4.5°C$ at 3 bar (gauge). It is recommended to use the dew point at operating pressure to avoid confusion.

The effect of condensation on the insulation of a large generator depends on the type and condition of the insulation system. Epoxy-bonded insulation systems have high resistance to moisture. However, if delamination has occurred in mica insulation, moisture could penetrate deeply into the insulation system. If water has penetrated into the insulation during an outage, the dielectric strength will drop significantly. A dryout is necessary under these conditions before returning the unit to service, to prevent a possible ground fault. It consists of replacing the moist gas in the generator with dry gas and recirculating it until the insulation becomes dry. Dryout techniques that rely on heating the conductors will not remove the moisture. They only move the moisture to the surface. Blowing heated air on the windings will cause additional condensation on the windings, thereby worsening the situation.

Vibration Monitoring

The turbine supervisory system monitors the vibration and other parameters of the turbines and the generator. Vibration transducers normally measure the velocities of the bearing pedestals.

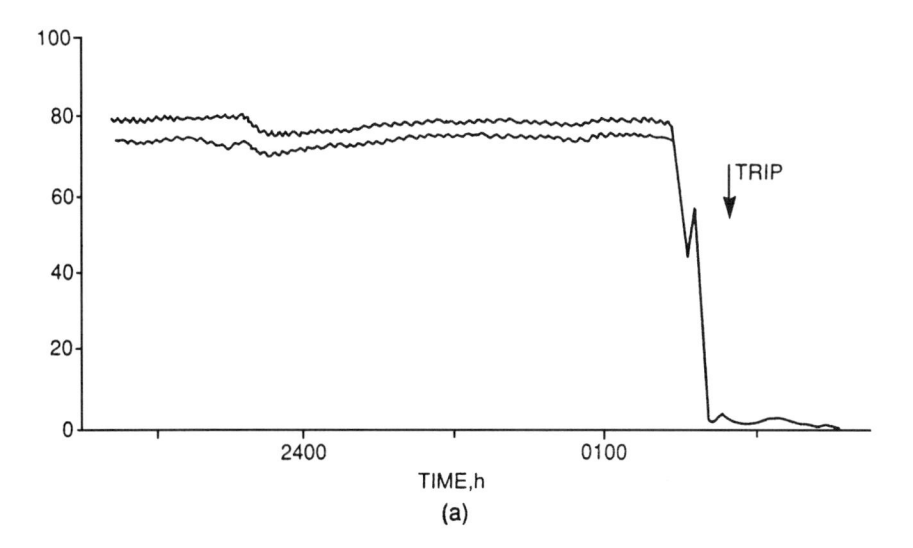

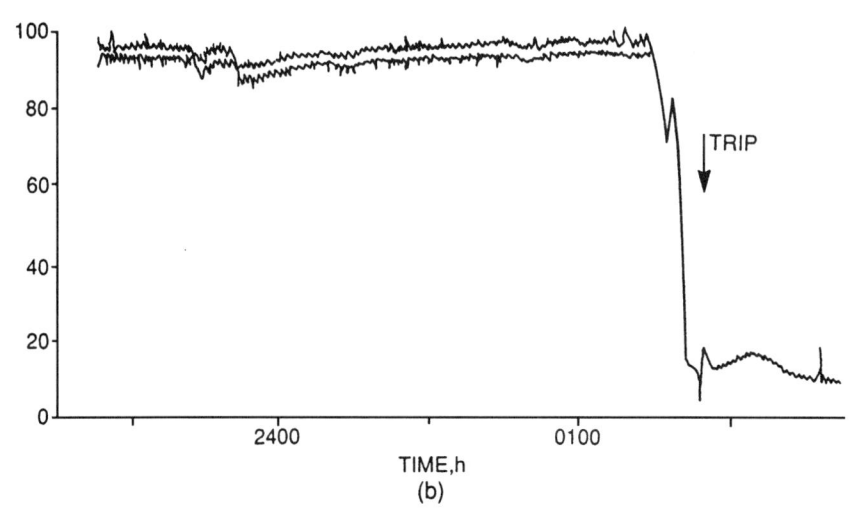

FIGURE 15.1 Generator core monitor response to generator core fault. (*a*) GCM (turbine end); (*b*) GCM (exciter end).

The corresponding levels of displacement are displayed and recorded. Alarms provide warning at a specified vibration amplitude. The unit is also tripped at a higher vibration amplitude to prevent damage to the turbines, connecting pipework, and bearings. Additional diagnostic equipment is normally installed at the bearing pedestals to monitor the vibration amplitude and phase.

The vibration readings are recorded during start-up and shutdown. They provide the signature of the turbine-generator over a wide range of frequencies. The phase and amplitude of these data are analyzed to identify if the vibration was caused by a mechanical imbalance, misalignment, oil whirl, or bent shaft. Thermal imbalances are also detected, and the rotor is balanced before returning the unit to service.

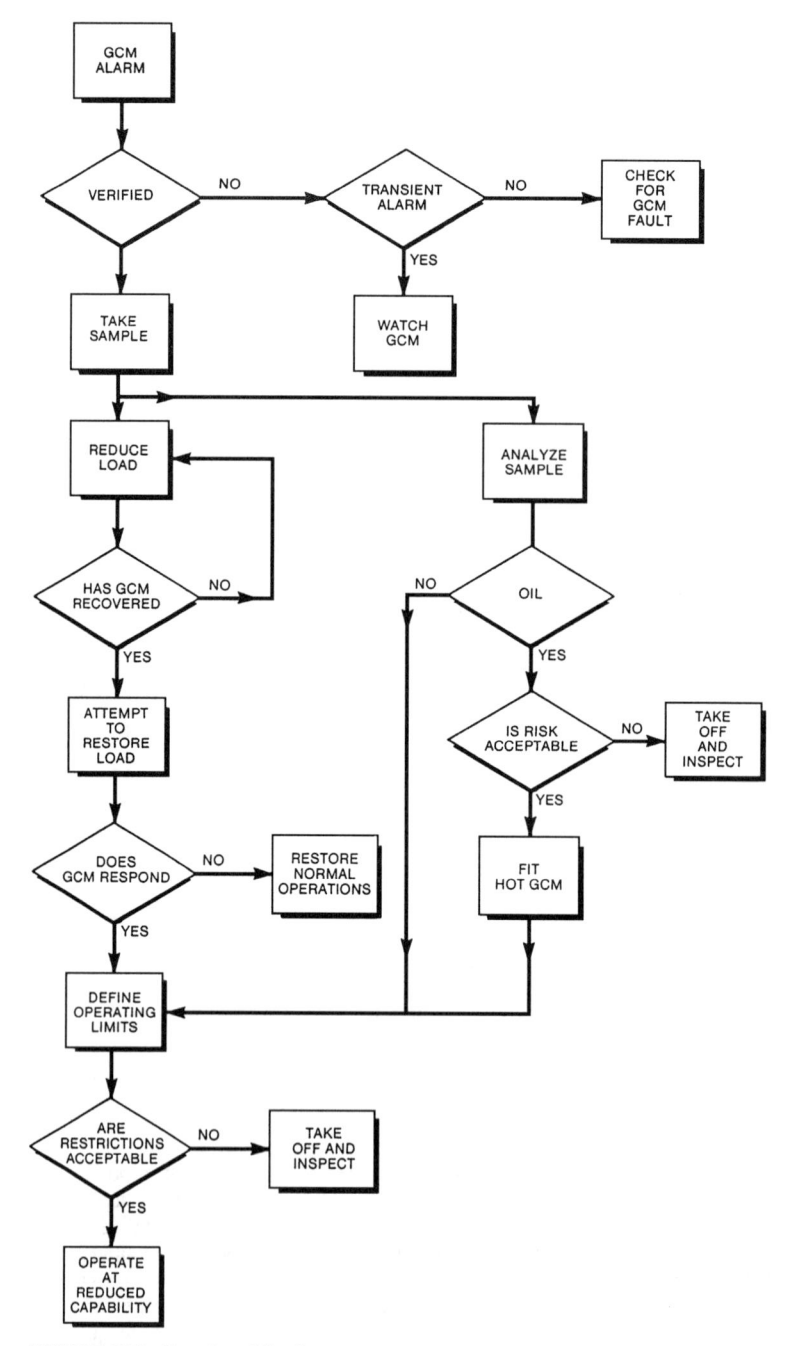

FIGURE 15.2 Procedure following a generator core monitor alarm.

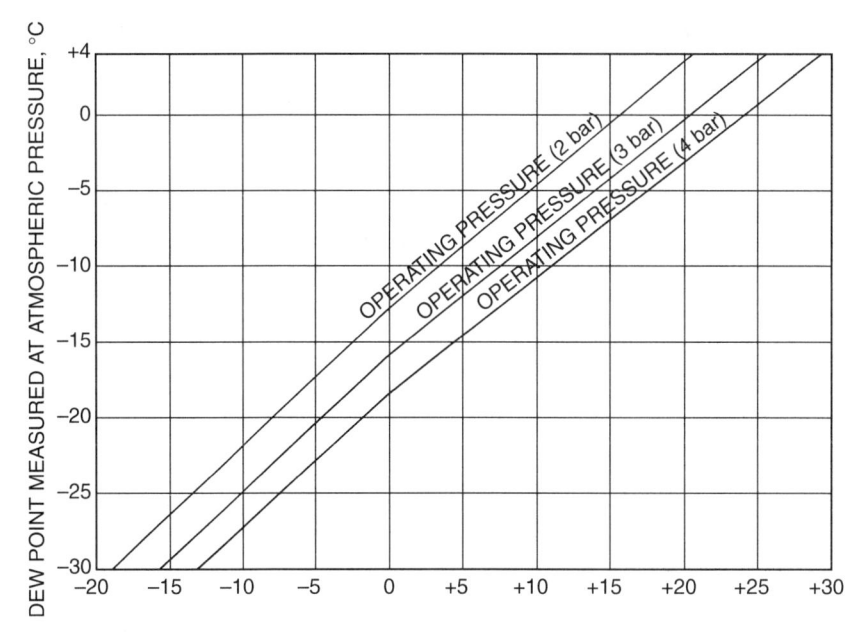

FIGURE 15.3 Relationship between dew point at pressure and the equivalent dew point measured at atmospheric pressure for operating pressure of 2, 3, and 4 bar.

OPERATIONAL LIMITATIONS

Temperatures

The temperature of the generator components and the cooling fluids' inlet and outlet temperatures are monitored and recorded. It is essential to maintain the stator coolant water below the boiling temperature. The highest core temperatures occur at the ends of the core. These temperatures are also monitored to prevent either core or conductor bar insulation damage.

The maximum temperatures occur at the highest leading power factor because the excitation current and resulting flux would be at their highest values. Thus, the core end temperatures can be reduced by operating the generator at a more lagging power factor. However, the fluxes and terminal voltage are reduced when the generator is operated in this mode. In summary, the leading power factor capability and the terminal voltage of the generator can be restricted if hot spots are identified, in order to keep the generator online.

Hydrogen Leakage

A hydrogen leakage from the generator can cause an explosion and fire. An explosive mixture is formed when the concentration of hydrogen in air is between 4 and 76 percent. Hydrogen leakage from the generator should not be tolerated. Hydrogen leakage detectors are used to alarm this condition. The unit should be shut down when the hydrogen leak exceeds a specified limit. The hydrogen pressure in the generator can be reduced under emergency conditions to reduce the hydrogen leakage from the casing. However, the hydrogen pressure should always be maintained at a higher value than the water pressure in the windings. Thus, the water flow should be decreased in some cases to maintain an acceptable pressure

TABLE 15.1 Dew Point Requirements for Hydrogen-Cooled Generators Operating at 4-bar Hydrogen Pressure

Generator operation	Stator coolant temperature and control	Hydrogen dew point and shutdown criteria	Liquid in, stator casing and hydrogen cooler drains
Starting up and after prolonged shutdown	Temperature maintained at or above 30°C and at least 5°C above the cold hydrogen temperature measured at the hydrogen cooler outlets	The dew point must be better than −18°C at 1 atm (equivalent to better than 0°C at frame pressure, 4 bar) with the set spinning (>2200 r/min) and immediately prior to excitation	Remove, identify, and log the volume of any liquid from stator and hydrogen cooler drains. Investigate cause. providing shutdown dew point was satisfactory, rely on dew point immediatley prior to excitation for decision on whether to load machine. *Note:* Rotor earth faults should be regarded as an indication of the presence of water in the frame until proved otherwise.
Running	Temperature maintained at or above 30°C and at least 5°C above the cold hydrogen temperature measured at the hydrogen cooler outlets	The target dew point should be better than −18°C at 1 atm (0°C at frame pressure, 4 bar). Remedial/investigative action should be taken to reduce moisture ingress if the dew point is worse than the above value. *Immediate* shutdown if the dew point reaches −5°C at 1 atm (e.g., to +18°C in frame)	Remove, identify, and log the value of all liquid from drains once per shift. Investigate the cause of any moisture ingress—ensure dew point measurement is correct since liquid accumulation would not be expected on load with dew point (at frame pressure) better than +18°C.
Running down	Temperature maintained at or above 30°C and at least 5°C above the cold hydrogen temperature measured at the hydrogen cooler outlets	Maintain dew point and remedial/investigative action as in "running"	No action
Shutdown for one week or less	Temperature maintained at or above 30°C and at least 5°C above the cold hydrogen temperature measured at the hydrogen cooler outlets	As above to avoid prolonged dryout on start-up	Drain, identify, and log the water, volume removed from the drains once per shift. Investigate the cause of any water ingress.
Shutdown for longer than 1 week	Isolated and drained. Seal oil system in service	Degassed	Drain, identify, and log the water volume removed from the drains once per shift. Investigate the cause of any water ingress.
Generator inspections	Isolated and drained. Seal oil system also shut down	Degassed and kept warm with hot-air blowers	Drain, identify, and log the water volume removed from the drains once per shift. Investigate the cause of any water ingress.

PERFORMANCE AND OPERATION OF GENERATORS 15.9

	Parameter to be monitored	
Drier operation and control	Drier blower operation	Moisture in oil concentration and control
In service and operating satisfactorily. Check efficiency during dryout after prolonged shutdown.	In service continously	The target moisture in oil concentration should be 0.05% w/w. The oil centrifuge should be in service and operating satisfactorily.
In service. Daily check for efficiency. Dew point measurements in and out. Increase frequency if dew point deteriorates from normal.	In service continously	In addition to the above weekly water in oil check providing H_2 dew point is normal. Increase test frequency if H_2 dew point deteriorates. Investigate the cause if H_2O in oil $>0.05\%$ since this will affect the H_2 dew point. Speed of action is determined by extent of deterioration of H_2 dew point.
In service	In service continuously	Maintain centrifuge in service and operating satisfactorily.
In service. Check daily for efficiency.	In service continuously	Maintain centrifuge in service and operating satisfactorily to keep seal oil dry.
Degassed and out of service. Dew point measurement in and out. Increase frequency if dew point deteriorates.	Degassed and out of service	Maintain centrifuge in service and operating satisfactorily to keep seal oil dry.
Degassed and out of service	Degassed and out of service	Maintain centrifuge in service and operating satisfactorily to keep seal oil dry.

differential between the hydrogen and the water. Therefore, the generator rating should also be decreased.

A generator with hydrogen-only cooling can be derated due to a decrease in hydrogen pressure to the extent that it can be operated in air at 1 atm (1.03 bar). Its apparent power rating in this condition will be around 60 percent of the rating in hydrogen. Figure 15.4 illustrates the derating of a generator due to a reduction in hydrogen pressure.

FAULT CONDITIONS

Faults can occur during operation of a generator. Protective systems connected to the generator will either provide an alarm or trip the unit when a fault occurs. The following are the most common faults that occur in generators.

Stator Ground (Earth) Faults

The generator stators are grounded through a distribution transformer. A resistor is installed on the secondary side of the transformer to limit the fault current. This arrangement will minimize the damage to the stator core when a fault occurs. Ground fault relays are installed on the generator system. They trip the unit upon a ground fault. Note that moisture or contaminated insulation can cause flashover and activation of these relays. Samples of the hydrogen in the stator and the oil in the generator transformer should be obtained following such an incident. These samples should be analyzed for hydrocarbons and other gases to help identify the cause of the fault and its location.

Stator Phase-to-Phase Faults

Large generators are normally connected to the grid via a step-up transformer. Phase-segregated bus bars connect the generator to the step-up transformers. Thus, the probability of a phase-to-phase fault is low. However, if a fault of this type occurs in the stator, it will cause extreme damage. The connections at the ends of the windings are the areas most prone to phase-to-phase faults. In these areas, nylon hoses provide cooling water to the end windings. If a conducting debris is left in the generator during maintenance, it can penetrate through the insulation, leading to a phase-to-phase fault. Hence, it is essential to take great care to prevent any debris from being left in the generator.

Stator Interturn Faults

Interturn faults are normally caused by debris on the winding, fatigue failure of the insulator, moisture condensation at the end windings, contamination of the stator cooling water, etc. Interturn faults generate high recirculating currents and do considerable damage at the location of the fault. There is no protection for this type of a fault. However, if the damage progresses, a ground (earth) or a phase-to-phase fault will occur. The protective systems will trip the generator in this case.

Negative Phase Sequence Currents

Negative sequence currents occur due to unbalanced loading or unbalanced faults. These currents flow in the rotor core or slot wedges (Fig. 15.5), causing unacceptable heating in

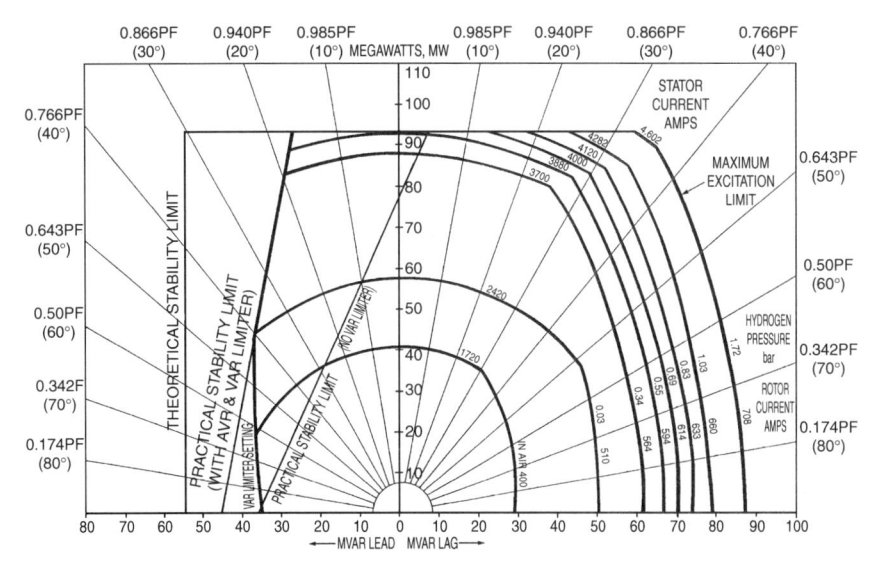

FIGURE 15.4 Derating of 94-MW generator with reduced hydrogen.

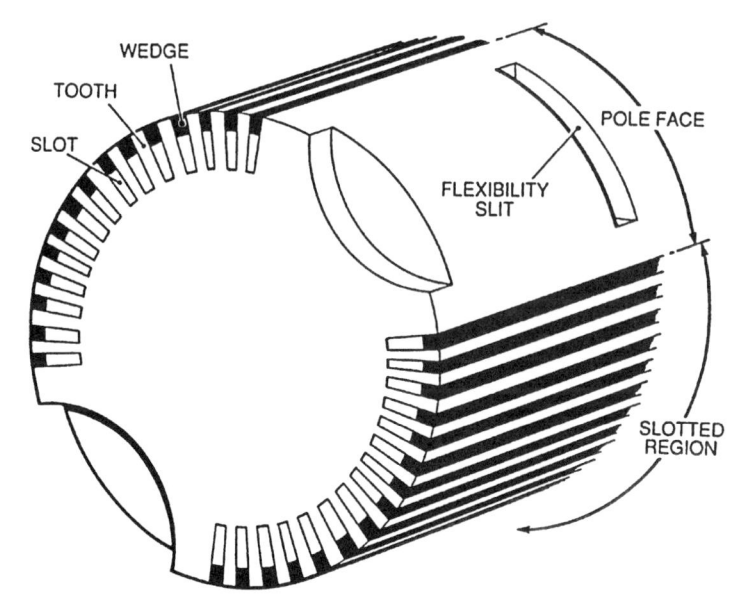

FIGURE 15.5 Typical two-pole turbine-generator rotor construction.

a localized area of the rotor. They can also cause cracks in the rotor in extreme cases. The protective systems will trip the rotor when potentially damaging negative sequence currents are generated. Alarms annunciate when low-level negative sequence currents occur, to warn operators that the load is unbalanced. It is necessary to identify and eliminate the cause of unbalance. The grid system should be notified of this condition because the cause of unbalance could be outside the power station. If the negative sequence currents are caused by an open circuit in a major interconnection, the unbalance will increase, leading to a unit trip. The negative sequence current will be shared by all the generators connected to the grid. Identical machines will share the current equally. Thus, they will all trip simultaneously at a specified value of the negative sequence current. This will lead to a further increase in the negative sequence current on the generators that are still connected to the grid.

Figure 15.6 illustrates the variation of the total rotor loss with negative sequence current I_2 per unit (i.e., portion of the rotor current) for a typical rotor. Figure 15.7 illustrates the variation of rotor temperature for two values of negative sequence currents I_2, suddenly applied at the end of a rotor flexibility slit. These results were obtained from manufacturers' tests.

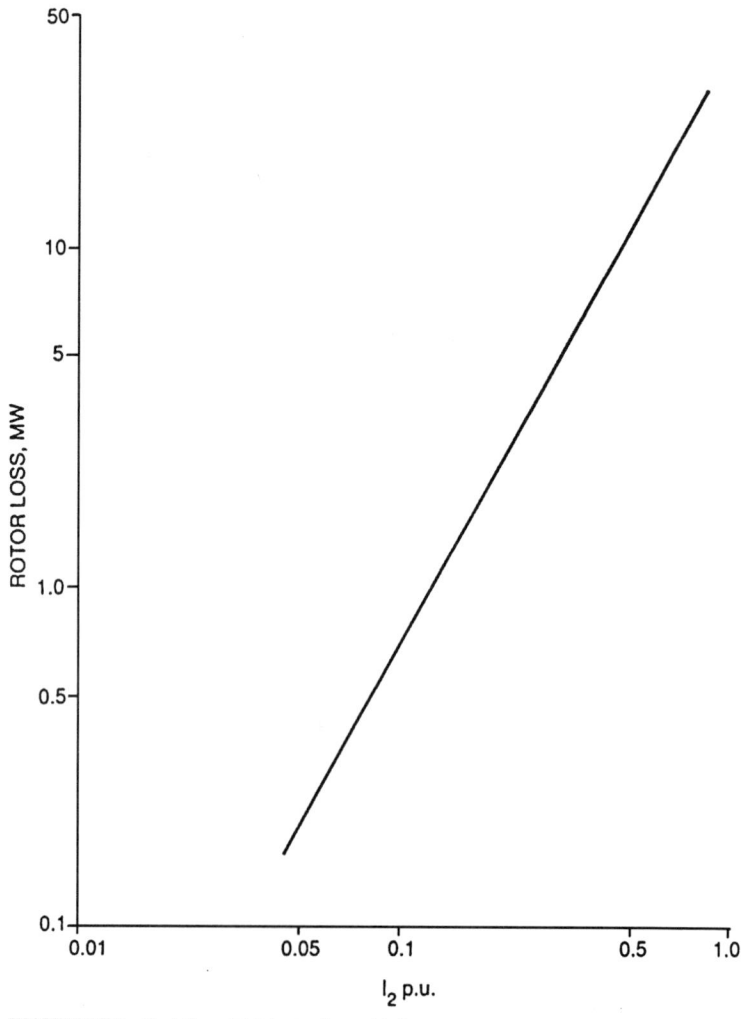

FIGURE 15.6 Variation of total rotor loss with I_2.

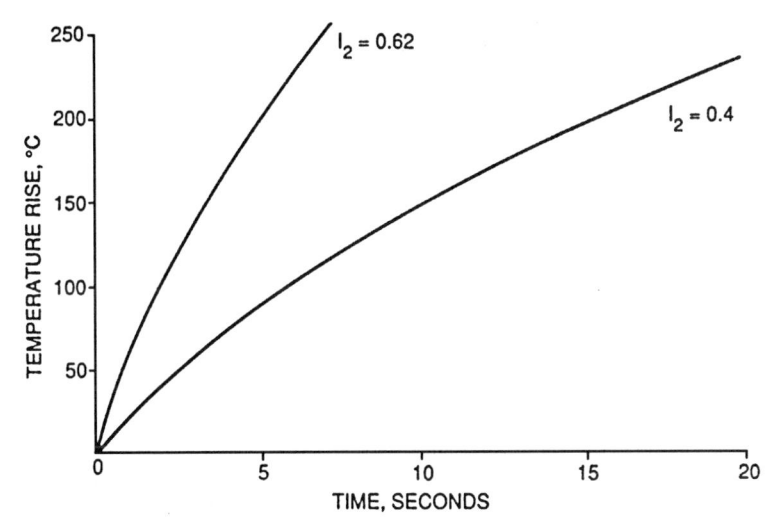

FIGURE 15.7 Transient temperature response for two values of suddenly applied negative sequence currents I_2 at the end of a flexible rotor slit.

Loss of Generator Excitation

The loss of excitation causes a decay in the rotor flux and an eventual loss of synchronism. The unit operates then as an induction generator. It draws reactive power from the system. The generator terminal voltage will drop. This could result in impairment of some of the auxiliary loads supplied from the generator. The protective systems trip the generator when this condition is detected. These systems rely on an impedance relay that monitors the generator impedance as seen from its terminals. The generator impedance changes when the excitation is lost. The impedance relay detects this change and trips the machine.

During recoverable transient conditions, there is a significant change in the generator impedance. Thus, a time delay is incorporated in the protective system, using the impedance relay to prevent tripping of the unit during recoverable transient conditions. Some generator protective systems trip the unit on loss of excitation when a large input of reactive power is detected. A time delay is also incorporated in this protective system to prevent unit trip during recoverable transient conditions.

Pole Slipping

Pole slipping in a generator is caused by loss of synchronism due to one of the following conditions:

- Inability of the excitation system to maintain the torque required for synchronism
- A system fault

This event causes large variations in voltage, active power, and reactive power delivered by the generator. It could also trip the auxiliary system motors, leading to a unit trip. This type of a shutdown could cause damage to the plant because it is nonsequenced. This event would also have an effect on the load and other generating plants connected to the grid. The transmission circuit protection could also operate indiscriminately during this event due to the large variation in power transmitted. This will lead to disconnection of some generating units from the grid. However, pole-slipping events are rare due to grid operating criteria. Thus, pole-slipping protection is not installed on most generating units.

Rotor Faults

A static or a rotating rectifier supplies dc power (excitation) to the windings of a generator rotor. The excitation voltage and current of a typical 600-MW plant are 550 V and 4500 A, respectively. The excitation circuit is insulated from the ground. The ground fault alarm annunciates when a ground fault having a resistance less than 20,000 Ω occurs. A ground fault can occur due to carbon dust on the slip ring insulation, or to moisture on the insulation of the excitation circuit. In this case, it is necessary to isolate the rotor windings by removing the carbon brushes to identify the location of the fault.

A ground fault can also occur due to a failure of the rotor winding insulation. However, one ground fault will not cause significant damage to the rotor. If a second ground fault occurs at a different location, a short circuit will develop in the winding, leading to flow of currents through the rotor forging or end rings. This current could cause extensive damage to the rotor. A generator having a high-resistance rotor ground fault can continue to operate with the aid of a ground fault protection system. This system monitors the resistance of the ground fault and its position. It trips the unit when the resistance of the ground fault drops below a specified value or when the location of the ground fault changes.

The three mechanisms of insulation failure are

- Bridging of the insulation by foreign material.
- Mechanical failure of the insulation.
- Copper dust generated by wear of the copper winding due to movement within the slots. Most of the copper dust is generated when the machine is rotated slowly by the turning gear. During this period, the copper windings contract because the machine is cold. There is more space for the windings to move within the rotor slots. A significant amount of copper dust could be generated if the machine is left on turning gear for an extended time.

A ground fault can occur by any of the mechanisms mentioned earlier. The current at the location of the fault could overheat the insulation and damage the windings. The damage could also spread to adjacent windings. The automatic voltage regulator (AVR) will increase the rotor current to compensate for the fault. Thus, a ground fault may not necessarily be noticed. The fault could also disappear when the unit is shut down due to movement of the insulation within the slot.

The effects of one set of shorted turns will be limited to one pole only. The rotor magnetic field will become unbalanced, leading to vibration. The magnitude of the vibration will vary with the rotor current. Since the current flowing in the shorted turns is lower, less heat is generated by them than by other turns. This will cause a thermal bend in the shaft that produces vibration. The magnitude of the vibration increases with the rotor current. These vibrations can limit the rotor currents, thereby limiting the capability of the generator. In this case, it is necessary to repair the rotor at the earliest opportunity. However, if continued operation of the unit is necessary until a suitable outage, then off-loading balance (known as offset balancing) can be done to maintain the vibration levels within acceptable limits. Figure 15.8a illustrates the vibration characteristics of a rotor prior to offset balancing. The cold vibration datum was at B (around 30 μm peak to peak). The vibration level changed to B4 (around 85 μm peak to peak) when the rotor current reached 1850 A.

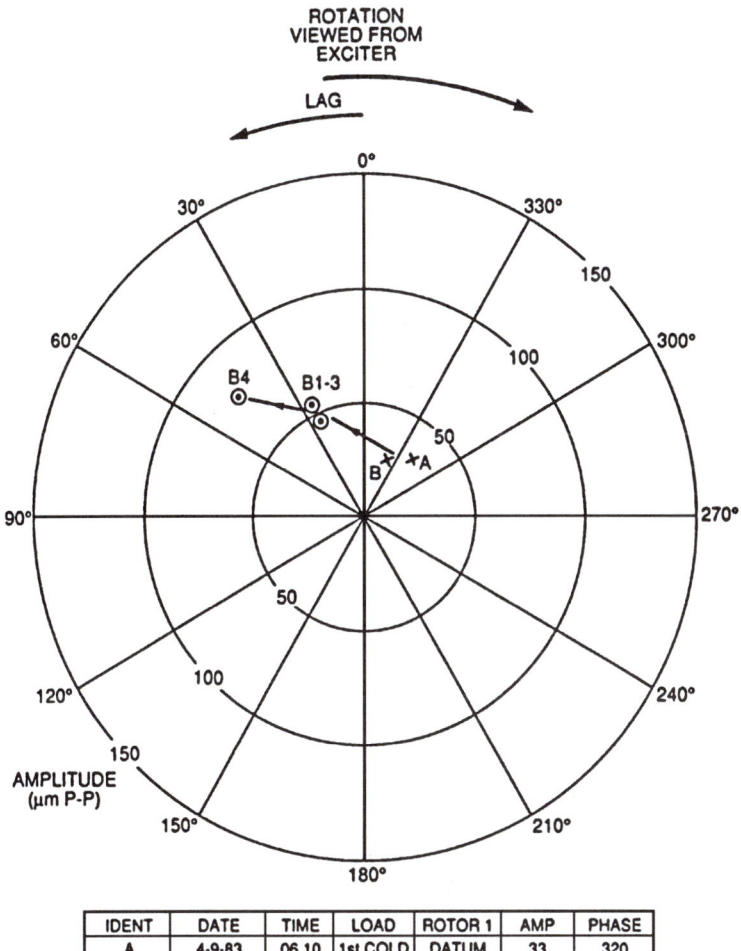

IDENT	DATE	TIME	LOAD	ROTOR 1	AMP	PHASE
A	4-9-83	06.10	1st COLD	DATUM	33	320
B	4-9-83	11.00	2nd COLD	DATUM	28	340
B1	4-9-83	23.25	80	1500	56	025
B2	5-9-83	02.20	150	1700	56	026
B3	5-9-83	05.50	260	1400	48	024
B4	5-9-83	12.00	270	1850	79	046

(a)

FIGURE 15.8 Thermal bending of rotor shaft, giving vibration phasor dependent on load and the effect of offset balancing. (*a*) Bearing 12-V initial response to loading.

Following offset balancing, the cold vibration datum (Fig. 15.8*b*) moved to G (around 55 μm peak to peak). The vibration level changed to H4 (around 70 μm peak to peak) when the rotor current reached 2470 A. Thus, the offset balancing has allowed the generator to deliver higher load.

The in-core leakage flux detector (known also as the search coil) detects shorted turns in the rotor. It is installed on the stator in the air gap between the stator and the rotor. Figure

15.16 CHAPTER FIFTEEN

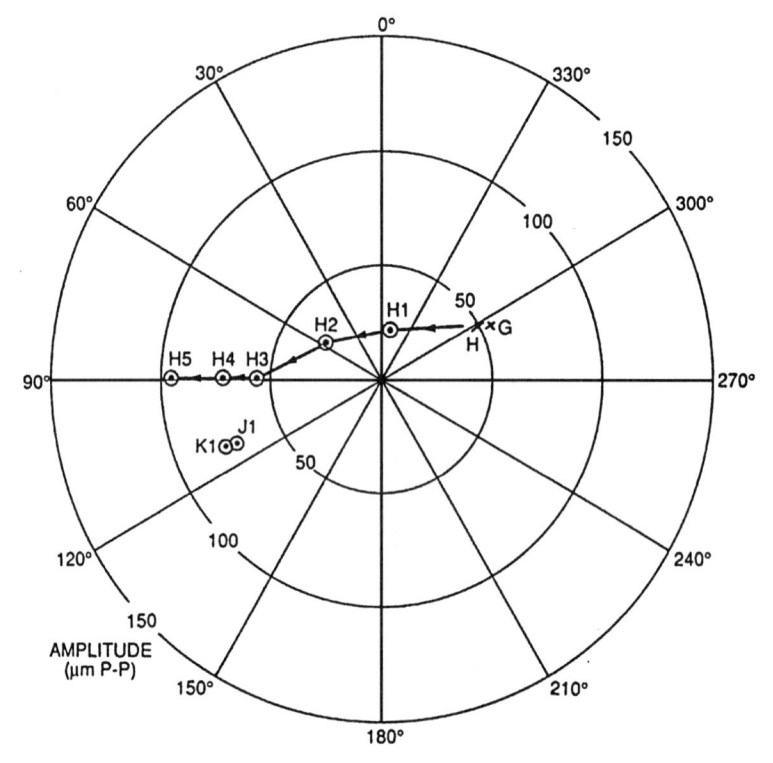

IDENT	DATE	TIME	MW	ROTOR 1	AMP	PHASE	
G	19-9-83		COLD	DATUMS	54	299	(MEAN)
H	19-9-83				49	300	(MEAN)
H1	20-9-83	04.50	280	1600	24	352	
H2	20-9-83	06.10	440	2100	33	055	
H3	20-9-83	11.50	470	2300	56	088	
H4	20-9-83	13.55	460	2470	72	088	
H5	20-9-83	15.36	460	2400	97	088	
J1	7-10-83	10.15	470	2220	71	112	
K1	21-10-83	14.30	460	2250	74	112	

(b)

FIGURE 15.8 (*Continued*) Thermal bending of rotor shaft, giving vibration phasor dependent on load the effect of offset balancing. (*b*) Response to loading following balance.

15.9 illustrates the voltage output waveform of a search coil. The tooth harmonic waveform is different for each pole. Figure 15.9*a* illustrates faults on coils D and F. Figure 15.9*b* illustrates additional shorted turns that developed on coil D.

REFERENCE

1. British Electricity International, *Modern Power Station Practice—Station Operation and Maintenance, Volume G*, 3d ed., Pergamon Press, Oxford, United Kingdom, 1991.

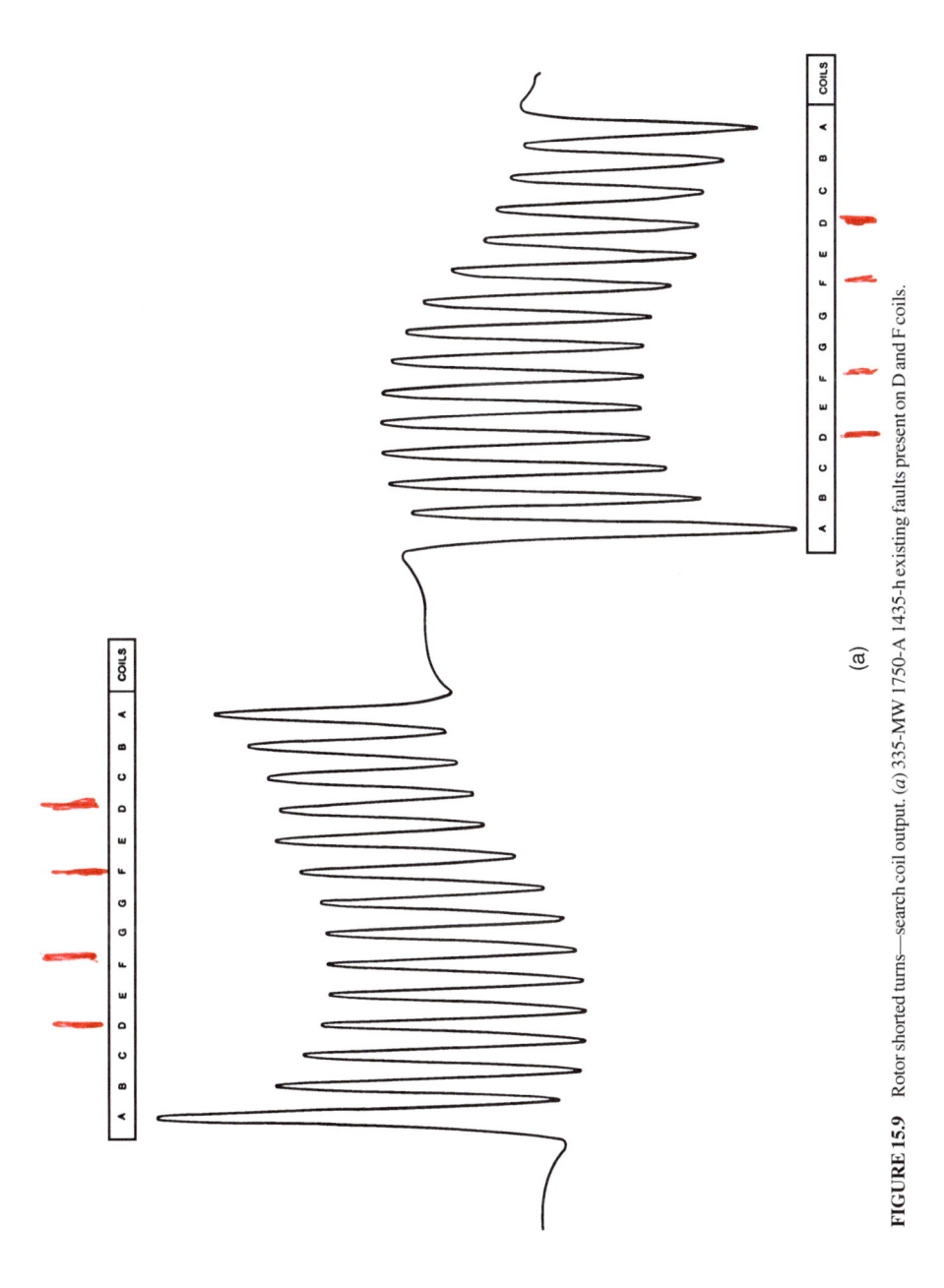

FIGURE 15.9 Rotor shorted turns—search coil output. (*a*) 335-MW 1750-A 1435-h existing faults present on D and F coils.

15.17

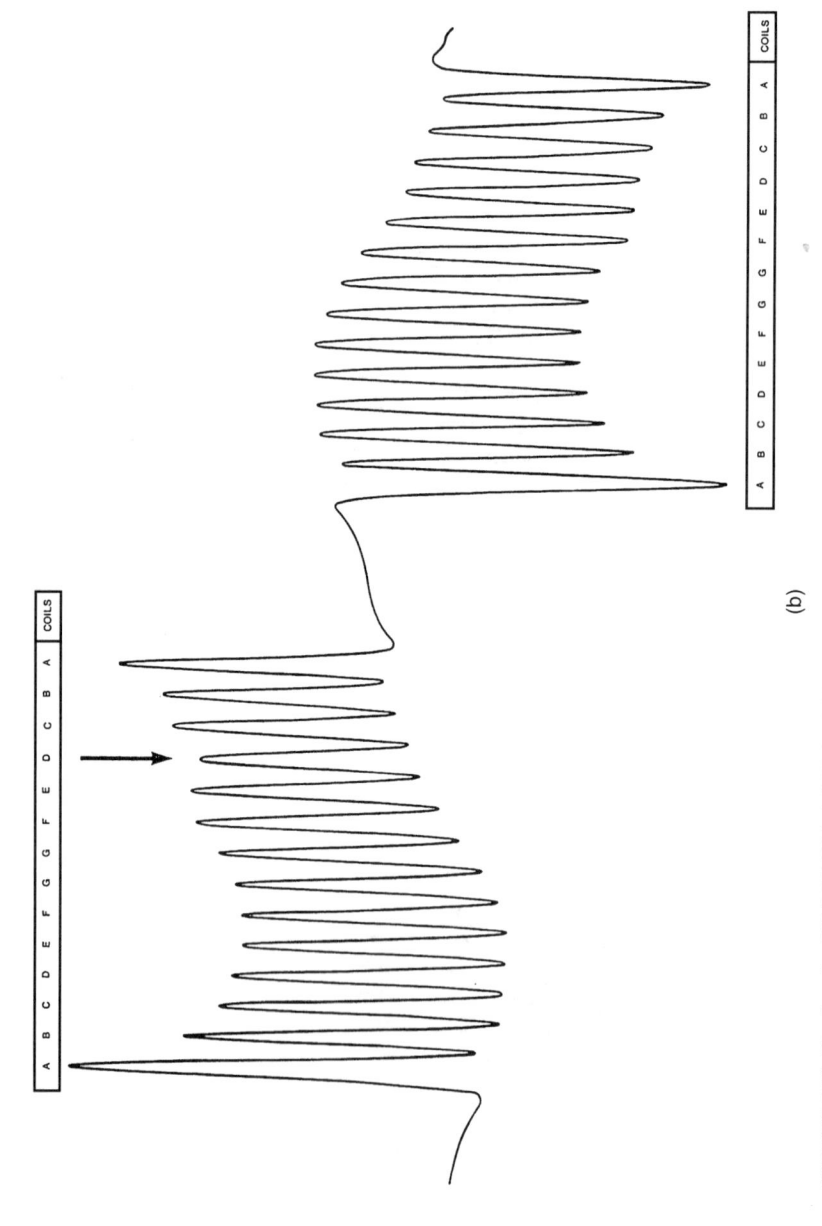

FIGURE 15.9 (*Continued*) 320-MW 2029-A 1600-h additional shorted turns, D coil. (*b*) At increased rotor current, fault on D coil worsens, and more turns are shorted.

CHAPTER 16
GENERATOR SURVEILLANCE AND TESTING

GENERATOR OPERATIONAL CHECKS (SURVEILLANCE AND MONITORING)

Regular monitoring of the following parameters is required:

1. Temperature of stator windings.
2. Core temperature.
3. Temperature of slip rings.
4. Vibration levels at the bearings.
5. Brush gear inspection (monthly or bimonthly). Remove the brush holder. Clean the carbon brushes by compressed air. Inspect the brushes for uneven wear (do not touch the brushes with bare hands). Replace worn brushes. Use a stroboscope to inspect the slip rings. Vary the frequency of the stroboscope to check for uneven wear of the slip rings. The temperature of the slip rings should be measured using an infrared detector. If the slip ring temperature is high, the brushes will overheat and wear quickly. Generator derating is required if the slip ring temperature is high.
6. On-line partial discharge activity.

Major Overhaul (Every 8 to 10 Years)

The electrical and mechanical tests required are described in Appendixes A and B, respectively. In summary, the work includes the following:

1. Perform insulation resistance and polarization index tests.
2. Investigate the causes of partial discharge.
3. Check the tightness of the stator wedges by tapping them with a hammer. The wedges must be tight to minimize the movement of the stator conductor bars during operation. Rewedging and adding packing may be required.
4. Perform the EL-CID test to determine if there is any loosening in the core laminations or deterioration in the core insulation. Apply penetrating epoxy if insulation is degrading.
5. Perform casing pressure and stator pressure and vacuum decay tests.
6. Refurbish the rotor including inspection of radial pins and end caps (including ultrasonic and dye penetrant testing), vacuum test, slip ring refurbishment, and check for copper

dusting. *Note:* Copper dusting occurs due to fretting of the rotor copper windings as they move in the slots when the machine is on turning gear. This problem does not occur during normal operation because centrifugal forces push the windings against the wedges. Copper dusting can cause shorts in the machine.
7. Calibrate protection equipment.

APPENDIX A: GENERATOR DIAGNOSTIC TESTING

The following factors affect the insulation systems in generators:

- High temperature
- Environment
- Mechanical effects such as thermal expansion and contraction, vibration, electromagnetic bar forces, and motor start-up forces in the end turns
- Voltage stresses during operating and transient conditions

All these factors contribute to loss of insulation integrity and reliability. These aging factors interact frequently to reinforce one another's effects. For example, high-temperature operation could deteriorate the insulation of a stator winding, loosen the winding bracing system, increase vibration, and cause erosion. At some point, high-temperature operation could lead to delamination of the core and internal discharge. This accelerates the rate of electrical aging and could lead to a winding failure.

Nondestructive diagnostic tests are used to determine the condition of the insulation and the rate of electrical aging. The description of the recommended diagnostic tests for the insulation system of motors and the conditions they are designed to detect are discussed.

Stator Insulation Tests

An electrical test is best suited to determine the condition of electrical insulation. The tests on insulation systems in electrical equipment can be divided into two categories:

1. High-potential (hipot), or voltage-withstand, tests
2. Tests that measure some specific insulation property, such as resistance or dissipation factor

Tests in the first category are performed at some elevated ac or dc voltage to confirm that the equipment is not in imminent danger of failure if operated at its rated voltage. Various standards give the test voltages that are appropriate to various types and classes of equipment. They confirm that the insulation has not deteriorated below a predetermined level and that the equipment will most likely survive in service for a few more years. However, they do not give a clear indication about the condition of the insulation.

The second category of electrical tests indicates the moisture content; presence of dirt; development of flaws (voids), cracks, and delamination; and other damage to the insulation. A third category of tests includes the use of electrical or ultrasonic probes that can determine the specific location of damage in a stator winding. These tests require access to the air gap and energization of the winding from an external source. These tests are considered an aid to visual inspection.

DC Tests for Stator and Rotor Windings Index

These tests are sensitive indicators to the presence of dirt, moisture, and cracks. They must be performed off-line with the winding isolated from ground, as shown in Fig. 16.1. Suitable safety precautions should be taken in performing all high-potential tests. When high-voltage dc tests are performed on water-cooled windings, the tubes or manifolds should be dried thoroughly, to remove current leakage paths to ground and to avoid the possibility of damage by arcing between moist patches inside the insulating water tubes. For greater sensitivity, these tests can be performed on parts of the windings (phases) isolated from one another.

The charge will be retained in the insulation system for up to several hours after the application of high dc voltages. Hence, the windings should be kept grounded for several hours after a high-voltage dc test to protect personnel from a shock.

Tests using dc voltages have been preferred over the ones using ac voltages for routine evaluation of large machines for these reasons:

1. The high dc voltage applied to the insulation during a test is far less damaging than high ac voltages due to the absence of partial discharges.
2. The size and weight of the dc test equipment are far less than those of the ac test equipment needed to supply the reactive power of a large winding.

Insulation Resistance and Polarization Index

The *polarization index* (PI) and insulation resistance tests indicate the presence of cracks, contamination, and moisture in the insulation. They are commonly performed on any motor and generator winding. They are suitable for stator and insulated rotor windings.

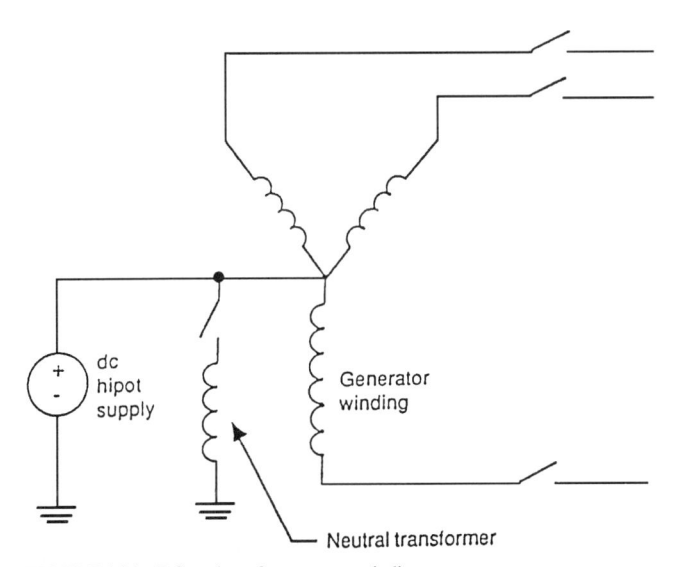

FIGURE 16.1 DC testing of a generator winding.

The insulation resistance is the ratio of the dc voltage applied between the winding and ground to the resultant current. When the dc voltage is applied, the following current components flow:

1. The charging current into the capacitance of the windings.
2. A polarization or absorption current due to the various molecular mechanisms in the insulation.
3. A "leakage" current between the conductors and ground (the creepage path). This component is highly dependent on the dryness of the windings.

The first two components of the current decay with time. The third component is mainly determined by the presence of moisture or a ground fault. However, it is relatively constant. Moisture is usually absorbed in the insulation and/or condensed on the end winding surfaces. If the leakage current is larger than the first two current components, then the total charging current (or insulation resistance) will not vary significantly with time.

Therefore, the dryness and cleanliness of the insulation can be determined by measuring the insulation resistance after 1 min and after 10 min. The polarization index is the ratio of the 10-min reading to the 1-min reading.

Test Setup and Performance. Several suppliers, such as Biddle Instruments and Genrad, offer insulation resistance meters that can determine the insulation resistance accurately by providing test voltages of 500 to 5000 V dc. For motors and generators rated 4 kV and higher, 1000 V is usually used for testing the windings of a rotor, and 5000 V is used for testing the stator windings.

To perform the test on a stator winding, the phase leads and the neutral lead (if accessible) must be isolated. The water must be drained from any water-cooled winding, and any hoses removed or dried thoroughly by establishing a vacuum (it is preferable to remove the hoses because vacuum drying is usually impossible).

The test instrument is connected between the neutral lead or one of the phase leads and the machine frame (Fig. 16.1). To test a rotor winding, the instrument should be connected between a lead from a rotor winding and the rotor steel. During the test, the test leads should be clean and dry.

Interpretation. If there is a fault or the insulation is punctured, the resistance of the insulation will approach zero. The IEEE standard recommends a resistance in excess of $(V_{LL} + 1)$ MΩ. If the winding is 13.8 kV, the minimum acceptable insulation resistance is 15 MΩ. This value must be considered the absolute minimum since modern machine insulation is on the order of 100 to 1000 MΩ. If the air around the machine had high humidity, the insulation resistance would be on the order of 10 MΩ.

The insulation resistance depends highly on the temperature and humidity of the winding. To monitor the changes of insulation resistance over time, it is essential to perform the test under the same humidity and temperature conditions. The insulation resistance can be corrected for changes in winding temperature. If the corrected values of the insulation resistance are decreasing over time, then there is deterioration in the insulation.

However, it is more likely that the changes in insulation resistance are caused by changes in humidity. If the windings were moist and dirty, the leakage component of the current (which is relatively constant) will predominate over the time-varying components. Hence, the total current will reach a steady value rapidly. Therefore, the polarization index is a direct measure of the dryness and cleanliness of the insulation. The PI is high (>2) for a clean and dry winding. However, it approaches unity for a wet and dirty winding.

The insulation resistance test is a very popular diagnostics test due to its simplicity and low cost. It should be done to confirm that the winding is not wet and dirty enough to cause a failure that could have been averted by a cleaning and drying out procedure.

The resistance testing has a pass/fail criterion. It cannot be relied upon to predict the insulation condition, except when there is a fault in the insulation.

The high potential tests, whether dc or ac, are destructive testing. They are not generally recommended as maintenance-type tests. For stator windings rated 5 kV or higher, a partial discharge (PD) test, which in the past has been referred to as *corona*, should be done. The level of partial discharge should be determined because it can erode the insulation and lead to insulation aging.

DC Hipot Test

A high dc voltage withstand test is performed on a stator or rotor winding to ensure that the ground wall insulation can be stressed to normal operating voltage. The outcome of the test is simply pass or fail. Thus, it is not classified as a diagnostic test. The DC hipot test is done sometimes following maintenance on the winding, to confirm that the winding has not been damaged. It is important to consider the consequences of a hipot failure. Spare parts and outage time should be available before proceeding with this test.

The DC hipot test is based on the principle that weakened insulation will puncture if exposed to a high enough voltage. The test voltage is selected such that damaged insulation will fail during the test and good insulation will survive. Insulation that fails during a hipot test is expected to fail within a short time if placed in service. The distribution of electrical stresses within the insulation during a dc test is different from normal ac operation because the dc electric field is determined by resistances rather than capacitances. Figure 16.1 illustrates how the test is done. The winding is isolated, and a high voltage is connected between the winding and ground. If the stator windings are water-cooled, they must be drained and the system dried thoroughly to avoid electrical tracking of the coolant hoses. The hoses should be removed to ensure they are not damaged by the test. The stator frame and all temperature sensors must be grounded. All accessories such as current and potential transformers must be disconnected or shorted. The suggested voltage test for a new winding is 1.7 times the rms ac voltage. A typical routine voltage used during maintenance is $2V_{L-L}$ kV dc. However, the test voltages used by the manufacturer and during commissioning are significantly higher than the maintenance test voltage level. The rotor windings do not have a standard test voltage level.

If a hipot test is successful, it confirms that there are no serious cracks in the ground wall and the insulation system. The insulation will most likely withstand normal operating stresses until the next scheduled maintenance test.

High-Voltage Step and Ramp Tests. The variation of current (or insulation resistance) should be monitored as the dc hipot test is performed. If there is a weakness in the ground wall, a sudden nonlinear increase in current (or decrease in insulation resistance) will precede a breakdown as the voltage is increasing. An experienced operator can interrupt the test when the first indication of warning occurs. If the voltage achieved is considered sufficient, the machine can be returned to service until the repairs can be planned. Following identification of a suspect phase, the location of the "weakness" must be found. The variations of voltage with current obtained during the test can be used in future comparisons on the same winding if the same conditions exist.

The winding must be completely isolated (Fig. 16.1). A special "ramp" or conventional high-voltage dc test set is used for this test. The leakage current must be calculated at the end of each voltage step. The test operator must make a judgment based on the increase in leakage current before increasing the voltage further. The test voltage can alternatively be increased slowly with a recorder plotting the variations of leakage current against voltage (Fig. 16.2*a* and *b*).

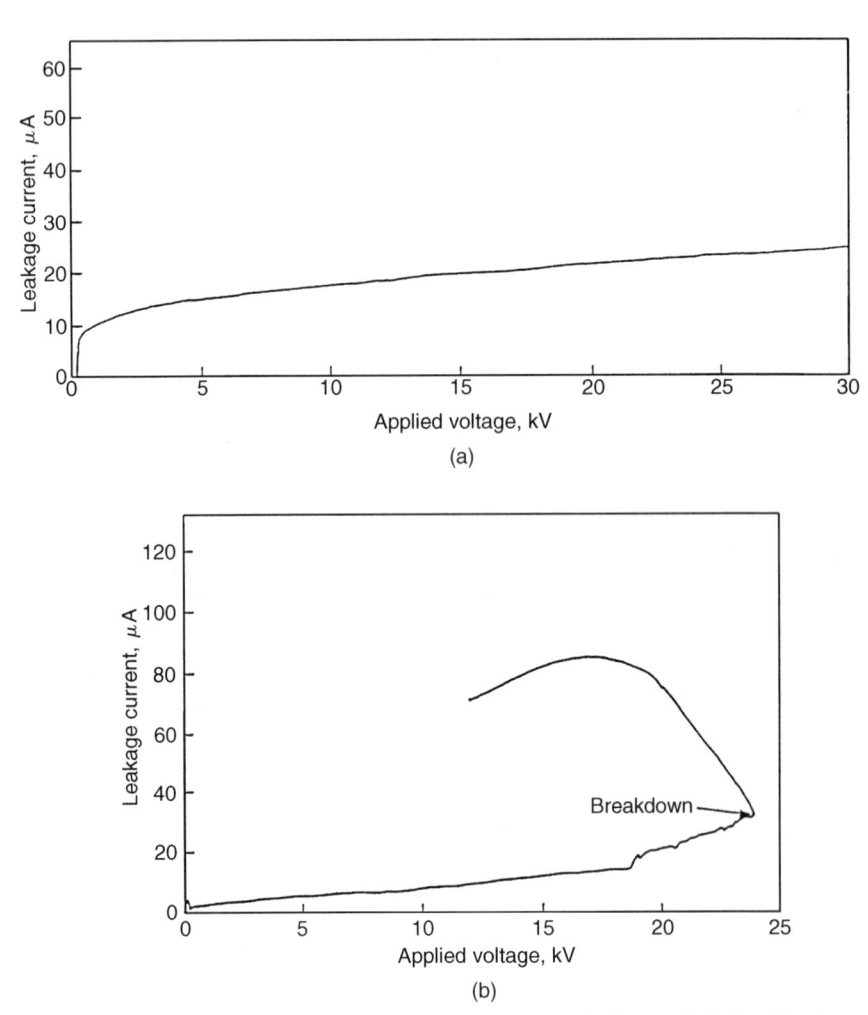

[Figure: (a) Leakage current vs Applied voltage showing smooth curve rising from 0 to ~25 μA over 0-30 kV. (b) Leakage current vs Applied voltage showing curve rising to ~85 μA at ~17 kV then dropping to Breakdown at ~24 kV.]

FIGURE 16.2 Ramp test characteristic output. (*a*) Voltage-current plot for a good winding; (*b*) voltage-current plot showing instabilities prior to insulation puncture.

A weakness in the ground wall can be detected by a sudden increase in leakage current. If the condition of the ground wall is questionable, the machine can be returned to service if the achieved voltage is considered sufficient. Further investigations can be scheduled at a more convenient time.

AC Tests for Stator Windings

The dc tests are only capable of measuring the conductivity of the insulation system. The ac tests are usually more revealing of the insulation condition. However, they are more onerous than dc tests. The ac tests are also capable of being sensitive to the mechanical condition of the system. For example, if delamination (air-filled layers) is present in the ground wall, the capacitance between the conductors and the core will decrease. However, if the winding is wet, the capacitance will increase.

Partial Discharge Tests. Partial discharges (known in the past as corona) are spark charges which occur in voids within high-voltage insulation (more than 5 kV). They occur between the windings and core, or in the end winding region. These are "partial" discharges because there is some remaining insulation. PD can erode the insulation and therefore contributes to its aging. However, PD is a symptom of insulation aging caused by thermal or mechanical stresses. The measurement of PD activity in a stator winding is an indication of the health of the insulation. PD tests provide the best means for assessing the condition of the insulation without a visual inspection. These tests should be done on stator windings in motors and generators rated higher than 5 kV.

Off-Line Conventional PD Test. The conventional PD test involves energizing the winding to normal line-to-ground ac voltage with an external supply. A PD detector is used to measure the PD activity in the winding. The sparks caused by PD are fast current pulses that travel through the stator windings. These pulses and the accompanying voltage pulses increase with the PD pulse. Figure 16.3 illustrates a high-voltage capacitor that can block the power frequency voltage and allows the high-frequency pulse signals to reach the PD detector. An oscilloscope is used to display the pulse signals, after further filtering.

The pulse magnitudes are calibrated in picocoulombs (pC) even though the actual measurements are in millivolts. The conventional test is done off-line. A separate voltage supply is used to energize the windings to normal voltage. The interference from high-frequency electrical noise in this test is a minimum.

Test Setup and Performance. The conventional test involves isolating the winding from the ground and energizing one phase of the winding by a 60-Hz power supply cable to rated line-to-ground voltage. This test is normally done on each phase separately while the remaining two are grounded. The phases are disconnected from one another at the neutral. Draining of the water-cooled winding is not required.

The test equipment includes a power separation filter (high-voltage capacitor and a high-pass filter to block the power frequency and its harmonics, Fig. 16.3). The oscilloscope displays the PD pulses (Fig. 16.4). A pulse height analyzer is used to process the pulse data. It gives the pulse counts, pulse magnitudes, and comparisons between positive and negative pulses.

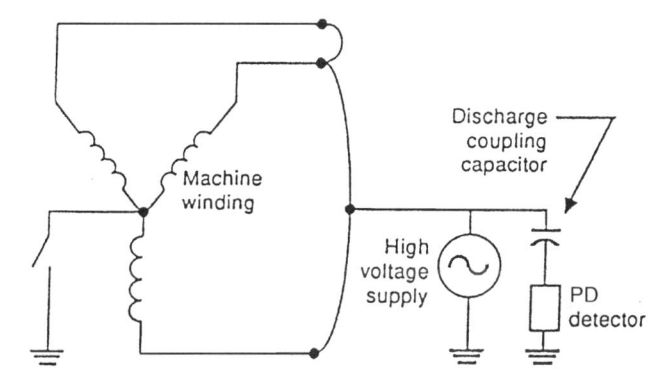

FIGURE 16.3 Test arrangement for conventional partial discharge test.

16.8 CHAPTER SIXTEEN

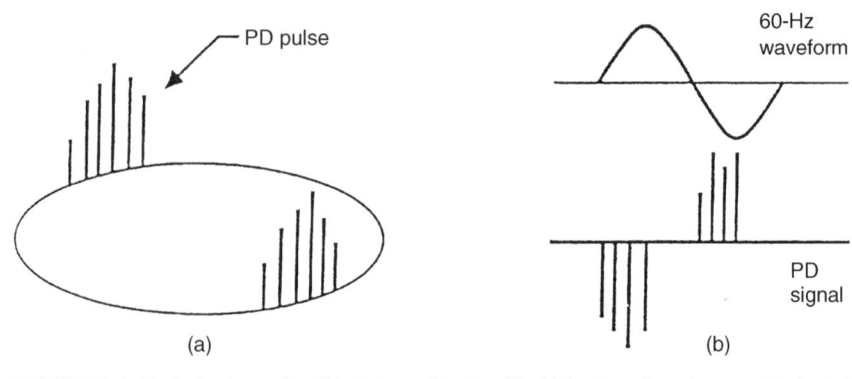

FIGURE 16.4 Typical outputs of partial discharge detectors. The higher the pulses, the more deteriorated the winding. (*a*) Shows the elliptical trace from many types of commercial detectors, where the pulse position on the ellipse indicates the phase position; (*b*) shows the display on a conventional dual-channel oscilloscope.

The ac voltage is raised gradually until PD pulses are observed on the oscilloscope. The voltage at which PD starts is called the *discharge inception voltage* (DIV). When the test voltage reaches the normal voltage, the magnitude of the pulses is read from the screen. The analysis of the pulse height is normally recorded. As the ac voltage is decreased, the voltage at which the PD pulses disappear is recorded. It is called the *discharge extinction voltage* (DEV). It is usually lower than the DIV. The actual test takes about 30 min normally. However, the setup and disassembly can take up to a day.

Interpretation. There is no general agreement on the acceptable magnitudes of PD, DIV, and DEV. The inductive nature of the windings makes the calibration of the measured PD magnitudes (conversion from millivolts on the screen of the oscilloscope into picocoulombs) difficult. Thus, the measurement of the pulses may not provide an accurate value of the PD activity. These measurements cannot be calibrated from machine to machine or among the different types of commercial detectors.

The most useful method for interpreting the PD test results is by performing the test at regular intervals and monitoring for trends. The recommended interval for air-cooled machines is once or twice per year and 2 years for hydrogen-cooled generators. As the condition of the insulation worsens, the magnitude of the PD will increase and the DIV and DEV will decrease. An increasing trend of PD activity indicates that the insulation is aging. Visual inspection of the winding condition may be required. PD results should be compared only if the same equipment and procedures are used during testing. This is due to the calibration problems mentioned earlier. Comparison of results can also be misleading if there are differences between the types or ratings of the windings. Comparison of PD results are valid if the windings and test methods are identical.

A PD magnitude of less than 1000 pC indicates that the winding should not fail during the next few years. A visual inspection is recommended if the magnitude of PD is more than 10,000 pC, especially if other identical machines have a PD less than 1000 and the insulation is made of epoxy-mica. The DIV in modern epoxy or polyester windings should be greater than one-half the operating line-to-ground voltage. The test indicates that slot discharge is occurring if the DIV value is very low in epoxy-mica windings. However, older asphaltic and micafolium windings may not be in danger even if the magnitude of the discharge is high and the DIVs are low. This is in contrast with newer machines that have synthetic insulation, especially Mylar. Their condition deteriorates quickly in the presence of PD. However, older windings should be inspected if there is an increasing trend of PD activity.

There are many disadvantages for off-line conventional PD tests. Since the entire winding, including the neutral end, is fully energized, sites which are not normally analyzed can generate pulses. Large discharges can occur in sites which are not normally subjected to high voltages. This is misleading because the operator may believe that the winding is deteriorating.

On-Line Conventional PD Test. This test is similar to the off-line test except that an external power supply is not used to energize the winding. The generator is driven at normal speed by the turbine, and sufficient field excitation is applied. Therefore, the stator is at the normal operating voltage. The test can be performed with the generator synchronized to the grid or not. When the test is done on a motor, the winding is energized by the normal power supply. Extreme caution is required when the test is performed due to the considerable risk to personnel and the machine if the capacitor fails.

The test is more realistic than the off-line one because the voltage distribution in the windings is normal. Also, slot discharges that are caused from bar or coil movement are present.

The equipment used in the off-line test can be used in this test. The blocking capacitors are connected to the phase terminals during an outage. *Dangerous events can occur if the capacitor fails during the test.* An experienced operator can distinguish true PD from electrical interference from brushes, thyristor excitation systems, and background. If the generator is not synchronized, some generators can handle variations in the field current. In these cases, the DIV and DEV can be measured.

Some utilities leave the test equipment connected during normal operation. The PD activity can then be measured at low and full power. Deterioration in the condition of the insulation is detected by an increasing trend of PD. Since the test is done during normal operation, it gives the most accurate indication of the true condition of the insulation. External interference (from power line carrier, radio station, etc.) can be severe during the test, especially in large generators. The interference can be misleading. The operator may believe that high PD activity is occurring while the winding is perfectly good.

Dissipation Factor and Tip-Up Tests. The condition of the insulation system in a high-voltage winding can be evaluated by treating it as a dielectric in a capacitor. The capacitance and dissipation factor (or power factor, or tan δ, see note*) of a winding can be measured during an outage. These measurements are normally made over a voltage range. Since partial discharges are initiated when there are voids in the ground wall insulation, the change in dissipation factor with voltage is a measure of the initiation of additional internal losses in a winding.

Machine manufacturers use dissipation factor and tip-up (change in tan δ as the voltage is increased or, Δ tan δ) as quality control tests for new stator bars and coils. A general weakness in the bulk insulation, normally caused by incorrect composition or insulation that is not fully cured, is indicated by abnormally high dissipation factor. Excessive voids within the insulation will discharge at high voltage. They are indicated by a higher-than-normal increase in dissipation factor when the voltage is increased (tip-up).

The winding must be isolated into coils or coil groups to obtain a sensitive measurement during this test. This consumes a significant amount of time. Thus, this test is not widely used for testing motors and generators in utilities.

Tip-Up Test. A half-day outage is normally required for a dissipation factor test. It can be performed on motors and generators of all sizes. A single measurement of dissipation factor on a complete winding has limited use. However, trends of measurements on coils or coil groups over years provide useful information. This test is most useful when done on low and high voltage. As the voltage is increased from a low to a high value, the dissipation

---
*Note: Angle δ is defined as $\delta + \theta = 90°$, where θ is the phase angle between the current and the voltage.

factor will increase, i.e., tip up. The PD activity within the insulation will increase with the voltage level. Dissipation factor is a measure of electrical losses in the insulation system. It is a property of the insulation. It is desirable to have a low dissipation factor. However, a high dielectric loss does not confirm that the insulation is poor. A capacitance bridge is normally used to measure the dissipation factor. For example, the dissipation factor of a good epoxy-mica and of asphaltic insulation is 0.5 and 3 percent, respectively.

The dissipation factor will not increase with the voltage in a perfect insulation. However, if air-filled voids are present in the insulation, partial discharge will occur at a high enough voltage. The electrical losses in the winding will increase due to energy consumption by heat and light generated by the discharges. The dissipation factor will increase with the voltage. As the partial discharge activity increases, the tip-up will increase and the condition of the winding will worsen. Therefore, the PD activity is measured indirectly by a tip-up test. A bridge is used to measure the dissipation factor. It effectively measures the ratio of the in-phase current in the sample to the capacitive (or quadrature) current. This ratio is determined over the total current in the sample. Thus, it represents the average loss over the entire winding being tested.

Stator Turn Insulation Surge Test. The surge tests are hipot tests used to check the integrity of the interturn as well as the capability of the ground wall insulation to withstand steep transients that are likely to be encountered in normal service. These surge tests are normally done on new windings in the factory to detect faults. A voltage is applied for a very short time to the turn insulation during the test, causing weak insulation to fail. Thus, the surge test is not a diagnostic test but a hipot test for the turn insulation. The impedances of two matching sections of the winding are compared by commercial surge testers. A voltage surge having around 0.2-μs rise time and adjustable magnitude is applied simultaneously to the two winding sections L_1 and L_2 (Fig. 16.5). The shape of the surges is superposed on an oscilloscope. A high-voltage transient is developed across the winding turns due to the short rise time. The two waveforms will be identical if both windings are free from faults (because the impedances are the same). Any discrepancy in the two waveforms may indicate a shorted turn in one of the windings. An experienced operator can identify the nature of the fault (Fig. 16.6) by comparing the magnitude and type of the discrepancy between the two waveforms.

In some cases, the surge voltage is applied to an exciter coil, which is placed over the stator coil to be tested (direct connection of the surge tester to a stator coil is not required). This allows testing of coils in complete windings without disconnecting each coil from the other. The voltage is induced into the stator coil by transformer action when the surge is applied. This produces the turn-to-turn stress in the stator coil. However, interpretation of

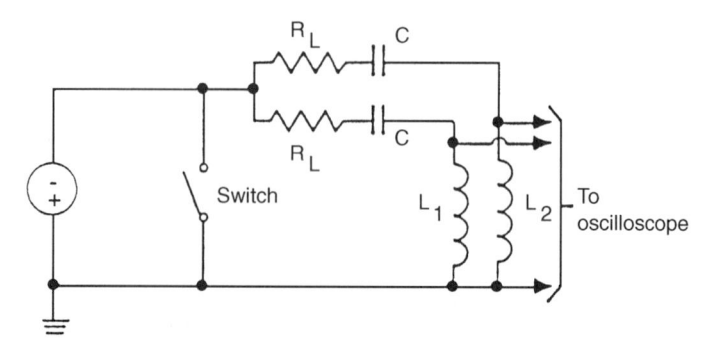

FIGURE 16.5 Simplified schematic for a turn insulation surge tester; L_1 and L_2 are either coils or phases in a winding.

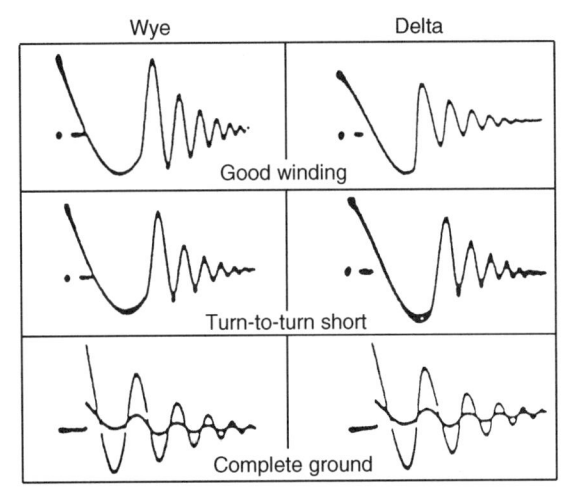

FIGURE 16.6 Typical waveforms from a surge tester. For good coils, both traces will be the same.

the result is even more difficult. This "induced" surge test should be carefully done because high voltages may also be induced into coils other than the coil being tested. The test is normally performed by connecting the high-voltage output of the surge tester to two of the phases. The third phase is grounded to the surge tester. The voltage is applied and increased to specified limits, which should not exceed the ground wall dc hipot voltage. If the surge waveforms are identical, the turn insulation is presumed sound.

This test is based on comparing the shapes of surges applied to two winding sections. The two surge shapes may not be identical if the two winding sections being tested have slightly different impedances due to having different dimensions of coils. This will suggest a fault even when the insulation system is in good condition. It is also difficult to detect a turn fault in a coil tested in a circuit parallel with more than 10 coils because a shorted turn will have a minor change in the total impedance of the winding. As the number of coils being tested increases, it becomes more difficult to determine if a defective coil is present. This test does not indicate the relative condition of the turn insulation in different coils. It only indicates if shorts exist. It is a go/no-go proof test like the ac and dc hipot test for the stator ground wall insulation.

Synchronous Machine Rotor Windings

the presence of faults in rotor winding insulation can sometimes be indicated by a change in machine performance rather than by the operation of a protective relay. For example, if a coil develops a short circuit, a thermal bend may develop due to an asymmetric heat input into the rotor. This could lead to an increase in shaft vibration with increasing excitation current. This change can be used in some cases to determine if the interturn fault is significant. The location and severity of a fault cannot always be found easily even when the rotor is removed. This is especially true in large turbine-generator rotors whose concentric field windings are embedded in slots in the rotor body and covered by retaining rings at the ends. Many ground and interturn failures disappear at reduced speed or at a standstill. This makes their detection very difficult and emphasizes the need for on-line detection technique. The

16.12 CHAPTER SIXTEEN

following tests are used to determine if faults exist in the rotor winding and/or indicate their location. Solid-state devices used in exciters should be shorted out before conducting any test involving the induction or application of external voltages to the rotor winding.

Open-Circuit Test for Shorted Turns. An open-circuit test can be used to confirm if shorted turns in rotor field winding exist when there are indirect symptoms such as a change in vibration levels with excitation. The machine should be taken out of service for a short while but does not need to be disassembled.

Figure 16.7 illustrates the open-circuit characteristic of a synchronous machine. It relates the terminal voltage to field current while the machine is running at synchronous speed with its terminals disconnected from the grid. The open-circuit curve can be used to verify shorted turns if an open-circuit test characteristic with healthy turn insulation was done previously. A higher field current will be required to generate the same open-circuit voltage if there are shorted turns in a rotor field winding. If the difference between the two curves in more than 2 percent, the possibility of a turn insulation fault will be confirmed. The difference in characteristics to indicate a shorted turn depends on the number of turns in the field winding and the number of shorted turns. For example, a single shorted turn cannot be detected by this test if the connected field winding has a large number of turns. This test is done while the machine is running at synchronous speed with its stator winding terminal open-circuited and the field winding energized. Generators can easily be driven at synchronous speed because their drivers are designed to operate at synchronous speed. Motors may need to be driven by ac or dc drive at synchronous speed.

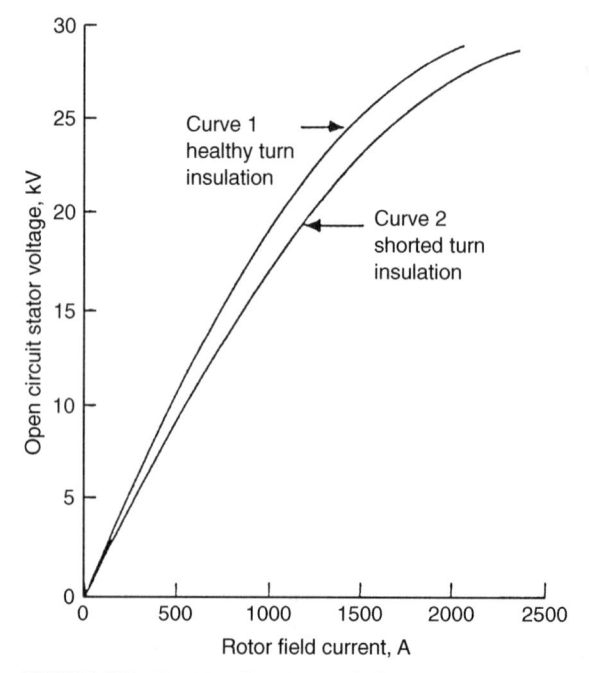

FIGURE 16.7 Detecting shorted rotor winding turns.

GENERATOR SURVEILLANCE AND TESTING   **16.13**

If the test indicates the possibility of shorted turns, further confirmation should be obtained by performing additional tests. This test has these limitations:

- It may not detect shorted turns if the machine has a large number of turns and/or there are parallel circuits in the field winding.
- Differences in the open-circuit curve will also be created when the machine's magnetic characteristics change, for example, when the rotor wedges are replaced with a different material.

Air Gap Search Coil for Detecting Shorted Turns. Interturn faults in rotors are detected by air gap search coils. Methods have been developed for on-line and off-line testing. This technique is especially useful for detecting faults present at operating speed which disappear on shutdown. The coils and slots having shorted turns as well as the number of turns shorted can be identified by this method. Permanent flux probes have already been installed on some machines. Each rotor slot has local fields around it. This leakage flux is related to the current in the rotor. The magnetic field associated with a coil will be affected if the coil is shorted. The search coil records the high-frequency waveform (known as slot ripple) generated in the air gap. Each rotor slot generates a peak of the waveform in proportion to the leakage flux around it. If an interturn fault occurs, the peaks associated with the two slots containing the faulted coil will be reduced. The recorded data are analyzed to identify the faulted coil and the number of (faults). Shorted turns also generate significant levels of even harmonics (multiples of the frequency) while a fault-free rotor generates only odd harmonics.

The search coil is normally mounted on a stator wedge. A gas-tight gland is required for the leads of the probe. Shorted rotor turns should not be a cause of grave concern if the rotor vibration is not excessive and the required excitation is maintained. A generator can operate adequately for a period of time under this condition. However, these shorted turns are normally caused by serious local degradation of the interturn insulation and possibly major distortion of the conductors. In some cases where static exciters are used, arcing damage and local welding have been found.

It is difficult to interpret the on-load test results from the search coil due to the effects of saturations and magnetic anomalies in the rotor body. More complex and time-consuming detection techniques are required. However, modern on-line monitors have overcome these difficulties. They are designed for use with turbine generators equipped with air gap search coil. The output from the search coil is continuously being processed. An alarm is initiated when a current-carrying shorted turn occurs in the rotor winding.

Impedance Test with Rotor Installed. Shorted turns in a field winding can also be detected by periodic measurement of rotor impedance using an ac power supply. These tests should ideally be performed while the machine is operating at synchronous speed because shorted turns may only exist when centrifugal forces are acting on the turn conductors. When the machine is shut down, there may not be any contact, or the fault resistance may be high. Shorted turns can be detected more accurately by impedance rather than resistance measurements. This is due to the induced backward current in a single shorted turn, which opposes the magnetomotive force (mmf) of the entire coil, resulting in a significant reduction in reactance. This technique is particularly effective in salient-pole rotors, where one short-circuited turn eliminates the reactance of the complete pole. There is a sudden change in impedance when a turn is shorted during run-up or rundown (Fig. 16.8). A sudden change of more than 5 percent is needed to verify shorted turns.

The highest field current used for this test should be significantly lower than the normal current required for rated stator voltage at open circuit. The voltage applied should not exceed the rated no-load stator voltage. A normal winding will exhibit a reduction in

16.14 CHAPTER SIXTEEN

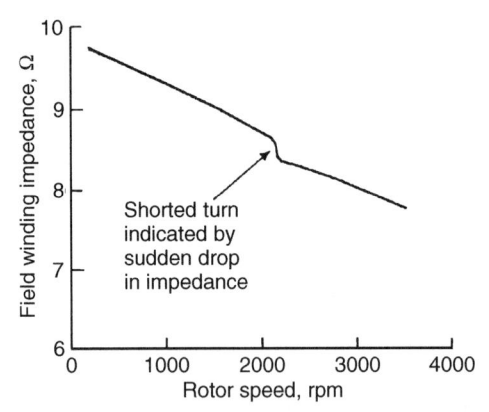

FIGURE 16.8 Detecting shorted rotor turns by impedance measurements.

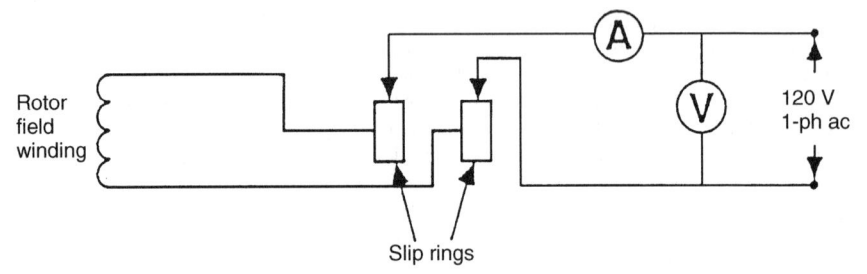

FIGURE 16.9 Test setup for impedance measurement with rotor installed.

impedance up to 10 percent between standstill and operating conditions due to the effects of eddy currents on the rotor.

This test can only be performed if the field winding is accessible through collector rings because the low-voltage ac power should be applied while the machine is running. A 120-V, 1-phase, 60-Hz ac power is applied. The voltage, current, and shaft speed are measured. The power supply should be ungrounded because the rotor could get damaged if the field winding has a ground fault.

The test includes these steps:

1. Perform an insulation resistance test on the field winding of the machine to be tested to check for ground faults. The impedance test should not be performed if a ground fault is found. The ground fault should be located using a different procedure.
2. Connect an instrumented and ungrounded power supply to the field winding (Fig. 16.9). The instruments used should be properly calibrated.
3. Take the reading from the local speed indicator to determine the relationship between impedance and speed.
4. Adjust the field winding voltage to give a maximum permissible current of 75 percent of the current required to achieve the rated open-circuit stator voltage.

5. Increase and decrease the speed of the machine while the stator windings are disconnected from the power supply. Measure the current, voltage, and speed starting at zero and increasing the speed at 100 rpm intervals until the rated speed is reached. Continuous measurements can also be recorded simultaneously on a multichannel strip chart recorder.

The values of the impedance ($Z = V/I$) should be plotted against the speed (Fig. 16.8). A sudden change in impedance of 5 percent or more or a gradual change of more than 10 percent will indicate a strong possibility of shorted turns in the winding. This test is not as sensitive as the previous two described earlier. It is also important to note that solidly shorted turns will not produce an abrupt change in impedance.

Detecting the Location of Shorted Turns with Rotor Removed. The exact location of a shorted turn should be found to minimize the disturbance to the winding when making repairs. One or a combination of the following procedures should be used:

Low-Voltage AC Test. When the field winding of a synchronous machine rotor having shorted turns is connected to a low ac voltage (typically 120 V), the tips of the teeth on either side of the slot(s) having the shorted turns will have significantly different flux induced in them. Figure 16.10 illustrates how the relative magnitudes of tooth fluxes can be measured. The teeth are bridged by a flux survey using a laminated-steel or air-core search coil, which is connected to a voltmeter and wattmeter. The voltage is measured by the voltmeter while the direction of the induced flux is given by the wattmeter. The search coil is moved across all the teeth of the rotor, and voltage and wattage readings are taken. The search coil readings depend on its axial location along the rotor. Therefore, all the readings should be taken with the coil located the same axial distance from the end of the rotor. Since the readings vary significantly near the end of the rotor, the coil should not be placed near the end of the rotor. It is important to note that core saturation may occur when a 60-Hz power supply is used. A higher frequency should be used, if possible, to reduce this problem.

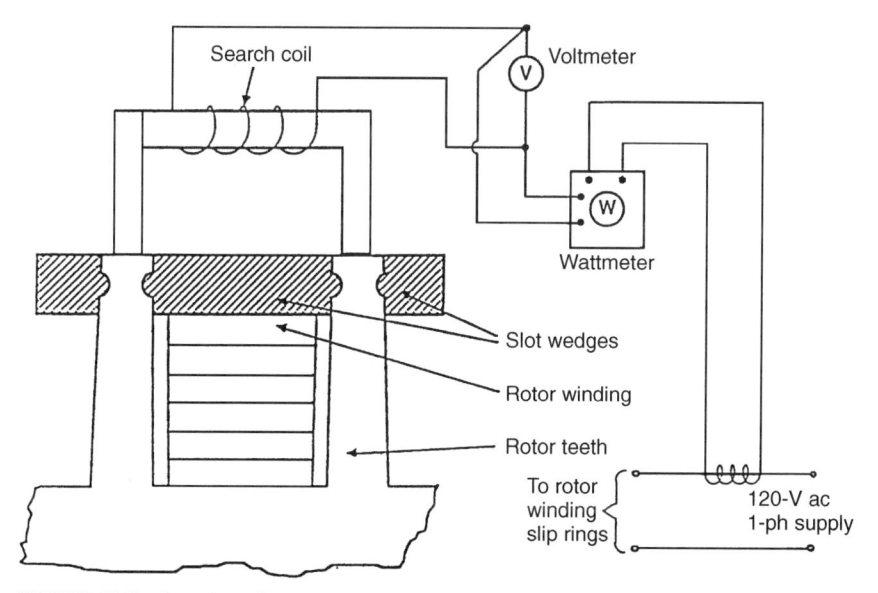

FIGURE 16.10 Setup for ac flux survey test with rotor removed.

16.16 CHAPTER SIXTEEN

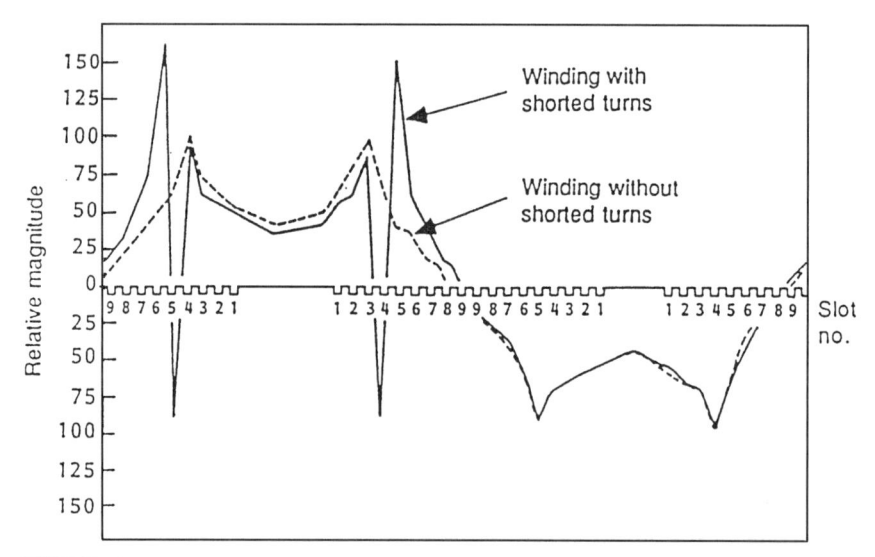

FIGURE 16.11 Flux distribution survey for two-pole turbine generator.

The equipment used for the EL-CID test, described later, can be used to detect the shorted field winding turns. This test can be done without removing the end-winding retaining rings if the rotor has steel wedges and no damper winding. If the rotor has a separate damper winding or aluminum alloy slot wedges (shorted at the ends) used as a damper winding, they must be open-circuited at the ends before the test can be done. In this case, the retaining rings should be removed. Since many shorts are created by the action of centrifugal forces, they may not appear at standstill.

Figure 16.11 illustrates the flux distribution for a rotor with and without shorted turns. The sharp change in direction of the induced flux indicates the slot containing the shorted turns.

Low-Voltage DC Test (Voltage Drop Test). This method is used to locate the shorts based on dc voltage drop between turns. The end rings should be removed to provide access to the turns. In some cases, the shorts should be induced by applying a radial force to the coils. This is normally done by tapping the wedges with a wooden block or clamping the coils at the corner.

The test is done by applying a dc voltage to the field winding and measuring the drop in voltage across the turns. If a short occurs, the voltage drop across the turn will be lower than normal.

Field Winding Ground Fault Detectors. A large generator rotor operates at 500 V dc and 4000 A normally. If the insulation between the winding and the body is damaged or bridged by conducting materials, there will be a shift of the dc potential of the winding and exciter. The part of the winding where the fault occurred becomes the new zero potential point. In most cases, this will not cause an immediate problem if there is no additional ground fault. A second ground fault in the rotor will be catastrophic.

A rotor ground fault detector is used to enunciate when a fault occurs. Some units are tripped automatically due to possible extensive damage to the rotor body by a dc arc across a separate copper connection. Ground fault detectors have various configurations. The rotor winding is grounded in simple dc schemes on one end through a high ohmic resistance.

However, these schemes become insensitive if the fault occurs close to this end. Ohm's law determines the magnitude of leakage current from the rotor winding to the ground fault relay. The shaft should also be grounded.

A sophisticated technique was developed to continue operation of a generator having a known ground fault (second ground fault detector). It uses a microprocessor and measuring resistors to determine whether the power dissipated by the leakage current exceeds a value that would cause a failure if there were two or more ground faults from the winding. A search coil mounted in the air gap has been used to detect interturn faults and a second ground fault.

If a fault is identified, measurements of slip ring to shaft voltages will give an indication of the location of the fault. (Is the fault at the middle or end of the winding?) After disconnecting the ground fault detector and while the generator is still on-line, the voltage readings between the brush holders and the shaft are taken. If one ground fault is present, the approximate location of the fault in percent of winding resistance is

$$\frac{\text{Lower ring-to-shaft voltage}}{\text{sum of two ring-to-shaft voltages}} \times 100\%$$

During rundown of the unit (when it is unloaded and tripped), an insulation resistance tester is used to test the fault resistance. The brushes are raised or the field circuit breaker is opened to determine if the fault is in the generator rotor, external bus, or exciter. The fault is also monitored as the speed drops. If the fault disappears, it will be impossible to find its location. The operator may decide to put the machine back in service. If the fault reappears when the unit is returned to service, the process should be repeated. If the fault is sustained, a low voltage is applied across the slip rings while the rotor is at standstill. It is usually provided from a 12-V car battery or from a 120-V ac variac. The voltage between the rotor body and each slip ring is measured. If the readings are full voltage with one and zero with the other, there is likely a low-resistance path at the slip rings. It could be caused by carbon dust or insulation failure. It may easily be corrected with a good cleanup. The rotor should be withdrawn if the fault is within the winding. When the rotor is removed, the low-voltage source is reapplied to the slip rings, and a voltmeter is installed between the rotor body and a long insulated wire. The insulation is removed from the last 5 mm of the wire, and a probe is used to contact the rotor winding metal through ventilation holes and under the retaining rings. This technique will identify the slot, bar, or ventilation hole having the closest voltage to the rotor body. The fault is usually located under the wedge near this location. The problem is rectified sometimes by cleaning the ventilation ducts. Otherwise, additional dismantling may be required.

If the ground fault is transient and needs to be found, a failure is forced with a moderate high-potential test and the same technique is used. The hipot test should be used as a last option.

Surge Testing for Rotor Shorted Turns and Ground Faults. This off-line method is used to detect rotor winding faults on stationary and rotating shafts. The location of the fault is identified. This method is very effective in finding ground faults and shorted turns. There is electrical symmetry in a healthy rotor winding. The travel time of an identical electric pulse injected at both slip rings through the winding should be identical. The reflection of the pulse back to the slip rings would also be identical. If there is a short or ground fault, some of the pulse energy will be reflected back to the slip ring due to the drop in impedance at the fault. The reflections will change the input pulse waveform depending on the distance to the fault. Therefore, a fault will generate different waveforms at each slip ring unless it is located exactly halfway in the winding.

Recurrent surge oscillography (RSO) is a technique based on the above principle. This test cannot be done on-line because the winding should be isolated from the exciter. Two identical, fast-rising voltage pulses are injected simultaneously at the slip rings. The potential at each injection point is plotted versus time. Identical records should be obtained if there is no fault due to the symmetry in the winding. Differences between the traces are indicative of the winding fault. The fault is located from the time at which irregularity occurred. Ground faults having a resistance less than 500 Ω will be detected by the RSO method. These faults are also normally detected by the generator protection systems. The RSO technique is used to confirm ground faults. Interturn faults having a resistance of less than 10 Ω will also be detected by RSO. Faults which have a resistance more than 10 Ω are more significant during operation and less severe off-load. These faults cannot be detected by RSO.

Low Core Flux Test (EL-CID)

The conventional method for testing for imperfections in the core insulation of motors and generators has been the *rated flux* test. This test requires high power levels for the excitation winding to induce rated flux in the core area behind the winding slots. The alternative low-flux test described in this section has been performed successfully across the world. Its main advantage is that it requires much smaller power supply for the excitation winding. Only 3 to 4 percent of rated flux is induced in the core. In reality, the power supply can be obtained from 120-V ac wall socket source. Also, the time required to perform this test is much shorter.

The *electromagnetic core imperfection detector* (EL-CID) identifies faulty core insulation. It is based on the fact that eddy currents will flow through failed or significantly aged core insulation even if the flux is a few percent of the rated flux. A *Chattock coil* (or Maxwell's worm) is used to obtain a voltage signal proportional to the eddy current flowing between the laminations. The solenoid coil is wound in a U shape.

Figure 16.12 illustrates how the coil is placed to bridge the two core teeth. The fault current I_F is approximately proportional to the line integral of alternating magnetic field along its length l (Ampere's law). Thus, if the effects of the field in the core are ignored, the voltage output in the Chattock coil is proportional to the eddy current flowing in the area encompassed by the coil (the two teeth and the core behind them).

The excitation winding that generates the test flux in the core induces an additional voltage across the coil due to the circumferential magnetic field. A signal processor receives the output voltage from the Chattock coil. It eliminates the portion generated by the excitation winding and gives a voltage proportional to the eddy current (Fig. 16.13).

The output milliamperes of the signal processing unit are proportional to the voltage in the Chattock coil generated by axial eddy currents. High milliampere readings are normally caused by faulty insulation in the core or interlamination shorts at the core surfaces. A reading higher than 100 mA indicates significant core plate shorting.

The Chattock coil is moved along the teeth of the core, and the current readings are recorded. Areas where the readings exceed 100 mA should be marked with a nonconductive substance and examined for defects.

APPENDIX B: MECHANICAL TESTS

These tests will help in determining the integrity of the windings. Loose stator windings can cause mechanical and electrical damage to the ground wall insulation. Large machines

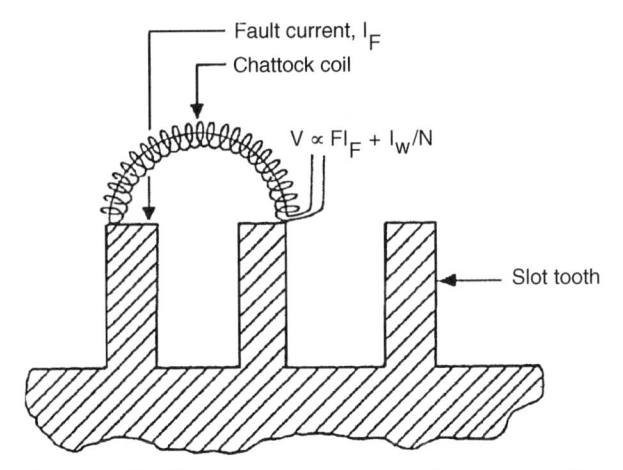

FIGURE 16.12 Chattock coil mounting configuration and output voltage.

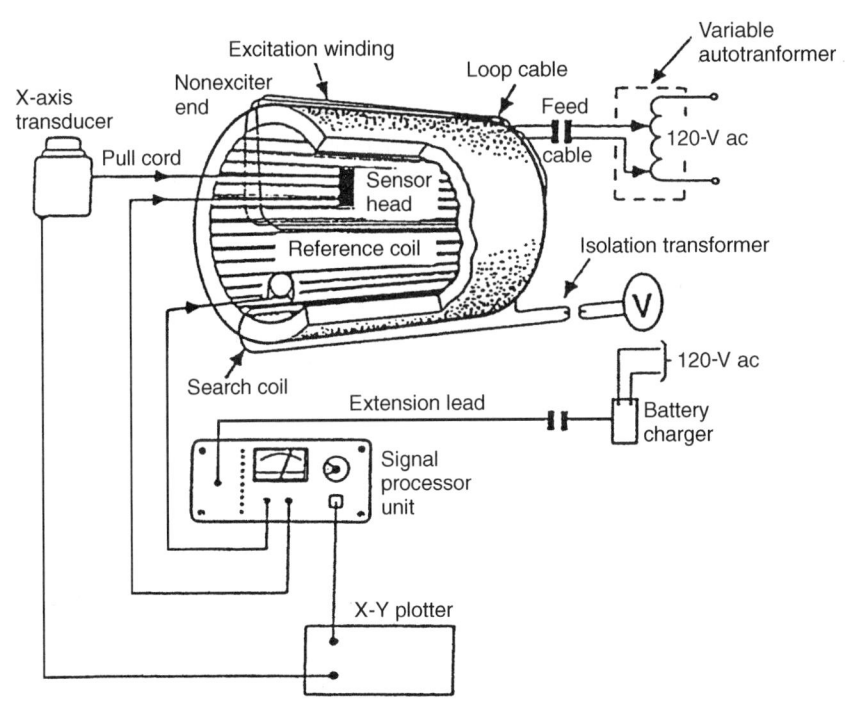

FIGURE 16.13 EL-CID test setup for large generator.

are more susceptible due to the increased forces and slot discharges. The following tests should be done during outages:

Stator Windings Tightness Check

The tightness of the stator wedges in the slot should be checked on a regular basis. The effective methods are wedge tapping and ultrasonic detection. The stator wedge is struck by a blunt object. The tightness of the wedge in the slot will determine the type of sound "ring." A tight wedge will give a dull sound, and a slack wedge a hollow sound. A ring between these extremes indicates that the wedge will become loose in the future.

A measuring instrument using an ultrasonic technique can also be used to determine the tightness of the stator wedges. This portable equipment uses a vibrator, accelerometer, and force gauge to excite the wedge. It identifies the natural vibration resonance and assesses the tightness of the wedge assembly.

Stator Winding Side Clearance Check

This test is done to ensure a tight fit between conductor bars and the slot sides. A feeler gauge is inserted to determine the tightness.

Core Laminations Tightness Check (Knife Test)

This test involves inserting a standard winder's knife blade (maximum thickness of 0.25 mm) between laminations at several locations in the core. If the blade penetrates more than 6 mm, then the core is soft. It should be retightened by packing.

Visual Techniques

If there is indication of insulation aging or a fault, the machine should be visually inspected. Flashlights and magnifying glasses are normally used. Small mirrors can be used for examining the inside edges of retaining rings. Boroscopes are used sometimes to gain access to stator core laminations or conductors. Magnetic strips are swept along the internal surfaces to pick up small magnetic fragments.

Ground Wall Insulation

Early detection of deterioration of insulation can help in extending the life of the machine. Dusting or powdering of the insulation along the slot wedges or in the ventilation ducts may indicate damage to the insulation by mechanical abrasion. This powder should not be confused with the reddish powder which is caused by core problems, or copper dusting (which occurs due to fretting of the rotor winding when the machine is on turning gear). Another white or gray powder is caused by partial discharge. This powder is only found in bars and coils near the line ends of the winding. This powder should not be confused with the one caused by abrasion which is found throughout the winding.

Other signs of mechanical distress are debris at slot exit, stretch marks and cracks in surface paint in the slot area or in the mechanical supports in the end winding area. Thermal aging is indicated normally by discoloration or undue darkening of the insulation surface.

Electrical effects are normally indicated by carbonized tracking paths to grounded components. External partial discharge, which normally occurs in line end coils, is confirmed by a gray or white powder. Chemical analysis of the dust will show a high percentage of salts.

Rotor Winding

Mirrors and boroscopes are used to perform rotor inspection without disassembly. In some cases, it may be necessary to disassemble the rotor partially by removing the retaining rings and some wedges.

Turn Insulation

The location of the turn faults in the rotor cannot usually be determined by visual means without some disassembly. Turn faults in the end windings become visible after removing the end caps. Those in the slot become visible by removing the wedges and lifting the turns.

Faults caused by copper dusting can be verified by small copper particles in the slots and vent ducts.

Slot Wedges and Bracing

The movement of slot packing under wedges in gas-cooled rotors can be detected without disassembly. This problem can be identified by examining the gas exhaust holes in the wedges to see if packing has been moved to block the flow of cooling gas out through them. The problem is also indicated by rotor thermal unbalance.

The rotor should be disassembled if there is evidence of early signs of deterioration in slot wedges.

Stator and Rotor Cores

Severe overheating and melting may occur at the surfaces of laminated cores due to insulation faults. These can easily be detected by a visual examination. Faults that occur in the slot region are normally hidden by the winding and slot wedges. These are normally indicated by signs of burning in the vicinity of the insulation and wedges.

CHAPTER 17
GENERATOR INSPECTION AND MAINTENANCE

The two types of generator maintenance are *on-load* and *off-load* maintenance. The on-load maintenance consists mainly of monitoring activities. This type of maintenance is very important because it detects the conditions that could lead to a catastrophic failure.

ON-LOAD MAINTENANCE AND MONITORING

This type of maintenance varies with the design of the machine and the instrumentation installed in it. The following paragraphs describe the standard maintenance and monitoring activities as well as the recent developments in monitoring techniques.

Stator

Temperature Measurement. Thermocouples are installed in the stator to monitor the temperature of the stator bar insulation, core teeth, core end plates, stator water connections, and hydrogen gas. Eighty thermocouples are used for a typical 660-MW generator. The readings of these thermocouples are displayed in the control room. Analysis of the information obtained from them will help to identify developing faults and deficiencies in the cooling circuit and the stator.

End Winding Vibration. A vibrations transducer module is installed in the end windings. It measures the vibration in all directions. It identifies deterioration in the structural integrity of the end winding and stresses in the conductor coil that approach the initiation of fatigue cracking.

Stator Water Flow. The flow rate of the stator water is monitored by sight-glass indicators having a flap in the flow. The angle of the flap provides a rough indication of the flow rate. Regular monitoring of the flow rate will give an indication of a change in the flow rate that requires investigation during an outage.

Hydrogen Dew Point. Hydrogen dew point is checked regularly to ensure that moisture condensation will not occur at the cooler parts inside the generator. An increase in the hydrogen dew point could have any of the following consequences:

- Moisture in oil
- Leaks from the hydrogen heat exchanger
- Leaks from the stator coolant system

- Deterioration in the performance of the hydrogen driers

Hydrogen Entrainment into Stator Water. A small amount of hydrogen enters the water circuit normally due to the pressure differential between the hydrogen and the water. Alarms indicate excessive ingress of hydrogen into the water. The volume of hydrogen detrained from the stator water should be monitored for early detection of a developing leak.

Hydrogen Leakage. Hydrogen leakage checks should be initiated over the whole generator when there is evidence of high hydrogen makeup. Nonflammable gas detectors are used to detect the presence of hydrogen. The exact location of the leak is then identified by using an aerosol foam or a soap solution. The search for hydrogen should include the slip ring cooling circuits and the lubricating oil.

Core Monitor. The core monitor is an online gas sampling device used to detect the decomposition products of overheated insulation. The products are normally released due to a developing fault. The monitor alarms at a specified level of decomposition products in the gas. A gas sample also can be collected in the monitor for future analysis of entrained decomposition products.

Radio-Frequency Monitor. Corona is discharged due to deterioration in the condition of the insulation. This discharge can be detected in the radio-frequency currents in the neutral conductors. Analyzing equipment is used to monitor the level of corona discharge.

Rotor

Vibration Monitoring. The vibration of the turbine-generator rotor should be monitored during normal operation and rundown. The vibration levels are monitored in the axial, vertical, and lateral directions at each bearing. The signals are recorded and subsequently analyzed to identify the cause of the vibration. Any change in the phase or amplitude of the vibration should be investigated. The investigation may include obtaining a rundown vibration plot as soon as the unit can be shut down. This plot should be compared with the rundown vibration plot that was obtained initially to fingerprint the machine. Any anomalies between the two plots should be identified to provide early indication of the possible cause of the problem.

Rotor Ground Fault Detection. The ground fault protection system should detect a serious rotor ground fault. The rotor insulation resistance should be checked periodically, using the measuring circuit of the ground fault protection system to confirm the integrity of the insulation.

Shaft Voltage Measurement. The shaft voltage readings should be taken regularly. Any deviation from the normal readings should be investigated. Some units have a shaft voltage alarm system. It is recommended to seek a specialist's advice on the cause of such abnormal voltage. However, consideration should be given to the following possible contributory factors:

- Changes in the condition of the bearing insulation
- Problems with the shaft ground brush
- Magnetized shaft

Flux Monitor. The search coil is a device used to identify changes in the leakage flux from the rotor due to interturn faults. It is installed on the stator in the air gap between the rotor and the stator. This device allows regular monitoring of the condition of rotor interturn insulation.

Shaft Ground Brush. Contamination buildup normally on the shaft ground brush renders it ineffective. It is recommended to inspect the brush frequently to maintain its cleanliness.

Bearing Pedestal Insulation. The insulation resistance of the generator outboard bearing, shaft seal ring, pipework, and exciter baseplate should be inspected regularly. This insulation can fail due to dirt accumulation, moisture ingress, or a short circuit

GENERATOR INSPECTION AND MAINTENANCE   17.3

caused by badly placed conduit. The insulation failure will allow significant currents to go through the bearings, causing damage in the bearing and journal surfaces. On-load measurement of the insulation resistance can be taken in some bearings integral with the generator end plate.

The on-load check of the pedestal bearing insulation can be done as follows:

1. Connect the rotor shaft to ground at the turbine end, using a portable copper gauze brush and lead on an insulated handle.
2. Measure the ac voltage differential between the shaft and ground (using a gauze brush) at the exciter end of the shaft.
3. Develop a short circuit between the rotor shaft and pedestal at the exciter end and measure the differential voltage between the pedestal and ground.

The two voltage readings will be similar if the insulation is in a good condition. If the insulation is faulty, the voltage measured in step 3 will be lower than that in step 2. Finally, measure the differential voltage between the exciter end of the shaft and the pedestal. This value should normally be less than 25 percent of that measured in step 2, if the insulation is in good condition.

Excitation System

Brushless System. The brushless excitation system requires minimal on-load maintenance. Leaks from the exciter and rectifier heat exchangers are a great hazard to the system. Thus, regular visual checks for moisture around the cooling circuit are recommended.

If the alarm indicating a diode failure in the rotating rectifier is received, the failed diode should be replaced at the earliest opportunity. This is done to prevent overloading of the remaining diodes and causing a bridge-arm failure.

Slip Ring Systems. Regular checks and maintenance are required to have a trouble-free operation of the rotor and exciter slip ring brush gear. This work is normally done on-load. It includes checks to determine

- Freedom of the brushes within the boxes
- Individual brush length
- Individual brush current
- Condition of brushes
- Evidence of sparking
- Brush vibration
- Brush spring tension
- Oil ingress from adjacent bearings

The condition of the slip ring, exciter cooling circuit including the filters should be checked.

The number of brushes that can be changed on-load within a given period should be limited. In general, when one brush reaches the minimum length, all the brushes in the box should be replaced. However, only the brushes in one box should be replaced within a 24-h period. This allows the brushes to settle and share current equally before additional changes are made.

Strict safety procedures must be followed while performing all on-load brush gear work. They include placing the unit on manual excitation, disconnection of the ground fault protection system, and adoption of ground-free work practice by using rubber mats, insulated tools, rubber gloves, etc.

OFF-LOAD MAINTENANCE

The following paragraphs describe the maintenance activities that will be done on every planned generator overhaul and as a result of a problem identified on the machine while operating.

Stator Internal Work.

The removal of hydrogen coolers or access doors on most machines will allow the inspection of the core back, end windings, and cooling connections. An effective *clean-conditions system* should be set up before entry into the stator to prevent debris from entering the stator and to ensure that no equipment is left in the machine when the work is completed.

The following maintenance activities should be performed when access to the generator internal parts is available:

- Inspect the end winding and check for tightness of the winding and its support structure.
- Inspect the stator bars and wedges.
- Inspect the stator core for evidence of hot spots and damage.
- Clean any contaminated surfaces with an approved solvent. The dirt should not be driven into the winding.
- Identify and remove any debris on the stator.
- Check the phase and neutral connections to bushings.
- Check the stator coolant pipework and hoses. If a leak is discovered, identify its location by one of the following two methods:

    Fill and pressurize the winding with water, and inspect for leaks. There is a possibility of contaminating the insulation with water when using this method. Thus, it is not recommended by some manufacturers.

    Inject a tracer gas into the dried winding, and locate the leak by using a gas detector.

Stator External Work

Stator Water System. A careful inspection of stator water system pipework, pumps, coolers, valves, and filters is needed before shutting down the generator to identify the areas requiring work during the outage. Following isolation of the system, the filter should be checked and the debris identified. This check will help in identifying the areas that need further investigation. For example, if gasket materials are found, the gaskets of the pipework could have deteriorated. Great care should be taken while working on the stator water system to prevent any contamination from entering it.

Following any work on the stator water system, the pipework should be flushed with the winding bypassed. This is done to prevent any debris from entering the winding.

Hydrogen Coolers. Any small water leakage inside the hydrogen coolers can prevent the system from reaching the required hydrogen dew point. If a cooler is suspected of leaking, it should be removed and pressure-tested. If a leak is detected, the unit should be repaired. A pressure-drop test should be performed following the repair.

Hydrogen Driers. The following checks should be done when there is evidence of deterioration in the performance of the hydrogen driers:

- Performance of the heater
- Operation of the valve
- Obstructions in the pipes

- Desiccant contamination

Main Connections. The condition of the main connection enclosures should be monitored to prevent the ingress of moisture, oil, or dirt and subsequent insulation damage. The insulators should be cleaned and inspected regularly for any evidence of damage. All the joints of the conductors should be inspected for any evidence of overheating. The tightness of the joint bolts should be maintained at the specified torque to ensure that the value of the joint resistance is within acceptable limits.

Rotor

The rotor of all generators should be withdrawn every 8 years for inspection and testing. The condition of end rings and winding should be carefully examined during this work. The rotor should also be withdrawn if the on- or off-load tests indicate a rotor fault.

Slip Rings and Brush Gear

The maintenance of the slip rings and brush gear includes the following activities:

- Remove the brushes from the boxes, mark their position on them, and store them. Great care should be given to prevent chipping the brushes during this work. If one or more brushes are near the minimum length, all the brushes in the brush box should be discarded.
- Disconnect and remove the brush gear enclosure.
- If the condition of the slip ring surfaces is satisfactory, wrap the slip rings with moisture-absorbent paper and polythene for protection.
- Remove carbon dust, oil, and other contamination from the enclosure, ventilation circuit and connections, slip ring sides, and shaft. However, do not clean the slip ring surface.
- Overhaul the ventilation system.
- Reassemble the brush gear enclosure and bed in any new brushes in situ by using a strip of fine abrasive cloth. Ensure that there are no abrasive particles in the brushes or between the brush and ring. All carbon dust should have been removed.
- Check the spring tension and freedom of the brushes in the boxes.

It is recommended that the brushes remain lifted off the ring until the end of the outage. Corrosion of the rings should also be prevented by keeping the enclosure dry and warm. The measures listed above are recommended for any prolonged outage.

Refurbishment of the slip ring surfaces is required occasionally to remove marks or uneven wear, or to recut the grooves. Grinding of the slip rings is normally done in situ by using one of the following methods:

- Machine while the shaft is on turning gear (barring). Grind the rings, using a motor-driven grindstone mounted on a sliding baseplate attached to the generator bedplate.
- Machine while the shaft is stationary. Grind the rings, using a specially designed collar with ring of grindstones assembled around the slip ring. A pony motor drives the collar, using a belt.

A lathe is used instead of the motor-driven grindstone for extensive in situ machining of slip rings or recutting of grooves.

Exciter and Pilot Exciter

Maintenance of this equipment includes inspection, cleaning, testing the windings, checking connections, and performing the maintenance described in the previous section on brush gear and slip rings.

Rectifier

The maintenance activities of rectifiers include testing of the insulation. The test voltage applied should be kept low to prevent damaging of the diodes. If it is necessary to apply a higher voltage, the diodes must be disconnected before performing the test.

Static Rectifiers. The maintenance activities of static rectifiers include cleaning and inspection of all components, diode testing, checking of fuses, insulation testing, and of checking of all connections.

Rotating Rectifiers. The maintenance activities of rotating rectifiers are similar to the ones for static rectifiers. However, particular care should be taken to the following points due to the high centrifugal forces experienced by the components of the rectifier during operation.

- A thorough inspection of the rectifier components is required to confirm their mechanical integrity. The failure of a component during operation could cause extensive damage to the machine.
- Each module should be accurately weighed following maintenance to ensure the assembly is balanced properly. The balance weights of a module should be adjusted, if required, to ensure it is within the weight tolerance (typically ±0.5 g).

Due to the complexity of some designs of rotating rectifiers, it is recommended to follow an inspection system during maintenance. It consists of an independent check following each stage of maintenance and subsequent rebuild. It is also recommended to test the performance of the measuring and alarm equipment associated with the rectifier following maintenance.

Field Switch

The switchgear maintenance should include the inspection and testing of the field switch and its operating mechanism. The field switch overcurrent trip unit contacts should be inspected for freedom of movement and operation.

Automatic Voltage Regulator

The off-load maintenance of the AVR includes the following activities:

- Cleaning of the equipment with a soft brush
- Vacuum cleaning of the equipment to remove dust
- Examining the components and conductors thoroughly for any sign of damage and overheating
- Checking and cleaning of the relays
- Checking and lubricating of the motorized potentiometers and their cam-operated switches; also checking of their traverse times
- Checking for correct operation of all tripping devices

GENERATOR INSPECTION AND MAINTENANCE   17.7

Supervisory and Protection Equipment

A schedule should be prepared to identify each supervisory and protection device. The schedule should provide a reference to the test, calibration procedures, and calibration date for each device. A check should be done to confirm that the tests and calibration of the devices were completed as required. A plan should also be made to schedule the tests and calibration of the devices in the future. This work is required to maintain the accuracy and reliability of the equipment. A check of the overall protection system should also be carried out annually.

GENERATOR TESTING

Tests should be done on the generator during the outage to monitor its condition and to ensure that it is in a satisfactory state for return to service, following maintenance. A major overhaul of the generator should be done every 8 years. Details about the tests were provided in Chap. 16. The following paragraphs provide practical information about these tests.

Insulation Testing

Insulation Resistance (IR) Measurement. The insulation resistance test consists of a measurement of the dc resistance between the winding and ground, or between two windings. It is normally performed using a battery- or mains-powered megohmmeter. The test voltage on rotors is typically 500 V dc (the diodes should be disconnected before performing the test). The test voltage for stators up to 6.6 kV is 1 kV dc. It is 5 kV dc for higher stator voltages. The IR test results should be corrected to a standard temperature of 20°C. Figure 17.1 illustrates a nomogram used for this purpose. The IR results are also affected significantly by humidity if the insulation system is open to the atmosphere. Dry air should be blown through water-cooled windings following draining to remove the water from all the hoses. This is necessary to obtain satisfactory IR readings.

Polarization Index Test. The polarization index (PI) test is an extension of the insulation resistance test. The dc voltage is applied across the insulation for 10 min. The PI is the ratio of the insulation resistance reading obtained after 10 min to the reading obtained after 1 min. The insulation condition is considered acceptable if the PI is greater than 2. Lower values of PI indicate the presence of surface contamination such as dirt or moisture on the insulation. If the PI value is lower than 2 and the 1-min IR reading is low, serious degradation of the insulation may have occurred. The extent of the degradation in the insulation can only be determined by the loss angle test.

Loss Angle Test. The loss angle (or tan delta) test is used to determine the condition of the insulation. Voids develop in the insulation as it deteriorates, resulting in a change in capacitance of the winding. The loss angle (δ) changes due to the change in capacitance of the winding (Fig 17.2).

The test is performed as follows:

- Confirm that the winding has a satisfactory IR reading.
- Apply ac voltage (60 or 50 Hz) to the winding.
- Increase the voltage from 20 to 100 percent of the line voltage V_1 in increments of $0.2V_1$.
- Record the capacitance and tangent of the insulation loss angle at each step, using an ac bridge.
- Discharge the winding following completion of the test. This step is necessary to ensure the safety of personnel working with this equipment.

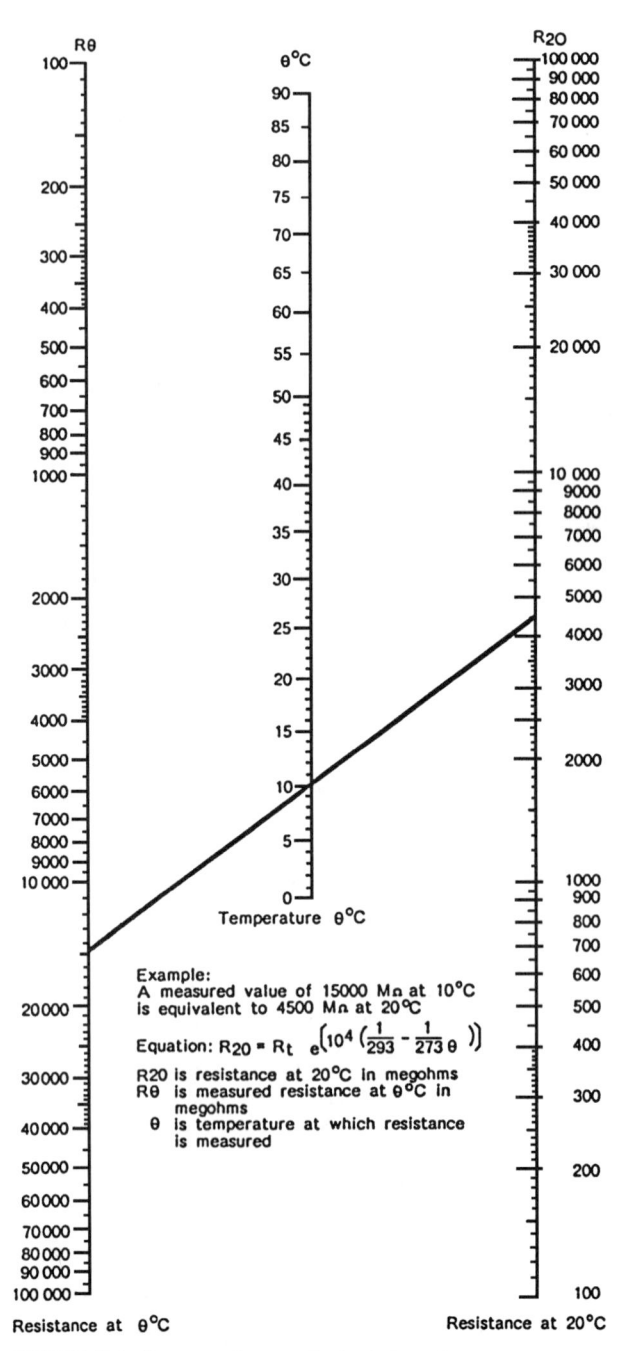

FIGURE 17.1 Nomogram for correcting insulation resistance measurements to a standard temperature.

Example:
A measured value of 15000 MΩ at 10°C is equivalent to 4500 MΩ at 20°C

Equation: $R_{20} = R_t \, e^{\left(10^4 \left(\frac{1}{293} - \frac{1}{273\,\theta}\right)\right)}$

R_{20} is resistance at 20°C in megohms
$R\theta$ is measured resistance at θ°C in megohms
θ is temperature at which resistance is measured

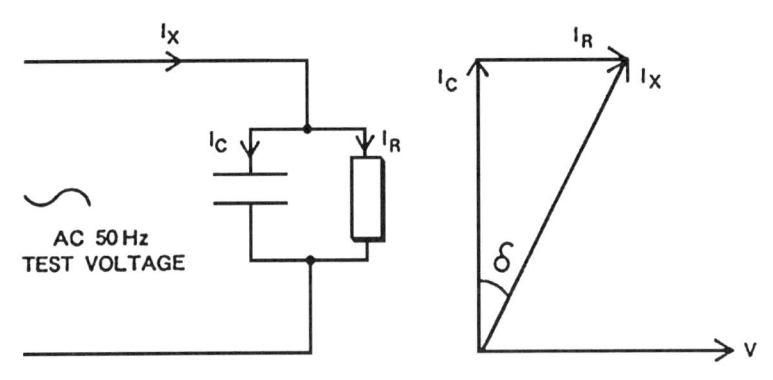

FIGURE 17.2 Insulation loss angle. The insulation may be considered as a capacitor and resistor in parallel. The ratio of the resistive leakage current to capacitive leakage current is represented by the tangent of the loss angle, δ.

The tan delta should be plotted against the line voltage. The test results are affected by the following factors:

- Presence of cooling water in the winding
- Type and pressure of gas inside the generator
- Amount of surface contamination on the windings

The condition of the insulation is determined by comparing the test results with other tests performed on the same and similar machines.

High-Voltage Tests. High dc voltage tests that subject the winding to a higher voltage than the normal operating voltage are rarely performed on a commissioned generator. However, proof voltage tests are performed normally following repair. The actual test voltage depends upon the repair and the condition of the machine. These tests should be performed in a CO_2, nitrogen, or pure hydrogen environment to remove the risk of fire.

Testing the Stator Core

Following removal of the rotor from the generator, the stator core should be examined for evidence of damage or breakdown in the core plate insulation. The following two methods are used commonly:

Ring Flux Test. A flux is induced in the core by using several turns of heavy high-voltage (HV) cable wound toroidally around the core and supplied from a voltage source of 3 to 14 kV. The pulsating magnetic flux induced in the core is similar to the rotating flux induced while the machine is operating. Hot spots will appear in the core (Fig. 17.3) due to the induced eddy currents in the regions where the interlaminar insulation (between the core laminations) is damaged. An infrared camera is used to detect these regions. This test is difficult to perform, expensive, and time-consuming.

The Electromagnetic Core Imperfection Detector (EL-CID) Test. One turn of a light conductor is wound in a similar fashion as above around the core to generate a small flux in the core. The conductor is normally supplied from 120 or 240 V ac. The conductor will carry a current in the range of 13 to 30 A (depending on the voltage source). The EL-CID detector contains a pickup coil which detects the magnetic flux induced in the air by the small fault currents flowing in the damaged region. The detector is scanned along the conductor slots inside the stator, and battery-operated electronic circuits are used to identify the faults.

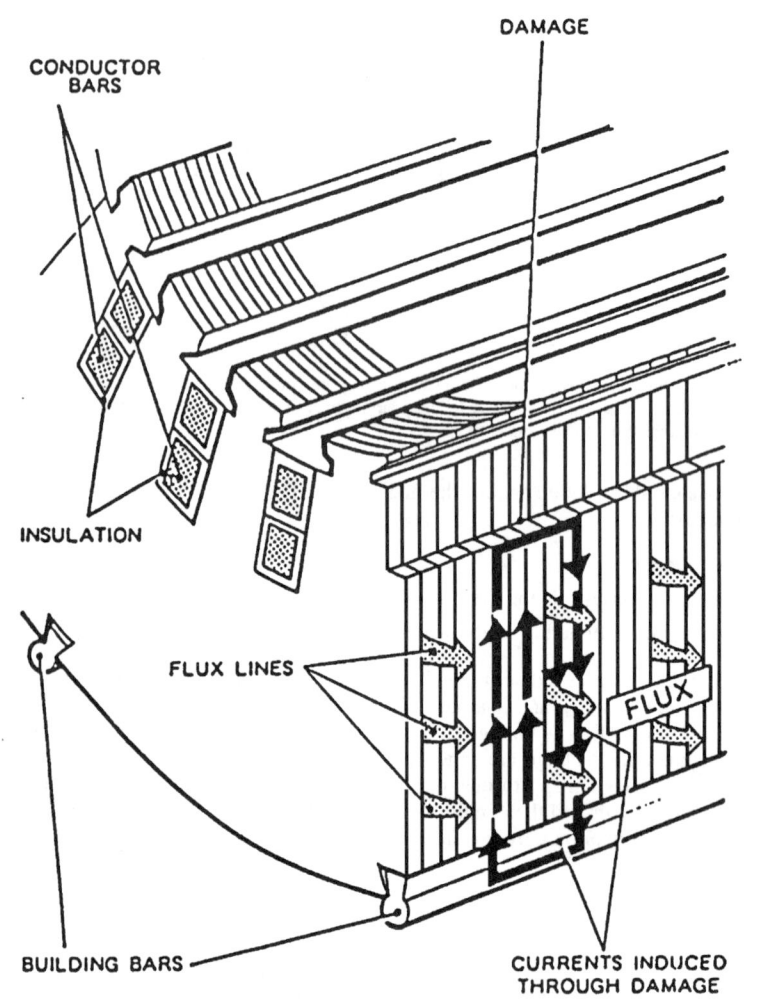

FIGURE 17.3 Induced electric currents in a damaged area of the stator. The hot spots created by these currents may be detected by infrared techniques in the ring flux test. Alternatively, the electromagnetic effects of such currents may be detected in the EL-CID test.

The EL-CID method can identify the majority of faults in the core. It has become the preferred method for core testing due to its simplicity and low cost. Fine grinding or etching techniques are used sometimes to rectify the local areas of core damage.

Stator Coolant Circuit Testing

Following maintenance work on the stator, it is necessary to prove that the windings, hoses, and manifolds within the stator frame are watertight. This can be done by performing a vacuum loss test. It consists of drawing a 700-mbar vacuum in the winding and observing the change in pressure over a 12-h period. The manufacturer would normally specify the max-

GENERATOR INSPECTION AND MAINTENANCE   17.11

imum permissible change in pressure. A typical 660-MW stator should drop around 10 mbar within 12 h. If the vacuum test results are unsatisfactory, a mixture of air and 20 percent (by volume) of a tracer gas should be pressurized in the winding. A leak search should then be performed, using an electronic gas detector.

Helium or refrigerant gases, such as Arcton or Freon, are used normally as tracer gases. If Freon is used, precautions must be taken to ensure that the concentration of the gas in the generator atmosphere is below the *long-term occupational exposure limit*. This is the limit of concentration of gas in the air that permits safe work for an 8-h period. The limits of refrigerant gases are listed in health and safety manuals.

Another important check should be done following work that involves opening of the stator water coolant circuit. It is the confirmation that the coolant is flowing through all coils. A *bubble check* is normally performed on hose-type machines (where coolant flows through translucent PTFE hoses in and out of the winding). Dry filtered air is injected into the stator water system after it has been placed in service. The flow should be examined in all the hoses, as indicated by the movement of entrained air bubbles.

An alternative method for checking the flow should be used in designs where the manufacturer does not recommend filling the winding with water unless the frame has been pressurized with gas. It consists of measuring the temperature increase across each conductor as the pump losses warm up the stator coolant. On other stator designs, it may be possible to measure accurately the total flow through each phase and to compare the values with the flow readings obtained during the commissioning phase of the machine.

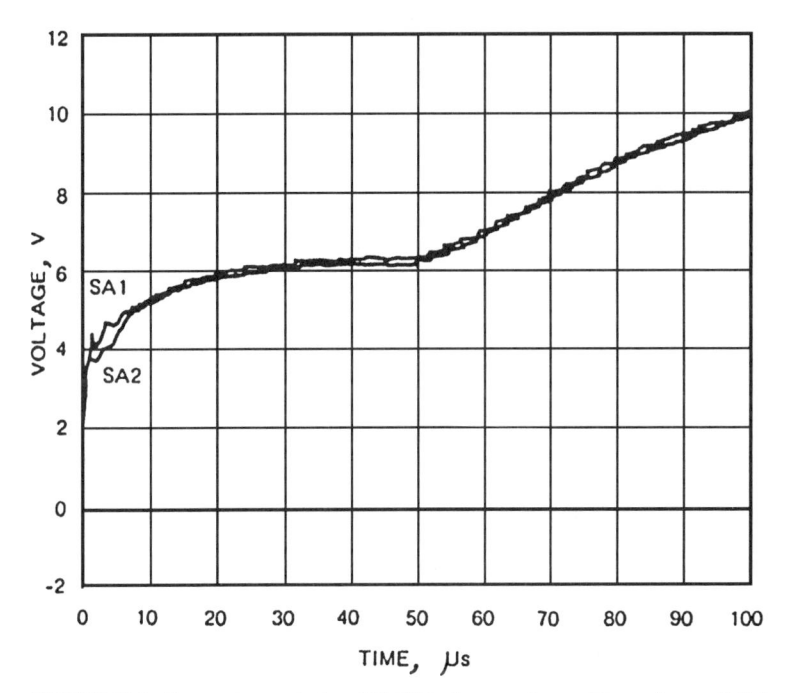

FIGURE 17.4 Recurrent surge test on 660-MW rotor; superimposed traces from a digital storage oscilloscope Trace SA1 are the input voltage at end −1 of the rotor winding as the 12-V square wave is applied with end −2 open-circuited. Trace SA2 is the converse. The divergence of the traces indicates an interturn degradation of the rotor winding between 2.5 and 12.5 percent of the rotor winding length from SA2. The traces, when compared with those from the rotor when new, also showed that the effective rotor winding length had reduced by 5 percent.

Hydrogen Loss Test

A pressure drop test should be performed using air, nitrogen, or CO_2 to ensure that the frame is gas-tight. Hydrogen is not used in the test, to avoid the safety hazard that would occur in the event of a leak.

The pressure drop test is performed normally over a 12-h period. The volume of gas lost during the test, at normal temperature and pressure (NTP), can be found using the following expression:

$$\frac{6800V}{\tau} \left(\frac{P_{gs} + P_{bs}}{\theta_s + 273} \right) - \frac{P_{gf} + P_{bf}}{\theta_f + 273} \quad m^3/day$$

where V = volume of generator voids and associated pipework, m^3
τ = duration of test
P_{gs} = gas pressure at start of test, bar gauge
P_{gf} = gas pressure at finish of test, bar gauge
P_{bs} = barometric pressure at start of test, bar
P_{bf} = barometric pressure at finish of test, bar
θ_s = temperature at start of test, °C
θ_f = temperature at finish of test, °C

The results of the test should be compared with the manufacturer's specifications. If the results are not acceptable, the test should be repeated, using a tracer gas such as helium. Gas detectors are used to identify the source of the leak. The leaks should be sealed and the test repeated until satisfactory results are obtained.

Rotor Winding Tests

The installed rotor ground fault detection equipment can be used to detect and measure rotor ground faults on-load. The insulation resistance test is used to identify rotor ground faults during outages. The detection of interturn faults [failure of the insulation between the turns (windings) of the rotor] is more difficult. However, if there is a solid short between the turns, the IR test can identify the interturn fault.

The *recurrent surge test* is used to identify ground and interturn faults. A 12-V dc square wave is injected through a matching resistor at each end of the rotor winding in turn. The input and output waveforms of both waves are stored and viewed on an oscilloscope. If there is fault in the winding, a reflected wave will travel back to the input, causing a change in the input voltage. The time of this change is proportional to the distance of the fault from the input. The input traces from the two ends of the winding are superimposed to show any divergence between the traces. This will ease the detection of the voltage change described earlier.

The number of shorted turns indicated by the test is proportional to the increase in total transit time of the pulse from the total transit time that was recorded when the rotor was new. This test can be performed while the rotor is rotating on machines having slip rings. This provides the ability to detect the faults that disappear at rest or change with speed. Figure 17.4 shows the traces of a recurrent surge test performed on a 660-MW rotor. It indicates the presence of shorted turns.

REFERENCE

1. British Electricity International, *Modern Power Station Practice—Station Operation and Maintenance, Volume G*, 3d ed., Pergamon Press, Oxford, United Kingdom, 1991.

CHAPTER 18
GENERATOR OPERATIONAL PROBLEMS AND REFURBISHMENT OPTIONS

TYPICAL GENERATOR OPERATIONAL PROBLEMS

The typical operational problems encountered with generator rotors are

- Shorted turns
- Field grounds
- Thermal sensitivity
- Negative sequence heating
- Contamination
- Misoperation
- Forging damage

Shorted Turns and Field Grounds

A shorted turn fault (known also as a winding short or an interturn fault) occurs due to failure of the insulation between the winding turns of the rotor. These faults are not desirable. However, rotors have operated with a limited number of short turns without suffering significant effects on generator operation. Shorts can occur anywhere in the rotor winding. However, they are frequently located in the end windings under the retaining rings.

Ground faults are caused by the failure of the ground wall insulation. The generator should not be operated following a grounded fault in the rotor. A single ground fault will not cause a circulating current in the forging (the rotor body) because the source of excitation is ungrounded. However, if a second ground fault occurs, the current will circulate in the forging, causing melting and serious damage to the rotor insulation.

Figure 18.1 illustrates a winding short and a ground fault in the slot of a generator rotor. Figure 18.2 illustrates a ground fault in the end winding between the top turn of the winding and the retaining ring.

The following conditions can cause breakdown in the rotor insulation:

Operation Time. During normal operation, the insulation is exposed to electrical, mechanical, and thermal stresses. Thus, the condition of the insulation is expected to degrade with operation time.

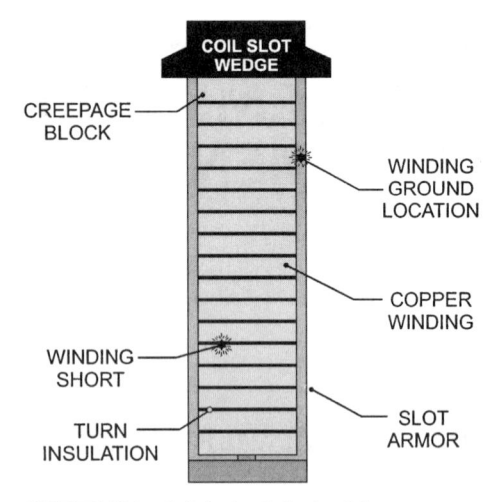

FIGURE 18.1 Coil slot insulation breakdown.

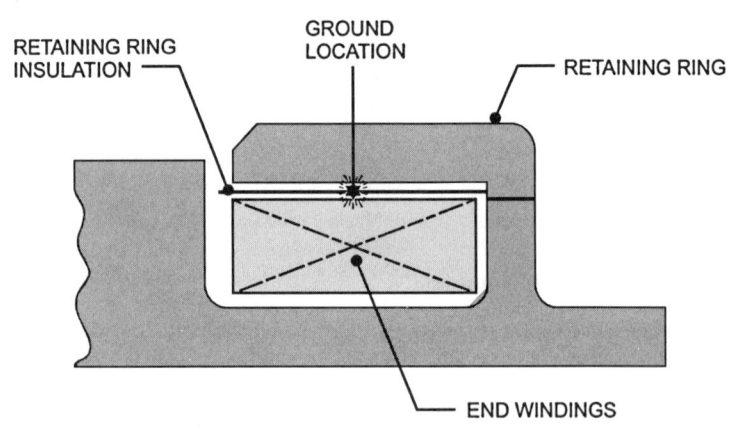

FIGURE 18.2 Field end winding insulation breakdown.

Type of Operation. The condition of the insulation degrades faster in a generator that experiences frequent shutdowns or frequent cycling in the field current (to change the amount of reactive power delivered from the generator) than in a generator that operates at base load.

Contamination. The condition of the insulation degrades due to the presence of contamination or burning inside the generator (burning produces conductive material that circulates throughout the generator).

Abnormal Operating Incidents. The rotor insulation deteriorates following an abnormal incident that results in heating, burning, arcing, high stress, etc. These incidents include generator motoring (i.e., the generator acts as a motor when it is synchronized to the grid and there is no steam or hot gas driving the turbine), a negative sequence event (unbalance in the three phases in the stator, which can be caused by closing the generator breaker when the voltage in the three phases is different from the corresponding voltages in the grid), burning inside the generator, or an overspeed incident.

As mentioned earlier, the generator can operate satisfactorily with interturn faults (shorted turns) in the rotor windings. However, the generator should not be operated with a ground fault.

Shorted turns do not subject the machine to high risk. The consequences of operating a rotor with shorted turns include the following:

1. Inability to reach the nominal rating of the machine
2. Possibility of developing an unbalance in the rotor that will result in high vibrations

In the worst case, a complete rewind of the rotor will be required.

When one ground fault occurs, the current will not flow in the forging (body of the rotor). However, when a second ground fault occurs, the current will start to flow between the two ground points. The resulting heat generation (from the current) could melt the forging within a few seconds.

The cause of the ground fault should be investigated immediately and corrective action taken. Operation with one ground fault could lead to serious damage, should a second ground fault occur.

Shorted turns will develop in the rotor after a long period of operation. Many generators have operated satisfactorily for years with shorted turns. The shorted turns should be repaired only when continued generator operation becomes unacceptable due to unit derating or high vibrations. The generator field should be inspected and tested regularly to confirm the integrity of the insulation system. Diagnostic tests can identify the location of an insulation fault and its severity. This ability for taking quick corrective action allows the rotor to be returned to service within a short time.

The following additional concerns affect rotor operation.

Thermal Sensitivity

The excessive rotor vibration caused by the heating effect resulting from an interturn fault is known as *thermal sensitivity*. The rotor tends to bow due to the temperature differential across it. This leads to imbalance and high vibrations.

The following factors can cause thermal sensitivity:

Shorted Turns. The windings of a pole having several shorted turns will have lower resistance than the windings of a healthy pole. Since the current flowing in the poles is the same, the pole having a higher winding resistance (healthy pole) will heat up and expand more than the other. This causes a rotor bow in the direction of the healthy pole.

Blocked Ventilation. The rotor will have uneven temperature distribution if a ventilation path becomes blocked. This will lead to a rotor bow similar to the one caused by short turns.

Uneven Insulation. Nonuniform field insulation can cause the field coils to bind in the slots and in the end windings (under the retaining ring). This will lead to transmitting uneven axial forces on the rotor, causing it to bow.

Uneven Wedge Tightness. Uneven tightness of the wedges can cause nonuniform axial force distribution around the rotor. This will also lead to a bow in the rotor.

Uneven End Winding Insulation. Unevenly spaced and fitted end winding insulation can cause nonuniform forces to be transmitted to the rotor. This will also cause bowing in the rotor, leading to high vibrations.

There are two types of thermal sensitivity: reversible and irreversible. The reversible type is characterized by a repeatable behavior. In this case, the rotor vibration varies linearly with the field current. The thermal sensitivity is considered irreversible, or slipstick, if the rotor vibrations increase with the field current and do not decrease when the current drops. The type of thermal sensitivity should be identified by performing tests. This will aid in determining the corrective action required to eliminate the problem.

Contamination

The cooling method of the generator determines the type and extent of contamination found in it. A hydrogen-cooled generator normally has little contamination because it is well sealed. A totally enclosed water to air cooled (TEWAC) unit will have some particles due to makeup air. An open ventilated (OV) generator will have large amounts of contamination in it.

Carbon is a common contaminant in generators. It originates from slip ring (collector) brush wear or the exhaust of gas turbines. Silicon or petroleum by-products are also found in some generators. They originate from nearby systems or processes. The inlet filter eliminates most of the air contaminants. However, since the airflow is very large, even a small percentage of contaminants passing through the filter generates significant deposits over time. Other contaminants come from the generator itself. They originate from worn insulation and wedges.

The problems caused by contamination buildup include low insulation resistance (low meggar resistance to ground) readings, overheating, and interturn faults (turn shorts).

Collector, Bore Copper, and Connection Problems

The excitation system includes the collectors, bore copper, and main leads. *Collector flashover* is a problem that occurs in this area. In this case, contamination causes the positive collector to flash to ground. A creepage path (a degradation of the collector shell insulation) is developed, resulting ultimately in a forced outage.

As the rotor ages, the bore copper, terminal studs, and copper coils become loose. The relative movement between the main lead and the number 1 coil will increase. This can lead to fatigue failure of the connection and a forced outage due to *loss of field current*. The insulation around the copper between the collectors and the number 1 coils can deteriorate, leading to a ground fault and a forced outage (Fig. 18.3).

Copper Distortion. The copper winding becomes distorted in some generators after a long period of operation (Fig. 18.4). This problem occurs most frequently in the end winding area. It is caused by the following:

- Frequent thermal cycling of the winding
- Overheating of the winding
- Use of soft or annealed copper
- Friction between the top turns and the retaining ring

The damaged copper should be repaired or replaced before returning the unit to service. Failure to do so can result in ground faults or interturn faults (shorted turns).

GENERATOR OPERATIONAL PROBLEMS AND REFURBISHMENT OPTIONS   18.5

FIGURE 18.3 Collector and brush holder neglect.

Forging Concerns

The rotor body is made of a single forging. It is not laminated because it experiences significant mechanical, electrical, and thermal stresses. The rotor forging should be inspected before rewind to determine the long-term structural integrity. This inspection becomes especially necessary in the following situations:

- If the unit has been exposed to negative sequence currents*
- If the unit has been operated as a motor
- If the forging was manufactured before 1960

Negative sequence currents can cause arcing between the wedges and the retaining rings, burning, hard spots, and cracking on the surface of the forging. The rotors manufactured before 1960 tend to have lower toughness than modern rotors. They also have higher levels of impurities. This makes them marginal for continued operation.

FIGURE 18.4 Moderate copper distortion.

Recent inspection of some large steam turbine generators identified cracks in the rotor teeth and damage from in-service negative sequence events. This damage included strikes on the rotor teeth and the area of the retaining rings that faces the slot wedges. The cracks that occurred in the teeth of the rotor were in generators that experienced extended operations on turning gear.

The generator manufacturer should be consulted if there are negative sequence events. The manufacturer will examine the severity of the damage and recommend an inspection program for the rotor. The rotors, which experience extended periods of operation on turning gear, should also be inspected by the manufacturer.

---

*Negative sequence currents occur when there is a load or a phase unbalance in the three phases.

The bore of the rotor forging should be inspected every 8 years if the generator experienced frequent start-stops. The rotor teeth and wedges should also be inspected using magnetic particle and fluorescent techniques, prior to 5000 start-stop cycles.

Retaining Rings

The main purpose of the retaining rings is to restrain the centrifugal force of the rotor winding end turns (known as end windings). This centrifugal force is extremely large during normal operation. It produces a hoop stress that stretches the ring into a slight elliptical shape for two-pole rotors due to the uneven weight distribution of the end turns and associated insulation. The retaining ring is shrunk-fit at the end of the rotor body. This ensures that the retaining ring will remain cylindrical at full speed and prevents the ring from moving with respect to the body. The retaining ring will be subjected to high circulating currents during unbalanced load conditions that lead to negative sequence currents in the rotor.

The retaining ring is made of two types of material: magnetic and nonmagnetic. The nonmagnetic rings are used in large generators to reduce the leakage flux. The magnetic rings that were exposed to moisture developed corrosion. This led to failure during normal operation that resulted in extended outages. These magnetic rings were made from Gannalloy or 18% manganese and 5% chromium (18 Mn, 5 Cr). Both of these materials had serious problems. Gannalloy became embrittled when used in hydrogen-cooled generators. It is highly recommended to replace the Gannalloy retaining rings with ones made from 18% manganese and 18% chromium (18 Mn, 18 Cr). This material has been proved to be highly resistant to hydrogen embrittlement. The retaining rings made from 18 Mn, 5 Cr were found to be subjected to stress corrosion cracking (SCC). Since the failure rate of the retaining rings due to SCC was high, these rings were replaced with 18 Mn, 18 Cr material. This material has also proved to be highly resistant to SCC. Catastrophic damage could occur if the retaining rings made from Gannalloy or 18 Mn, 5 Cr were not replaced with 18 Mn, 18 Cr (Fig. 18.5).

Misoperation

Misoperation of the generator rotor can cause catastrophic consequences on the rotor and secondary damage to the stator and prime mover. The reasons of misoperation include

- Operator error

FIGURE 18.5 Catastrophic retaining ring failure.

- Equipment failure
- Abnormal system conditions

Figure 18.6 shows the most common modes of misoperation that can affect the rotor.

GENERATOR ROTOR RELIABILITY AND LIFE EXPECTANCY

The expected life of a generator rotor depends on its design, mode of operation, and operating incidents. Generators that are operated as peaking units with a large number of start-stop cycles and high field current (to deliver large amounts of reactive power) will generally have lower life expectancy than base load units having a low number of start-stop cycles and operating near unity power factor with relatively low field current. The frequent start-stop cycles increase insulation wear and distortion of the copper conductor. This is due to the high centrifugal forces that the insulation and copper experience when the rotor accelerates to full speed. When the field current is applied to the copper winding at operating speed, large forces are exerted on the windings in the axial direction. This can cause deformation and distortion of the copper winding and abrasion of the insulation, leading to premature failure.

The rotor forging and retaining rings will be damaged during an operating incident such as motoring or negative sequence operation. The insulation will also be damaged following a high-voltage spike incident. This could lead to an interturn fault or ground fault. The operating life of a generator rotor and its maintenance interval can be extended if the number of operating incidents is minimized.

Generator Experience

A generator operating at base load and having minimal operating incidents will have a life expectancy for its rotor forging of around 35 to 40 years. It may be necessary to repair or replace the insulation and copper during this period. The rotors that experience frequent start-stop cycles will have a much shorter expected life span. The frequent start-stop cycles can reduce the expected life of the insulation by 30 to 50 percent of that of a base load generator.

Knowing the rotor life expectancy will help in determining the inspection and maintenance intervals. It also minimizes the number of forced outages. An inspection and test

■ Field overheating	■ Abnormal frequency and voltage
■ Loss of excitation	■ Breaker failure
■ Rotor or stator vibration	■ Voltage surges
■ Synchronizing errors	■ Transmission line switching
■ Motoring	■ Electrical faults
■ Reduced seal oil pressure	■ High-speed reclosing
■ Unbalanced armature currents	■ Subsynchronous resonance
■ Loss of synchronism	■ Accidental energization

FIGURE 18.6 Common modes of misoperation.

program will help in diagnosing the condition of the insulation. It also assists in planning future repairs and rotor rewinds.

GENERATOR ROTOR REFURBISHMENT

Generator Rotor Rewind

The rotor is the most susceptible component in the generator to operating incidents such as motoring or negative sequence currents. It is also subjected to very high centrifugal forces during normal operation. Thus, it is the component that requires the most maintenance in the generator. The rotor rewind involves a replacement of all field winding insulation.

TYPES OF INSULATION

There are three types of winding insulation used for generator rotors. The first type consists of taped winding turns (Fig. 18.7). Every other turn including the end turns is taped with mica mat tape. This system requires the most labor due to the hand taping required. However, it is the least expensive and provides the best contamination protection.

The second insulation system involves the use of strip turn insulation in the slots and taped ends. Every other turn is taped with mica mat tape. This system permits the use of wider copper in the slots, while still providing protection against contamination in the end winding region.

The third system involves the use of "all strip turn insulation." It is normally used in applications requiring improved cooling in the end winding. The insulation strips are normally made of Nomex or glass laminate (Fig. 18.8).

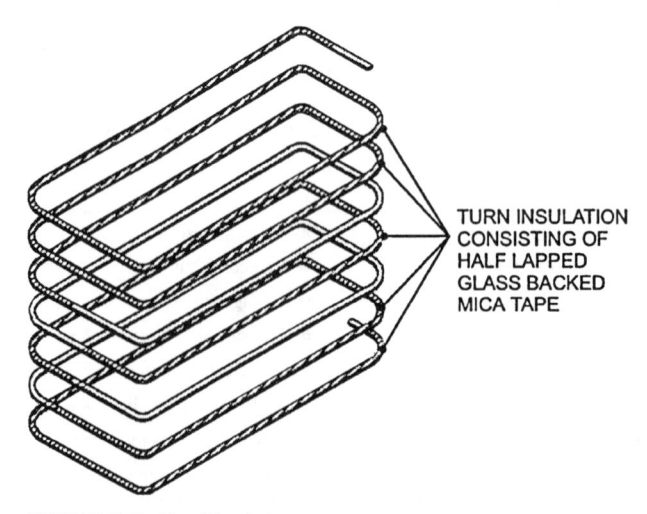

FIGURE 18.7 Taped insulation system.

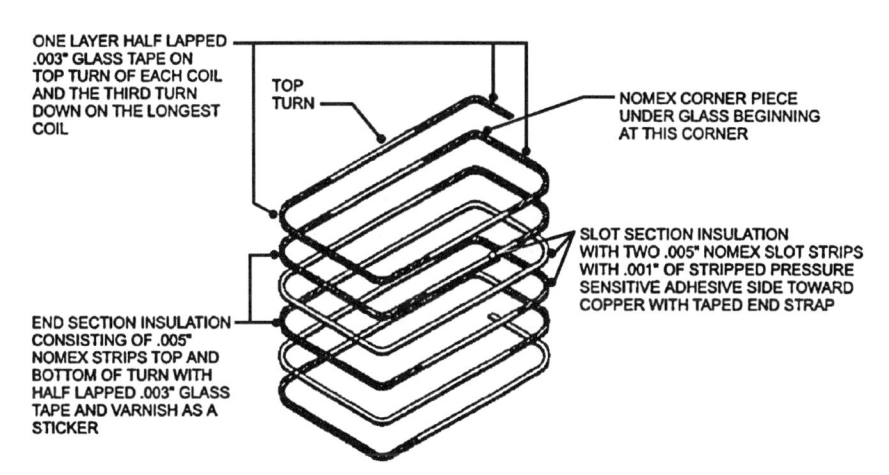

FIGURE 18.8 Taped and strip insulation system.

There are two types of material used for ground insulation (or slot armor). The first is a rigid armor made of a glass base. This material has high mechanical strength. The second type is Nomex. This material is tough and flexible.

The retaining ring insulation provides the ground insulation in the end winding region (Fig. 18.9). It is strong to be able to withstand the centrifugal forces. It normally has an outer layer of glass and an inner layer of Nomex.

The end winding blocking supports the winding and prevents distortion while allowing for thermal expansion (Fig. 18.10). The material used in modern rotors is epoxy glass laminates.

GENERATOR ROTOR MODIFICATIONS, UPGRADES, AND UPRATES

A generator uprate (increase in rating) can be achieved during a rotor rewind. This can be done by installing new class F insulation. This allows the rotor to have higher field current. Thus, the real and reactive power of the generator can be increased.

Generators used with gas turbines, steam turbines, and combined-cycle power plants have been operating at higher power factor than originally designed, to increase the real power output from the machine. For example, these generators would operate at a power factor of 0.95 lag rather than 0.9 lag. This reduces the reactive power (MVAR) output from the machine. This mode of operation does not require any modifications to the generator, as long as it operates within its capability limits. This uprating practice is being used more commonly due to energy shortages.

HIGH-SPEED BALANCING

The high-speed balancing of the generator rotor is normally done at 120 percent of the operating speed. It maximizes the dynamic stability and provides a comprehensive evaluation of the suitability of the rotor to return to service. The quality of the refurbishment work

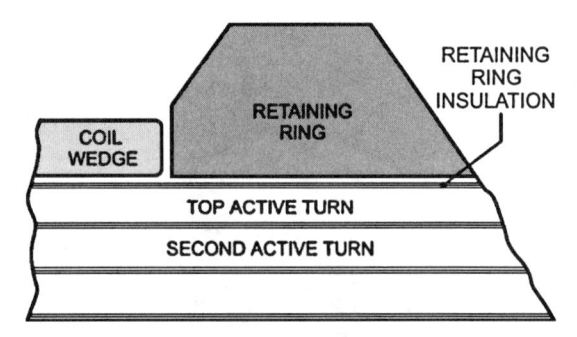

FIGURE 18.9 Retaining ring insulation.

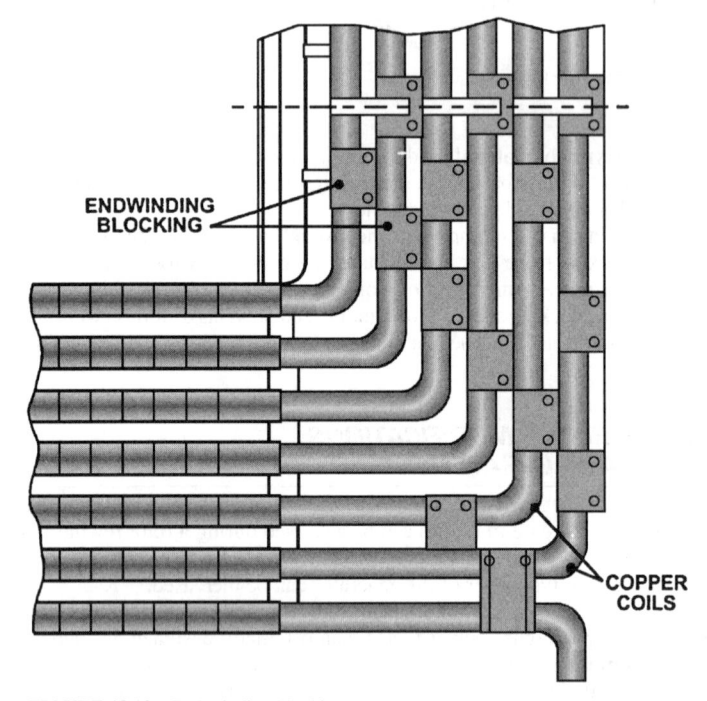

FIGURE 18.10 End winding blocking.

done on the rotor can be fully assessed. This reduces the chance of a costly forced outage. This work is normally done at a high-speed balancing facility.

The high-speed balance will eliminate the requirement of "trim" balancing on site, which is a process that can be very time-consuming (more than 14 days in some cases). This is due to the work required to disassemble and reassemble the generator to gain access to the rotor. It is recommended to perform a high-speed balance when new coils are used in rewinding the rotor. However, if the rewind was done using the same coils, the high-speed balance may not be required.

FLUX PROBE TEXT

The flux probe test determines if there is an interturn fault (shorted turn) in the rotor. Excitation is applied to the rotor while it is rotating at operating speed. The flux probe measures the leakage flux from each coil. If there is a shorted turn in a coil, its leakage flux will be different from that of the neighboring coil.

The flux probe test should be done during the high-speed balancing. It is also recommended that a flux probe be installed permanently in large generators. It provides the advantage of identifying a shorted turn as soon as it occurs.

REFERENCE

1. R. J. Zawoysky and K. C. Tornoos, *GE Generator Rotor Design, Operational Issues, and Refurbishment Options,* GE Power Systems, New York, 2001.

CHAPTER 19
CIRCUIT BREAKERS

THEORY OF CIRCUIT INTERRUPTION

A circuit breaker is a device that switches on and switches off electric circuits during normal as well as abnormal operating conditions. During the making or breaking of the switching contacts, there is a transition stage of arcing between the contacts. The study of this phenomenon is of great importance for understanding the design and operational characteristics of circuit breakers.

PHYSICS OF ARC PHENOMENA

Discharge in ac circuit breakers, generally in the form of an arc, occurs in the following ways:

1. When the contacts are being separated, arcing is possible even when the circuit emf is considerably below the minimum cold electrode breakdown voltage, because of the large local increase in voltage due to the circuit self-inductance.

   Note: This way of drawing arc is common to both dc and ac circuit breakers.

2. In an ac circuit breaker, the arc is extinguished every time the current passes through zero and can restrike only if the transient recovery voltage across the electrodes already separated and continuing to separate reaches a sufficiently high value known as the breakdown voltage.

The arc phenomenon depends upon

- The nature and pressure P of the medium
- The external ionizing and deionizing agents present
- The voltage V across the electrodes and its variation with time
- The nature, shape, and separation of electrodes
- The nature and shape of the vessel and its position in relation to the electrodes

An ideal gas is a pure dielectric because it consists of molecules which are electrically neutral. It can be made to conduct only when some means are employed to create free electrons and ions in the gas.

When the gas temperature increases significantly, the molecules start to break down at the most severe collisions and *dissociate* into their atoms. Energies of 9.7 eV and 5.1 eV, respectively, are needed to dissociate an N_2 and an O_2 molecule.

At higher temperatures, some molecules and atoms are deprived of an electron and the hot gas called *plasma* becomes a conductor.

Figures 19.1 and 19.2 show the variation of the degree of dissociation x_d, the degree of ionization x_i, thermal conductivity K, and electrical conductivity σ with temperature.

Note: The free electrons which are caused by high temperature can carry an electric current under an electric field in roughly the same way as do the free electrons in a metal (the electrical conductivity of a metal is due to the existence within it of *free* conductivity electrons).

Arc Interruption Theory

In circuit breakers, the modes of arc interruption are high-resistance interruption and low-resistance or current zero interruption.

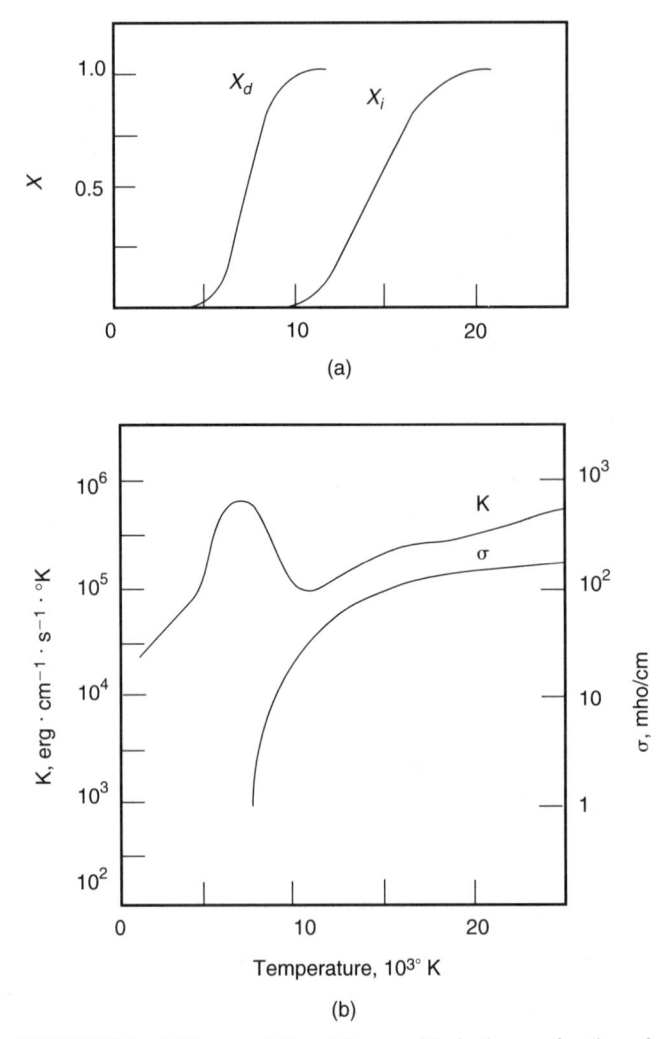

FIGURE 19.1 (*a*) Degrees of dissociation x_d and ionization x_i as functions of temperature. (*b*) Thermal conductivity K and electrical conductivity σ as functions of temperature.

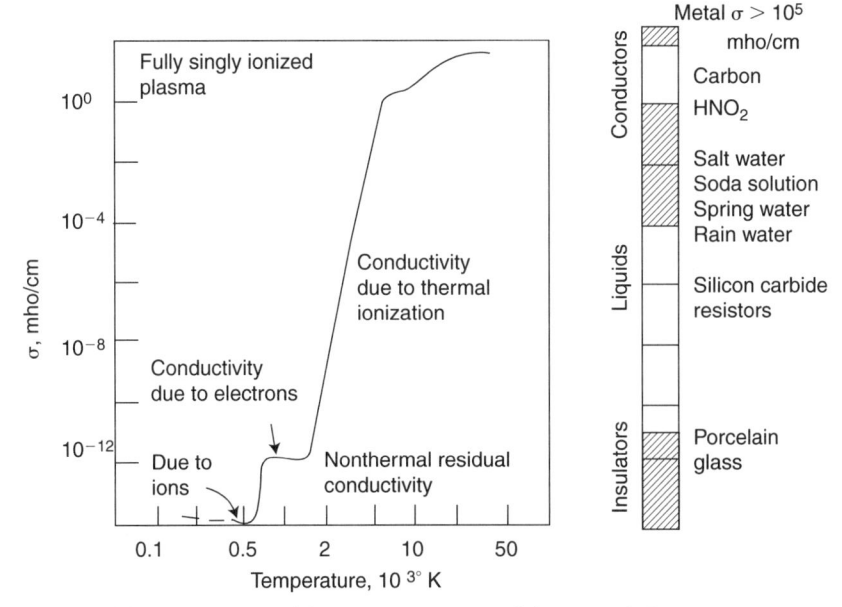

FIGURE 19.2 Electrical conductivity σ versus temperature of air at atmospheric pressure.

High-Resistance Interruption

The arc is controlled in such a way that its effective resistance increases with time, resulting in a decrease in current until it cannot be maintained. The arc resistance can be increased by lengthening, cooling, and splitting the arc.

Low-Resistance or Current Zero Interruption

In an alternating current, every time the current passes through zero, the arc extinguishes for a brief moment and again restrikes with the rising current. The reestablishment or interruption of the arc is an energy balance process.

If the energy input to the arc, subsequent to the current zero, continues to increase, the arc restrikes; if not, the circuit is interrupted.

CIRCUIT BREAKER RATING

The circuit breaker must be capable of carrying continuously the full load current, without excessive temperature rise, and should be capable of interrupting fault currents safely.

An ac circuit breaker has the following ratings:

1. Rated voltage, rated current
2. Rated frequency
3. Rated breaking capacities
4. Rated making capacities

5. Rated short-time current or rated maximum duration of short circuit
6. Rated operating duty

CONVENTIONAL CIRCUIT BREAKERS

Automatic Switch

The simplest circuit interruption device is the knife switch shown in Fig. 19.3. By closing the switch against the action of a spring, energy is stored. By applying a small force on the latch, the stored energy is released and the contacts open within a short time.

Air-Break Circuit Breakers

Air-circuit breakers cool the gases to naturally deionize them, causing arc interruption. The arc can be stretched. Its resistance can be increased by increasing its length. The increase in resistance is significant so that the current and voltage are brought into phase.

If the phase difference between the system voltage and the short-circuit current is reduced, when the current is interrupted at its zero value, the recovery voltage is very low at that instant.

The application of high-resistance interruption is limited to low- and medium-power ac circuit breakers. It is also used for low- and medium-power dc circuit breakers.

METHODS FOR INCREASING ARC RESISTANCE

The following methods increase the arc resistance:

1. *Arc lengthening.* The resistance is approximately proportional to the arc length.
2. *Arc cooling.* A decrease in temperature increases the voltage required to maintain ionization. Therefore, cooling effectively increases the resistance.

PLAIN BREAK TYPE

The plain break type is the simplest type of air-break circuit breaker. The contacts are made in the shape of two horns, as illustrated in Fig. 19.4. The arc initially strikes across the shortest distance between the horns. Then it is driven steadily upward by the convection currents which are caused by heating the air during arcing. The arc extends between the horns. This results in lengthening and cooling of the arc.

The arc interruption process is relatively slow. It limits the application of these circuit breakers to 500 V and low-power circuits.

MAGNETIC BLOWOUT TYPE

This type of circuit breaker extinguish the arc by means of a magnetic blast. It is limited to circuits up to 11 kV. The arc is subjected to the action of a magnetic field set up by coils

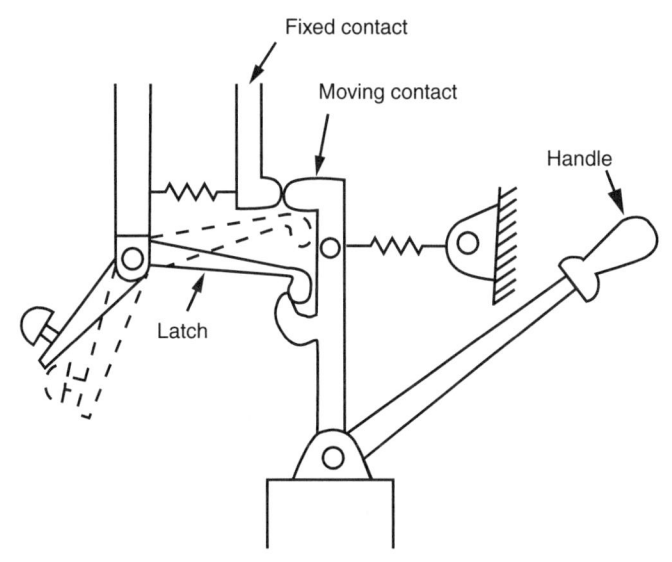

FIGURE 19.3 Automatic switch.

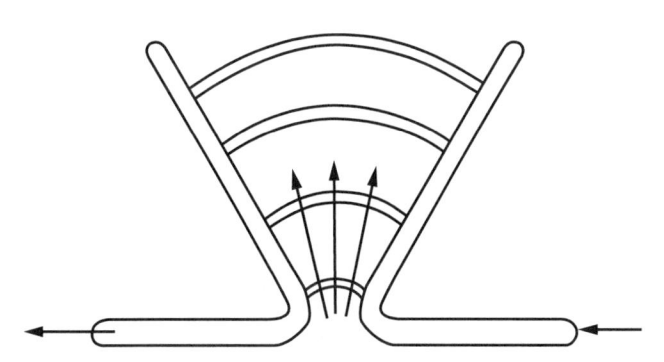

FIGURE 19.4 A simple air-break breaker.

connected in series with the circuit being interrupted (Fig. 19.5). These coils are called *blowout coils* because they help blow out the arc magnetically. The arc is magnetically blown into arc chutes where the arc is lengthened, cooled, and extinguished.

The breaking action becomes more effective with heavy currents. This results in higher breaking capacities for these breakers.

The arc chute is an efficient device for quenching the arc in air. It performs the following interrelated functions:

1. It confines the arc within a restricted space.
2. It provides control of the movement of the arc to ensure extinction occurs within the device.
3. It cools the arc gases to ensure extinction by deionization.

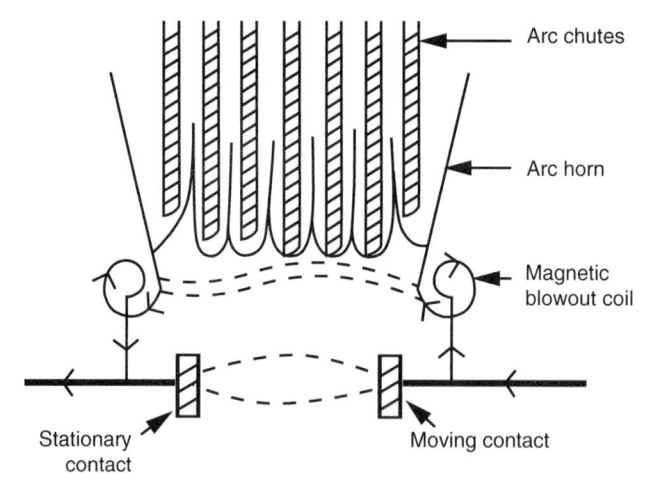

FIGURE 19.5 Principal scheme of a magnetic-blast breaker.

ARC SPLITTER TYPE

The blowouts in these breakers consist of steel inserts in the arcing chutes. These inserts are arranged so that the magnetic field induced in them by the current in the arc helps move the arc upward. The steel plates divide the arc into a number of short arcs in series. Figure 19.6 illustrates an air-break arc splitter-type circuit breaker.

APPLICATION

In general, air-break circuit breakers are suitable for the control of power station auxiliaries and industrial plants. They combine a high degree of safety with minimum maintenance. They do not require any associated equipment such as compressors. Since they have no oil, they are recommended for locations where fire or explosion hazards are feared.

OIL CIRCUIT BREAKERS

In oil circuit breakers, the arc energy is used to crack the oil molecules to generate hydrogen gas. The hydrogen is used to sweep, cool, and compress the arc plasma. This deionizes the arc plasma in a self-extinguishing process.

Even if the breaker contacts are immersed in oil, arcing still occur during contact separation. The heat from the arc evaporates the surrounding oil and dissociates it into carbon and a substantial volume of gaseous hydrogen at high pressure. The heat conductivity of hydrogen is high, resulting in cooling of the arc and the contacts. This increases the ignition voltage and extinguishes the arc.

CIRCUIT BREAKERS

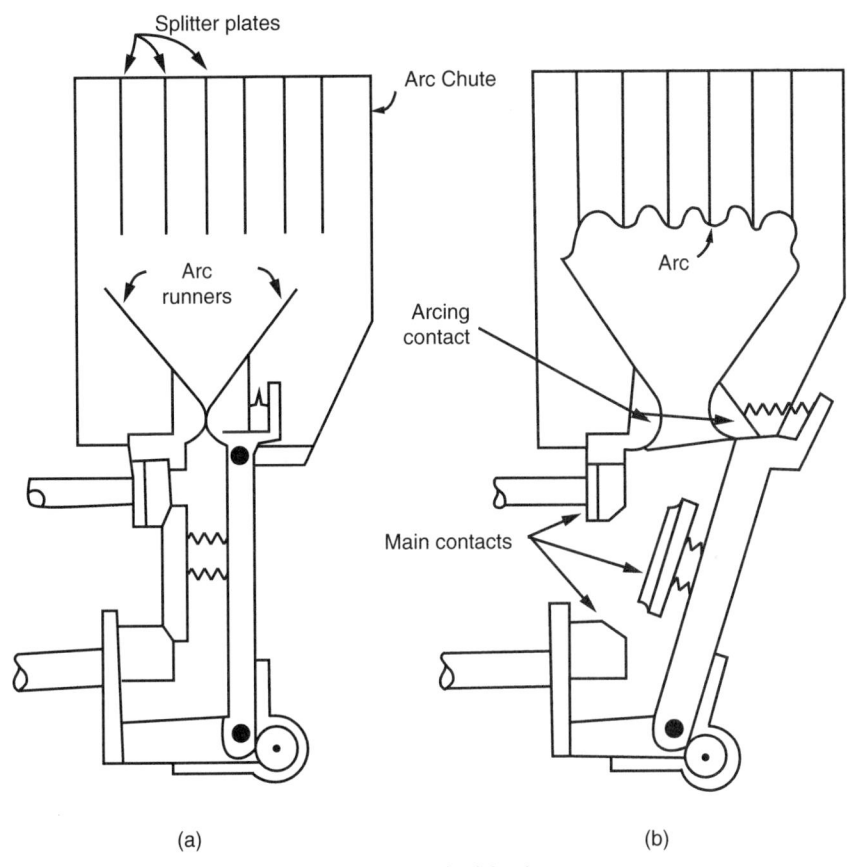

FIGURE 19.6 Elementary arrangement of air-break circuit breaker.

The cooling caused by the hydrogen (due to its high conductivity) is very effective. It increases the voltage required for reignition significantly (5 to 10 times higher than the reignition voltage required for air).

Hydrogen is produced spontaneously in arcs under oil.

Interruption of heavy short-circuit currents generates extremely high pressures which should be released safely or controlled properly. These high pressures are used to extinguish the arc which generated them.

Advantages of Oil

As an arc extinguishing medium, oil has the following advantages:

1. Oil produces hydrogen during arcing. The hydrogen helps extinguish the arc.
2. The oil provides insulation for the live exposed contacts from the earthed portions of the container.
3. Oil provides insulation between the contacts after the arc has been extinguished.

Disadvantages of Oil Circuit Breakers

Oil has the following disadvantages when used as an arc extinguishing medium:

1. Oil is inflammable and may cause fire hazards. When a defective circuit breaker fails under pressure, it may cause an explosion.
2. The hydrogen generated during arcing, when combined with air, may form an explosive mixture.
3. During arcing, oil decomposes and becomes polluted by carbon particles which reduces its dielectric strength. Hence, it requires periodic maintenance and replacement.

Plain Break Oil Circuit Breakers

In this type of circuit breaker the arc is confined only within the oil tank. Deionization of the arc is due to turbulence and increase of pressure (Fig. 19.7).

To ensure successful interruption, a long arc length is needed so that turbulence in the oil (caused by the pressures generated by the arc) helps in quenching the arc.

The tank should be weathertight to prevent moisture ingress (moisture will reduce the dielectric strength of the oil significantly).

Caution

1. If there is an air cushion on top of the oil, the hydrogen gas generated may mix with the air to form an explosive mixture. The air cushion should be kept to a minimum.
2. If the speed of contact movement is slow during short-circuit interruption, welding of the contacts may occur, resulting in a very dangerous situation and possible explosion.

The main features that affect the performance of a plain break oil circuit breaker are

1. Length of the break
2. Speed of contact movement
3. Head of oil above the contacts (to ensure adequate oil flow due to high pressure) during arcing
4. Clearance to earthed metal adjacent to the contacts

This type of circuit breaker is considered satisfactory up to 11 kV and 250 MVA.

Arc Control Circuit Breakers

Most oil control circuit breakers have an arc control mechanism. In these breakers, the gas produced during arcing is confined to small volumes by the use of an insulating, rigid arc chamber surrounding the contacts. Therefore, higher pressures can be generated during arcing to force the oil and gas through and around the

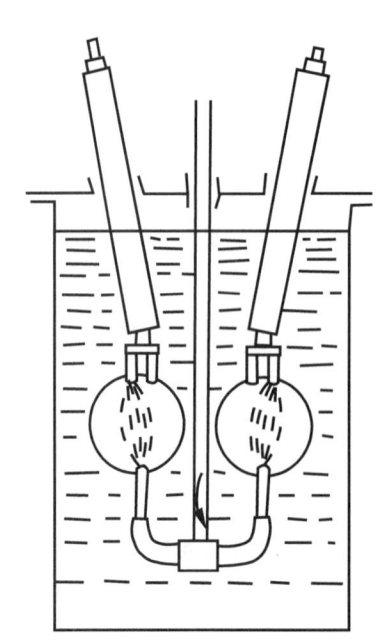

FIGURE 19.7 Plain break oil circuit breaker.

arc to extinguish it. These small high-pressure-resistant chambers are called *arc control pots* or *explosion pots*. These explosion pots have improved the efficiency of arc interruption and decreased significantly the risks of fire hazards. This improvement has resulted in significant reduction in arc duration time.

RECENT DEVELOPMENTS IN CIRCUIT BREAKERS

Vacuum Circuit Breakers

The two outstanding properties of high vacuum are that

1. It has the highest insulating strength known.
2. When an ac circuit is opened by separating the contacts in vacuum, interruption occurs at the first current zero. The dielectric strength across the contacts builds up at a rate thousands of times higher than that obtained with conventional circuit breakers.

These properties make the vacuum circuit breakers more efficient, less bulky, and cheaper.

The service life of these breakers is much longer than that of conventional circuit breakers. The maintenance required for these breakers is minimal. Low vacuum pressures are measured in terms of torr, where 1 torr = 1 mm of mercury. Pressures as low as 10^{-7} torr can be achieved.

When contact separation occurs in air, the ionized molecules are the main carriers of electric charges. They are responsible for the low breakdown value. The absence of air molecules increases the dielectric strength of vacuum significantly. Figure 19.8 illustrates the breakdown strength of various insulating materials.

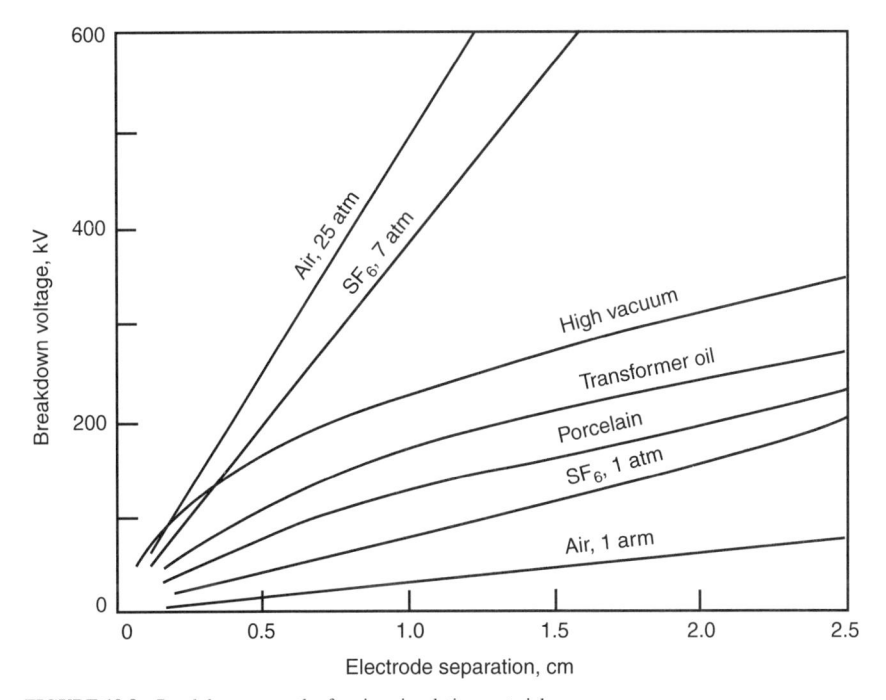

FIGURE 19.8 Breakdown strength of various insulating materials.

19.10 CHAPTER NINETEEN

Construction of Vacuum Switch/Circuit Breakers. The vacuum circuit breaker is a very simple device when compared with oil or air-blast circuit breakers. Two contacts are mounted inside an insulating vacuum-sealed container. One is fixed and the other could be moved through a very short distance. The contacts are surrounded by a metallic shield that protects the insulating container. A typical assembly of a vacuum switch is shown in Fig. 19.9.

Vacuum Chamber. It consists of the following subassemblies:

1. The vacuum chamber
2. The operating mechanism

The vacuum chamber is usually made of synthetic material such as urethane foam. It is enclosed in an outer glass-fiber-reinforced plastic tube or of simple glass or porcelain. It has two contacts, a metal shield and metal bellows, which are sealed inside the chamber. The metallic bellows are usually made of stainless steel. The metallic bellows are used to move the lower contact and provide a gap of 5 to 10 mm depending upon the application of the switch.

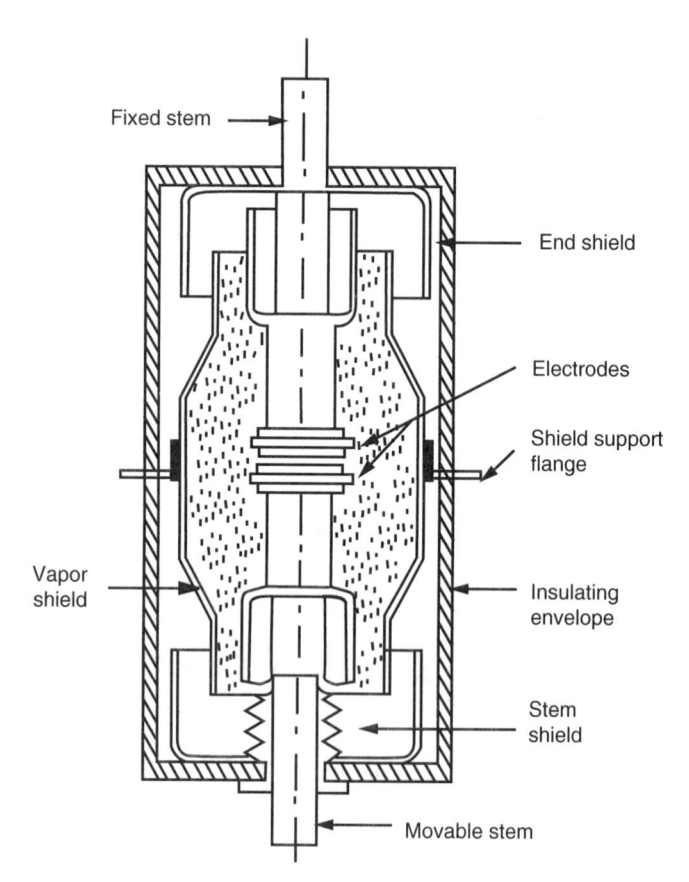

FIGURE 19.9 Vacuum interrupter schematic diagram.

Application of Vacuum Switches. The vacuum switches have a promising application in the high-voltage switchgear field. They are distinguished for their low cost, low fault interrupting capacity, and capability of a large number of load switching operations without maintenance. They are very high-speed switches and can be used for many industrial applications. They are well suited up to 35-kV and 100-MVA applications.

Sulfur Hexafluoride (SF$_6$) Circuit Breakers

The SF$_6$ circuit breaker is the most recent development in the field of high-voltage switchgear. SF$_6$ is a gas that is 5 times heavier than air. It is chemically very stable, odorless, inert, noninflammable, and nontoxic.

SF$_6$ and its decomposition products are electronegative. They capture electrons at relatively high temperature. Therefore, the dielectric strength rises rapidly and enables the breaker to withstand the recovery voltage even under extreme switching conditions.

SF$_6$ circuit breakers do not allow the gas to escape, as do air-blast circuit breakers following the quenching operation. The circuit breaker chambers are hermetically sealed. The gas pressure remains practically constant over long periods.

The SF$_6$ gas has low contact erosion and negligible decomposition in arc. The breaker can be operated for several years without having to be opened for the purpose of overhauling. At atmospheric pressure, the dielectric strength of SF$_6$ is about 2.5 times that of air. Figure 19.10 illustrates the variations of the dielectric strength with pressure for air, oil, and SF$_6$.

The dielectric strength of SF$_6$ is less than that of oil at atmospheric pressure. However, it increases rapidly with pressure. The gas is strongly electronegative. Free electrons are readily removed from a discharge by the formation of negative ions.

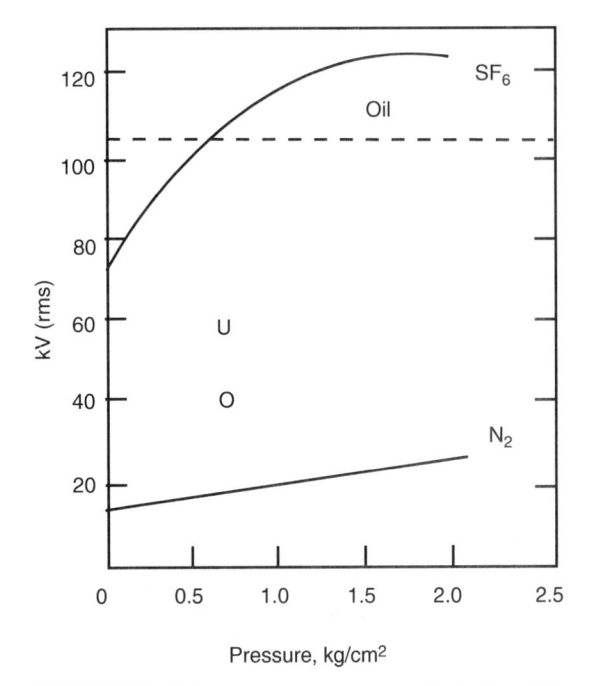

FIGURE 19.10 Dielectric strength versus pressure for air, oil, and SF$_6$.

The following are the processes by which a free electron is attached to a neutral gas molecule:

1. A direct attachment as in

$$SF_6 + e^- = SF_6^-$$

2. As dissociative attachment in

$$SF_6 + e^- = SF_5 + F^-$$

The resulting ions are heavy and ineffective carriers of current. Therefore, the ionized SF_6 has also a high dielectric strength.

Maintenance of SF_6 Breakers

Storage. Keep the breaker in a room protected from adverse conditions such as excessive humidity, dust, or corrosive chemical agents and sudden temperature changes.

Description. The SF_6 is an arc extinguishing and insulating medium. The chemical properties of the gas make SF_6 breakers suitable for indoor installation. SF_6 breakers have separate poles (one per phase). The active parts of the poles are housed in insulating cylinders (one per pole).

Each insulating cylinder is filled with SF_6 gas (electronegative gas), which is used because of its dielectric properties and its capacity to extinguish the arc quickly. The gas pressure is usually around 3.4 bar absolute referred to a temperature of 20°C.

During the breaker opening operation, the combined action of mechanical devices with that of the arc itself, which develops between the arcing contacts in each pole, creates a forced circulation of gas that blows the arc, causing its extinction.

The SF_6 gas is neither flammable nor explosive. Therefore, its use does not cause fire or explosion risks. The SF_6 circuit breakers have proved to be highly reliable under different circuit conditions. They are suitable for the most stringent duties (including high-humidity environment). The electrical life of the breaker depends on its usage and is in relation to the arcing contact wear. The SF_6 gas is not notably affected by current interruptions.

Pole Characteristics.
The pole is made of an insulating cylinder from which the upper and lower terminals come out. At the base of the cylinder, there is

1. The pole operating shaft
2. The gas filling valve
3. The pressure switches

The moving part is composed of

1. The arcing contact
2. The blowing nozzle
3. The moving tubular contact
4. The insulating tie-rod

The guiding contact provides the current passage between the moving tubular contact and the lower fixed main contact. The tubular contact is moved by an insulating tie-rod which is linked to the pole operating shaft with a lever. The pole operating shaft is provided with a lever externally which is connected to the output shaft of the operating mechanism via transmission tie-rods.

The gas is sealed on the shaft by a double gasket with a suitable liquid interposed. The pressure switches monitor the gas pressure in the pole. If the gas pressure drops below the preset value of the pressure switch, the built-in changeover contact will operate to annunciate an alarm.

CIRCUIT BREAKERS 19.13

Note:

1. The preset limit to provide an alarm is 2.5 bar absolute.
2. The breaker metal frame should be connected to ground.

Maintenance

Caution. Before carrying out any maintenance work, workers must ensure the following actions have been taken:

1. Open (off) the circuit breaker.
2. For draw-out or disconnectable circuit breaker, the work must be done only when they are withdrawn from the compartment.
3. With circuit breakers in fixed version or for maintenance on the enclosures or fixed parts, deenergize the following:
   a. The main voltage circuits connected to them.
   b. Auxiliary circuits.
   Also, connect all main voltage terminals (on both the supply and the load side) to ground in a clearly visible manner.

General. During normal service, only limited maintenance is required for the circuit breakers. The burden of service duty determines the frequency of inspections and maintenance. It is advisable to follow these rules:

1. Equip the circuit breaker with an operation counter.
2. For circuit breakers that will operate only a few times or that will remain closed (on) or open (off) for long periods, the breaker should be operated from time to time to prevent a tendency to clog which may cause reduction in the closing or opening speed.
3. During service, visually inspect the circuit breaker from the outside to detect any dust, dirt, or damage of any kind.

Maintenance Program. Table 19.1 illustrates the maintenance program for the SF_6 circuit breaker.

TABLE 19.1 Maintenance Program

	Interval	
Maintenance operations	Normal ambient installation	Dusty or polluted ambient installation
Carry out the general inspection (Table 19.2)	1 year	6 months
Carry out visual inspection from the outside and the inspection of the main voltage parts	1 year	6 months
Measure the insulation resistance (refer to instruction and maintenance manual of the breaker)	3 years	6 months
Lubricate the sliding points	1 year	6 months
Carry out the operating mechanism maintenance (see relevant section in maintenance manual)	5 years or every 5000 operations	3 years or every 5000 operations
Complete overhaul (see maintenance manual)	5 years or every 10,000 operations	3 years or every 10,000 operations

TABLE 19.2 Inspection

Part inspected	Negative inspection	Remedies
Operating mechanism (see relevant instruction manual)	• Presence of dust on the internal devices • Distorted or oxidized springs • Locking rings or clips out of place, loose nuts or screws • Detached wires and relative fasteners	• Clean with a dry brush or rag • Replace the damaged springs • Refit the locking rings or clips in their place and tighten the nuts or screws adequately • Replace the wire fasteners and reconnect the wires correctly
Main voltage part	• Presence of dust or dirt on the insulating parts • Locking rings or clips out of place, loose nuts or screws • Distortion or crackings of the insulating parts • Oxidized isolating contacts (only for draw-out circuit breaker) • Traces of overheating or of loose screws on the connections to the breaker terminals (only for fixed version circuit breaker)	• Clean with a dry brush or rag • Refit the locking rings or clips in their place and tighten the nuts or screws • Ask SACE for replacement of the damaged parts • Clean with a rough rag soaked in a suitable solvent and lubricate moderately with neutral grease • Clean the connections and breaker terminals with a rough rag soaked in a suitable solvent. Cover them with neutral grease and tighten the screws adequately
Grounding isolating contact (only for draw-out circuit breaker)	Presence of oxidation	Clean with a rough rag soaked in a suitable solvent and lubricate moderately with neutral grease
Grounding connection (only for fixed version circuit breaker)	Presence of oxidation and/or loose nut	Clean with a rough rag soaked in a suitable solvent. Tighten the grounding connection fully and cover it with neutral grease
Insulation resistance	See instruction manual	If the insulation resistance decreases with regard to the values measured during putting into service, search for the cause (dirt, humidity, or damage to the insulating parts) and restore the normal value
Auxiliary circuit feeding voltage	Check the feeding voltage of the operating mechanism electrical accessories	The releases (shunt opening – shunt closing) and the electrical locking devices must work correctly with a feeding voltage between 85 and 100% of the rated value
Operating and signaling devices	Carry out the operation tests	Replace the damaged or faulty devices (if necessary, ask SACE)
SF_6 gas pressure within each pole	See instruction manual	See instruction manual
Pole arcing contacts	See instruction manual	See instruction manual

General Inspection of the Circuit Breaker. Table 19.2 shows the general inspection required for the circuit breaker.

Lubrication. All sliding parts subject to friction must be lubricated at the required frequency with MU-EPIAGIP grease or equivalent.

Checking the Gas Pressure within the Poles. Refer to the appropriate section in the maintenance manual.

REFERENCE

1. R.W. Smeaton, *Switchgear and Control Handbook*, McGraw-Hill, New York, 1987.

CHAPTER 20
FUSES

A *fuse* is an overcurrent protective device with a circuit-opening fusible part that is heated and severed by the passage of current through it. Fuses can meet most of the protection requirements for good system operation.

The fuse is a thermal device. Heat will melt the fuse element regardless of its source. The fundamental features of fuses are as follows:

1. The fuse combines the sensing and interrupting elements in one unit.
2. It is a single-phase device. Only the fuse in the affected phase will melt to isolate the fault.

    Three-phase motors will continue to run on single-phase power for extended periods. This may result in overheating and damage to the motors.

3. The fuse response is a function of I^2T, where I is the current and T is the time the current exists. It has an inverse-time characteristic—the higher the current, the faster the fuse blows.
4. Most fuses require considerably more current than their amperage rating to operate. For example, NEMA standards require that E-rated fuses of 100E and below melt in 300 s at 200 to 240 percent of their rating. Fuses above 100E must melt in 600 s at 220 to 264 percent of their rating. These durations are considered extremely long for short-circuit protection.
5. Fuses should be coordinated with downstream devices to ensure faults are cleared within reasonable times (1 to 5 s or faster). A fault magnitude of 5 or more times the current rating of the fuse is required to clear the fault within this range of operating times.

    The application of fuses is a little difficult in some situations due to this current-magnitude requirement. Another protective scheme must be used in some critical applications, usually at higher cost.

TYPES OF FUSES

The most common types of fuses are single-element, dual-element, and current-limiting.

Single-Element Fuses

Single-element fuses have a high-speed response to overcurrents. They are usually used for the protection of nonmotor loads. Motor inrush currents are usually 6 times the normal full-load current.

Single-element fuses could cause nuisance opening during the starting period. Therefore, they are not normally suitable for motor controllers.

Note: The single-element fuses could cause nuisance openings if used in other inductive load applications such as transformers or solenoids.

Dual-Element Fuses

The dual-element fuse has two distinct series-connected sections. The first provides instantaneous operation for short circuits, and the second provides time-delayed operation for normal overloads. These fuses are ideally suited for motor controllers. The manufacturers of these fuses recommend their use for short-circuit and running overload protection.

However, this is not a common practice in industry. The dual-element fuses are not highly desirable for running overload protection because of the lengthy downtime required to obtain and install a new fuse. Most controllers are equipped with overload relays. These relays are manually or automatically resettable after an overload that caused the devices to open the circuit.

Current-Limiting Fuses

Current-limiting fuses are designed to open the circuit in less than $1/4$ cycle (based on 60 Hz). Figure 20.1 illustrates the performance of a current-limiting fuse. The current-limiting fuse will react as any other fuse to low and medium magnitudes of fault current (Fig. 20.2). The current-limiting action will occur at high magnitudes of fault current (Fig. 20.3).

A current-limiting fuse will not produce external arcing. The special quartz sand inside the fuse container is transformed to glass by the energy from the fault current. The glass creates an insulating material that results in circuit opening.

The circuit-limiting fuse has the highest capability to interrupt short circuits of any fuse available. The current-limiting fuse operates to drive the fault current of a voltage surge to zero. These fuses keep a voltage surge to a minimum to prevent equipment damage.

Figure 20.4 illustrates the differences in time-current characteristics of two different types of current-limiting fuses. The two fuses shown are both E-rated fuses at the 5.0-kV level. The fuse on the right is a "slower" fuse. This feature makes it more suitable for transformer primary protection than the "fast" fuse on the left.

EXAMPLE 20.1 *Consider a 500-kVA transformer connected delta-wye on a 4800-V system. The full-load current is a 60 A. The magnetizing inrush current is 10 to 12 times the full-load current. It lasts 0.1 s. At a current of 12 times full load, or 720 A, the "fast" fuse on the left in Fig. 20.4 may have blown on magnetizing inrush current. The fuse on the right had plenty of time to withstand the inrush current.*

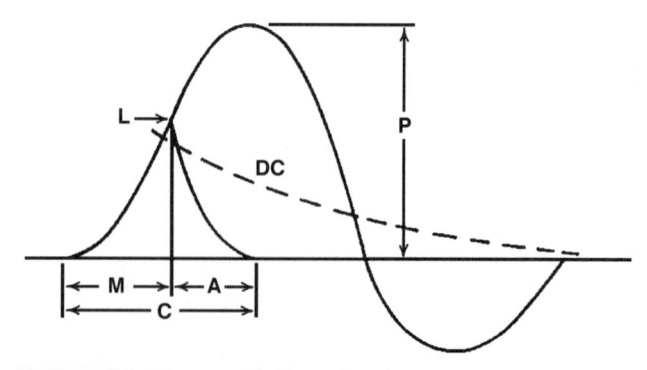

FIGURE 20.1 The current-limiting action of current-limiting-type fuses. *A*, arcing time; *M*, melting time; *C*, total clearing time; *P*, available peak current; *L*, let-through current; DC, dc component.

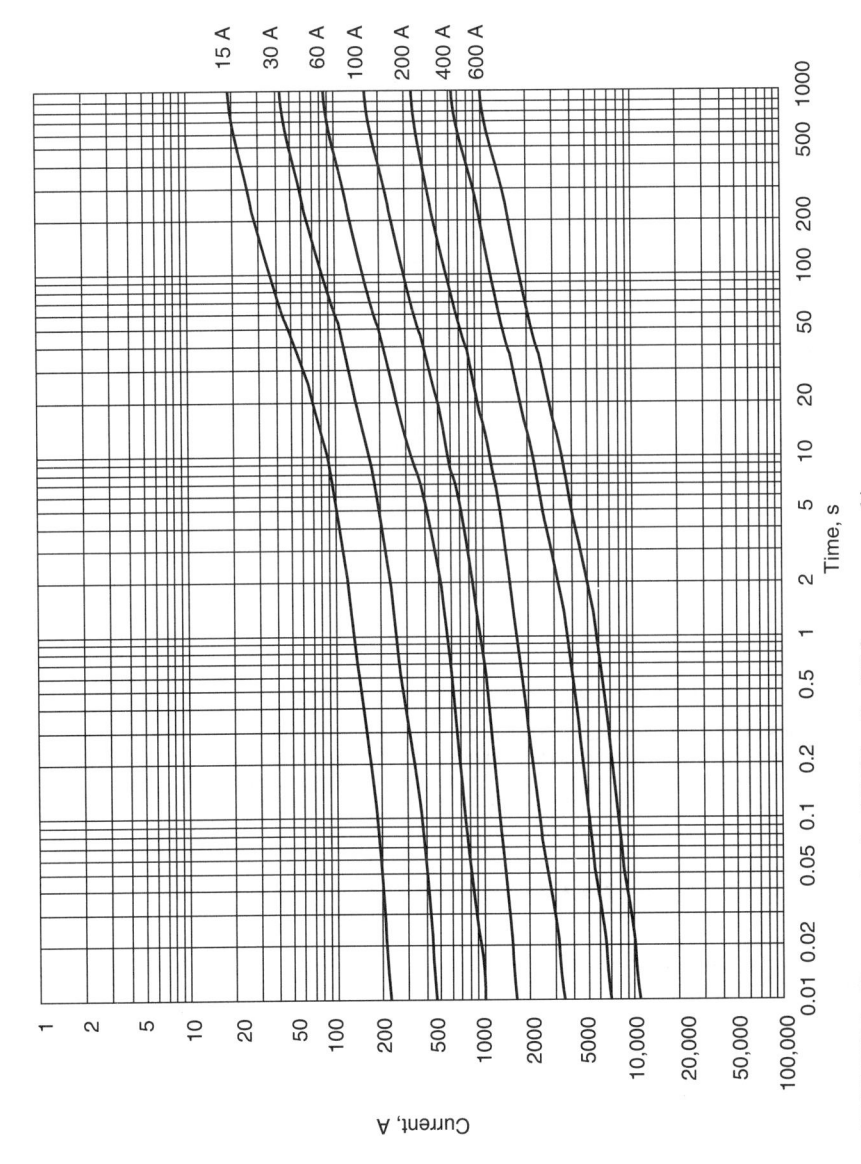

FIGURE 20.2 Time-current data for 600-V Duralim 6JD fuses, average melting.

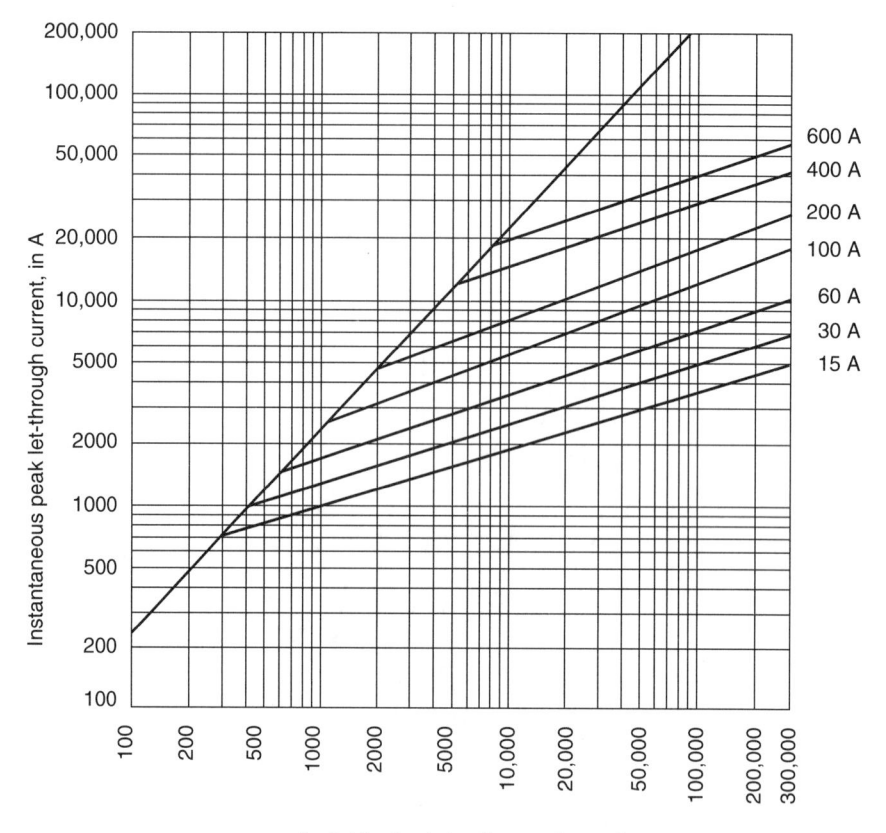

[Graph: Instantaneous peak let-through current (A) vs Available short-circuit current, rms A, for fuse ratings 15 A, 30 A, 60 A, 100 A, 200 A, 400 A, 600 A]

FIGURE 20.3 Current-limiting data for Duralim 6JD fuses at 600 V ac. (Tests are conducted at 600 V, one phase, with one fuse. In actual field applications, current limitation may be greater due to either the involvement of two or three fuses for line-line faults or reduced voltage for line-neutral or line-ground faults. Therefore, the above chart provides worst-case conditions with regard to current limitation.)

FEATURES OF CURRENT-LIMITING FUSES

Current-limiting fuses are fully rated to withstand moderate overloads without damage or a change in characteristic. Current-limiting fuses are generally used in motor starters and low-voltage circuit breakers to protect the motor contacter or circuit breaker from destruction when subjected to fault-current magnitudes in excess of their interrupting rating.

When current-limiting fuses are used, the NEC requires that "Fuseholders for current-limiting fuses shall not permit insertion of fuses that are not current-limiting" (Art. 240-60). Therefore, the fuse holder which will take class R fuses must be used. Class R fuses provide high degree of current limitation with interrupting current capability of up to 200,000 A.

This interrupting current rating distinguishes class R fuses, from class H fuses, which have an interrupting rating of only 10,000 A. Class K fuses (K-1, K-5, and K-9) have an interrupting rating as low as 50,000 A. Figures 20.5 and 20.6 illustrate the differences in construction of these fuses.

FUSES   20.5

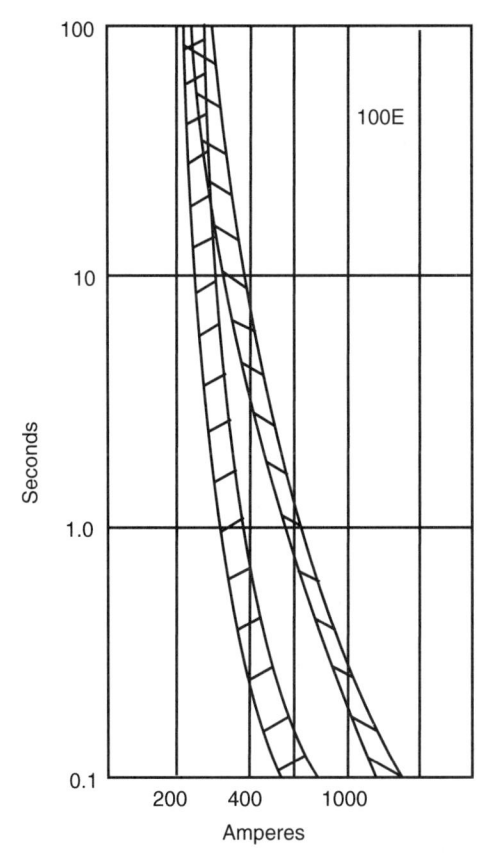

FIGURE 20.4 Plot of two different types of current-limiting fuses showing the differences in time-current characteristics.

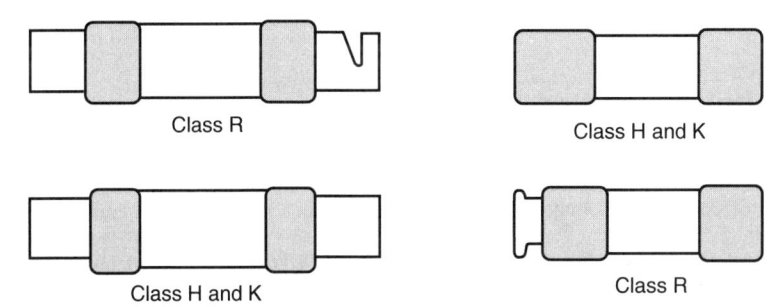

FIGURE 20.5 The two knife-blade fuses have the same physical dimensions. The class R fuse, however, has a rejection slot in one blade; the other fuse does not. Both fuses can be installed in a standard non-rejection-type fuseholder. Only class R fuses can be inserted in a rejection-type fuseholder.

FIGURE 20.6 The two ferrule-type fuses have approximately the same physical dimensions. The class R fuse, however, has a rejection groove on one ferrule; the other does not. Both fuses can be installed in a standard non-rejection-type fuseholder. Only the class R fuse can be installed in a rejection-type fuseholder.

Medium-voltage motor controllers are fault-protected by high-interrupting-capacity, current-limiting fuses. Backup R-rated fuses are especially designed for motor service with capability of carrying high starting currents during prolonged acceleration without fuse deterioration or nuisance blowing. These fuses do not have actual current ratings. However, they have application designations with typical melting time, total arc clearing time, and current-limiting characteristics.

ADVANTAGES OF FUSES OVER CIRCUIT BREAKERS

Fuses are preferred over circuit breakers for the following reasons:

1. Fuses do not have moving parts. They are maintenance-free and do not require periodic checking. They can be relied upon to protect a circuit for an indefinite time.
2. In general, fuses are considered more accurate and reliable than circuit breakers.
3 A blown fuse usually provides greater incentive to correct the cause of a failure than a tripped circuit breaker.

Note: A detailed comparison between current-limiting fuses and circuit breakers is presented in the appendix. This comparison was conducted by an independent firm (Noram). The comparison describes the numerous advantages that current-limiting fuses have over circuit breakers.

APPENDIX: ELECTRICAL SYSTEM PROTECTION CONSIDERATIONS

Why Noram Chose to Develop Duralim HRC Fuses (over CBs)

Noram embarked down the "protection road" by developing either circuit breakers (CBs) or current-limiting fuses (CLFs). After careful review of its users' and specifiers' list of desired core performance requirements for electrical protective devices, the Duralim HRC fuse was born.

Core Performance Requirements

General Purpose—One Device for All Applications. Both CBs and CLFs can be used as general-purpose devices if the proper types and ratings are chosen. Certain application decisions or compromises may have to be made in order to achieve a general-purpose capability.

For example, a CB's frame size may have to be increased to ensure that adequate interrupting rating (IR) is achieved for systems with high fault levels. A CLF with time-delay overload characteristics could be employed to permit the same fuse to be used for not only feeder protection, but also full backup motor protection.

Noram came to the conclusion that for this requirement (general purpose), a time-delay CLF with high interrupting capacity (IR 200 kA) provided an excellent means to protect main, feeder, and branch circuits with both noninrush and inrush circuit characteristics.

High Level of Protection—Overload and Short-Circuit. In the overload region, both the CB and the CLF can provide protection, assuming that the proper device and rating are provided for the specific application. In the short-circuit region, certain advantages arise with the *high-rupturing capacity* (HRC) fuse over that of the CB, due to the CLF's high level of current limitation. Although current-limiting CBs are available, they are generally larger than standard breakers, more expensive, and significantly less current-limiting than HRC class J fuses.

CHAPTER 21
BEARINGS AND LUBRICATION

The function of bearings is to keep the shaft or rotor properly aligned with the stationary parts under axial and radial loads. Bearings that provide radial positioning to rotors are known as *line* or *radial bearings*. Bearings that position the rotor axially are called *thrust bearings*. Thrust bearings usually serve as thrust and radial bearings.

TYPES OF BEARINGS

The main types of bearings are journal and rolling-contact (ball and roller) bearings. The ball and roller bearings are also known as antifriction bearings. The most common bearings used on centrifugal pumps are the various types of ball bearings. Roller bearings are used less often. However, spherical roller bearings are used frequently for large shaft sizes, for which there is a limited choice of ball bearings.

In horizontal pumps with bearings on each end, *inboard* bearings are located between the casing and the coupling. The bearing on the outside is the *outboard*.

The bearings are usually mounted in a housing supported by brackets which are attached or integral to the pump casing. The lubricant is usually contained within the housing. The bearings must be kept within proper temperature limits due to the heat generated within the bearings. For bearings that are provided with a forced-feed lubrication system, cooling is achieved by circulating the oil through a separate water-to-oil cooler. Otherwise, cooling liquid enters a jacket located inside the housing to remove the heat.

There are rigid and self-aligning bearings. A self-aligning bearing will adjust itself automatically to a change in the shaft angular position.

Ball and Roller Bearings

Since the coefficient of rolling friction is less than that of sliding friction, ball bearings offer a significant advantage over sleeve bearings. In ball bearings, the load is carried on the point of contact between the ball and the race. During normal operation (constant speed), the point of contact does not rub or slide over the race. The heat generated is minimal. Also, the point of contact is constantly changing because the ball rolls in the race. The bearing operation is almost *frictionless*.

In sleeve bearings, the surfaces are always rubbing against each other. Lubrication is used to reduce the friction.

The starting friction is only *slightly* larger than the operating friction in rolling-contact bearings. This is a significant advantage of rolling-contact (ball and roller) bearings. It makes them suitable for machines that are started and stopped frequently under load.

Rolling-contact bearings require little maintenance and lubrication. However, they are noisier than journal bearings and more expensive. The lifetime of rolling-contact bearings is limited due to fatigue failure of the raceways caused by the repeated high stresses as the shaft rotates. However, most failures are caused by lubrication failure or improper mounting.

Several types of rolling-contact bearings are capable of handling radial load and axial thrust. Although they are called rolling-contact bearings, the balls and rollers do considerable sliding during normal operation due to changes in speed.

Stresses during Rolling Contact

The stresses during rolling contact are high due to the small area of contact. The sliding motion (frictional force) causes tensile stress near the surface, which results in pitting. Since the contact stresses are usually greater than the yield strength, *residual* compressive stresses are induced in the races and the balls. These stresses cause deformations which increase with repetitions.

Since the area of contact between the balls and the races is small, the stresses in a particular point in a bearing race are very high. Fatigue failure is expected due to the repeated nature of the stresses. Figure 21.1 illustrates a single-row, deep-groove ball bearing.

Rolling bearings have a limited life because any normal loading creates stresses higher than the fatigue strength involved. The bearing life is limited by the number of stress repetitions to cause fatigue failure. For example, doubling the speed of rotation reduces the bearing life by one-half. The centrifugal forces within the bearings become significant as the speed increases. A bearing has a shorter life at very high rotating speed.

The load has a significant impact on the bearing life. The bearing life B was found experimentally to vary inversely as the power k of the load F; B is proportional to $1/F^k$, where k is 3 for ball bearings and is 10/3 for roller bearings. In equation form,

$$\left(\frac{F_1}{F_2}\right)^k = \frac{F_1^k}{F_2^k} = \frac{B_2}{B_1} \quad \text{or} \quad \frac{F_1}{F_2} = \left(\frac{B_2}{B_1}\right)^{1/k}$$

Note that for ball bearings ($k = 3$), the life increases 8 times when the load decreases by one-half. Corrosion from acid or water reduces the life of bearings.

The bearing life is usually measured in millions of revolutions (MR). However, some catalogs give the bearing life by the number of hours at various rotating speeds.

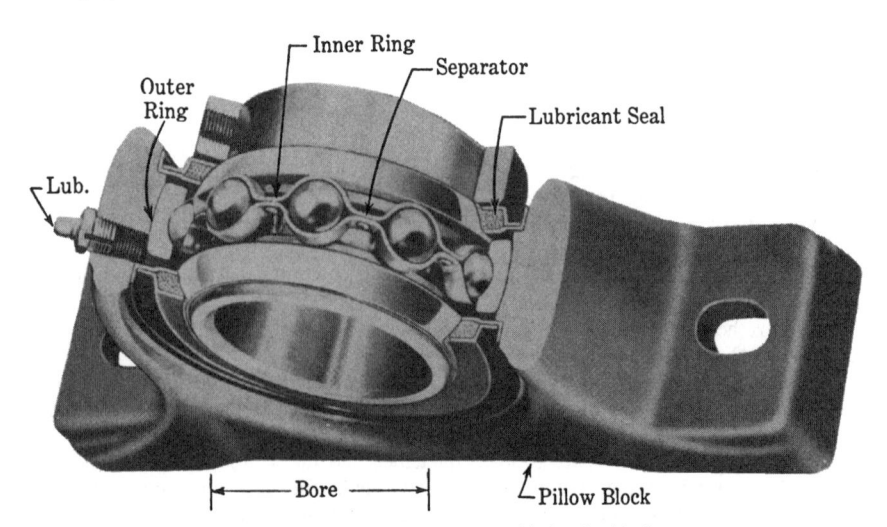

FIGURE 21.1 Single-row, deep-groove ball bearing mounted in bearing block.

The signs of fatigue are spalling or flaking of small particles of material from the surface of a ring or a ball (roller) caused by the high shearing stress below the contact surfaces. This mechanism of failure is preceded by noisy operation. Continued operation will eventually lead to fatigue failure.

STATISTICAL NATURE OF BEARING LIFE

The life of rolling bearings is rated on the basis that 90 percent of a large number of bearings in particular surroundings will survive a specific life. This life is usually designated as B_{10}, which is interpreted as a failure rate of 10 percent expected within B. However, other catalogs give the rated load for 50 percent survival B_{50} (median life). A commonly accepted relationship is

$$\text{Median life} = (5)(90\% \text{ life}) \qquad B_{50} = 5B_{10}$$

The design life of bearings varies considerably, depending on the machine and service. Table 21.1 illustrates the variations in design life depending on the machine and service.

MATERIALS AND FINISH

The most common material used in bearings is SAE 52100. It is an alloy steel of nominally 1 percent C and 1.5 percent chromium, hardened to Rockwell C58-65. Alloys such as nickel and molybdenum are used with chromium.

The bearing life depends heavily on the hardness. The average life of a bearing when the hardness is $R_c = 50$ is about one-half of that when $R_c = 60$.

The operating temperature of SAE 52100 and similar steels is usually held to 300°F. However, the maximum temperature for the installation is 200°F.

SIZES OF BEARINGS

Table 21.2 shows several series of bearings and their key dimensions that have been standardized. The bearing is classified as heavier by increasing the outside diameter as illustrated in Fig. 21.2.

TABLE 21.1 Design Life for Rolling Bearings

Type of service	Hours (90% life)
Infrequent use—instruments, demonstration apparatus, sliding doors	500
Aircraft engines	500–2,000
Intermittent use, service interruptions of minor importance—hand tools, hand-driven machines, farm machinery, cranes, household machines	4,000–8,000
Intermittent use, dependable operation important—work moving devices in assembly lines, elevators, cranes, and less frequently used machine tools	8,000–12,000
8-h service, not fully utilized—gear drives, electric motors	12,000–20,000
8-h service, fully utilized—machines in general, cranes, blowers, shop shafts	20,000–30,000
24-h service, continuous operation—separators, compressors, pumps, conveyor rollers, mine hoists, electric motors	40,000–60,000
24-h service, dependable operation important—machines in continuous-process plants, such as paper, cellulose; power stations, pumping stations, continuous-service machines aboard ships	100,000–200,000

TABLE 21.2 Dimensions of Rolling Bearings

This table does not give all standard dimensions. The maximum fillet radius r is the maximum radius at the shoulder on the shaft which is cleared by the corner radius on the bearing. Conversion factors: 0.03937 in/mm; 25.4 mm/in

Brg. No.	Bore mm	Bore in	Outside diameter, mm 200 Series	Outside diameter, mm 300 Series	Outside diameter, mm 400 Series	Width of races, mm 200 Series	Width of races, mm 300 Series	Width of races, mm 400 Series	Max. fillet r 200 Series	Max. fillet r 300 Series	Max. fillet r 400 Series
00	10	0.3937	30	35		9	11		0.024	0.024	
01	12	0.4724	32	37		10	12		0.024	0.039	
02	15	0.5906	35	42		11	13		0.024	0.039	
03	17	0.6693	40	47		12	14		0.024	0.039	
04	20	0.7874	47	52		14	15		0.039	0.039	
05	25	0.9843	52	62	80	15	17	21	0.039	0.039	0.059
06	30	1.1811	62	72	90	16	19	23	0.039	0.039	0.059
07	35	1.3780	72	80	100	17	21	25	0.039	0.059	0.059
08	40	1.5748	80	90	110	18	23	27	0.039	0.059	0.079
09	45	1.7717	85	100	120	19	25	29	0.039	0.059	0.079
10	50	1.9685	90	110	130	20	27	31	0.039	0.079	0.079
11	55	2.1654	100	120	140	21	29	33	0.059	0.079	0.079
12	60	2.3622	110	130	150	22	31	35	0.059	0.079	0.079
13	65	2.5591	120	140	160	23	33	37	0.059	0.079	0.079
14	70	2.7559	125	150	180	24	35	42	0.059	0.079	0.098
15	75	2.9528	130	160	190	25	37	45	0.059	0.079	0.098
16	80	3.1496	140	170		26	39		0.079	0.079	
17	85	3.3465	150	180		28	41		0.079	0.098	
18	90	3.5433	160	190		30	43		0.079	0.098	
19	95	3.7402	170	200		32	45		0.079	0.098	
20	100	3.9370	180	215		34	47		0.079	0.098	
21	105	4.1339	190	225		36	49		0.079	0.098	
22	110	4.3307	200	240		38	50		0.079	0.098	

BEARINGS AND LUBRICATION 21.5

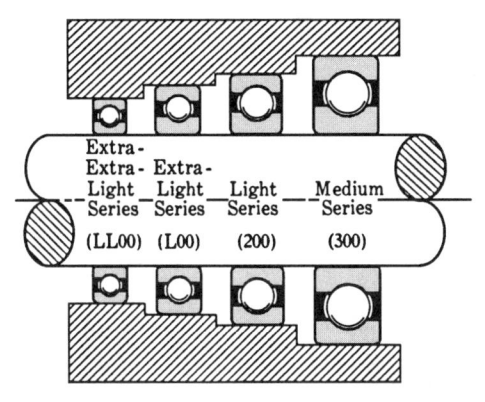

FIGURE 21.2 Various series of ball bearings. All these bearings have the same basic number as 09.

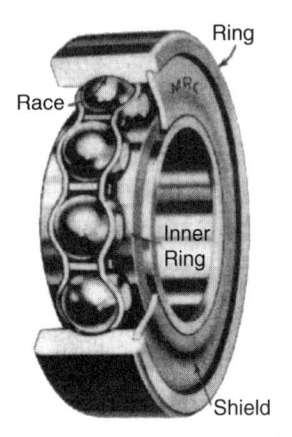

FIGURE 21.3 Single-row ball bearing with shield. The balls in this bearing were loaded with the aid of a filling slot. Notice their spacing. The shield aids in keeping out foreign matter, important in rolling bearings. (*Courtesy Marlin-Rockwell Corp., Jamestown, N.Y.*)

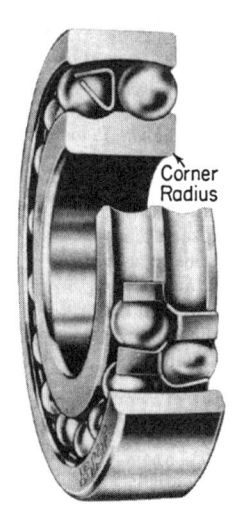

FIGURE 21.4 Self-aligning ball bearing. Self-aligning roller bearings are similar. (*Courtesy SKF Industries, Inc., Philadelphia.*)

TYPES OF ROLLER BEARINGS

Figure 21.1 has illustrated the *deep-groove ball bearing*. Figure 21.3 illustrates a filling-slot type of ball bearing. The slots or notches permit the assembly of more balls, giving the bearing larger radial load capacity. Both bearings are used for radial loads only.

Figure 21.4 illustrates a *self-aligning ball bearing*. It compensates for angular misalignments that arise from the deflection of the shaft or foundation or from errors in mounting. They are recommended for radial loads and moderate thrust in either direction.

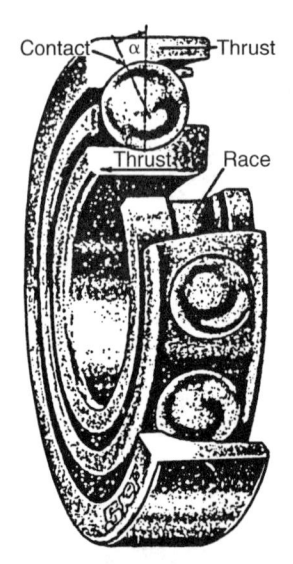

FIGURE 21.5 Angular contact bearing. Observe that the action of the thrust is such as to move the surface of contact away from the centerline plane of the balls. Compare the shape of the grooves in this figure with those of Figs. 21.1 and 21.3. (*Courtesy SKF Industries, Inc., Philadelphia.*)

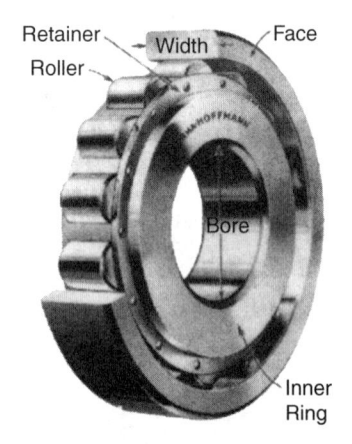

FIGURE 21.6 Cylindrical roller bearing. The rollers run in a groove in the inner ring. (*Courtesy Norma-Hoffman Bearings Corp., Stamford, Conn.*)

Figure 21.5 illustrates an *angular-contact bearing*. There is an angle between the line through the areas carrying the load and the plane of the bearing face. These bearings are used to carry heavy thrust loads. They are usually used in opposed pairs. They are also suited for *preloading*. Preloading a bearing involves placing it under an initial axial load which is independent of the working load to maintain a constant alignment of parts and reducing the axial movement as well as radial deflection under working loads.

Double-row ball bearings (not self-aligning) are similar to the single-row ball bearings. The only difference is that each ring has two grooves. The two rows of balls give a bearing capacity slightly less than twice that of a single-row type.

Figure 21.6 illustrates a cylindrical roller bearing. The contact is a line instead of a point, as is the case in ball bearings. This results in a greater area to carry the load. Hence, for a particular size, the cylindrical roller bearing has a larger radial capacity than a ball bearing. The main function of the retainer is to keep the roller axes parallel.

A popular type of cylindrical roller bearing is the *needle bearing* shown in Fig. 21.7. This type does not have a retainer to hold the rollers in alignment. Their main advantage is the small diametral dimension. They are used where the speed is relatively low and where there is oscillating motion.

Figure 21.8 illustrates a *tapered roller bearing*. The rolling elements are arranged as frustums of a cone. All their axes intersect at a point on the axis of the shaft. These bearings are capable of carrying significant axial loads.

FIGURE 21.7 Needle bearing. (*Courtesy of The Torrington Co., Torrington, Conn.*)

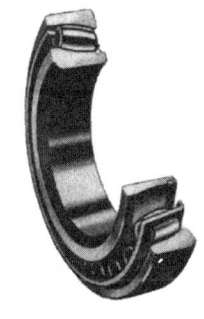

FIGURE 21.8 Tapered roller bearing. (*Courtesy Timken Roller Bearing Co., Canton, Ohio.*)

FIGURE 21.9 Spherangular roller bearing. (*Courtesy Hyatt Roller Bearing Co., Harrison, N.J.*)

Figure 21.9 illustrates a *"spherangular" roller bearing*. This is an angular contact bearing with rollers instead of balls. The bearing is self-aligning because the outer race has a spherical surface. It has also a high load capacity due to its large contact area.

THRUST BEARINGS

The rolling type of thrust bearings uses the following as rolling elements: balls, short cylindrical rollers, tapered rollers, and spherical rollers that run in spherical races (self-aligning). These bearings can be rigidly supported, or one of the races can be supported in a spherical seat to make it self-aligning. Figures 21.10 and 21.11 illustrate thrust bearings that use balls and tapered rollers as rolling elements, respectively.

LUBRICATION

The Viscosity of Lubricants

The *viscosity* is defined as the resistance of a liquid to flow or deformation. In general, the slower a fluid flow under a given pressure drop, the higher its viscosity. The ability of a lubricant to maintain an oil film (which reduces friction and wear) between the working parts of machines is determined by viscosity.

Figure 21.12 illustrates two parallel plates separated by a film of oil. The top plate is moving; the bottom plate is stationary. A force is applied to the top plate to keep it moving at a steady speed.

$$\text{Shear stress } S_s = \frac{\text{force}}{\text{area}}$$

FIGURE 21.10 Thrust ball bearing. (*Courtesy Aetna Ball & Roller Bearing Co., Chicago.*)

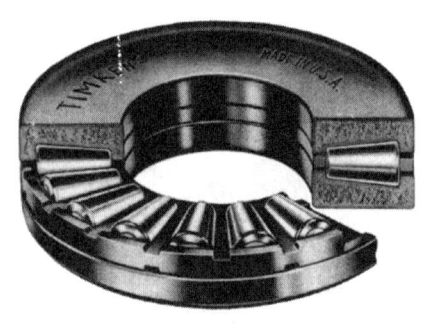

FIGURE 21.11 Tapered roller thrust bearing. (*Courtesy Timken Roller Bearing Co., Canton, Ohio.*)

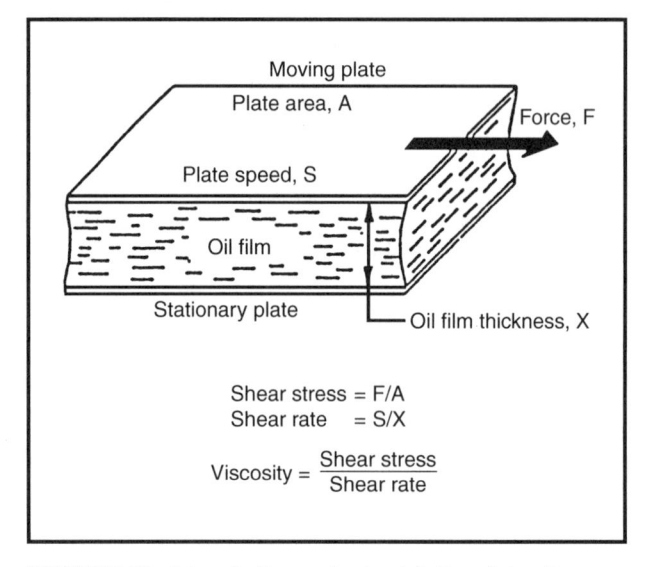

FIGURE 21.12 Schematic diagram showing definition of viscosity.

$$\text{Shear rate } R = \frac{\text{speed}}{\text{thickness}}$$

$$\text{Viscosity } V = \frac{S_S}{R} \quad \text{(Newton's equation)}$$

Water and most industrial oils follow this equation (newtonian fluids). The shear stress increases in direct proportion to the shear rate. The viscosity is independent of the shear stress or shear rate at which it is measured.

Nonnewtonian fluids exhibit a nonlinear relationship between the shear stress and the shear rate. Greases and oils below their solidification temperature are nonnewtonian lubricants.

The equation of viscosity can be rearranged to show that the force is proportional to the viscosity and the speed is inversely proportional to the viscosity.

Viscosity Units. The physical quantity in the equation above is the dynamic viscosity (also called *absolute* viscosity). However, the kinematic viscosity is traditionally used when referring to lubricants. The two measures of viscosity are related by density, as shown:

$$\text{Dynamic viscosity} = \text{kinematic viscosity} \times \text{density}$$

The friction in bearings or pressure loss in horizontal pipe flow is not affected by the density. Therefore, the *dynamic viscosity* is the controlling measure.

However, density has a significant effect on gravitational flow, such as flow from a reservoir through a hose. The driving force is proportional to the density in this situation, and the *kinematic viscosity* is the controlling measure.

The lubrication industry has traditionally used the kinematic viscosity. This viscosity is obtained by measuring the flow rate from a reservoir through a small capillary. The unit of kinematic viscosity in the International System (SI) is millimeters per second (mm/s). However, the lubrication industry has traditionally used the term *centistokes* (1 cSt = 1 mm/s).

The unit of dynamic viscosity in SI is the pascal-second (Pa·s). However, the lubrication industry has traditionally used the term *centipoises* (1 cP = 1 mPa·s). This equation is used to convert from kinematic viscosity to dynamic viscosity:

$$\text{Viscosity (cP)} = \text{viscosity (cSt)} \times \text{density (g/mL)}$$

The density is given at the temperature of the viscosity measurements.

The Saybolt universal second (SUS) is an older measure of the viscosity which is still used occasionally in the United States. Figure 21.13 illustrates the relationship between the kinematic viscosity and the SUS.

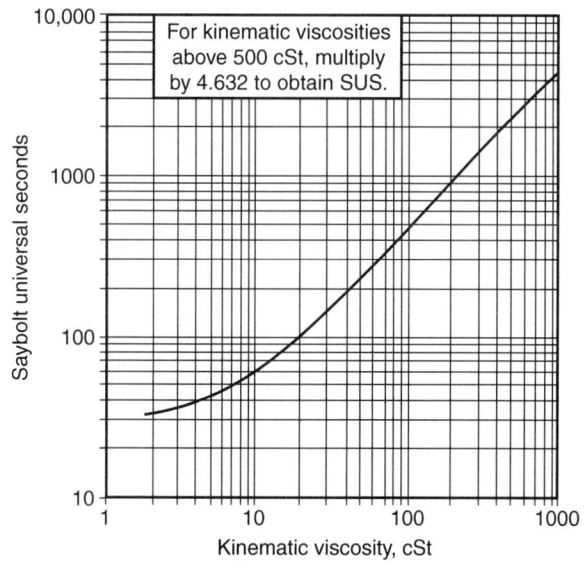

FIGURE 21.13 Relationship between kinematic viscosity (cSt) and Saybolt universal seconds (SUS).

A reference point for viscosity is water at room temperature which has a viscosity of 1 cSt approximately. The viscosity of a typical engine oil at 280°F is 5 to 6 cSt. The viscosity of the same oil at $-25°C$ is between 3000 and 3500 cP (temperature has a significant effect on viscosity).

Significance of Viscosity. Viscosity is the most important property of lubricating oils. In bearings, oil prevents wear between mating surfaces by keeping them separated, minimizes friction between rubbing surfaces, and removes heat generated by surface contact. Viscosity is important in all these functions.

The role of viscosity is illustrated in plain journal bearings. When the viscosity is high, the bearing load is completely separated by the *oil film*. The journal does not contact the bearing, and the friction is low. When the viscosity is low, the metal-to-metal contact and the friction are high. In this case, the oil viscosity has a minor effect. The antiwear properties of the oil play a major role. The antiwear properties of the oil play a minor role when the viscosity is high enough to support the bearing load completely.

The friction between the moving parts is reduced by lubricants even if the oil film does not carry the load. The effect of the lubricant viscosity is to "cushion" the asperity-to-asperity (microscopic hills) contact between moving surfaces to reduce the local stresses. The contact between the parts in roller bearings and gear teeth is along a line (one gear tooth against another). A very thin film is trapped between the surfaces due to the high contact pressure. The pressure in these contact zones increases to several hundred thousand pounds per square inch, which results in significant increase in viscosity.

The viscosity of the oil becomes high enough to keep the moving surfaces separated. This mechanism of lubrication is called *elastohydraulic lubrication.* The oil viscosity at these high pressures has a major role in determining the film thickness in such contacts.

The oil viscosity should be carefully selected to ensure proper lubrication. If the viscosity is low, high wear rates and frequent failures will result. If the viscosity is too high for the application, a significant amount of heat would be generated. This may lead to failures due to high temperature. Most manufacturers specify the viscosity required for their machines.

Flow-Through Pipes. In laminar flows through pipes, viscosity plays an important role. In this case, the pressure loss is proportional to the fluid viscosity. The flow rate is inversely proportional to the viscosity. When the flow through pipes becomes turbulent, the fluid friction increases by a factor of 6. The viscosity has a minor effect on pressure loss while the pipe roughness has a major effect.

Variation of Viscosity with Temperature and Pressure

Temperature Effect. The viscosity of all fluids is inversely proportional to temperature. The oil viscosity is reported at two standard temperatures, 40 and 100°C. MacCoull's equation describes the variation of viscosity with the temperature of simple lubricants:

$$\ln \ln (V_k + 0.7) = A + B \ln T$$

where V_k = kinematic viscosity
 T = absolute temperature
 A, B = constants for a particular oil

Note: MacCoull's equation is not accurate if the oil has additives or if the viscosity is lower than 1 cSt.

Viscosity Index. The *viscosity index* (VI) characterizes the change of viscosity with temperature. A high VI indicates a small change in viscosity with temperature. A low VI indicates a large change.

Effects of Pressure on Viscosity. The viscosity of all fluids increases with pressure.

Nonnewtonian Fluids

In newtonian fluids, the shear stress (force/area) is proportional to the applied shear rate (speed/thickness of oil film). Thus, the viscosity, which is the ratio of these properties, is constant.

Nonnewtonian fluids exhibit a nonlinear relationship between the shear stress and the shear rate. The viscosity of some nonnewtonian fluids changes with time after a flow starts. Nonnewtonian fluids are characterized by a graph of shear stress versus shear rate. The three classes of nonnewtonian fluids in the lubrication industry are greases, VI improved oils, and oils at low temperature (near the pour points).

Greases. The viscosity of greases is time-dependent. The viscosity is relatively high when the flow starts. Then it decreases as the flow progresses. The yield stress of a grease is the minimum stress required to cause a flow.

VI Improved Oils

VI improvers are often included in oils to expand the useful range of operating temperature to permit low viscosities for low-temperature start-up, while still providing adequate viscosities at operating temperatures.

Organic polymers are the active ingredients in VI improvers. They are extremely large molecules containing many thousands of carbon atoms. The concentration of polymer in VI improved oils is about 1 percent. However, this is sufficient to cause major effects on viscosity.

The viscosity of VI improved oil at 210 and 0°F is about 1.8 and 1.2 times the viscosity of the base oil, respectively. The viscosity of polymer-thickened oils (VI improved oils) decreases while the oil is in service. This loss is called the *permanent viscosity loss* (PVL). The *shear stability* is the resistance of the oil to this loss. In engine service, the oil viscosity decreases 7 to 14 percent of its 100°C kinematic viscosity after 500 to 1500 mi of service.

Oils at Low Temperatures

The viscosity of the oil increases rapidly as the temperature decreases. If the lubricant is used below its pour point, the machinery may experience severe problems. The working parts of the machinery may be covered fast enough by oil. This leads to high wear and could result in total failure.

The pour point is usually lowered by adding certain additives called *pour point depressants.*

Variation of Lubricant Viscosity with Use

Oxidation Reactions. The viscosity of the oil increases when it is placed in service. This phenomenon is significant when the oil is subjected to excessive heat. Some of the oil mol-

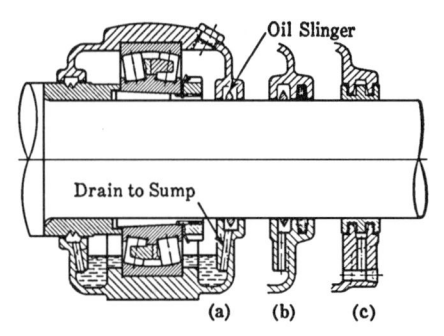

FIGURE 21.14 Typical housing, with various seals. Sometimes an oil slinger only is sufficient, as at (*a*). At (*b*), a slinger and a seal; (*c*) labyrinth seal. (*Courtesy SKF Industries, Inc., Philadelphia.*)

ecules react with oxygen to form oxygenated molecules. This increases the oil viscosity and darkens its color. In service, heat can result in a serious problem. Some lubricants contain antioxidant additives to control oxidative thickening.

Physical Reactions. The mineral oil is composed of a broad range of molecular types and sizes. The lower-viscosity molecules tend to evaporate in a hot environment while the heavier, higher-viscosity molecules remain. This results in an overall increase in viscosity.

Housing and Lubrication

Cleanliness has a major effect on the life of a rolling bearing. For this reason, these bearings must be protected from airborne particles and expected sources of dirt. Figure 21.14 illustrates typical bearing housing with various seals.

Interference fits for the rings must be tight to prevent relative motion. The bearings are usually mounted against a shoulder which makes it possible to move them off without transmitting the force through the rolling elements.

Grease is usually used at low and medium speeds because it provides a better seal against dirt. It is suitable for temperatures from -70 to $210°F$. It is preferable not to pack tightly the bearing and the housing with grease. Two-thirds full is usually enough.

Oil is used at higher speeds. The oil level should be kept below the center of the lowest rolling element, as shown in Fig. 21.14. Otherwise, excessive churning of the oil at high speed will cause overheating.

For a particular method of lubrication, the coefficient of friction is almost constant. Hence, the temperature rise under constant load is almost proportional to the speed.

Some bearings are cooled by circulating and cooling the oil due to excessive generation of heat.

Some prelubricated bearings having integral seals to run for the life of the bearing (or machine) without any required maintenance. Figure 21.15 illustrates a typical

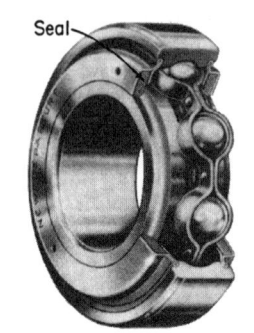

FIGURE 21.15 Ball bearing with seals. This type is relubricated by means of a hypodermiclike needle through the small holes on the side. (*Courtesy New Departure, Bristol, Conn.*)

ball bearing with seals. The seals are usually found on one or both sides.

Sealed prelubricated bearings require attention if they are not operated for a long time. The shaft should be rotated occasionally (once every 3 months) to move the lubricant and maintain a film coating of the balls.

The self-aligning ball bearings have been operated satisfactorily at high speeds. They also have a long life and very minimal thrust capacity.

The self-aligning spherical roller bearings are used for larger shafts and for applications having a considerable thrust component.

The single-row deep-groove ball bearing is used most commonly on centrifugal pumps having a relatively small diameter. It is capable of handling radial as well as axial loads. This type of bearing sometimes has seals to exclude dirt and retain lubricant.

The angular-contact bearing is good for heavy thrust loads. The single-row type is good for thrust in one direction only (commonly used on vertical pumps). The double-row type is capable of handling thrust in both directions (Fig. 21.16).

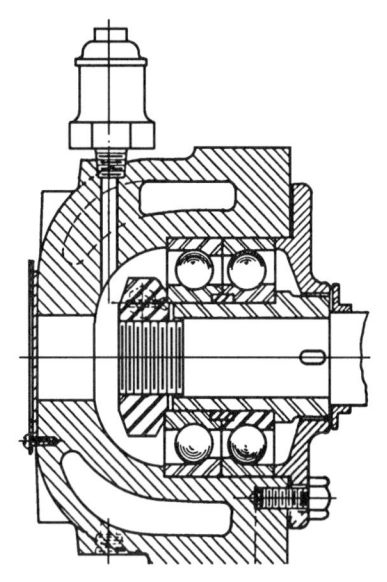

FIGURE 21.16 Double-row angular-contact ball thrust bearing that is grease-lubricated and water-cooled.

Lubrication of Antifriction Bearings. In general, the choice of lubricant is dictated by the application and the cost. For example, water is the preferred lubricant for vertical wet-pit condenser circulating pumps (Fig. 21.17). If oil or grease were used and the lubricant leaked, the condenser operation might be seriously affected because the lubricant will coat the tubes.

Most ball bearings used for centrifugal pumps are grease-lubricated bearings.

Figure 21.17 illustrates the seals required within the housing of some bearings to prevent the lubricant from escaping.

The proper grease circulation within the bearing and the housing is prevented if the bearing is overfilled with grease. Only one-third of the void spaces in the housing should be filled. Any additional amount of grease will cause the bearing to overheat, and the grease will flow out of the seal to relieve the pressure. If the grease cannot escape through the seal or through the relief cock, the bearing will fail early.

The oil is supplied to the bearing from the bearing housing reservoir by oil rings, as shown in Fig. 21.18. A constant-level oiler is used sometimes, as shown in Fig. 21.19.

It is advantageous to have the capability to interchange the lubricant. Some pump lines are built with bearing housing that is capable of using oil or grease lubrication with minimal modifications (Fig. 21.20).

REFERENCES

1. I. J. Karassik, W. C. Krutzsch, W. H. Fraser, and J. P. Messina, *Pump Handbook*, 2d ed., McGraw-Hill, New York, 1986.

2. V. M. Faires, *Design of Machine Elements*, 4th ed., Collier-MacMillan, Toronto, 1965.

3. D. L. Alexander, "The Viscosity of Lubricants," Texaco magazine *Lubrication,* vol. 78, no. 3, 1992.

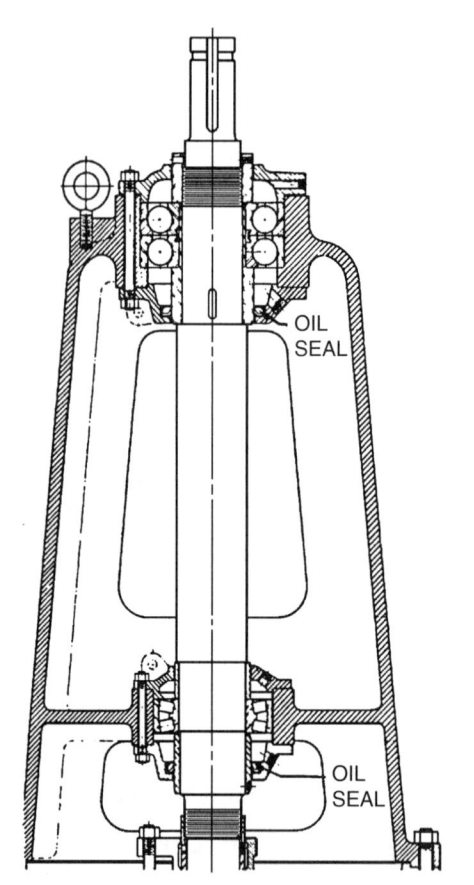

FIGURE 21.17 Ball bearing with seal in vertical pump. The seal guards against escape of grease.

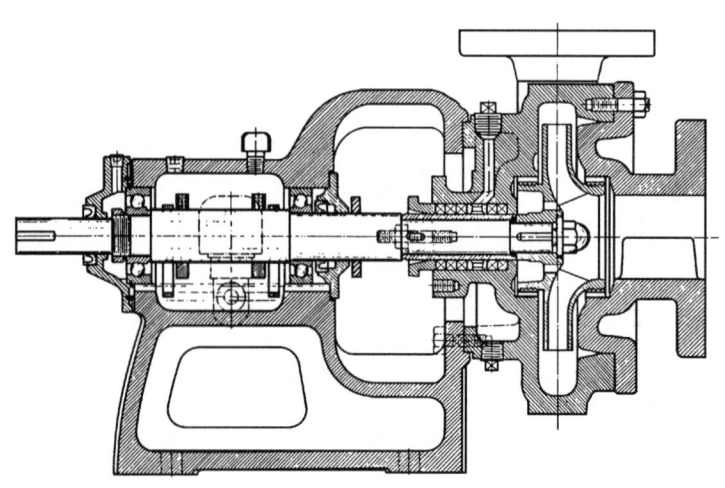

FIGURE 21.18 Ball bearing pump with oil rings.

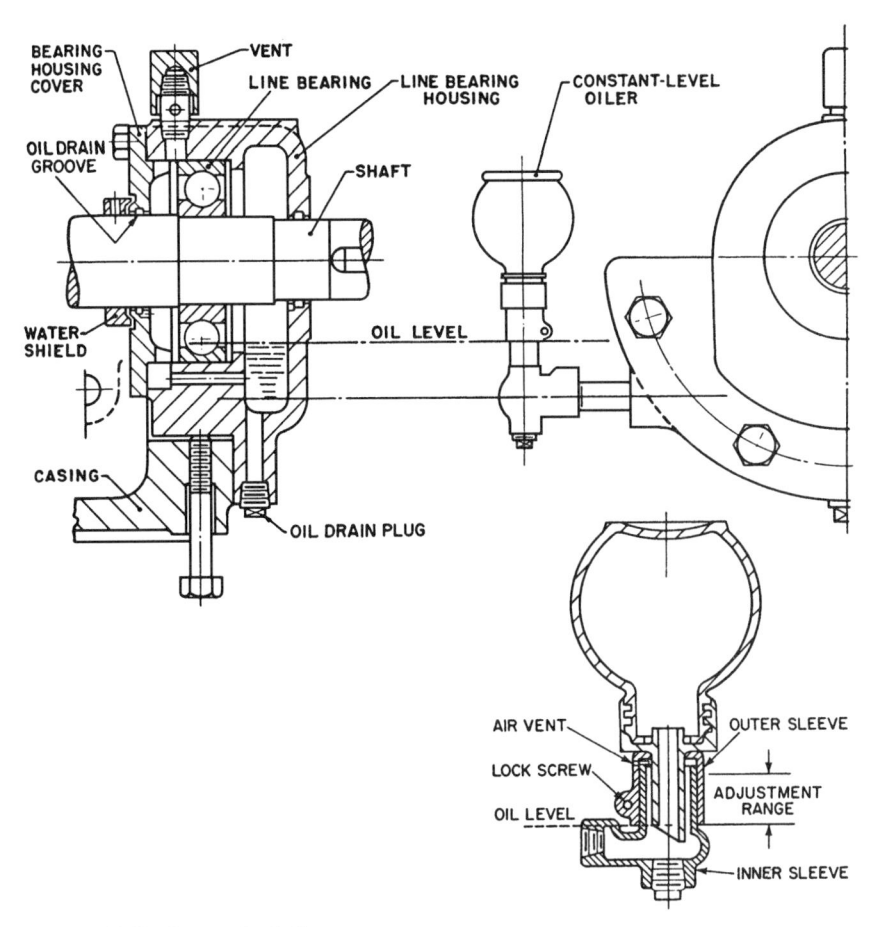

FIGURE 21.19 Constant-level oiler.

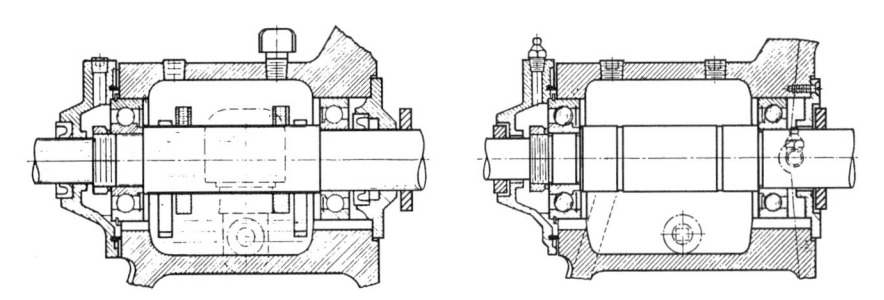

FIGURE 21.20 Ball bearings arranged (left) with oil rings in the housing and (right) for grease lubrication.

CHAPTER 22
USED-OIL ANALYSIS

PROPER LUBE OIL SAMPLING TECHNIQUE

All sample containers must be clean and dry. A used-oil analysis program is started by sending a sample of the new, unused oil to the analysis laboratory. This sample will be used as a reference standard. The analysis is based on deviation of properties relative to the new, unused oil. The used-oil sample should be collected when the oil is warm and well mixed. If a machine is down, it should be started to circulate the oil through the system.

The sample line should be purged to remove stagnant oil and debris (the sample must not be contaminated). The sample should be taken from the line before the filter. The sample should be representative of the system. If the sample is taken from a sump or a gearbox, it should be taken from the middle every time. Hand pumps can be used to withdraw a sample from a sump or crankcase into a bottle. The sample must properly identify the oil and equipment type, collection date, time since last oil change, and oil makeup rate. The sample should be taken to the laboratory as soon as possible.

TEST DESCRIPTION AND SIGNIFICANCE

Table 22.1 lists the test schedule for used industrial oils.

Visual and Sensory Inspections

An experienced observer can determine if the oil has contamination or deterioration. Engine oils in short service or those exhibiting little degradation will have a bland or an additive odor similar to that of unused oil. Oils with longer service periods under favorable operating conditions have a normal "used" odor.

In the examination of compressor, turbine, and hydraulic systems, appearance and odor are particularly useful. These oils are bright and clear and have bland odor under favorable operating conditions. The oil becomes "hazy" in the presence of a small amount of suspended water. The oil will have a cloudy appearance when it has a larger amount of water (Fig. 22.1). The reference line behind the oil to the left is sharply defined. This confirms that the oil is dry and clear. The oil is *hazy* in the middle bottle containing 200 parts per million (ppm). The reference line behind the oil to the right can hardly be seen because the oil is so cloudy (containing 400 ppm of water). Larger quantities of water coalesce and form free water droplets which will settle at the bottom of the container. When contaminants increase in oil, its water separation properties gradually deteriorate. When there is contamination or oxidation in the oil, its color becomes significantly darker. The oil has characteristic odors under these conditions. Severe oxidation is characterized by a sharp or burnt odor. Some chemical contaminants have specific odors typical of fuel, chlorinated solvents,

TABLE 22.1 Basic Test Schedule for Used Industrial Oils

Property	ASTM method	Compressor oil	Gear oil	Turbine oil	Hydraulic oil
Appearance	—	X	X	X	X
Odor	—	*	X	X	X
Water					
Crackle	—	—	X	—	X
Karl Fischer	D 1744	†	X	†	†
Viscosity					
@40°C	D 445	X	—	X	X
>100°C	D 445	—	X	—	—
Toluene insolubles	D 893	—	X	—	—
Emission spectrometry	—	X	X	X	X
Infrared spectrometry	—	X	—	X	X
Total acid number	D 664	‡	‡	‡	‡
Particle count	F 661	§	—	§	§

*Use caution when examining oils from ammonia or other noxious gas systems.
†Determined if sample is hazy or if water content is critical.
‡Determined when sensory or infrared methods indicate need.
§Determined if cleanliness is a major criterion or to meet equipment manuacturers' recommendations.

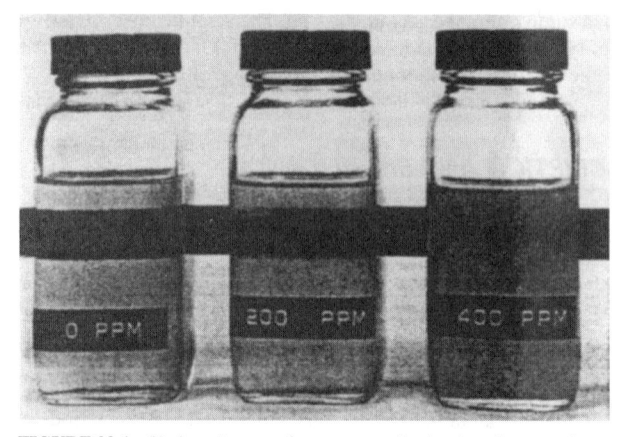

FIGURE 22.1 Various degrees of water contamination in a clear petroleum product.

or sour gas (i.e., a gas containing sulfur compounds). The oil analyst confirms these conditions by requesting additional tests.

Chemical and Physical Tests

Water Content. The presence of water in oil is confirmed by a crackle test. A few drops of oil are placed in a small aluminum dish and heated rapidly over a small flame or on a laboratory hot plate (Fig. 22.2). The test can be conducted by using a hot electric soldering iron which is immersed in oil. If the water concentration is 0.01 percent or less, an audible crackling sound will be heard. When a crackle test confirms the presence of water, a quantitative test is conducted by distillation (ASTM D95). The test involves dissolving a specific quantity of oil in a water-immiscible hydrocarbon solvent, such as xylene, and heated in a distillation flask. The water present in the oil co-distills with the solvent also. A condenser is used to cool the vapors. The water settles in the bottom of the trap where its volume can be measured, while the solvent flows back to the distillation flask.

Figure 22.3 illustrates an automated instrument used for the Karl Fisher method (ASTM D 1744). The advantage of this method is the ability to determine the amounts of free and dissolved water content in oil. However, it is normally used for relatively clean oils. This method is especially useful for refrigeration and hydraulic system lubricants which cannot have water content due to the potential damage of internal components. This is essential in refrigeration applications where free water droplets can freeze and plug the lubricant lines. The presence of water confirms lubricant contamination through leaking seals, coolant seepage, improper storage, or incorrect application of the lubricant. The water can cause rust, sludge, and deterioration in the lubricant properties. The source of water should be identified and eliminated as soon as possible.

Viscosity.   Viscosity is a measurement of the lubricant's resistance to flow at a specific temperature. It is the most important property of the lubricant. The flow characteristics of the lubricant must be adequate to ensure that all the parts of the system are properly lubricated. The viscosity varies with the type of the lubricant, degree of oxidation, and contamination in the lubricant. As the viscosity of the oil increases, it becomes harder to pipe it or take it off lubricated surfaces.

The film-forming properties of the lubricant depend heavily on viscosity. However, if the lubricant's viscosity is very high, it cannot flow fast enough at low temperatures to perform its function. Unnecessary heat would be generated in this case, resulting in damage to the equipment. The lubricant should flow easily to perform its other functions, such as carrying foreign material to system filters for removal, cooling, and transmitting force in hydraulic applications. The viscosity of the lubricant will increase with oxidation, foreign material contamination, degradation of additives, or evaporation of light base oil components. An increase of 10 percent in the viscosity of the lubricant would be a cause of concern for an operating machine. A decrease in viscosity is expected with fuel dilution and thermal or shear degradation of high-viscosity lubricant components such as viscosity index improvers. Sudden changes in viscosity provide indication of equipment malfunction or a change in operating conditions.

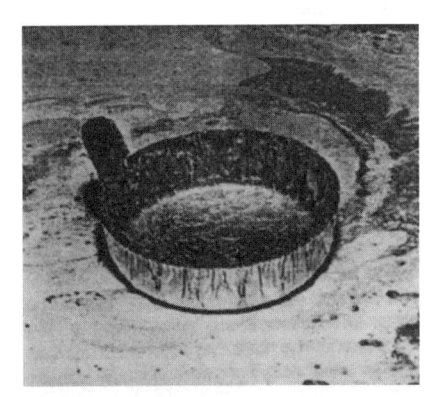

FIGURE 22.2   Crackle test for water using a small aluminum foil cup on a hot plate. Sputtering indicates water contamination.

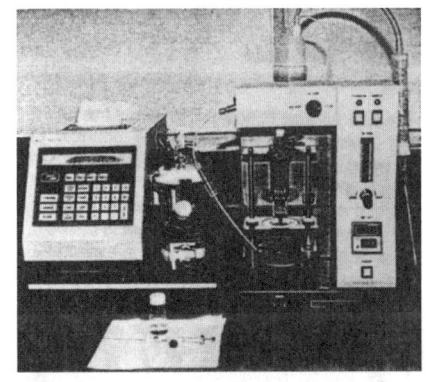

FIGURE 22.3   An automated apparatus for determination of water in petroleum products using the Karl Fischer method (ASTM D 1744).

Emission Spectrographic Analysis. Emission spectrometers determine the presence of inorganic contaminants and metalloorganic oil additive elements. A film of oil is carried in this instrument on a rotating graphite wheel to a narrow gap between the wheel and a sharpened graphite rod electrode. The film is subjected to a high voltage arc. The energy of the arc excites the metallic elements in the oil. Each element emits a characteristic spectrum of light. The individual wear metal elements are characterized by selected lines of this spectrum. A diffraction grating is used to separate or isolate the selected lines, permitting their light intensities to be measured by photomultiplier tubes. A computer is used to convert the intensities of the light to element concentration. The results are printed for examination by the analyst.

Other types of spectrometers are the *inductively coupled plasma* (ICP) and *direct-coupled plasma* (DCP) spectrometers, shown in Figs. 22.4 and 22.5, respectively. The advantage of these instruments is the ability to utilize automatic samplers for the introduction of the sample into the instrument.

Emission spectrographic analysis is a powerful tool for determining the levels of wear metal (trace amounts). The significance of wear metals in used oils depends on the make and model of the equipment, type of service (including working environment, drain frequency, filter change interval, etc.). The concentration of metals increases with time. A sudden increase in concentration of metallic elements such as copper, lead, or iron indicates an increase in wear rate and possibly abnormal operating conditions. Immediate attention is required (maintenance or resampling is needed).

Trends should be developed over a period of time to determine normal operating conditions for wear metals. A judgment should not be made on a single sample analysis. The emission spectrograph provides the levels of metalloorganic additive in the used oil. This provides information about the depletion of additive, oil makeup, and suitability of the oil being used.

This is the significance of common metals found:

Copper indicates wear in rocker arm bushings, wrist pin bushings, thrust washers, and other bronze or brass components.

Iron indicates wear from rings, liners, crankshaft, gears, pistons, cylinder walls, or valve train. It can appear as rust after storage.

Lead indicates wear from babbit or copper/lead bearings.

Tin indicates bearing wear when babbit overlays are used.

Silicon indicates the presence of sand, dust, or dirt entering the system. Major abrasion can occur, causing wear in ring, liner, bearing, and crankshaft.

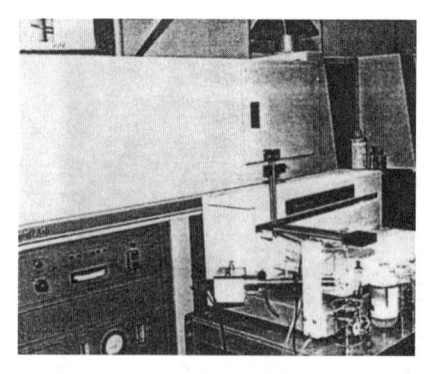

FIGURE 22.4 An inductively coupled plasma spectrometer.

FIGURE 22.5 A direct-coupled plasma spectrometer.

Aluminum indicates bearing or piston wear. Attention is needed even for a slight increase because rapid wear generates larger particles which may not be detected by emission spectrography.

Particles larger than 10 to 15 μm cannot be detected by a typical spectroscopy analysis. Particles smaller than 40 μm cannot be discerned by the naked eye. This is the reason for having emission spectrographic analysis done in conjunction with other analysis techniques such as ferrography (wear particle analysis) or vibration analysis.

Infrared Analysis. Infrared (IR) spectrometry is another powerful method for detecting low levels of organic contaminants, water, and oil degradation products. It provides a simple and rapid technique for establishing the following:

1. The general lubricant type (paraffinic or naphthenic)
2. The presence and often the quantity of certain contaminants such as alcohols, polar solvents, and free water (but not normally dissolved moisture)
3. The depletion of additives such as antioxidants
4. The presence of lubricant degradation products resulting from oxidation and nitration

Normally, a double-beam IR spectrometer is used in the differential mode (DIR). The used oil is placed in the sampling cell of the instrument, and an unused oil is placed in the reference cell. The instrument traces a curve representing the difference between the sample and the reference. It defines clearly the spectral regions characteristic of the organic contaminant or the oil degradation products. A newer technique for DIR spectrometry uses a Fourier transform infrared (FT/IR) analyzer. It involves placing the sample in a single beam cell. The spectrum is compared to the one of the reference oil stored in the instrument computer (Fig. 22.6). The method is particularly useful in analyzing oils from natural gas–fueled engines which have high combustion temperature. These engines promote the fixation of nitrogen by combining nitrogen and oxygen from air to form NO or NO_2

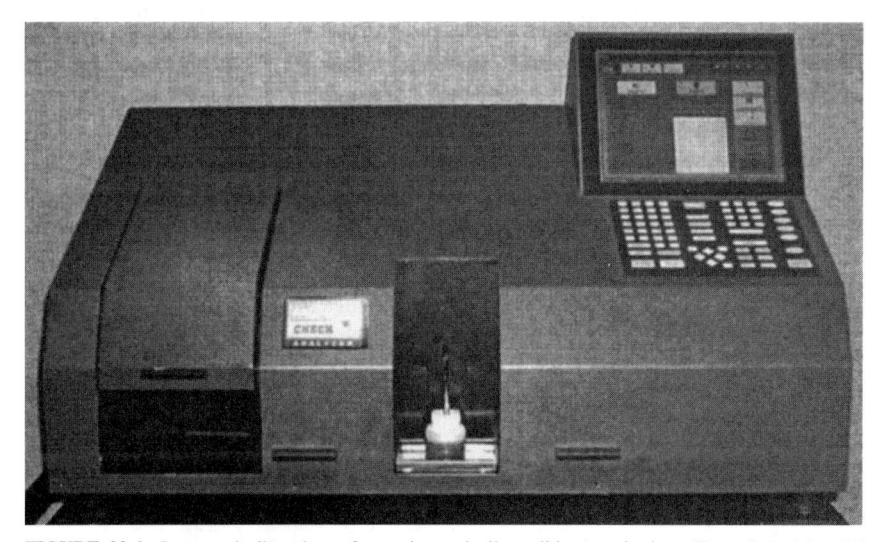

FIGURE 22.6 Integrated oil analyzer for engine and oil condition monitoring. (*From Practicing Oil Analysis 2001, Conference Proceedings, Noria Corp., Tulsa, Okla.*)

(termed NO_x), especially in naturally aspirated four-cycle gas engines when fuel/air ratios are lean.

The FT/IR technique calculates the *nitration number,* which measures the nitrogen compounds in the oil resulting from fixation of nitrogen. A sudden increase in the nitration number relative to the *oxidation number* results in the formation of acidic materials. This increases the rate of oxidation and thickening of the oil, which results in the formation of varnish and sludge deposits. A buildup in the concentration of nitration is an indication that the engine needs tuning.

The *oxidation number* is also obtained from the FT/IR analyzer. It is a measure of the amount of oxidation of the oil. This is an important parameter that determines the service life of a lubricating oil. The rate of oxidation increases significantly with high-temperature operation and contamination with water and glycol antifreeze. An increase in the level of oxidation (as measured by the oxidation number) requires attention to prevent a significant increase in viscosity, organic acid formation, filter plugging sticky rings, or piston deposits and lacquering.

FT/IR spectroscopy also indicates contamination due to the presence of water, glycol antifreeze, fuel soot, or gasoline and diesel fuel, depending on the type of lubricant. When any of these contaminants are detected, the analyst will perform other tests to quantify the amount of contamination present.

Total Base Number. The *total base number* (TBN) indicates the total alkalinity of a lubricant. It determines the ability of the oil to neutralize harmful acidic by-products. The TBN is measured by perchloric acidic titration (ASTM Method D2896) or by potentiometric titration with hydrochloric acid (ASTM Method D4739). The latter method replaced another potentiometric titration with hydrochloric acid (ASTM D 664 for TBN) because it is more accurate and is in agreement with ASTM D2896.

The main use of the TBN is for monitoring the remaining alkalinity of overbased detergent additives present in crankcase engine oils.

Total Acid Number. The *total acid number* (TAN) is measured by ASTM Method D664. It is used for industrial lubricants such as turbine, hydraulic, and refrigeration oils to monitor acid buildup in service. The TAN is used as a guide to track the oxidative degradation of an oil. When the TAN reaches a predetermined level, the oil should be replaced. A sudden increase in TAN indicates abnormal operating conditions that should be investigated.

Particle Count. Many systems such as hydraulic and turbine applications circulate lubricants through narrow passageways that must be kept clean and open to allow the lubricant to flow. These systems require fine filtration of the oil in service. The particle count analysis should be included in the used-oil monitoring program to confirm the proper cleanliness of the lubricant for the service. The level of oil cleanliness for an application depends on its precision of operation, required reliability, and the system clearances. Figure 22.7 shows a modern particle counter. It uses a photometric cell to determine the classification of particles in an oil sample. The cleanliness of the lubricating oil is defined by particle size ranges of 5 to 10, 10 to 25, 25 to 50, 50 to 100, and over 100 μm. The particle count is measured in the size ranges of concern for the system being monitored. Proper sampling technique should be followed to avoid false results. The samples should also be mixed well prior to analysis.

The international standard method for classifying the size distribution of solid particles in the lubricant is the ISO Solid Contaminant Code. It typically reports particle size ranges of >5 and >15 μm. Particles larger than 5 μm are generally caused by foreign contamination, while particles larger than 15 μm are wear particles in the system.

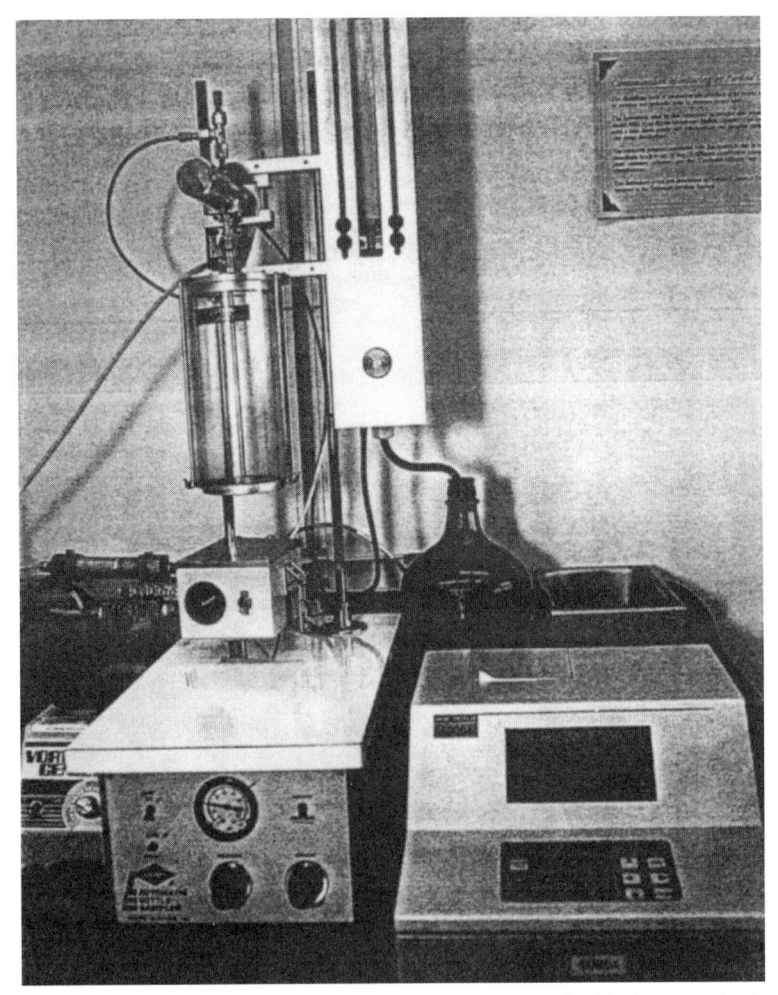

FIGURE 22.7 The modern particle counter uses a computer to classify cleanliness levels (ASTM F 661).

SUMMARY

The used-oil analysis program describes the condition of the oil in service. It is used to determine the suitability of the lubricant for continuing service and optimum intervals for preventive maintenance activities. Decisions regarding maintenance requirements and oil change intervals should be made based on trends, not on the results of one sample. The first step in the program is to determine which equipment should be monitored and which tests will give the most useful information for the application. If improper sampling or testing procedures are used, the results can be misleading. The used-oil analysis program is a useful technique if implemented properly. It is a significant contribution to the overall success of the maintenance program.

REFERENCE

1. E. Marshall, "Used Oil Analysis—A Vital Part of Maintenance," Texaco technical publication *Lubrication*, vol. 79, no. 2, 1993.

CHAPTER 23
VIBRATION ANALYSIS

Vibration is defined as small oscillations about some equilibrium point. The main characteristics of vibration are amplitude and frequency.

Figure 23.1 illustrates the movement of a piston driven by a crankshaft moving up and down in a cylinder. The position of the centerline of the piston above the centerline of the cylinder (equilibrium position) is assumed positive. When the piston is at top dead center, the piston centerline will be at the maximum positive distance above the cylinder centerline.

The plot of the piston versus time is given by a sine curve. The period of the sine wave is 1 s (the frequency is 1 Hz). The maximum peak (or amplitude) of the motion is D.

The energy produced by the motion is related to the area under the curve (Fig. 23.1c). The rms value of the amplitude is equal to 0.707 times the peak value of the amplitude. It represents the steady value of the amplitude that gives the same energy as the sine curve.

THE APPLICATION OF SINE WAVES TO VIBRATION

Any vibration can be described by a sine wave. The crankshaft/piston arrangement shown in Fig. 23.1 can be described by the spring mass system shown in Fig. 23.2.

Figure 23.3 illustrates the variation of velocity with time. The velocity curve is actually a cosine wave. It leads the displacement curve by 90°.

The maximum acceleration occurs at top dead center. The largest force exerted on the connecting rod is the force required to change the direction of the piston. Figure 23.4 illustrates the variation of acceleration with time. The acceleration curve is 180° out of phase with the displacement curve and 90° out of phase with the velocity curve.

The amplitude of displacement is proportional to the amplitude of acceleration (with a minus sign). Since the accelerometer sensors (measuring acceleration) are inexpensive and easy to use, the amplitude of displacement is often in units of acceleration, known as "G's" of acceleration. It is convenient to express vibration acceleration in terms of gravitational acceleration g.

$$\text{Acceleration } (g) = \frac{\text{acceleration (in/s}^2)}{386}$$

$$= \frac{\text{acceleration (mm/s}^2)}{9815}$$

MULTIMASS SYSTEMS

Any real machine can be assumed to be made of many springs and masses which are vibrating at different frequencies. The total vibration level of the machine is the sum of the motions of all the parts of the machine (Fig. 23.5).

23.2 CHAPTER TWENTY-THREE

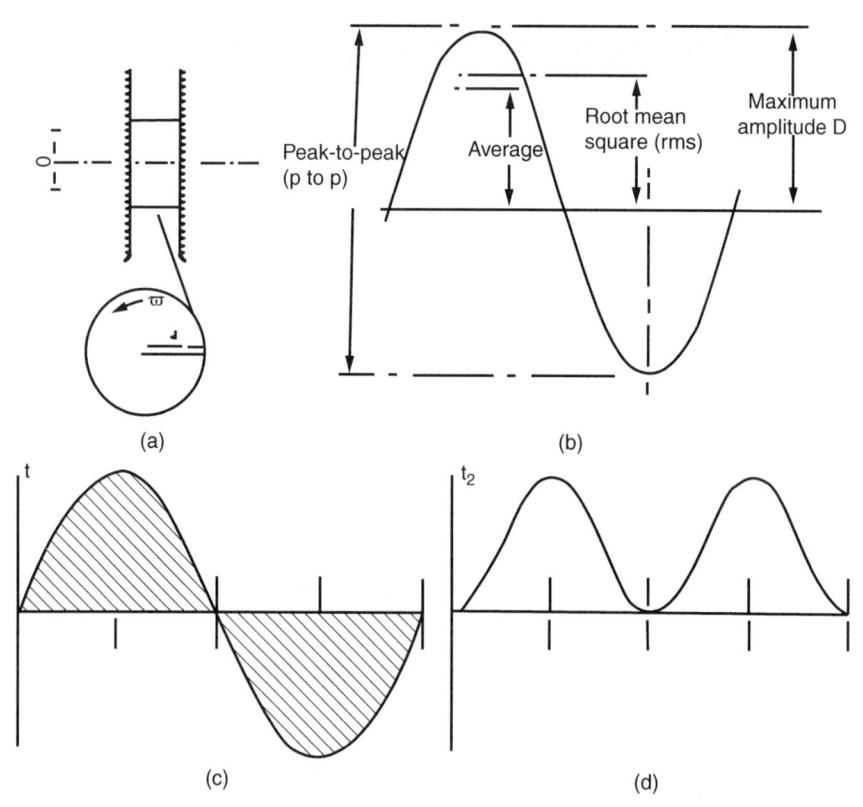

FIGURE 23.1 Vibration of a piston driven by a crankshaft.

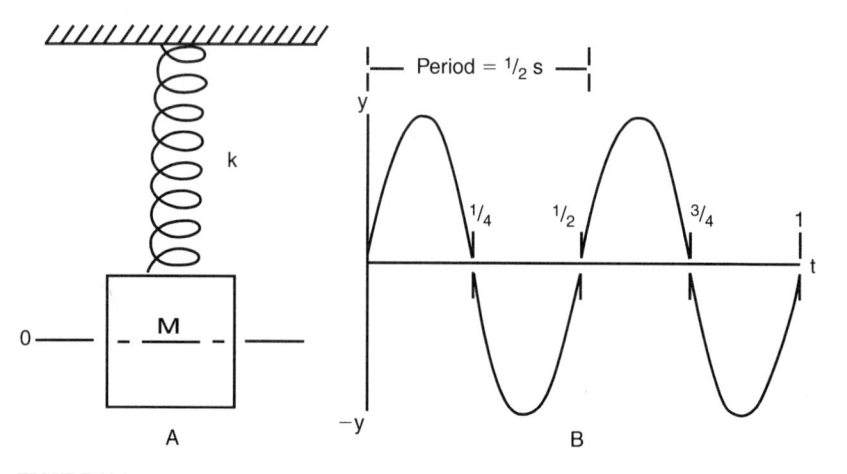

FIGURE 23.2 Vibration of a spring mass sytem.

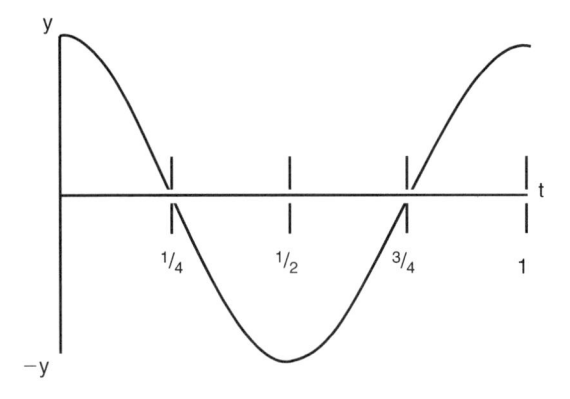

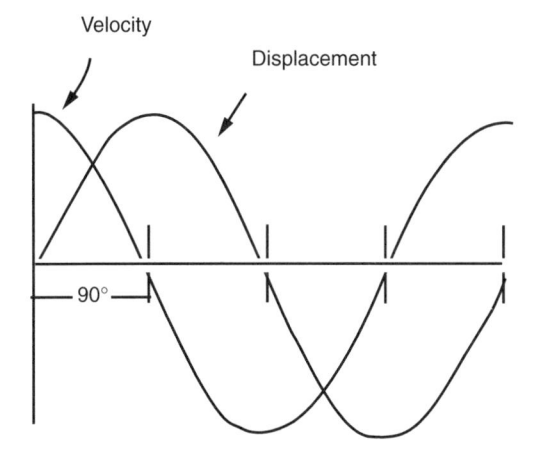

FIGURE 23.3 Variation of the vibration velocity with time.

RESONANCE

Every system has one or more *natural frequencies* where it "likes to vibrate." If a system is excited by a small force at a natural frequency, it will exhibit very large magnitude of vibration. This wild vibration at the system natural frequency is known as *resonance*. A system is more likely to fail when it is in resonance due to the high internal forces imposed on the springs by the masses. The values of the natural frequencies of a system can be calculated if its masses, spring constants, and damping constants are known.

By applying a set of known forcing frequencies to the system, the natural frequency can be measured by determining the frequencies at which the system exhibits large magnitudes of vibration (a two-channel real-time spectrum analyzer is usually used to obtain these measurements).

The natural frequencies of the *machine-support-piping* system should be determined when one is analyzing the vibration of a machine. This is necessary to ensure that the forcing frequencies are not near the natural frequencies.

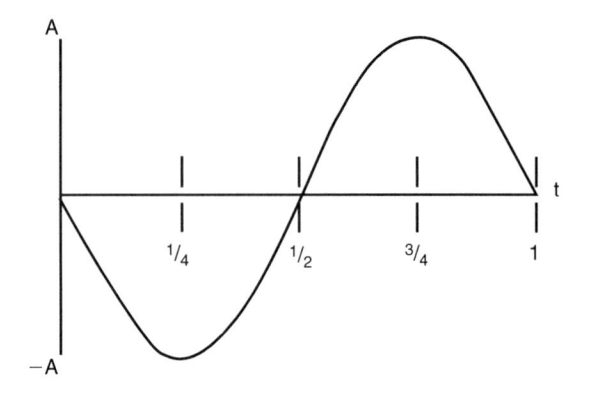

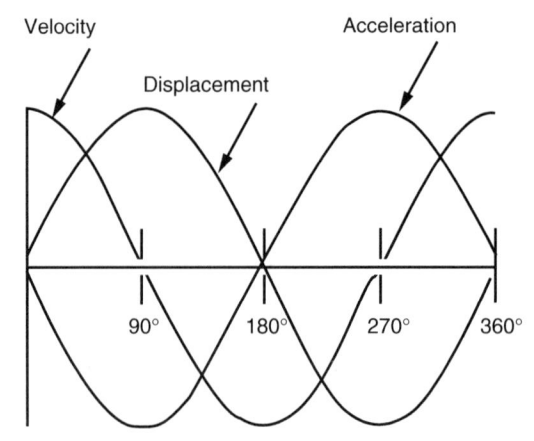

FIGURE 23.4 Variation of the vibration acceleration with time.

LOGARITHMS AND DECIBELS

A logarithmic scale is often used for vibration velocity and acceleration values due to the very wide range of vibration levels encountered with industrial machinery. Since the measurements of both sound and vibration vary from very low to very high levels, the decibel (dB) is often used. The definition of decibel is

$$\text{Decibel, dB} = 20 \log \frac{\text{amplitude}}{\text{reference amplitude}}$$

The standard reference for vibration is 10^{-3} cm/s^2.

For sound, the standard reference is a pressure of 0.0002 µbar. By definition, doubling the amplitude is equal to an increase of 6 dB.

Two machines each generate 100 dB when operating alone; if they operate in phase, they generate a maximum total noise of 106 dB (not 200 dB).

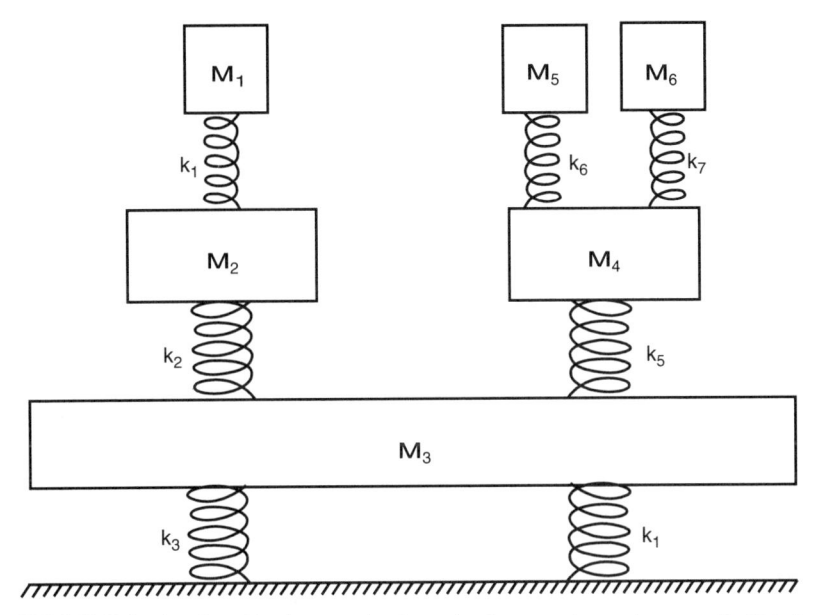

FIGURE 23.5 A real machine is assumed to be made of many springs and masses vibrating at different frequencies.

THE USE OF FILTERING

Filters are used to observe only the frequencies which are of immediate interest.

VIBRATION INSTRUMENTATION

The most common transducers in vibration monitoring are

- Displacement transducer
- Velocity transducer
- Acceleration transducer
- Key phasor

Displacement Transducer (Proximity Probe)

The displacement transducer is commonly known as the *proximity probe*. An electromagnet is placed in the tip of the proximity probe. It generates an electric eddy current proportional to the gap between the probe and the metallic object to be measured.

The proximity probe (unlike velocity and acceleration transducers) must be mounted on a stationary support relative to the observed object. The proximity probes are useful for monitoring shaft orbits because they do not have to contact the moving surfaces.

Velocity Transducer

The velocity transducer is a movable permanent-magnet core with an enclosed coil. The relative motion between the magnet and the coil creates a voltage change which is proportional to velocity.

Figure 23.6 illustrates two velocity transducers. The displacement can be obtained by integrating the output of the transducer.

Acceleration Transducer

Most accelerometers use the characteristic of piezoelectric crystals to measure acceleration. By applying a force to two faces of the crystal, a voltage is generated by the crystal that is proportional to the force. The acceleration levels can be integrated to obtain velocity and integrated again to obtain displacement.

Accelerometers are used at higher frequencies more than velocity transducers or proximity probes.

Transducer Selection

The amount of energy dissipated by a vibrating machine is proportional to the displacement, velocity, and acceleration. They all must be considered. If the machine is exhibiting high acceleration at high frequency, the associated displacement is probably *very low*. If low acceleration was observed at low frequency, the associated displacement could be *very high*. Figure 23.7 illustrates the relationship between acceleration, velocity, and displacement measurements.

If the velocity of the vibrating machine remains constant over a frequency range, the acceleration will increase with frequency whereas the displacement will decrease with frequency. Therefore, the accelerometer would indicate large values at high frequencies and very low values at low frequencies. A displacement transducer would indicate large values at low frequencies and very low values at high frequencies.

In general, displacement transducers are useful from zero to several hundred hertz (10,000 r/min). Velocity transducers are useful from 5 to 2000 Hz (300 to 120,000 r/min). Accelerometers are useful from 1 Hz (60 r/min) to more than 50 kHz (3,000,000 r/min).

The accelerometer is the most versatile transducer available. However, it may not have adequate sensitivity at low frequencies. Also, the accelerometer may be oversensitive at high frequency. These problems can be corrected by selecting the proper transducers and by the addition of external filters to eliminate the high-frequency response of the accelerometer and enhance the sensitivity at low frequency.

TIME DOMAIN

Figure 23.8 shows vibration due to an unbalanced rotor in the time domain. The amplitude of vibration is proportional to the amount of imbalance, and the cycle repeats itself once per revolution. Analysis in the time domain becomes more difficult when there is more than one vibration component present.

Figure 23.9 illustrates a situation where two sine wave frequencies are present. The individual components of this combination are difficult to derive from a time domain display.

The impulsive signal from bearings and gear defects can be detected in the time domain. It is also useful for analyzing the phase relationships of the vibration signals. However, it is difficult to determine the individual components of complex signals. The frequency domain is used to analyze these components.

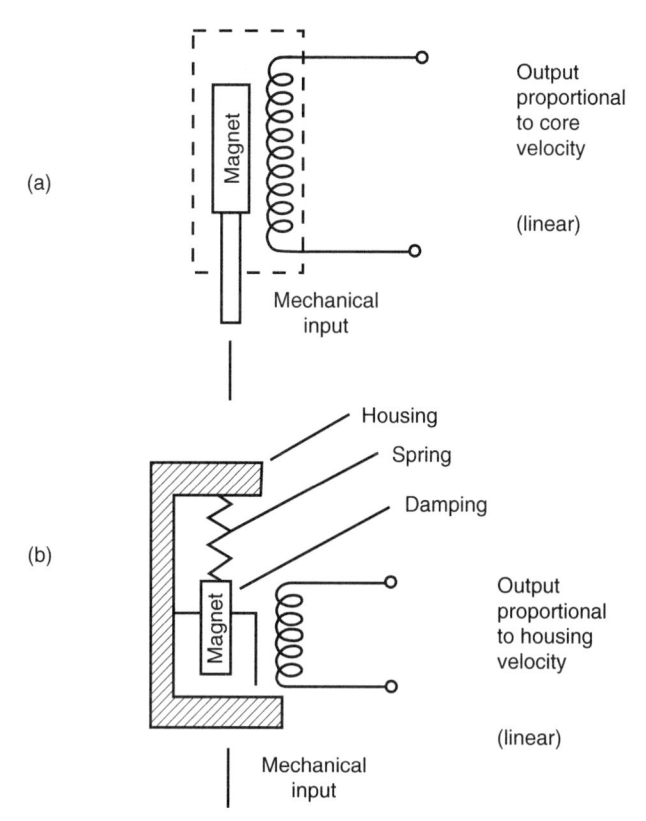

FIGURE 23.6 Velocity transducers. (*a*) LVT (contracting); (*b*) seismic (mounted on object to be measured).

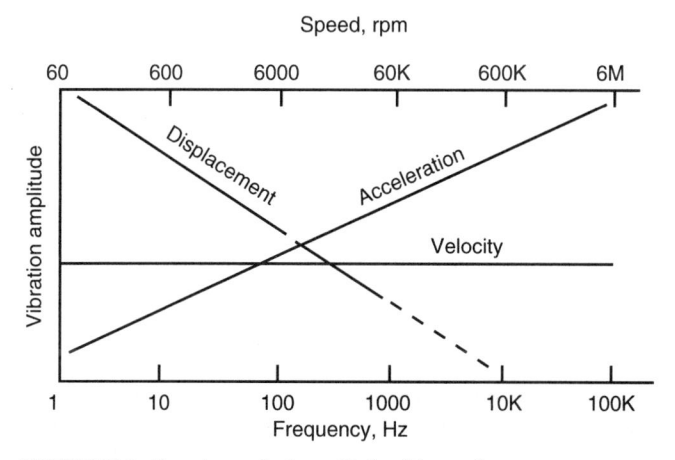

FIGURE 23.7 Transducer selection guide (useful ranges).

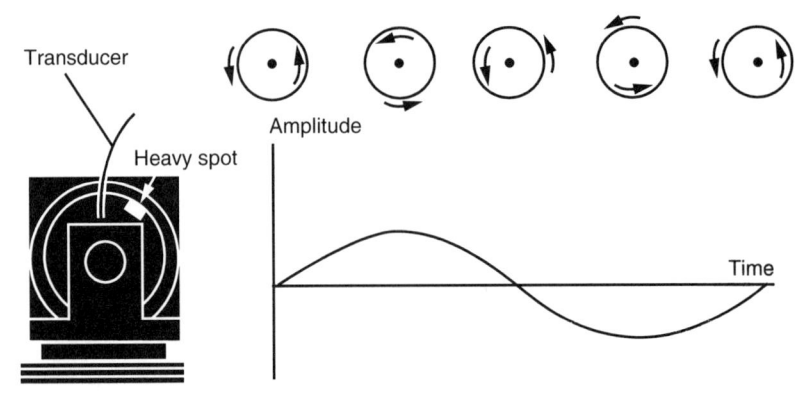

FIGURE 23.8 Vibration due to an unbalanced rotor in the time domain.

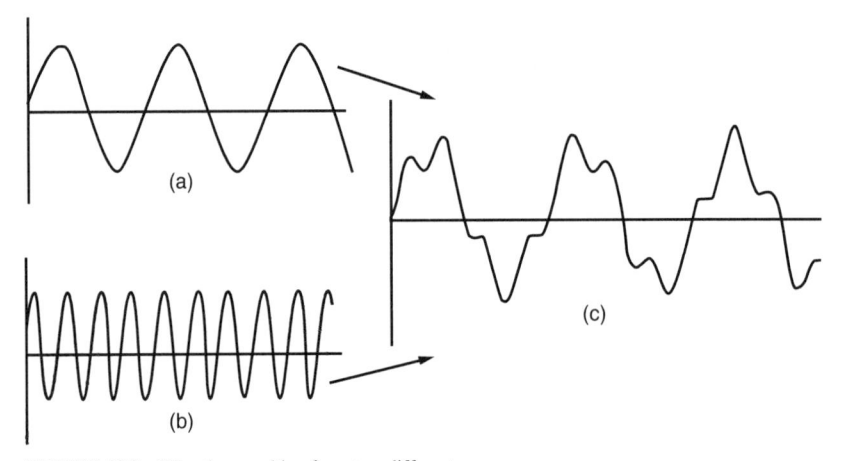

FIGURE 23.9 Vibration resulting from two different causes.

FREQUENCY DOMAIN

Figure 23.10 illustrates a three-dimensional graph of the signal shown in Fig. 23.9. The frequency domain permits separation of the components in the waveform. The frequency domain illustrates the same time domain that was shown in Fig. 23.9. It is the summation of two sine waves, which cannot be recognized in the time domain.

Each frequency component appears as a vertical line in the frequency domain. The height represents the amplitude of the vibration. The representation of the signal in the frequency domain is called the *spectrum* of the signal. The frequency domain is powerful because any real signal can be generated by adding up sine waves (Fourier's law). Therefore, the frequency domain can separate the sine wave components of any vibration signal.

The frequency spectrum of a vibration signal defines the vibration completely. There is no loss of information by converting to the frequency domain, provided that the phase information is included.

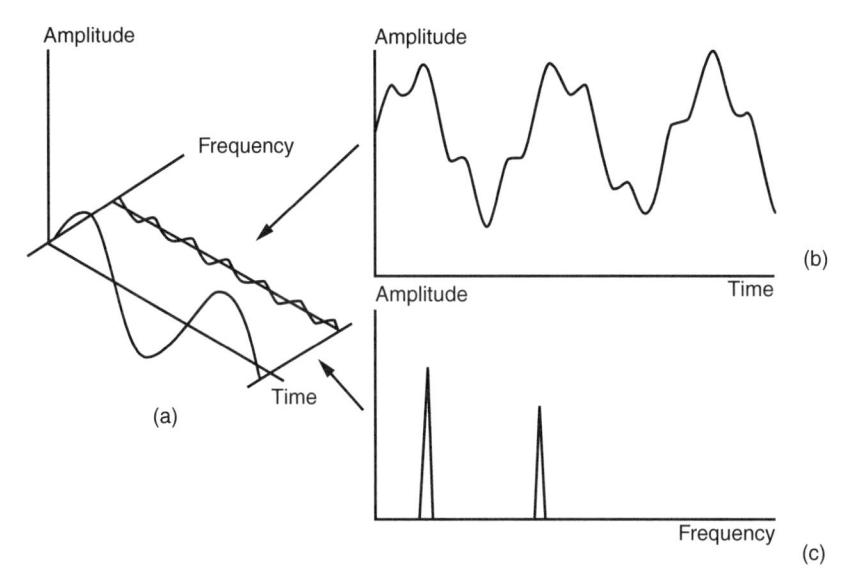

FIGURE 23.10 Vibration in the frequency domain.

MACHINERY EXAMPLE

Figure 23.11 illustrates the usefulness of the frequency domain analysis for a vibrating machine. The internal sources of vibration are rotor imbalance, a ball bearing defect, and reduction gear meshing.

VIBRATION ANALYSIS

The first step in vibration analysis is the understanding of the machinery (expensive diagnostic equipment is useless if the machine is not well understood). A study of the machine dynamics should precede the taking of vibration readings.

Vibration Causes

A wide variety of faults that result from wear, damage, or poor installation may lead to vibration in machinery. Vibration may also be caused by an inherent and unavoidable mechanical feature such as the reaction of turbine blades passing an opening.

However, serious vibration problems occur when one of the forcing frequencies approaches one of the natural frequencies of vibration. This is known as *resonance*. It amplifies the effect of the forcing frequency.

Forcing-Frequency Causes

The first step in performing vibration analyses is to investigate the forcing frequencies which result from any of the following causes.

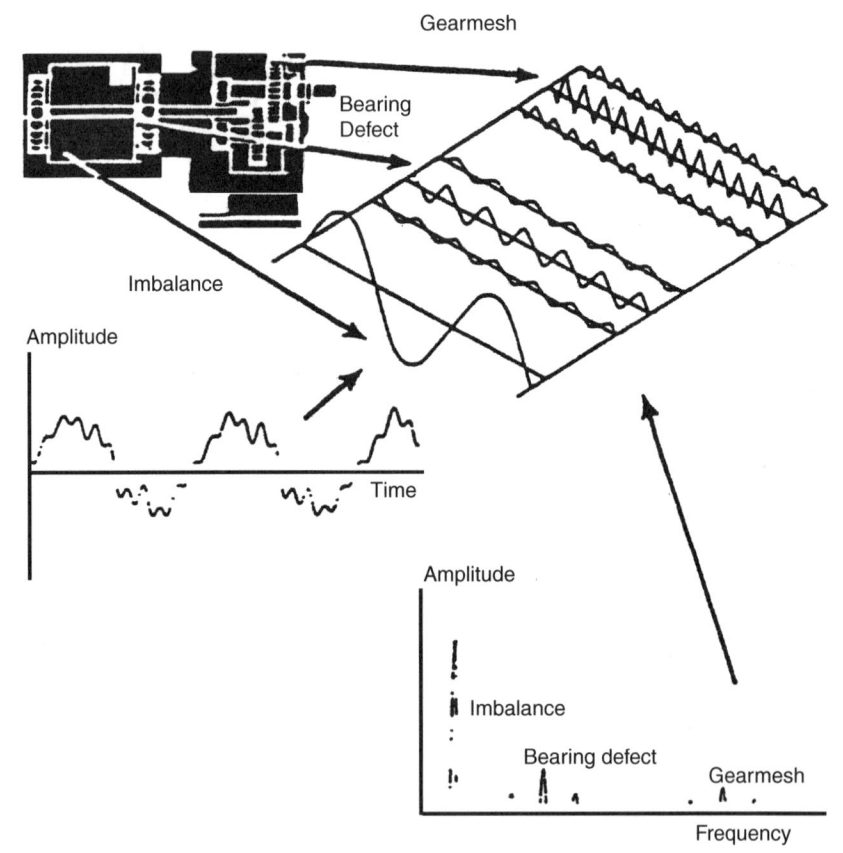

FIGURE 23.11 Identification of the vibration causes using the frequency domain analysis.

Unbalance

Unbalance is characterized by sinusoidal vibration at the machine's running speed. It occurs when the center of mass of a component does not coincide with its center of rotation.

The majority of vibration problems are caused by unbalance. The features of vibration from unbalance are that

1. It is a single-frequency vibration which has a constant amplitude in all radial directions.

2. It increases with speed and does not contain harmonics (higher-frequency vibrations).

Misalignment

Vibration is generated when there is a misalignment between shafts of connected machines. It can be mistaken for unbalance. Its distinguishing feature is that its principal frequency is at 1, 2, and 3 times the shaft rotational speed. Also, it contains a large axial component. Special cases of misalignment are improperly seated bearings and bent shafts.

Mechanical Looseness

Mechanical looseness vibrations occur at 1 × rotational speed. However, it always contains higher harmonics. Mechanical looseness can often be located by taking several velocity readings at different points on the machine. The measured vibrations will be the highest in the vicinity and direction of looseness.

Bearing Defects

The most common cause of failures in small machinery is due to rolling-element bearings. The vibration is usually caused by a defect in the inner race, outer race, or rolling elements. The vibrations are characterized by high frequency and low energy.

Formulas are used to determine the exact frequency generated by a fault in a race if the dimensions of the inner and outer races and the rolling elements are known.

Gear Defects

High-frequency vibrations are also generated by gear meshing. They are usually low-energy signals that are easy to recognize but difficult to interpret.

The vibration frequency is equal to the rotational speed of the gear times the number of teeth. The vibration amplitude varies with the load.

Oil Whirl

Oil whirl in bearings is caused by instability of the rotor which is supported by the liquid film. Changes in oil viscosity, pressure, or external preloads produce conditions that prevent the film from supporting the shaft, and bearing wipe occurs.

Blade or Vane Problems

Blade or vane passing generates single-frequency vibration or a large number of harmonics close to the blade passing frequency. The amplitude varies considerably with load. A cracked or missing blade usually increases the number of harmonics without changing the fundamental frequency.

Electric Motor Defects

Motor vibrations are caused by mechanical or electrical defects. The frequency generated by a broken rotor bar (mechanical defect) is distinguished from an electrical defect by the amount of motor slip. A simple technique is used to differentiate mechanical from electrical problems in motors. If the vibration remains unchanged when the motor is turned off, the deficiency is mechanical in nature.

Uneven Loading

Vibrations are generated by uneven loading (such as a belt drive on an eccentric pulley). The vibration is generated at running speed. It could be mistaken for unbalance. The vibration is distinguished by being unidirectional and usually varies with the load.

Driveshaft Torsion

Torsional vibration is becoming a problem with the advent of variable-frequency drives which may contain many electric harmonics. Torsional vibrations do not generate an externally measurable effect on the machine. They cannot be detected by conventional vibration sensors.

RESONANT FREQUENCY

If the running speed is close to one of the natural frequencies, the small amount of unbalance causes unacceptable vibration. Balancing the machine would not eliminate the vibration (because unbalance will return). The solution in this case is not to remove the forcing frequency, but to change the vibration characteristics of the machine.

This can be done by any of the following:

1. Increase the stiffness of the machine to raise the natural frequency.
2. Add mass to lower the natural frequency.
3. Add damping such as shock absorbers to reduce the machine's response to the forcing frequency.

Here is a general rule: If the forcing frequency is below resonance, damping is the most effective way to reduce the response. If the forcing frequency is higher than resonance, the mass should be increased.

VIBRATION SEVERITY

Figure 23.12 illustrates the severity of vibration depending on displacement, frequency, and velocity.

Vibration Diagnostic Chart. Table 23.1 illustrates a vibration diagnostics chart. It includes comments about the amplitude, frequency, phase, and features of each vibration cause.

A CASE HISTORY: CONDENSATE PUMP MISALIGNMENT

Problem

Machinery diagnostic program baseline data were collected for a power station's condensate pump. The data showed that levels recorded at the inboard pump axial and horizontal positions exceeded alarm values defined for these points.

Test Data and Observations

Velocity spectra from the inboard pump bearing showed the majority of the vibration was at 2× rotational speed in the axial direction and at 1 times rotational speed in the horizontal

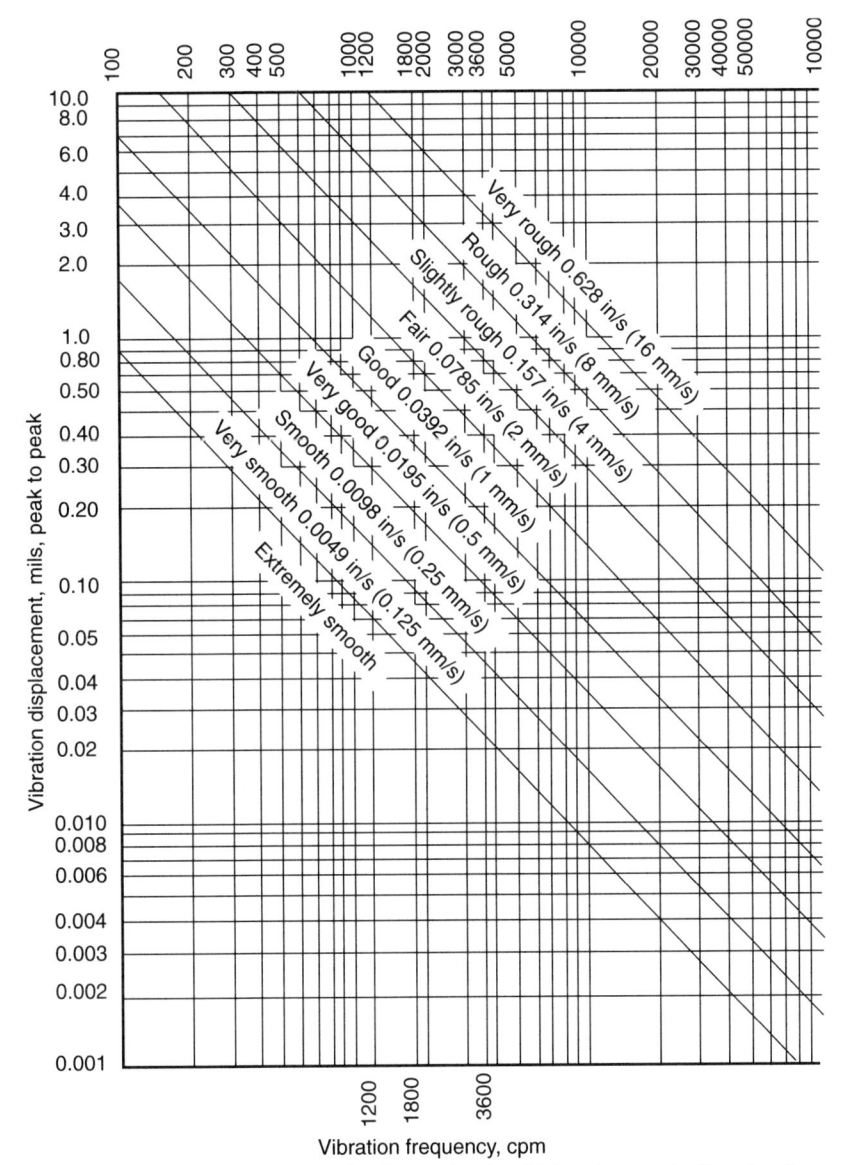

FIGURE 23.12 General machinery—vibration severity chart. (Values shown are for filtered readings taken on the machine structure or bearing cap. Vibration velocity, in/s peak.)

direction. Misalignment between the motor and pump was suspected, and a coupling inspection was performed. The coupling was found to be out of alignment, as the following readings indicate: parallel side to side was out 0.009 in; parallel top to bottom was out 0.018 in; and the motor was 0.018 in lower than the pump. Two of the four upper bearing housing bolts were also discovered to be loose.

TABLE 23.1 Vibration Diagnostics Chart

Cause	Amplitude	Frequency	Phase	Remarks
Unbalance (worn or lost parts, bent shaft)	Proportional to unbalance; largest in radial direction	1 × rpm	Single reference mark	One of the most common causes of vibration; appears after maintenance or after long periods of use
Bent shaft	Steady	1 × rpm; 2 × rpm also if bent at coupling	180° out of phase axially	
Thermal bow	Varies during operation	1 × rpm	Single reference mark	Increasing vibrations during variations in load or during start-up from cold condition
Misalignment of couplings or bearings	Large in axial direction; 50% or more of radial vibration	1 × rpm usual; 2 and 3 × rpm sometimes	Single, double, or triple	One of the most common causes of vibration; best found by appearance of large axial vibration; appears after maintenance or after long periods of use
Mechanical looseness		1, 2, and 3 × rpm	Two reference marks; slightly erratic	
Soft foot	Variable with load	1 × rpm; sometimes 2 × rpm		Check mountings for variations in vibration amplitude
Bad bearings of antifriction type	Unsteady—use velocity, or acceleration, or spike energy measurements, if possible	Very high; several times rpm	Erratic	Bearing responsible; most likely the one nearest point of largest high-frequency vibration
Eccentric journals	Usually not large	1 × rpm	Single mark	If on gears, largest vibration in line with gear centers; if on motor or generator, vibration disappears when power is turned off; if on pump or blower, attempt to balance

Electrical	Disappears when power is turned off	1 × rpm or 1 or 2 × line frequency	Single or rotating double mark	If vibration amplitude drops off instantly when power is turned off, cause is electrical
Rubs		0 to 50% of 1×, 1×, higher harmonics	Erratic	Many frequencies over entire frequency spectrum
Oil or steam whirl	May change rapidly	40 to 50% or 1 × rpm	Unsteady	Small changes in radial steam forces, or bearing loading may result in large vibration changes; may excite critical
Aerodynamic hydraulic forces		1 × rpm or number of blades on fan or impeller × rpm and harmonics	Erratic	
Bad gears or gear noise	Low—use velocity or acceleration measure if possible	Very high; number of gear teeth times rpm		
Structural resonance	Steady or variable	1 × rpm but often odd nonsynchronous frequency	Erratic	May be flow excited in vertical pump; if problem starts on old machine, check for soft feet or cracked supports
Bad drive belts	Erratic or pulsing	1, 2, 3, and 4 × rpm of belts	One or two depending on frequency; usually unsteady	Strobe light best tool to freeze faulty belt
Reciprocating forces		1, 2, and higher orders × rpm		Inherent in reciprocating machines; can only be reduced by design changes or isolation
Cracked shaft	Variable during transients	1, 2 × rpm	Change in 1 × phase often occurs	2 × excitation of critical during coastdown; *note* other faults can also cause change in 2 × rpm

23.15

Corrective Actions

The coupling was realigned to the following tolerances: 0.002 in parallel side to side and 0.006 in parallel top to bottom. The motor also was raised to a position 0.002 in lower than the pump. New bolts were installed on the upper inboard pump bearing housing.

Final Results

A new set of baseline vibration spectra was collected for this equipment at a later date. The data showed that alignment greatly reduced the vibration levels. Inboard pump axial vibration dropped from 0.42 to 0.09 in/s. The inboard pump horizontal vibration likewise was reduced from 0.18 to 0.12 in/s. Similar reductions were observed at other motor and pump measurement points.

Conclusion

Had this misalignment condition continued unabated, it is very likely that the pump and motor bearings would have been damaged. This, in turn, could have led to a catastrophic failure of the condensate pump, consequently derating the unit. It took two workers 12 h to align the coupling. If the bearing needed replacement, it would have taken 3 or 4 workers about 48 h to complete the job. In summary, correcting this condition before the bearings or coupling was damaged saved the station many worker hours as well as the expenses of new parts and lost production.

CHAPTER 24
POWER STATION ELECTRICAL SYSTEMS AND DESIGN REQUIREMENTS

INTRODUCTION

The electrical systems in a power station includes those parts associated with connecting the station to the grid and the ones that distribute power to the auxiliary systems inside the station. The reliability of the power supply depends on the importance of the equipment. For example, if the impairment of a piece of equipment will result in a unit trip, its power supply should be more reliable than a sump pump used occasionally. It is therefore necessary to determine the importance of each piece of equipment to the power plant when the degree of reliability of the power supplies is required. The sources of power supplies in the plant vary from grid-derived ac supplies to battery-backed ac and dc supplies.

The electrical supply to the auxiliary loads in the power plant can be taken from the following sources:

- Grid: derived ac supplies through a station system transformer (SST)
- Unit: derived ac supplies through a unit service transformer (UST)
- A combination of grid- and unit-derived ac supplies

Most modern power plants are choosing the second option while relying on the grid-derived ac supplies as backup power. The complete details of the electrical loading, rating, and duty information for all station loads including reactor, boiler, and turbine-generator should be established at the onset to enable accurate design of the electrical systems. The design ratings of the transformers, switchgear, cables, etc., can be determined when all the electrical system loadings are assembled. The electrical loading, rating, and duty schedules of all plant equipment should be included in the specifications. The most suitable equipment should be selected based on a comparison between the competitive tenders. The system designer should select the equipment based on the lowest lifetime cost and suitability for duty. The selection criteria of the most suitable electrical systems for nuclear and fossil-fired (coal, oil, and natural gas) power stations will be presented in this chapter.

SYSTEMS REQUIREMENTS

Station Operation Systems

The designers of the electrical systems should have a clear definition of the operating criteria required. *Station technical particulars* (STPs) should be developed during the early planning stages. They contain the requirements for the availability, operating flexibility, and control systems of the plant. They also include the technical requirements for the generator transformers, auxiliary systems, and protection arrangements.

The Appendix will detail the specifications that should be met. The minimum features of the electrical systems are determined based on the STPs.

Additional documentation is required to cover the safety aspects of nuclear power stations. It is normally presented in the form of a safety report. The designer will determine the electrical requirements based on the STPs. These requirements will generally include the following:

- The station-rated output is required over a frequency range of 59.5 to 60.5 Hz.
- A fault or a fire in any auxiliary system should not trip more than one generator.
- The electrical supplies to the auxiliary systems should remain stable for three-phase faults having duration up to 200 ms.
- The electrical supplies to the auxiliary systems should be designed to meet all the operating modes including part-loading and load rejection (sudden disengagement of the generator from the grid).
- The electrical supplies to the auxiliary systems should be designed to withstand any switching or transient overvoltages.
- The auxiliary systems should operate adequately within a voltage range of 95 to 105 percent of the rated value.
- The designer may supplement the STP requirements with additional features to improve the reliability of the electrical supply to the auxiliary system.

For example, the addition of alternate power supplies to selected equipment will reduce the outage time. This will consequently increase the revenues from power sales. The designer should provide economic justification for this design feature.

Grid Criteria

The grid system has criteria associated with frequency, voltage, and the partial or total loss of grid connections near the power plant. The frequency deadband of the turbine governing system is around 36 mHz. An automatic trip should be initiated when the frequency reaches 57.5 Hz for 12 s. The auxiliary systems are also expected to withstand without tripping for periods not exceeding 15-min transient overfrequency excursions (up to 8 percent) that would occur following a load rejection. These overfrequency excursions will result in an increase in the current to the motors and a possible trip on overload. The grid also expects the power plant to continue delivering power while the grid voltage varies within ±5 percent of the nominal grid voltage.

The auxiliary equipment should be rated to withstand this variation in voltage. It is important to mention that when the grid voltage drops, the current drawn by the motors of the auxiliary equipment will increase.

This could lead to tripping of the motors on overload. These motors will not reset. Thus, the decrease in grid voltage could lead to impairments of many auxiliary systems. Therefore, the protection devices of the auxiliary equipment should be designed to tolerate the increase in current drawn by this equipment following a decrease in grid voltage.

The impact of partial or total loss of the grid on the power plant varies with the type of the plant. Nuclear power plants are most affected by these events. These plants rely on diesel generators or gas turbines to provide power to their safety systems following the loss of power from the grid. They also rely on large battery banks to provide power to the essential safety systems.

Safety Requirements

The designer should abide by the standards, codes of practice, and rules of the safety acts to ensure the safety of the plant and personnel. All the electrical systems should incorporate features that allow the operation and maintenance of the equipment while complying with all safety requirements.

Operationally, the main design considerations are to ensure that the prospective system fault capabilities are not exceeded during normal and abnormal duties. The designer should rate the circuit for the required voltage and for normal and fault currents. The switchgear should be designed for making and breaking currents that occur during normal and fault conditions. The designer should also include interlocking and monitoring schemes to ensure that the equipment ratings are not exceeded due to an operational error.

ELECTRICAL SYSTEM DESCRIPTION

The output voltage of a 660-MW (776-MVA) generator is normally around 23.5 kV. It is connected to the grid via a *main output transformer* (MOT). The grid voltage varies from 230 to 800 kV. Insulated metal-clad SF_6 circuit breakers are used to connect the generator to the grid in modern power plants. The station auxiliary loads should always be supplied with electric power. Figure 24.1 illustrates a typical electrical distribution of a power-generating station.

The Generator Main Output System

Main Generator. The generator is connected to the main output transformer via phase-isolated connections. These connections are air-cooled and rated around 20,000 A for a 660-MW (776-MVA) unit.

Generator Transformers. The generator MOT connects the generator to the grid. It is fitted with on-load tap changers to accommodate the variations in grid voltage. It could be a three-phase unit or a three single-phase unit. It is rated around 800 MVA (for a 776-MVA generator). This rating is based on possible additional 44 MVA from a gas-turbine generator, less than a unit auxiliary load of 20 MVA. The MOT of a 900-MW unit is rated around 1145 MVA. It is normally made of three single-phase transformers, which are fitted with on-load tap changers.

Electrical Auxiliary Systems

The electrical auxiliary systems provide power for the unit auxiliaries. These systems must be designed to prevent a fault including a fire from affecting more than one generating unit. Similarly, a fault on a system that does not have an immediate effect on the operation of the unit should be prevented from propagating into the electrical system and affecting the unit output. This is achieved by segregating the switchgear and cables between the units.

Type of Stations

There are a wide variety of power stations including coal-fired, nuclear, oil-fired, hydraulic, gas turbines, and combined cycles. The following sections describe the different features associated with fossil-fired power stations.

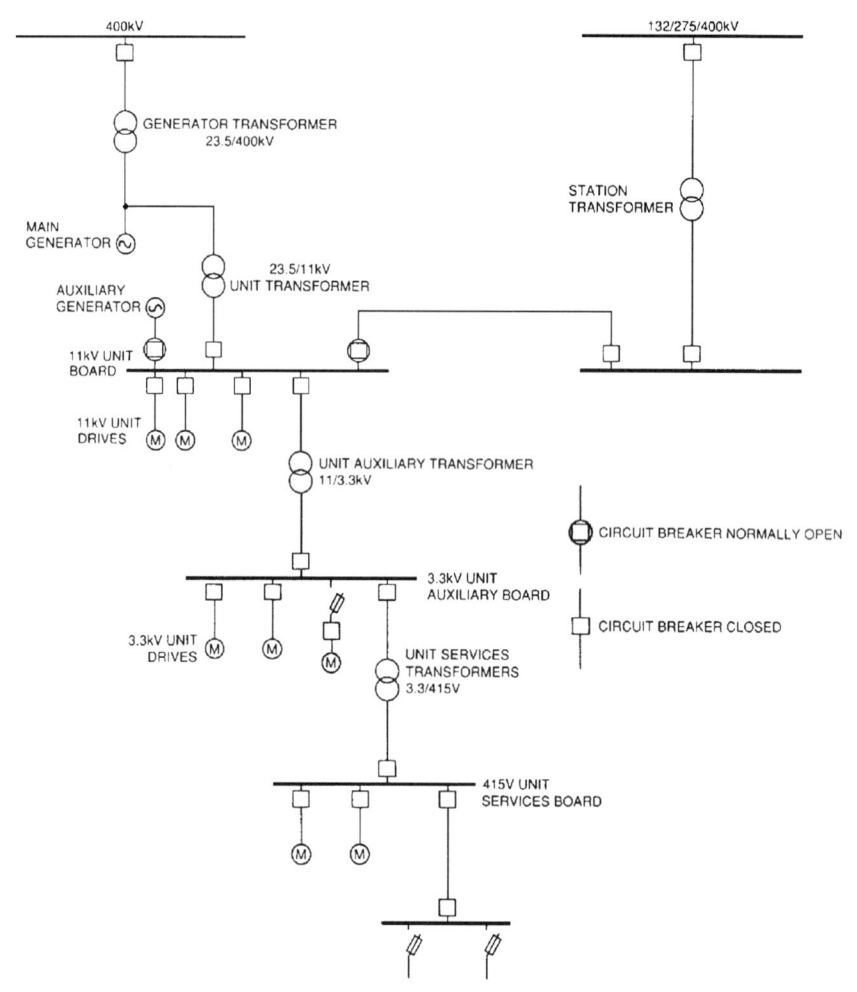

FIGURE 24.1 Typical unit station system for a 660-MW generator.

Fossil-Fired Power Stations. The fossil-fired power stations are fueled by coal, oil, or natural gas. The main difference in the electrical auxiliaries between a coal-fired and an oil-fired station is the fuel handling and combustion systems. Figures 24.2 and 24.3 illustrate the electrical auxiliaries system for Littlebrook *D* oil-fired station.

SYSTEM PERFORMANCE

Unit Start-up

During start-up, all types of power plants require sufficient electric power from a grid or a larger power source such as a gas turbine. The high-voltage (HV) circuit breaker of the

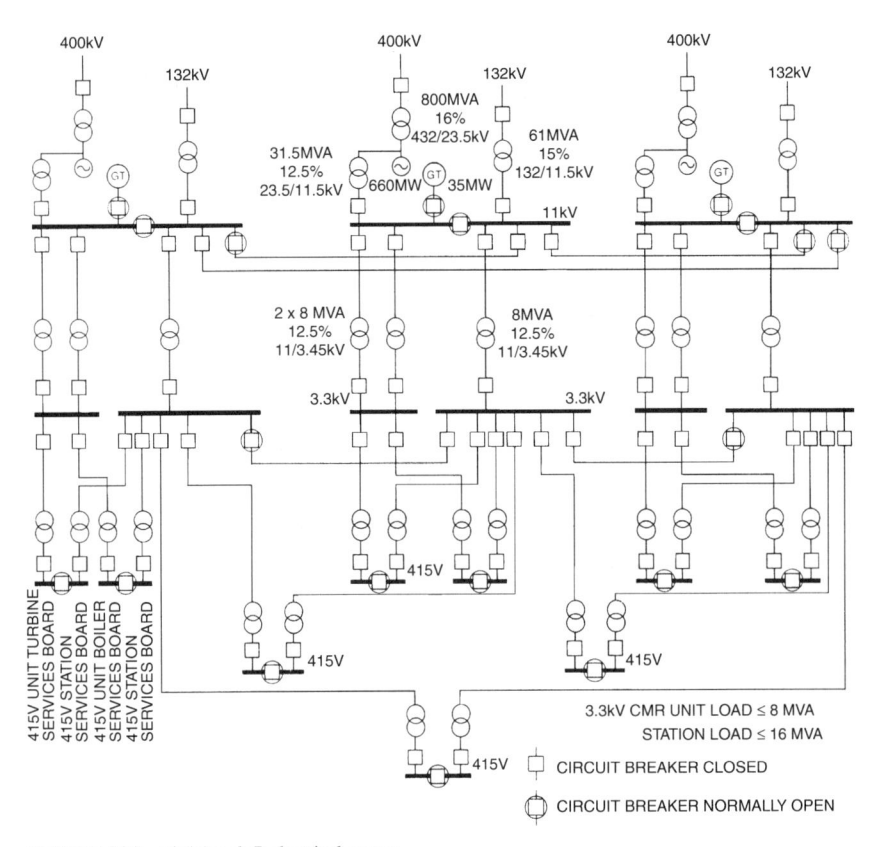

FIGURE 24.2 Littlebrook D electrical system.

generator transformer and the low-voltage (LV) circuit breaker of the unit transformer are held open during start-up (Fig. 24.4). The unit auxiliaries are supplied from the grid through the station transformer.

Plant Requirements. During start-up, all power plants require a minimum of one cooling water (CW) pump and one 50 percent electric boiler feed pump. Fossil plants require also coal mills or oil pumps and a draught system, e.g., forced-draft (FD) and induced-draft (ID) fans. When steam is generated at a specified enthalpy, the unit supporting auxiliaries powered from the station transformers will run up the turbine-generator to speed.

Synchronizing to the Grid. When the turbine-generator reaches the operating speed, it is synchronized to the grid via the HV circuit breaker of the generator transformer (Fig. 24.5). Synchronization is normally done using dedicated automatic equipment, which is controlled, from the main control room (MCR). It can also be done manually using portable check synchronizing trolleys in the MCR.

Synchronizing the Unit to the Station. When the generator is lightly loaded, the unit loads should be transferred to the unit transformer. This is done by paralleling the unit and station boards for a short time. Check synchronizing equipment is needed because the two sources could be out of phase and frequency.

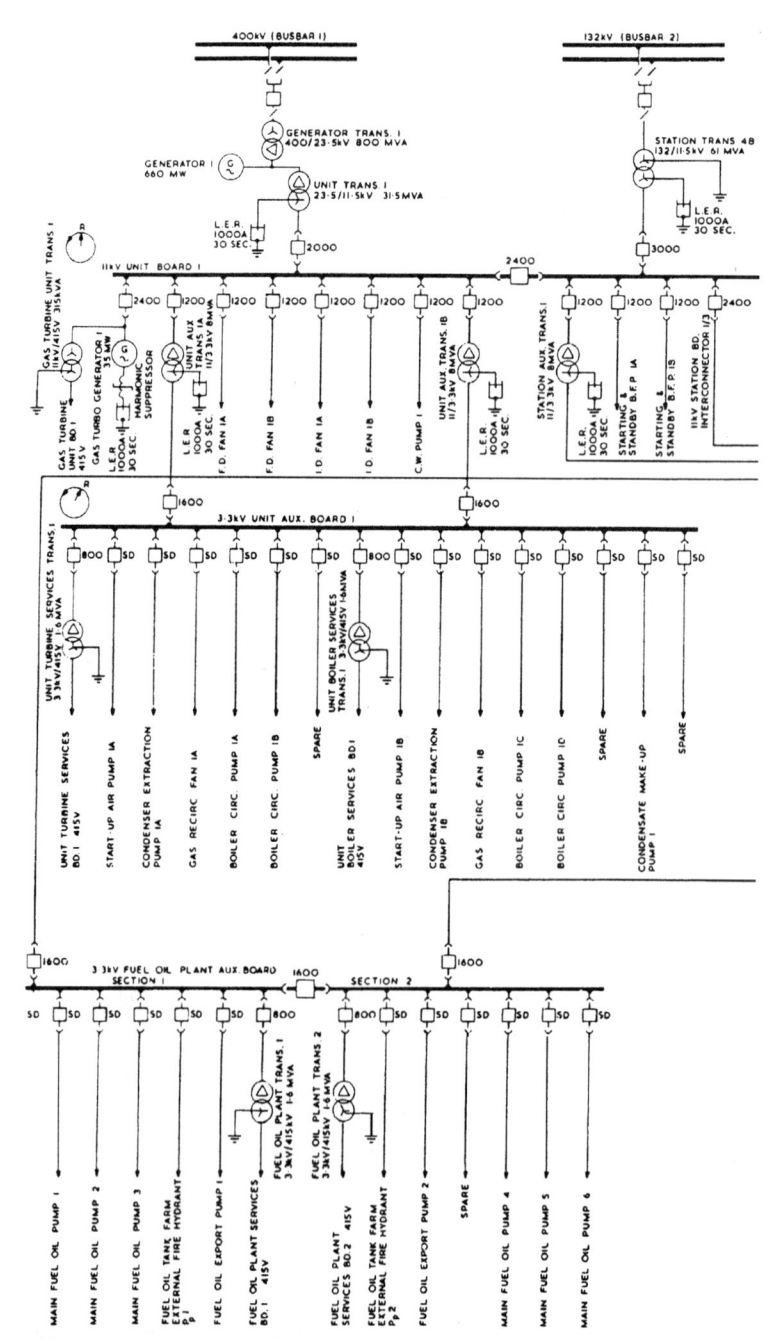

FIGURE 24.3 Littlebrook *D* electrical auxiliaries system.

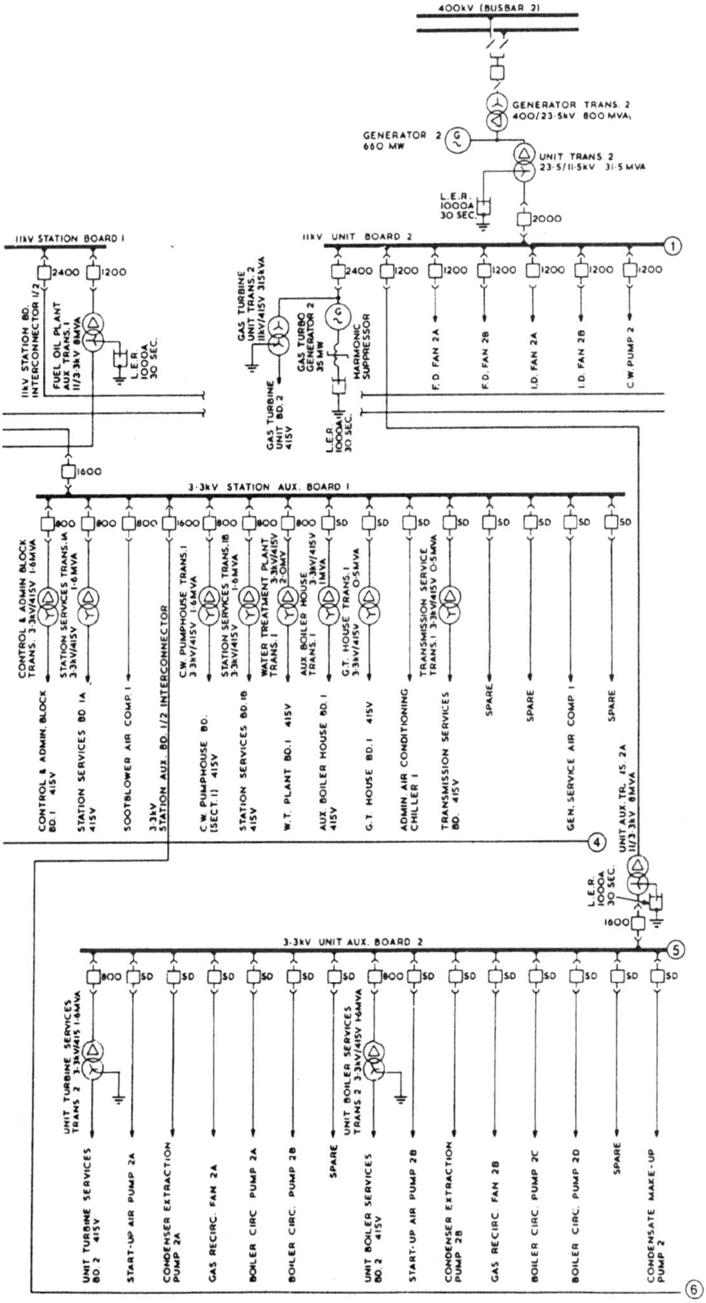
400kV (BUSBAR 2)
GENERATOR TRANS. 2
400/23·5kV 800 MVA
GENERATOR 2
660 MW
UNIT TRANS 2
23·5/11·5kV 31·5 MVA
L.E.R.
1000A
30 SEC.
2000
11kV STATION BOARD 1
11kV UNIT BOARD 2
2400
1200
11kV STATION BD. INTERCONNECTOR 1/2
FUEL OIL PLANT AUX TRANS.1 11/3·3kV 8MVA
GAS TURBINE UNIT TRANS. 2 11kV/415V 315kVA
GAS TURBINE UNIT BD. 2 415V
GAS TURBO GENERATOR 2 35 MW
HARMONIC SUPPRESSOR
F. D. FAN 2A
F. D. FAN 2B
I. D. FAN 2A
I. D. FAN 2B
C.W. PUMP 2
1600
3·3kV STATION AUX. BOARD 1
800
SD
CONTROL & ADMIN BLOCK TRANS. 3·3kV/415V 1·6MVA
STATION SERVICES TRANS. 1A 3·3kV/415V 1·6MVA
C.W. PUMPHOUSE TRANS.1 3·3kV/415V 1·6MVA
STATION SERVICES TRANS. 1B 3·3kV/415V 1·6MVA
WATER TREATMENT PLANT TRANS. 1 3·3kV/415V 2·0MV
AUX BOILER HOUSE TRANS. 1 3·3kV/415V 1MVA
G.T. HOUSE TRANS. 1 3·3kV/415V 0·5MVA
TRANSMISSION SERVICE TRANS. 1 3·3kV/415V 0·5MVA
CONTROL & ADMIN. BLOCK BD. 1 415V
STATION SERVICES BD 1A 415V
SOOTBLOWER AIR COMP. 1
3·3kV STATION AUX. BD. 1/2 INTERCONNECTOR
C.W. PUMPHOUSE BD. (SECT. 1) 415V
STATION SERVICES BD. 1B 415V
W.T. PLANT BD. 1 415V
AUX. BOILER HOUSE BD. 1 415V
G.T. HOUSE BD. 1 415V
ADMIN AIR CONDITIONING CHILLER 1
TRANSMISSION SERVICES BD. 415V
SPARE
GEN. SERVICE AIR COMP. 1
UNIT AUX. TR. 45. 2A 11/3·3kV 8MVA
3·3kV UNIT AUX. BOARD 2
UNIT TURBINE SERVICES TRANS 2 3·3kV/415 1·6MVA
UNIT BOILER SERVICES TRANS. 2 3·3kV/415V 1·6MVA
UNIT TURBINE SERVICES BD. 2 415V
START-UP AIR PUMP 2A
CONDENSER EXTRACTION PUMP 2A
GAS RECIRC. FAN 2A
BOILER CIRC PUMP 2A
BOILER CIRC. PUMP 2B
UNIT BOILER SERVICES BD. 2 415V
START-UP AIR PUMP 2B
CONDENSER EXTRACTION PUMP 2B
GAS RECIRC. FAN 2B
BOILER CIRC. PUMP 2C
BOILER CIRC. PUMP 2D
CONDENSATE MAKE-UP PUMP 2
1
4
5
6

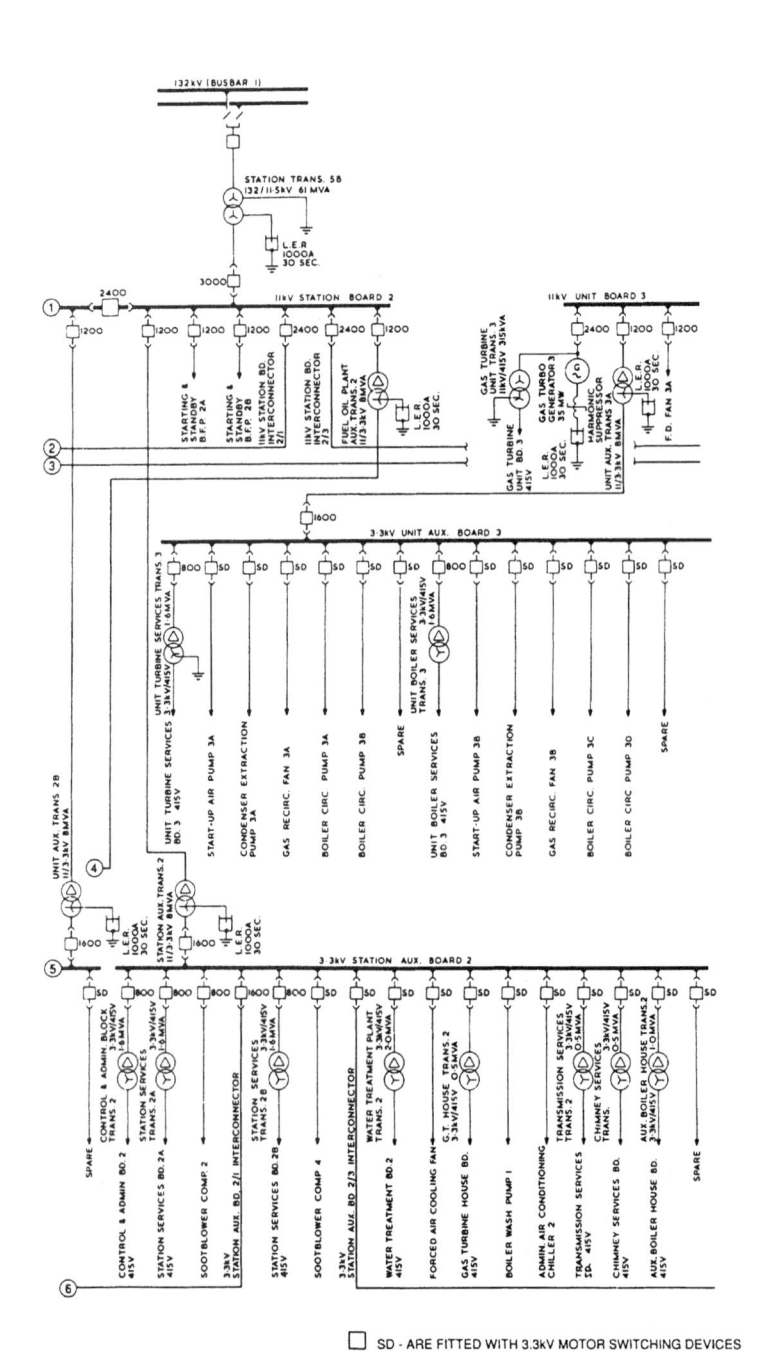

FIGURE 24.3 (*Continued*) Littlebrook *D* electrical auxiliaries system.

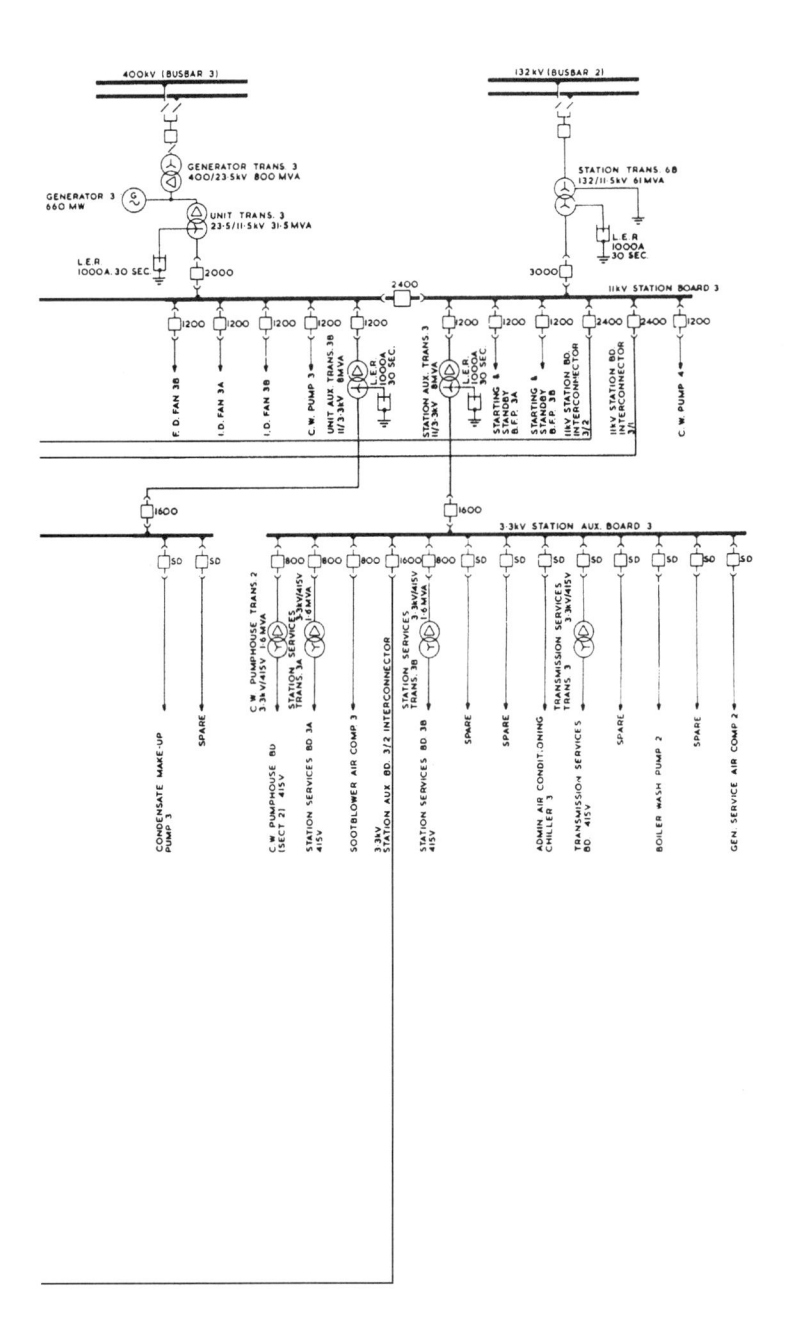

POWER STATION ELECTRICAL SYSTEMS AND DESIGN REQUIREMENTS   24.9

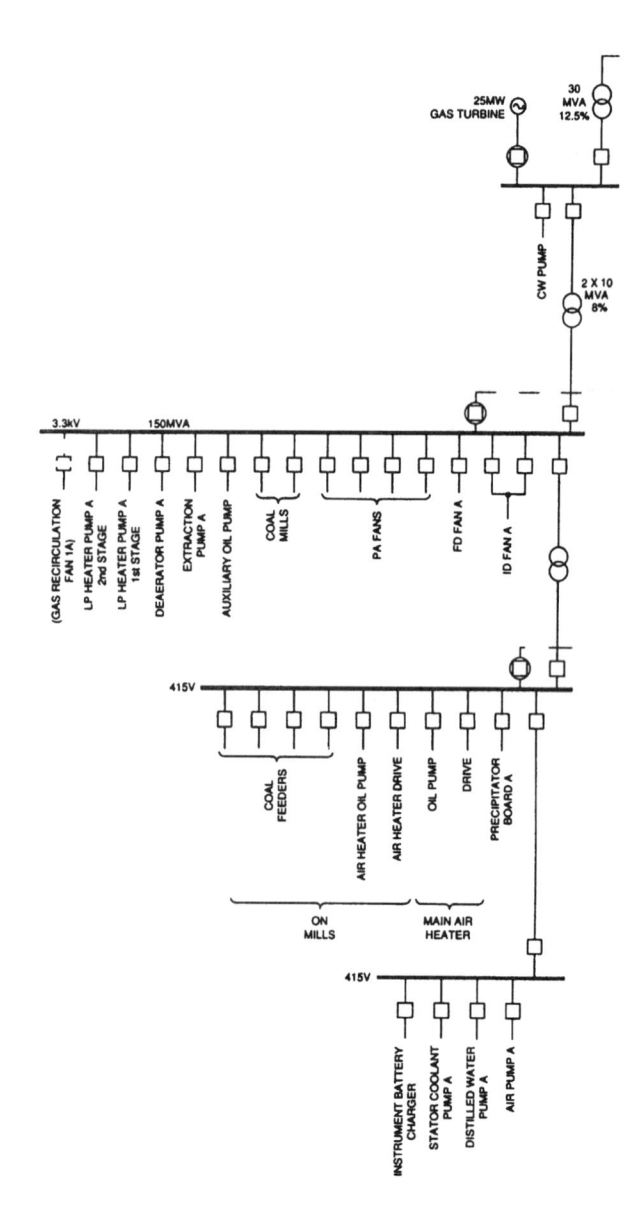

FIGURE 24.4 Typical electrical system for four 500-MW coal-fired units.

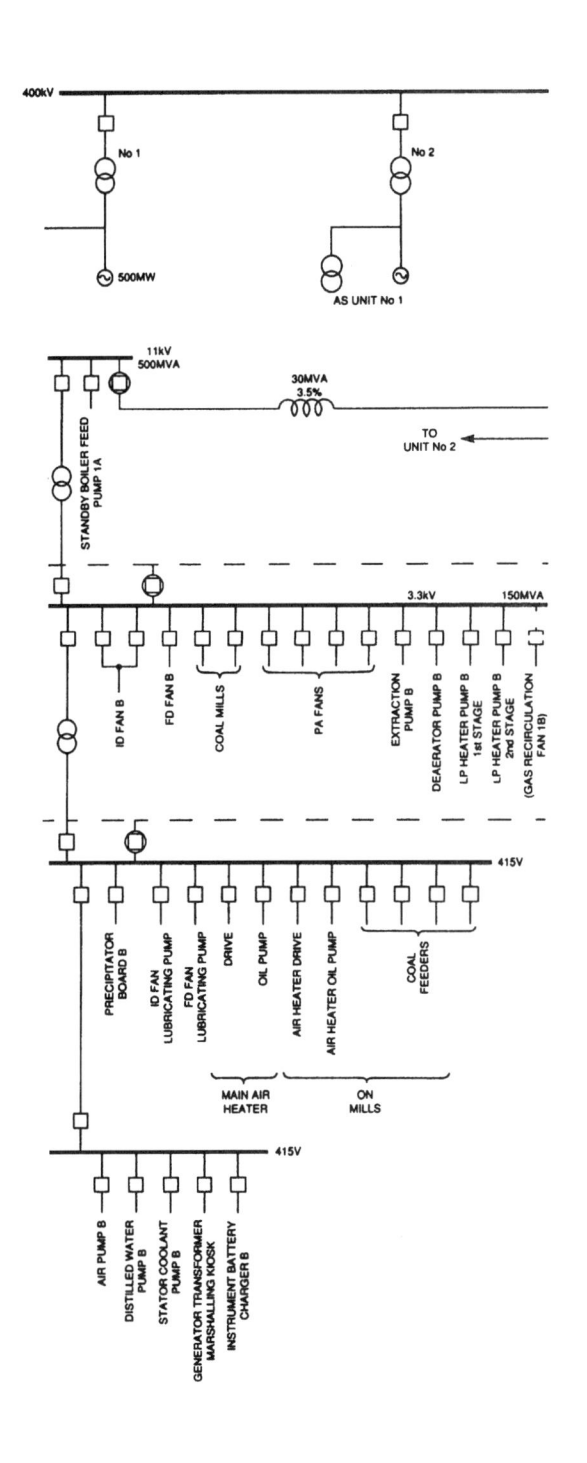

24.12 CHAPTER TWENTY-FOUR

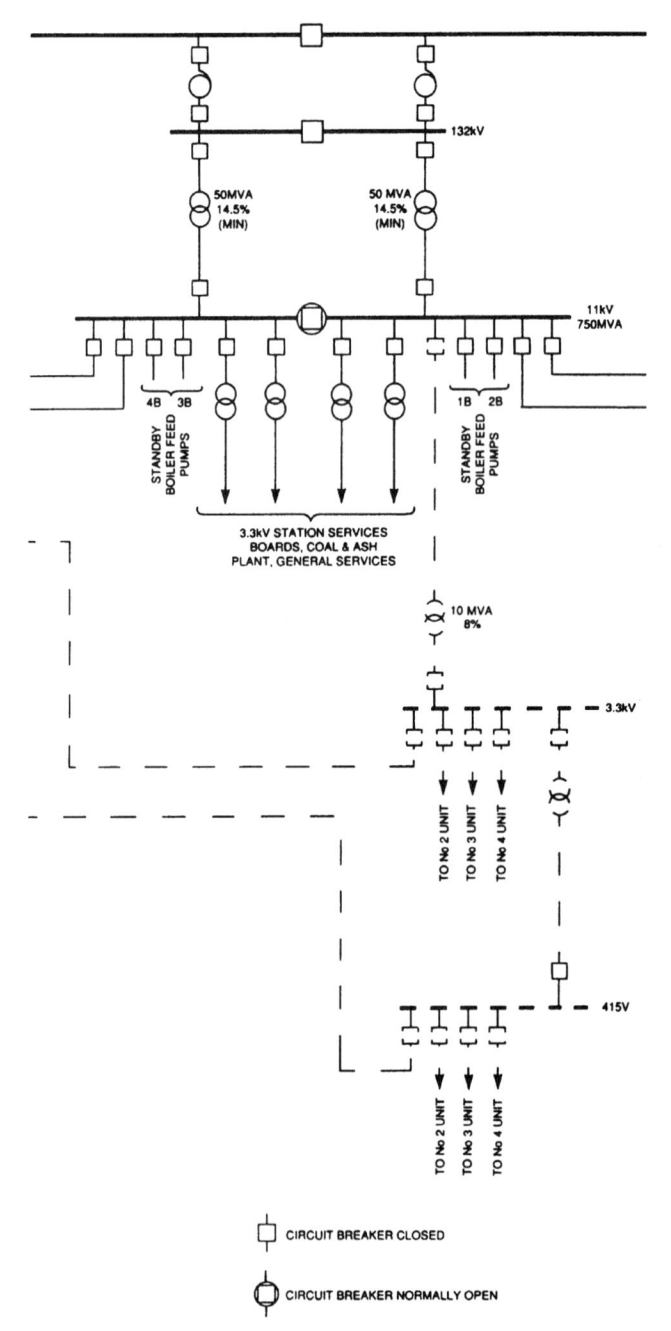

FIGURE 24.4 (*Continued*) Typical electrical system for four 500-MW coal-fired units.

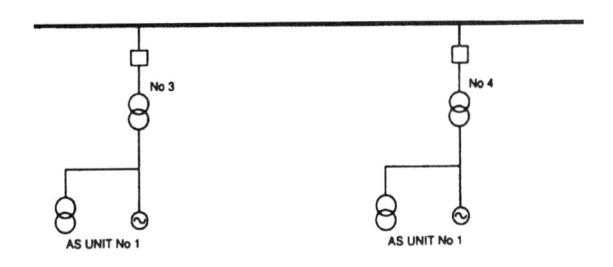

POWER STATION ELECTRICAL SYSTEMS AND DESIGN REQUIREMENTS 24.13

No 3   No 4

AS UNIT No 1   AS UNIT No 1

→ TO UNIT No 3
→ TO UNIT No 4

STANDBY
SUPPLIES

STANDBY
SUPPLIES

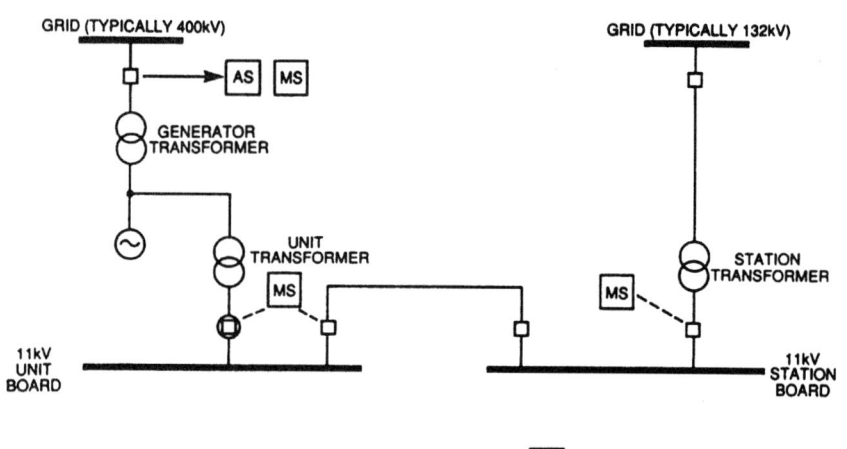

FIGURE. 24.5 Synchronizing points for a direct-connected generator.

Shutdown and Power Trip

There are two types of shutdown:

- A controlled shutdown, due to a requested reduction in power generation or prior to an outage.
- An emergency shutdown, due to an internal or external fault. The unit must be disconnected from the grid following this type of shutdown.

Controlled Shutdown. A controlled shutdown consists of the reverse of the start-up sequence. The unit power is reduced to an acceptable level, and the unit auxiliaries are transferred to the station transformer source.

Power Trip. There are five types of power trips in a power plant:

1. Grid faults causing disconnection of a single unit or the whole station.
2. Electrical faults in the generator system including the generator, unit transformers, and the main connections. These faults require the instantaneous opening of the generator HV and unit transformer LV circuit breakers.
3. Mechanical faults in the generator or the turbine including loss of lubrication oil or control fluid, loss of condenser vacuum, etc., requiring unit shutdown. Instantaneous disconnection of the unit from the grid may not be required following any of these faults. This is so because the turbine-generator overspeed can be prevented when the generator remains connected to the grid for a short time following a fault of this type.
4. Electrical auxiliaries system faults resulting in impairment of equipment leading to a unit trip.
5. Consequential unit trip due to unavailability of steam at the required enthalpy.

The unit is normally subjected to 100 percent load rejection following a fault in group 1. The turbine-generator will overspeed following disengagement of the unit from the grid due to the excess energy in the steam entering the turbines. The turbine governing system and the automatic voltage regulator of the generator are designed to handle this situation. A backup overspeed protection system is used to trip the steam valves when the governing system fails to limit the overspeed to 10 percent. It consists of bolts driven by centrifugal force against springs. They trip a lever when the overspeed reaches 10 percent, resulting in the shutdown of the steam valves.

The generator remains electrically connected to the grid following a fault in group 3. This is done to limit the overspeed due to steam flowing in the turbines. The unit is disconnected from the grid when the power measurement relay detects low forward power.

The auxiliary systems are normally subjected to voltage and frequency transients following some of the trips listed above. For example, following a load rejection that occurs immediately after a fault in group 1 above, the turbine-generator will overspeed. This will increase the frequency of the power generated and transmitted through the unit transformer to the auxiliary systems. The motors of the auxiliary systems will start to pull higher current due to this increase of frequency. This will lead to the tripping of some of these motors on overload and subsequent impairment of the auxiliary systems.

The faults in groups 1 through 5 will cause a loss of normal electrical supply to the unit. Essential systems within the unit will be supplied from battery-backed systems, local gas turbines, or diesel generators.

The Effects of Loss of Grid Supplies

The loss of grid supplies is a generic term used to cover events that result in a grid voltage and/or frequency outside the operating limits. It can be caused by the loss of power stations or an ice or a wind storm.

From a system operations perspective, it is desirable to keep all the power generating plants connected to the grid during a disturbance that involves a frequency or a voltage transient. This is desirable because the loss of power generating plants reduces the stability of the power transmission system. However, a variation in grid voltage or frequency outside the specified limits can cause impairment of systems inside power plants. For example, a decrease in grid voltage will result in an increase in the current supplied from the grid to the motors inside the power plant. These motors could trip on overload, resulting in impairments of essential systems. These systems will remain impaired until the motors are manually reset.

System impairments will also occur when the grid frequency increases or decreases outside the specified range. Thus, the designer of the power station should consider deliberate disconnection of the units from the grid when the grid voltage or frequency drifts outside the specified limits. This issue is of great importance to nuclear power plants because the impairment of their safety-related systems could result in meltdown of the fuel and release of radioactive material. However, it is also of importance to conventional power plants because forced outages have significant economic impacts on the station.

POWER PLANT OUTAGES AND FAULTS

The following are the two types of outages that occur in power plants:

- *Planned outage.* These outages are planned regularly to perform maintenance on the units.
- *Forced outage.* These outages occur due to failure of equipment.

Sufficient redundancy is normally incorporated in the design of the power plant to minimize the frequency of forced outages. The power supplied to redundant equipment should originate from different sources to achieve the required level of reliability. The electrical supply routes to this equipment should also be separated or segregated.

UNINTERRUPTIBLE POWER SUPPLY (UPS) SYSTEMS

Introduction

The UPS systems, also known as *guaranteed instrument supplies* (GIS) or *no-break supplies*, provide battery-backed ac and dc power supplies having higher reliability, quality, and continuity than the normal supplies provided from the electrical auxiliary system. They supply essential instruments, controls, and computers required for safe and reliable operating conditions. The loads supplied from the UPS systems include the following:

- Those required for postincident monitoring and recordings following a unit trip and loss of normal ac supplies.
- Those requiring very high reliability of power supply.
- Those required for the shutdown of the unit. These systems are normally supplied with dc power because it offers higher reliability than the ac power supply does.
- Those required for "black start," i.e., when the normal ac power supplies are not available.

Battery-backed motor-generator (MG) sets and static inverters are used to provide UPS in power stations. Figure 24.6 illustrates a typical motor-generator scheme. The following are its main disadvantages.

- Excessive maintenance required, e.g., shutdown every few months for inspection and maintenance.
- Frequent replacement of the brushes due to excessive wear.
- Difficult parallel operation when the load is low.
- Unreliable frequency control.
- High failure rates due to large number of components. The *mean time before failure* (MTBF) of an MG set is around 2 years.

Figure 24.7 illustrates a UPS system that uses static inverters. This system has proved to be more reliable than the one that uses MG sets. It consists of one inverter for a 500-MW generating unit. The unit electrical auxiliary system normally supplies the load via a single-phase step-down 415 V/110 V transformer. The 240-V dc supply to the inverter was derived from the 240-V dc station system. This system supplies also the emergency dc motors (e.g., emergency turbine lube oil pump motor and emergency generator seal oil pump motor), emergency lighting, switchgear and solenoids, etc.

DC SYSTEMS

Introduction

The dc supply systems are considered the most reliable power source in a power plant. They are normally supplied from a reliable ac power source through a rectifier. They are also battery-backed to supply essential loads when the normal ac power supply fails.

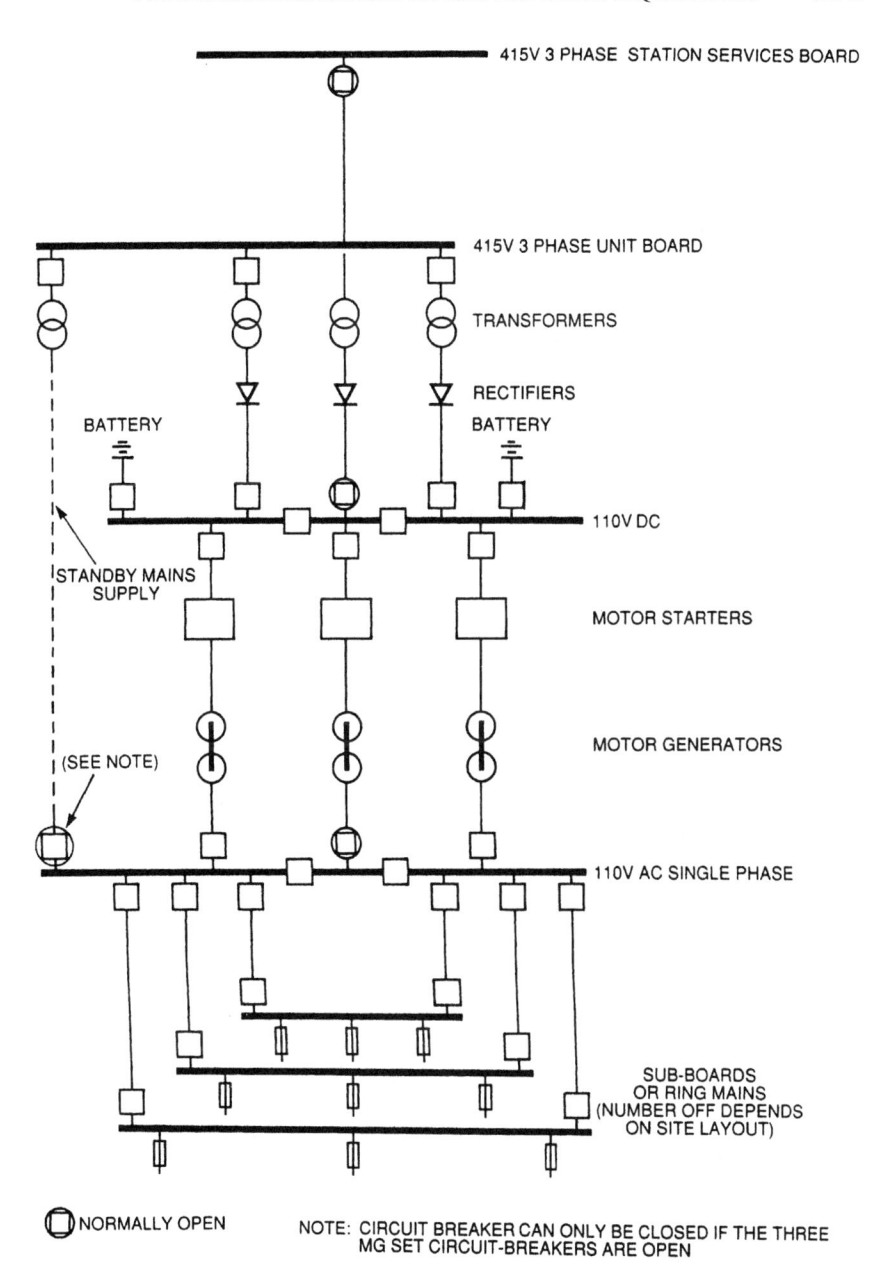

FIGURE. 24.6 Motor-generator scheme—single line diagram of connections.

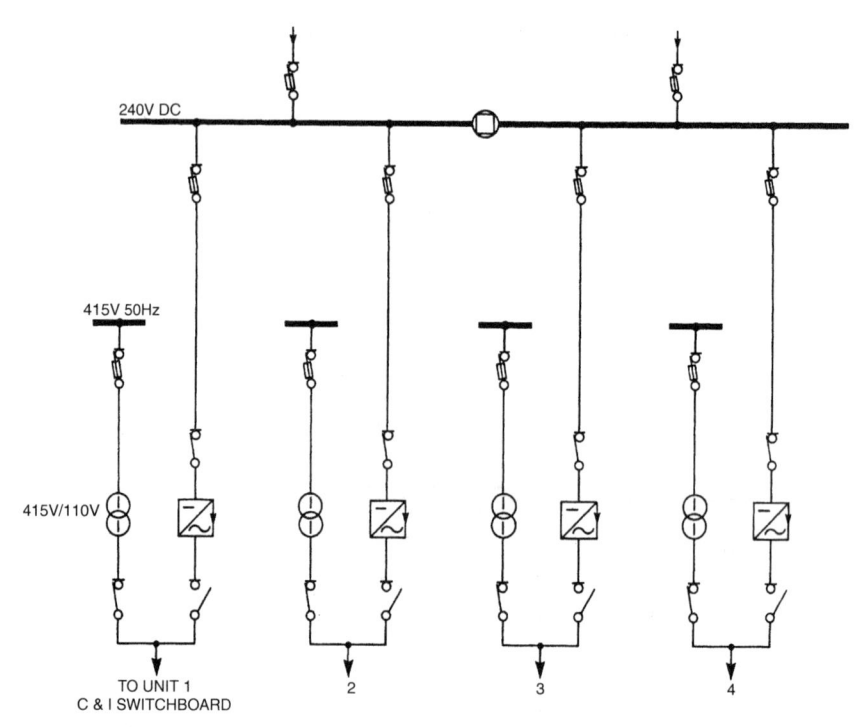

FIGURE 24.7 Static inverter scheme (one inverter per main unit).

DC System Functions

The dc systems have the following functions:

- To supply equipment requiring dc power during normal operation and which is also required to operate following loss of normal ac supplies, e.g., essential instruments, control equipment, switchgear closing and tripping, telecommunications, protections, interlocks, and alarms
- To supply standby equipment required to operate following loss of normal ac power supplies, e.g., emergency lighting, emergency turbine lubricating oil pumps, and emergency generator seal oil pump
- To supply the equipment required to start the standby power generators, which consist of gas turbines or diesel generators

Mission Time of DC Systems

The dc systems are designed to operate from batteries for a least 30 min following a complete loss of ac power supplies. This period is known as the *mission time* for these systems. The standby batteries must have sufficient capacity to meet the mission time. A period of 30 to 45 min is considered acceptable for the mission time because the ac supplies are expected to be restored during this period. The nominal voltage for these systems is 48, 110, 220, or 250 V dc.

The dc systems provide reliable power to the following equipment:

- Emergency lighting
- Emergency auxiliary drives, e.g., turbine emergency lubrication oil pump and generator emergency seal oil pump
- Emergency valve operation
- Fire sirens
- Switchgear and control gear tripping mechanism
- Interlocks and protection
- Control equipment
- Sequence equipment
- Alarm and indication
- Remote control equipment

Figure 24.8 illustrates the dc scheme for the Heysham 2 power station which has two 660-MW generating units. A battery and a charger are provided for each unit. A standby battery and charger can also be made available to either unit when necessary. The charger is rated to supply the entire dc load for one unit plus 25 percent capacity to cover unknown and future loads.

REFERENCE

1. British Electricity International, *Modern Power Station Practice—Electrical Systems and Equipment, Volume D*, 3d ed., Pergamon Press, Oxford, United Kingdom, 1992.

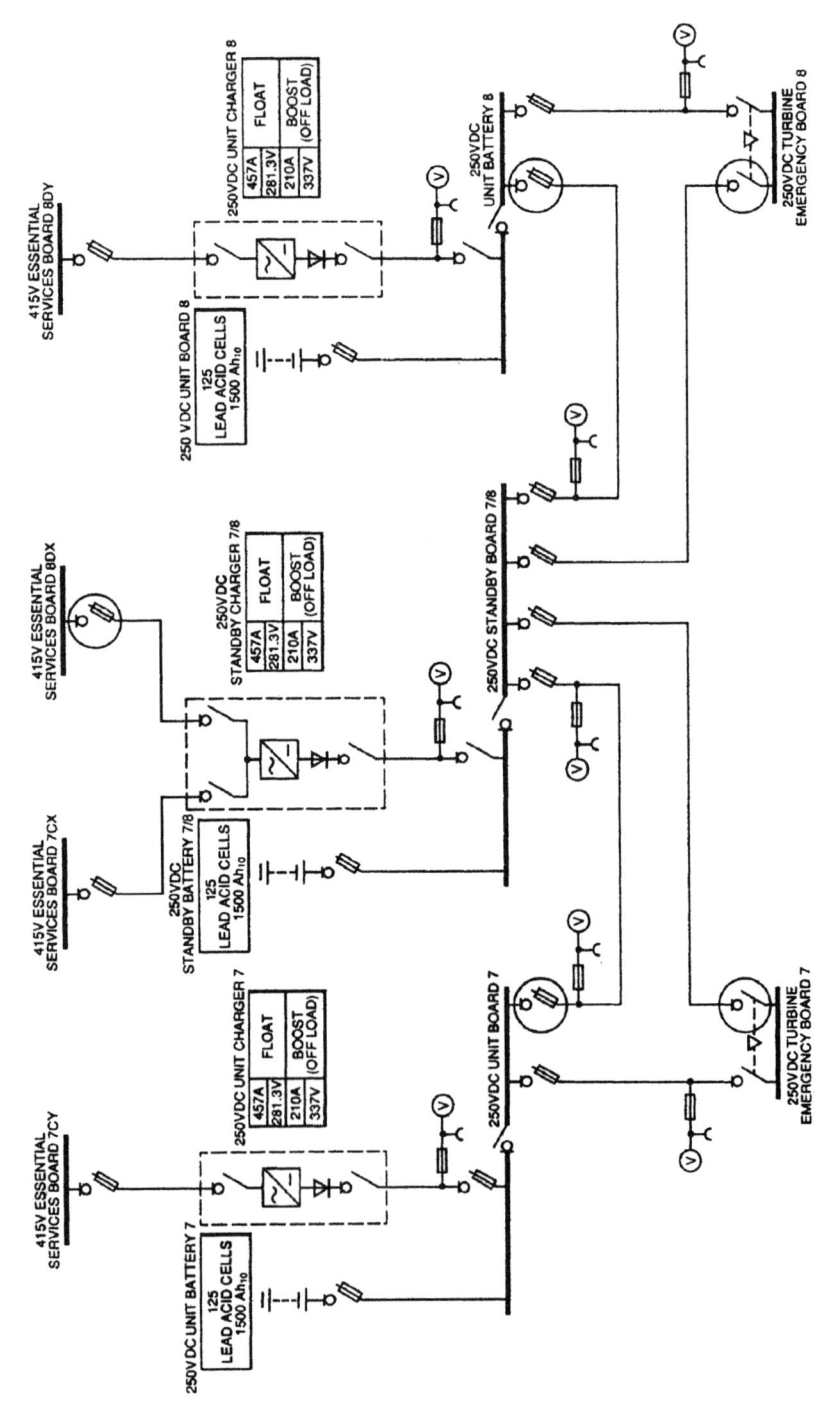

FIGURE 24.8 Heysham 2 power station, a 250-V dc system.

CHAPTER 25
POWER STATION PROTECTIVE SYSTEMS

INTRODUCTION

Figure 25.1 illustrates the overall protection scheme in a generating station. The auxiliary equipment inside the plant requires 5 to 10 percent of the power generated. The remainder goes to the grid. Since the plant is closely interconnected, a single failure requires more than the electrical and mechanical disconnection of the faulted system. This chapter covers the following topics:

1. The main unit protection. It includes the electrical protection of the plant equipment for which faults result in the tripping of one of the main plant systems.
2. The methods used to initiate the tripping of other associated plant systems.

DESIGN CRITERIA

The protective systems are designed to disconnect the faulty system when faults occur. The following are the main design requirements for the protective systems in power stations:

1. A system must be disconnected as quickly as possible when a fault occurs on it.
2. A secondary or a backup protection must be able to clear the fault if the main system protection did not clear it.
3. The system protection must be designed to match as closely as possible its operating characteristics; e.g., the generator negative sequence protection must be designed to match its thermal withstand to negative phase sequence currents.

The protection systems should generally be designed to prevent failure of a single protective device from causing a trip or allowing a fault to remain connected to the system. The exception occurs where the probability of failure of the protective device is very low such that the event is not considered credible, i.e., electromagnetic relays having a very low failure rate. These relays are used in electrical protection systems. They are normally installed in a fully dust-proofed environment where the temperature and humidity are controlled. All other protection systems using plant-mounted tripping devices rely on a "two out of three" logic to initiate a trip.

The protective relay output contacts must be allocated such that the operation of any relay will not trip more than one protective system. This is done to facilitate diagnostic testing while the unit is operating and one protective system is isolated. The trips which would remain active following a unit shutdown should be removed, e.g., the mechanical trips of the turbine and generator.

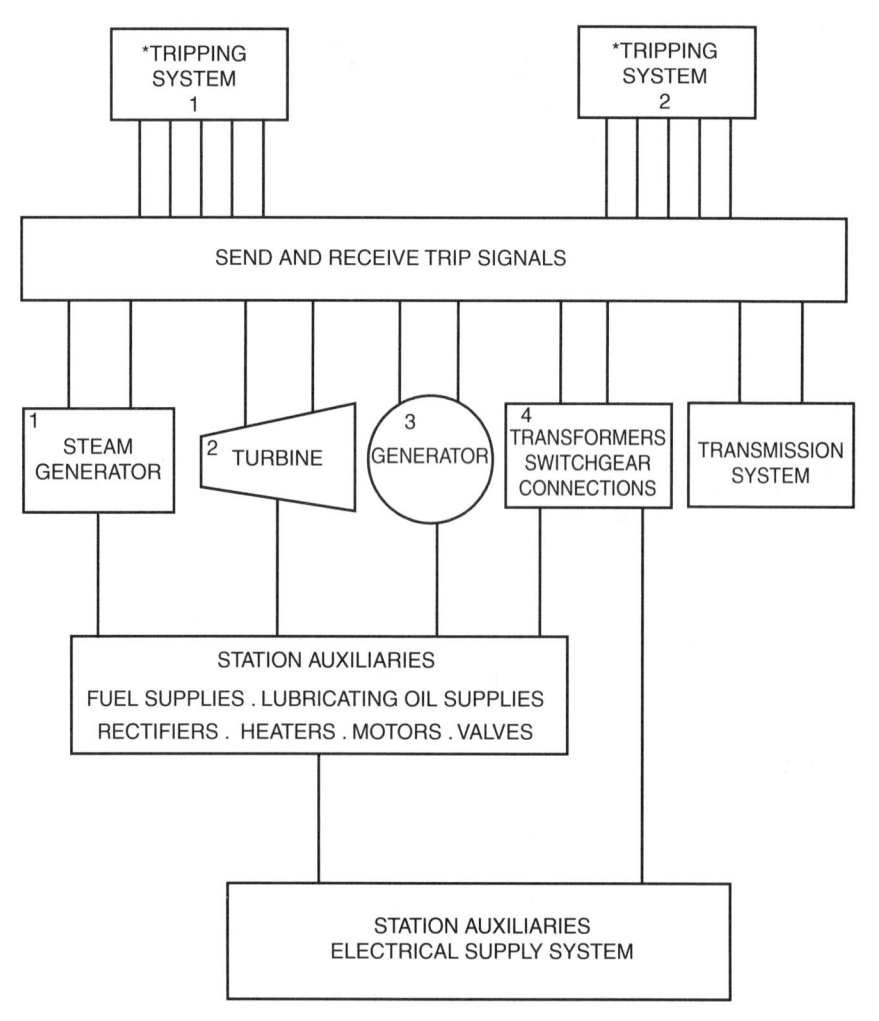

```
                    *TRIPPING                              *TRIPPING
                    SYSTEM                                 SYSTEM
                       1                                      2

              ┌─────────────────────────────────────────────────────┐
              │           SEND AND RECEIVE TRIP SIGNALS             │
              └─────────────────────────────────────────────────────┘

   ┌──────────┐  ┌──────────┐  ┌──────────┐  ┌─────────────┐  ┌─────────────┐
   │ 1        │  │          │  │    3     │  │ 4           │  │             │
   │  STEAM   │  │ 2 TURBINE│  │GENERATOR │  │TRANSFORMERS │  │TRANSMISSION │
   │GENERATOR │  │          │  │          │  │ SWITCHGEAR  │  │   SYSTEM    │
   │          │  │          │  │          │  │ CONNECTIONS │  │             │
   └──────────┘  └──────────┘  └──────────┘  └─────────────┘  └─────────────┘

                  ┌─────────────────────────────────────────┐
                  │           STATION AUXILIARIES           │
                  │  FUEL SUPPLIES . LUBRICATING OIL SUPPLIES│
                  │  RECTIFIERS . HEATERS . MOTORS . VALVES │
                  └─────────────────────────────────────────┘

                          ┌──────────────────────────┐
                          │    STATION AUXILIARIES   │
                          │ ELECTRICAL SUPPLY SYSTEM │
                          └──────────────────────────┘
```

*TRIPPING SYSTEMS 1 AND 2 CONTAIN UNIT TRIPPING LOGIC SHOWN IN FIG. 25.2

FIGURE 25.1 Overall generating protection scheme.

The equipment of a tripping system should be physically segregated from the other tripping systems as much as possible. This can be done by using separate relay panels and terminal blocks. The routing of the tripping system cables should also be different. The electrical connections for the protective relay circuits should be in accordance with the appropriate standards and codes.

Figure 25.2 illustrates a typical turbine protection system in a power station. There is a protection circuit for each of the main plant systems. A trip relay trips the unit when any of the protection devices shown operates. All fault conditions initiate the same signal to trip the unit. Category A faults are those requiring immediate disconnection of the unit from the grid when the fault occurs. Category B faults are those requiring the circuit breaker connecting the unit to the grid to remain closed following the fault to prevent rotor overspeed. The circuit breaker will eventually open on reverse power or low forward power.

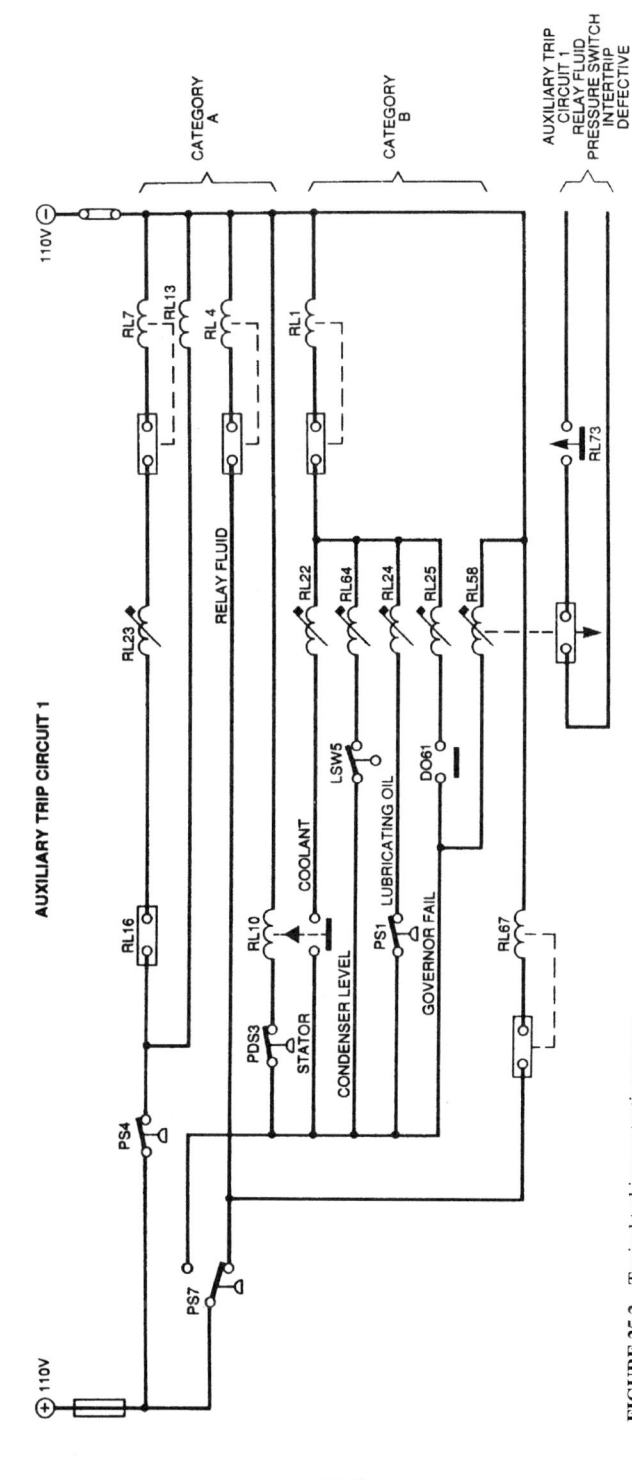

FIGURE 25.2 Typical turbine protection system.

25.4

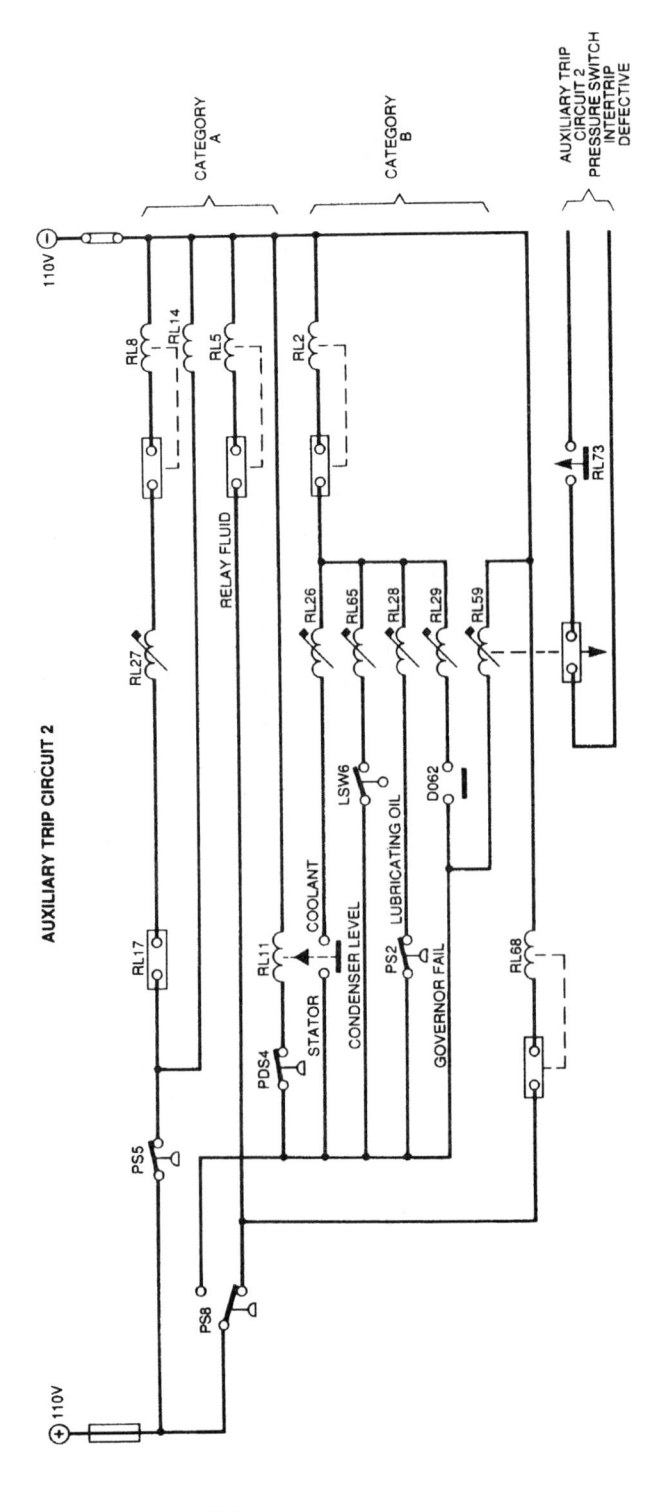

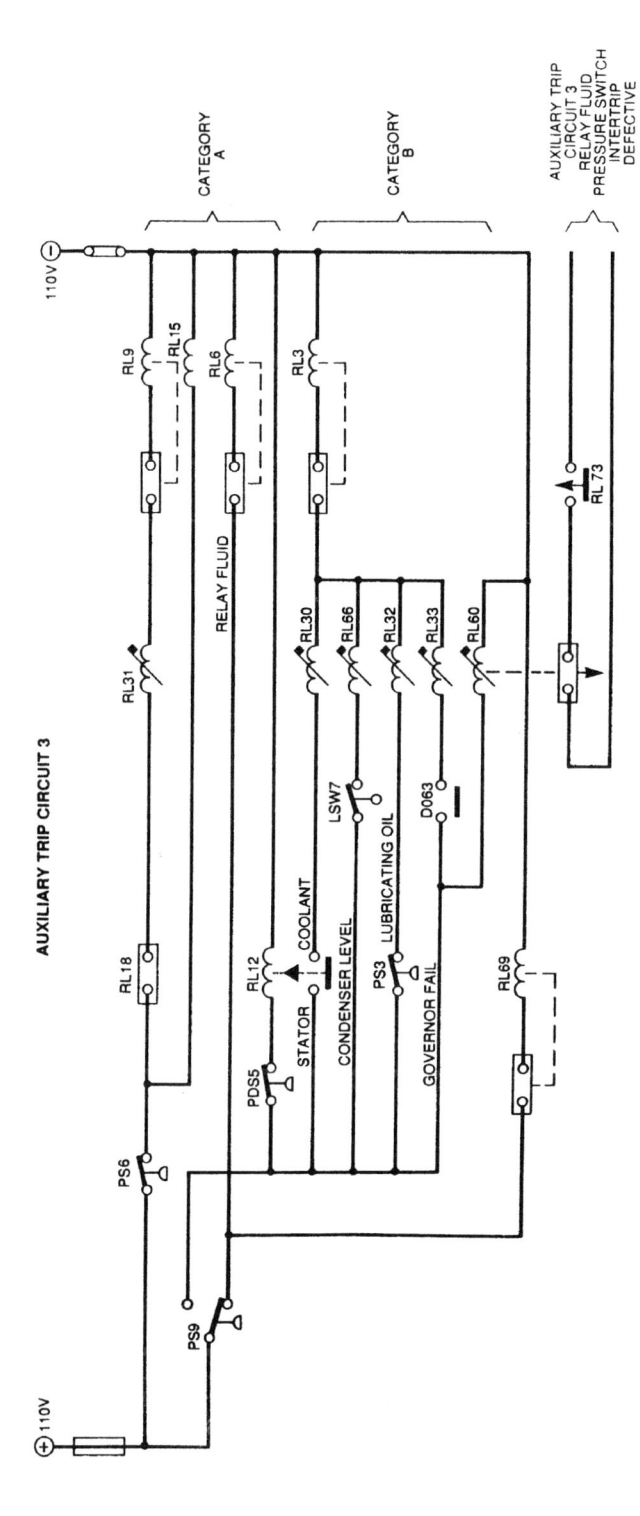

FIGURE 25.2 (*Continued*) Typical turbine protection system.

25.6

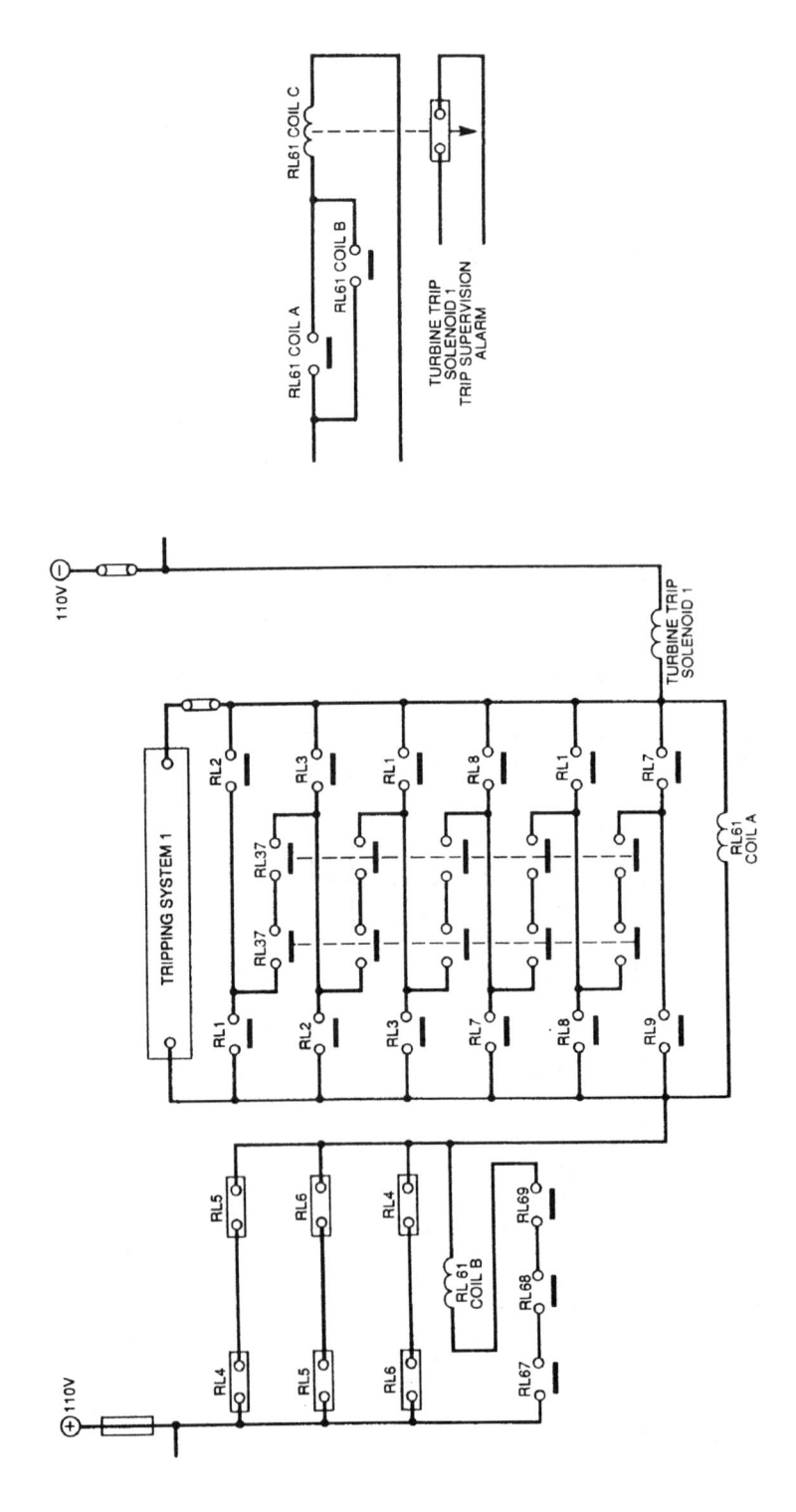

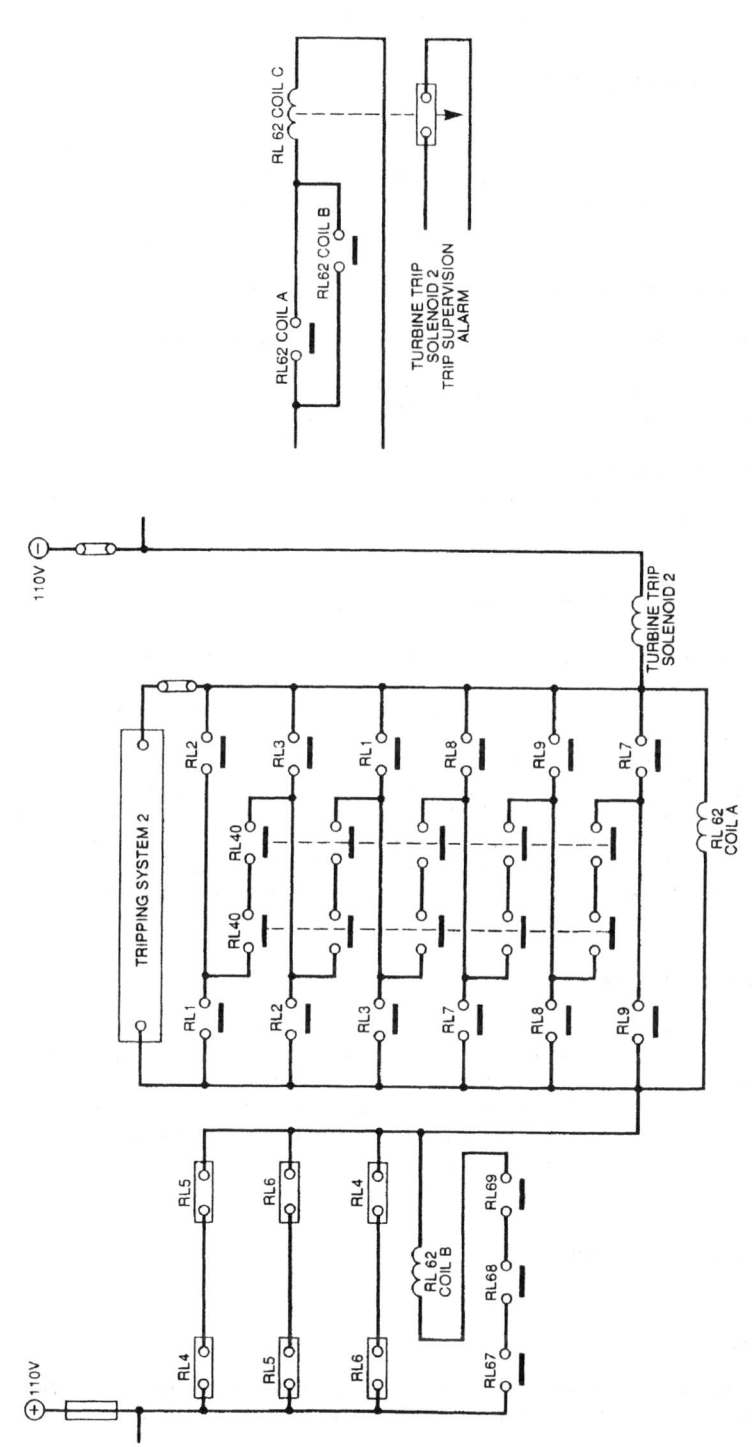

FIGURE 25.2 (*Continued*) Typical turbine protection system.

25.7

GENERATOR PROTECTION

The generator protection scheme includes the main connections and the windings of the generator transformer. There is an overlap between the protection arrangement of the generator and its transformer. Figure 25.3 lists the generator protective systems.

Stator Ground (Earth) Faults—Low-Impedance Grounding

The generator star point is grounded through a low-resistance liquid grounding (earthing) resistor, as shown in Fig. 25.4. The objective is to limit the fault current to 300 A for all ratings of generators. A single 300-A current transformer (CT) in the generator neutral line supplies two relays connected in series. These relays have an instantaneous and inverse time characteristic, respectively. The inverse time characteristic relay is set at 5 percent and the instantaneous relay at 10 percent. This method was replaced with the one described in the following section due to core burning events that occurred at a ground fault current up to 300 A.

Stator Ground Faults—High-Resistance Grounding

The modern practice for generators rated around 660 MW is to design the stator ground fault protection system for neutral currents of 10 to 15 A. The previous practice for designing for 2 to 3 A has led to problems in locating the ground faults which activated the protection system.

Figure 25.5 illustrates the ground fault protection system using resistance grounding through a distribution transformer. It includes two protection methods. Relay R1 is operated by a current transformer and relay R2 by a voltage transformer (VT). Both methods use the same relay which consists of an induction disk with an adjustable inverse time/voltage characteristic (Fig. 25.6). The interposing voltage transformer limits the maximum voltage across the relay during fault conditions to a value lower than its continuous voltage.

Stator Phase-to-Phase Faults

Phase-to-phase faults occur normally in the end windings. However, they can also occur in the slots if the slots contain windings of different phases. In this case, the phase-to-phase fault will cause a ground fault within a short time. The ground fault protection will clear the fault after a time delay due to the inverse time characteristic of the relay. The circulating current principle is used to provide fast protection against phase-to-phase faults. It relies on a differential protection using either biased relays or high-impedance relays.

DC TRIPPING SYSTEMS

Logic Diagram

Figure 25.7 illustrates the overall protection logic diagram for a generating system. The group definitions are listed on the left-hand side of the drawings. They identify which

GROUP DEFINITIONS

1. A DIRECT TRIP TO SHUT DOWN THE UNIT (SEE NOTE (a)) WITH DISCONNECTION FROM THE SYSTEM BY OPENING THE HV CIRCUIT BREAKER (PREPARES CB FAIL PROTECTION).
2. A DIRECT TRIP TO SHUT DOWN THE UNIT (SEE NOTE (a)) WITH DISCONNECTION FROM THE SYSTEM BY OPENING THE GENERATOR VOLTAGE CIRCUIT BREAKER.
3. AN INDIRECT TRIP TO SHUT DOWN THE UNIT (SEE NOTE (a)) BY TRIPPING THE TURBINE/BOILER (FROM LOSS OF POWER FLUID) & COMPLETING A UNIT TRIP VIA THE LOW FORWARD POWER RELAY WITH DISCONNECTION FROM THE SYSTEM BY OPENING THE HV CIRCUIT BREAKER.
4. AN INDIRECT TRIP TO SHUT DOWN THE UNIT (SEE NOTE (a)) BY TRIPPING THE TURBINE/BOILER (FROM LOSS OF POWER FLUID) & COMPLETING A UNIT TRIP VIA THE LOW FORWARD POWER RELAY WITH DISCONNECTION FROM THE SYSTEM BY OPENING THE GENERATOR VOLTAGE CIRCUIT BREAKER.

FIGURE 25.3 Overall protection logic diagram for main generating units.

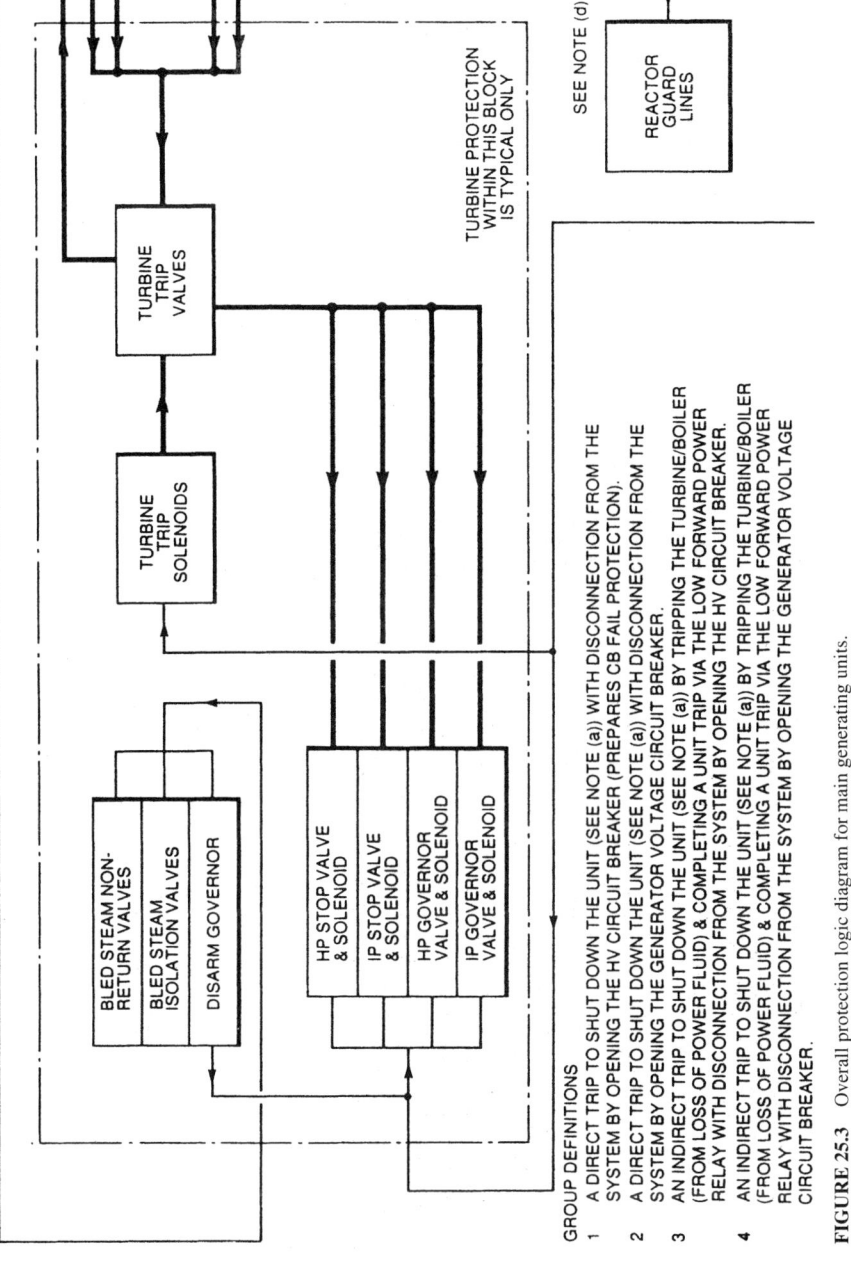

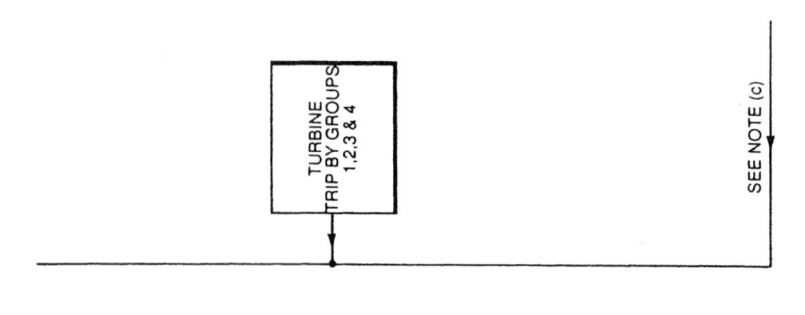

TURBINE TRIP BY GROUPS 1, 2, 3 & 4

SEE NOTE (c)

5   A TRIP TO OPERATE THE MAIN EXCITER FIELD CIRCUIT BREAKER & GENERATOR VOLTAGE CIRCUIT BREAKER WHEN THE GENERATOR HV CIRCUIT BREAKER IS OPEN & THE GENERATOR VOLTAGE CIRCUIT BREAKER IS CLOSED.

6   A TRIP TO UNIT TRANSFORMER CIRCUIT BREAKERS.

7   A TRIP OF A SECTION OF HV BUSBARS WHEN THE GENERATOR HV CIRCUIT BREAKER FAILS TO OPEN FOR A GROUP 1 TRIP.

NOTES

(a) A DIRECT TRIP MEANS A TRIP OF THE TURBINE TRIP SOLENOIDS HP & IP GOVERNOR & STOP VALVES SOLENOIDS SENDING A SIGNAL TO THE REACTOR FROM LOSS OF POWER FLUID (THIS ENSURES THE TURBINE IS TRIPPED FIRST FOR ALL EXCEPT REACTOR PROTECTION, MAIN EXCITER FIELD CIRCUIT BREAKER & FOR GROUPS 1 & 3 UNIT TRANSFORMERS LV BREAKERS). AN INDIRECT TRIP IS A TRIP OF THE TURBINE, A SIGNAL TO THE BOILER/REACTOR & A TRIP OF THE OTHER PLANT ITEMS FOLLOWING OPERATION OF THE LOW FORWARD POWER RELAY.

(b) WHERE TEN MINUTES RUN-THROUGH IS POSSIBLE, BUSBAR PROTECTION & BACK-TRIP RECEIVE, TRIP THE HV CIRCUIT BREAKER ONLY

(c) AN ADDITIONAL TRIP INTO THE STOP, GOVERNOR & TURBINE TRIP VALVES SOLENOID CIRCUITS.

(d) INTERLOCKED BY CIRCUIT ISOLATORS & CIRCUIT BREAKER POSITION SWITCHES SO THAT WHEN THE GENERATOR IS ON OPEN CIRCUIT THE BOILER OR REACTOR IS NOT TRIPPED. THE SIGNAL MUST BE TIME DELAYED SO THAT THE LOSS OF POWER FLUID PRESSURE SWITCHES OPERATE BEFORE THE CIRCUIT BREAKER OPENS.

KEY DIAGRAM

GRID VOLTAGE
GENERATOR TRANSFORMER
UNIT TRANSFORMERS
MAIN GENERATOR
EARTHING TRANSFORMER
11kV UNIT BOARD A
11kV UNIT BOARD B

25.10

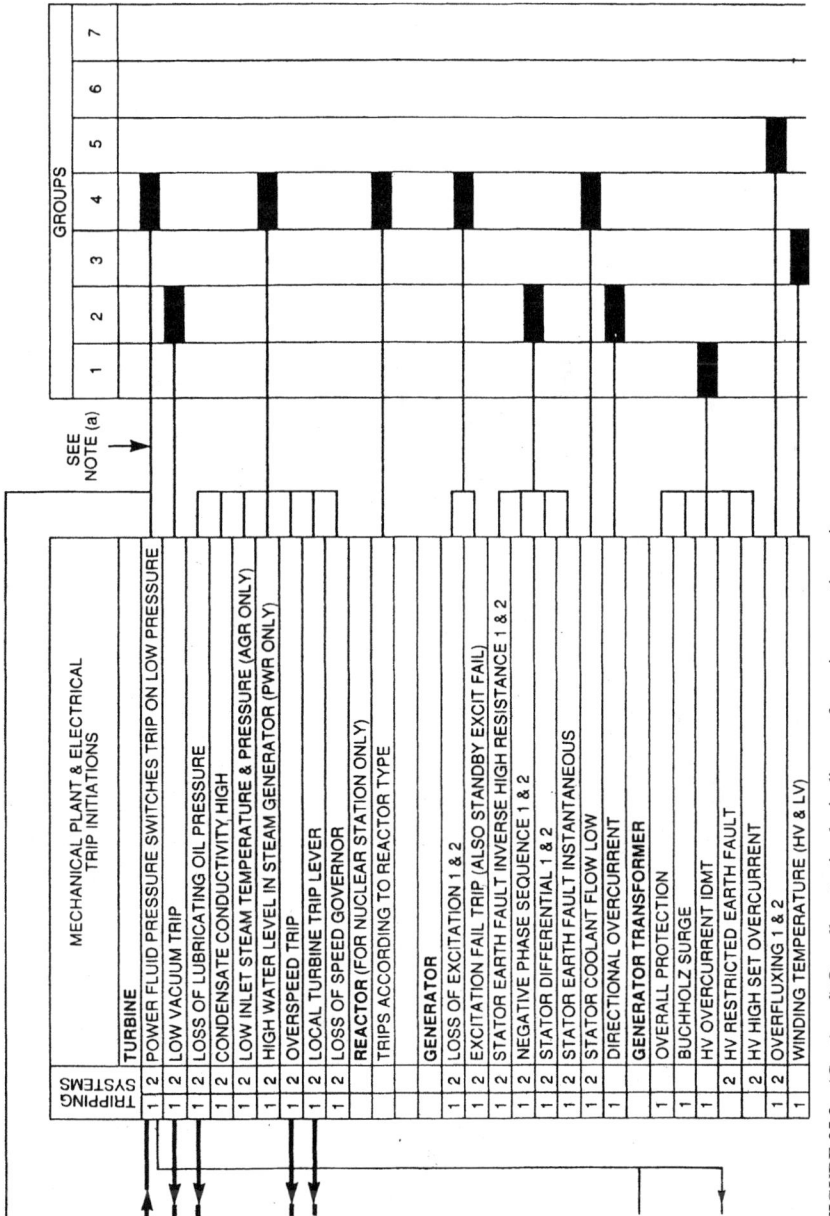

FIGURE 25.3 *(Continued)* Overall protection logic diagram for main generating units.

	UNIT TRANSFORMER(S)
2	HS OVERCURRENT
1	OVERALL PROTECTION
1	BUCHHOLZ SURGE
1	HV IDMT OVERCURRENT (2nd STAGE)
2	LV RESTRICTED EARTH FAULT
2	LV STANDBY EARTH FAULT (2nd STAGE)
1	HV IDMT OVER CURRENT (1st STAGE)
2	LV STANDBY EARTH FAULT (1st STAGE)
1	WINDING TEMP (NOT REQD ON AN TRANSFORMERS)
	HV CONNECTIONS (POWER STATION)
1	FIRST MAIN FEEDER PROTECTION
2	SECOND MAIN FEEDER PROTECTION
1	FIRST INTERTRIP RECEIVE
2	SECOND INTERTRIP RECEIVE
	GENERATOR HV FEEDER (TRANSMISSION STN)
1	HV BUSBAR PROTECTION
2	HV BUSBAR BACK TRIP RECEIVE
1	FIRST MAIN FEEDER PROTECTION
2	SECOND MAIN FEEDER PROTECTION
1	FIRST INTERTRIP RECEIVE
2	SECOND INTERTRIP RECEIVE
2	TRIP INITIATION HV CIRCUIT BREAKER FAIL
1	EARTHING TRANSFORMER
1	EARTHING TRANSFORMER BUCHHOLZ
2	EARTH FAULT INVERSE HIGH RESISTANCE 1 & 2
	GENERATOR VOLTAGE CIRCUIT BREAKER
2	FAIL PROTECTION
	COMMON EQUIPMENT
2	EMERGENCY STOP BUTTON - CCR
2	EMERGENCY STOP BUTTON - LOCAL
1	LV CONNECTIONS (POWER STATION)
2	LV CONNECTION PROTECTION 1 & 2

FIGURE 25.3 (*Continued*) Overall protection logic diagram for main generating units.

POWER STATION PROTECTIVE SYSTEMS 25.13

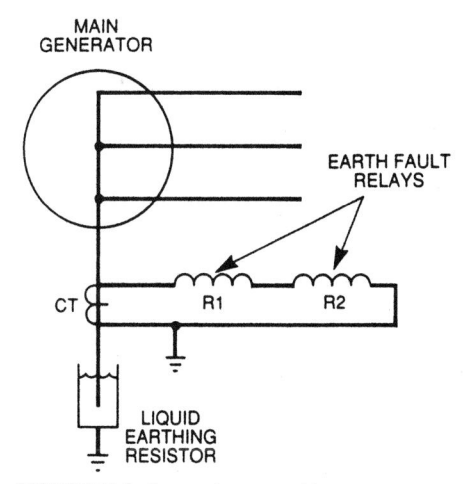

FIGURE 25.4 Low resistance earthing.

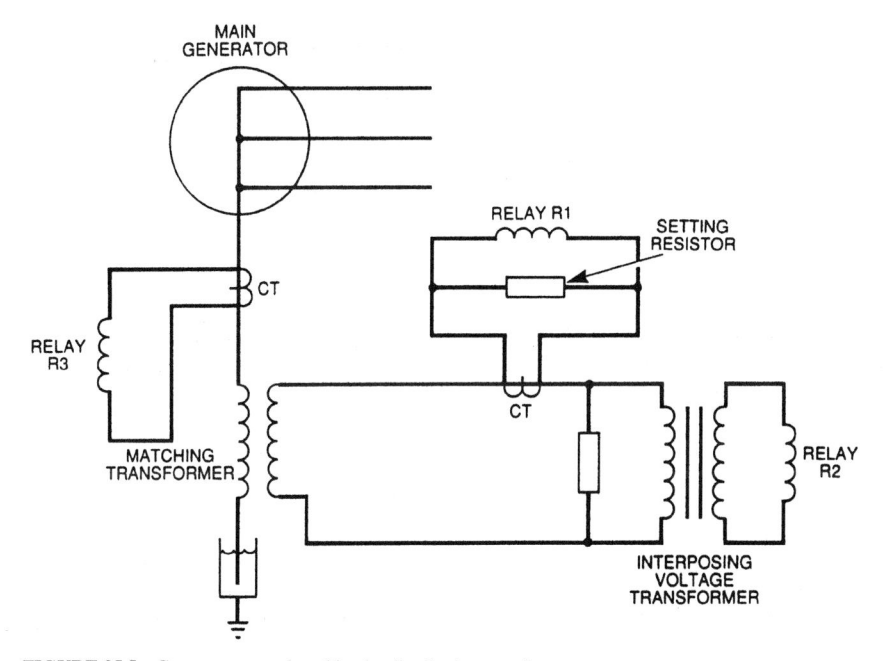

FIGURE 25.5 Generator neutral earthing by distribution transformer.

system should be tripped to bring the unit into a safe state. The logic diagram has the following format:

- The group 1 trips are category A. They require immediate opening of the generator high-voltage circuit breaker. This circuit breaker is located between the main generator and the generator transformer.

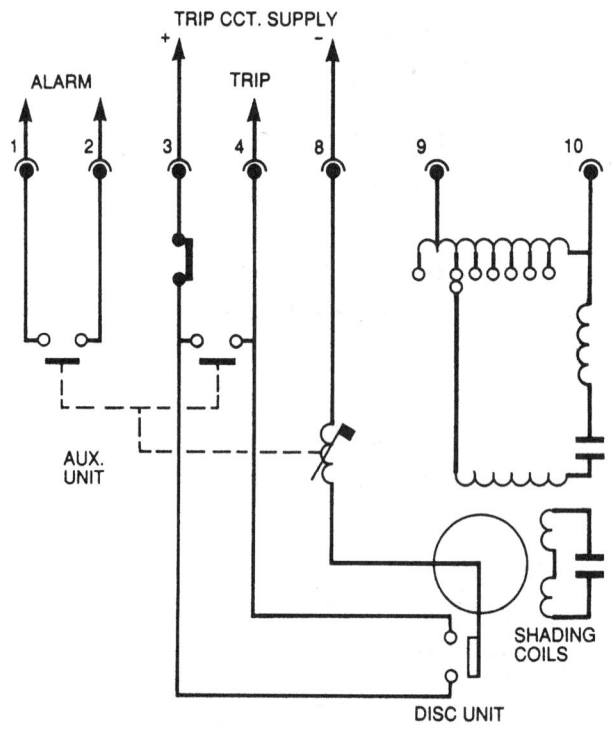

FIGURE 25.6 Earth fault protection relay.

- The group 2 trips are also category A. They require immediate opening of the generator voltage circuit breaker to clear the fault. This circuit breaker is located between the generator transformer and the grid. In this configuration, the generator will remain connected to the unit transformer.

- The group 3 trips are category B. They require opening of the generator high-voltage circuit breaker on low forward power.

- The group 4 trips are also category B. They require opening of the generator voltage circuit breaker on low forward power.

Figure 25.7 illustrates the overall protection schematic diagram for the generating station. A number is associated with each trip-initiating device. This number is cross-referenced to Fig. 25.3. The group numbers are shown on both the tripping schematic and the logic diagram.

REFERENCE

1. British Electricity International, *Modern Power Station Practice—Electrical Systems and Equipment, Volume D*, 3d ed., Pergamon Press, Oxford, United Kingdom, 1992.

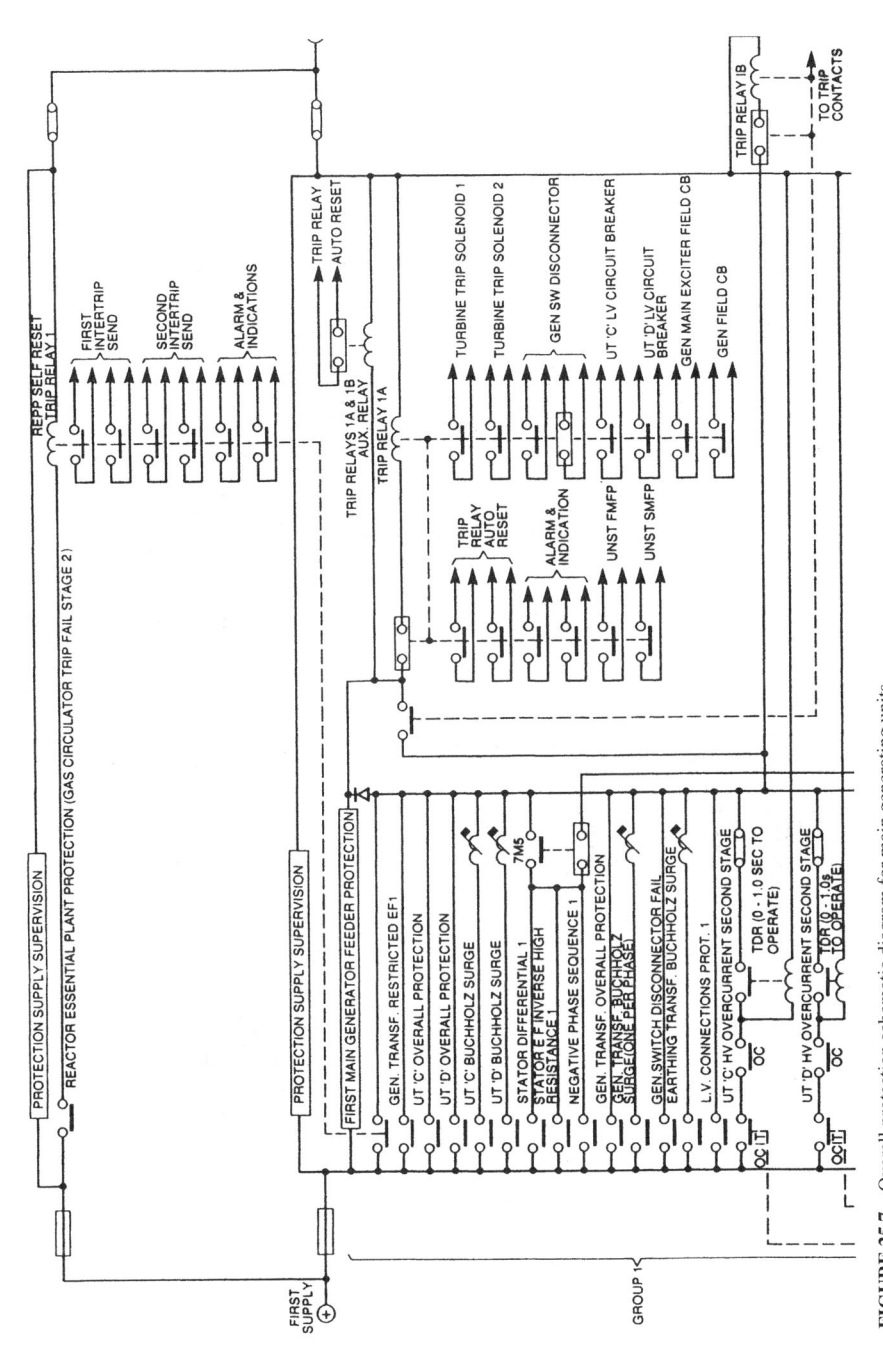

FIGURE 25.7 Overall protection schematic diagram for main generating units.

25.15

25.16

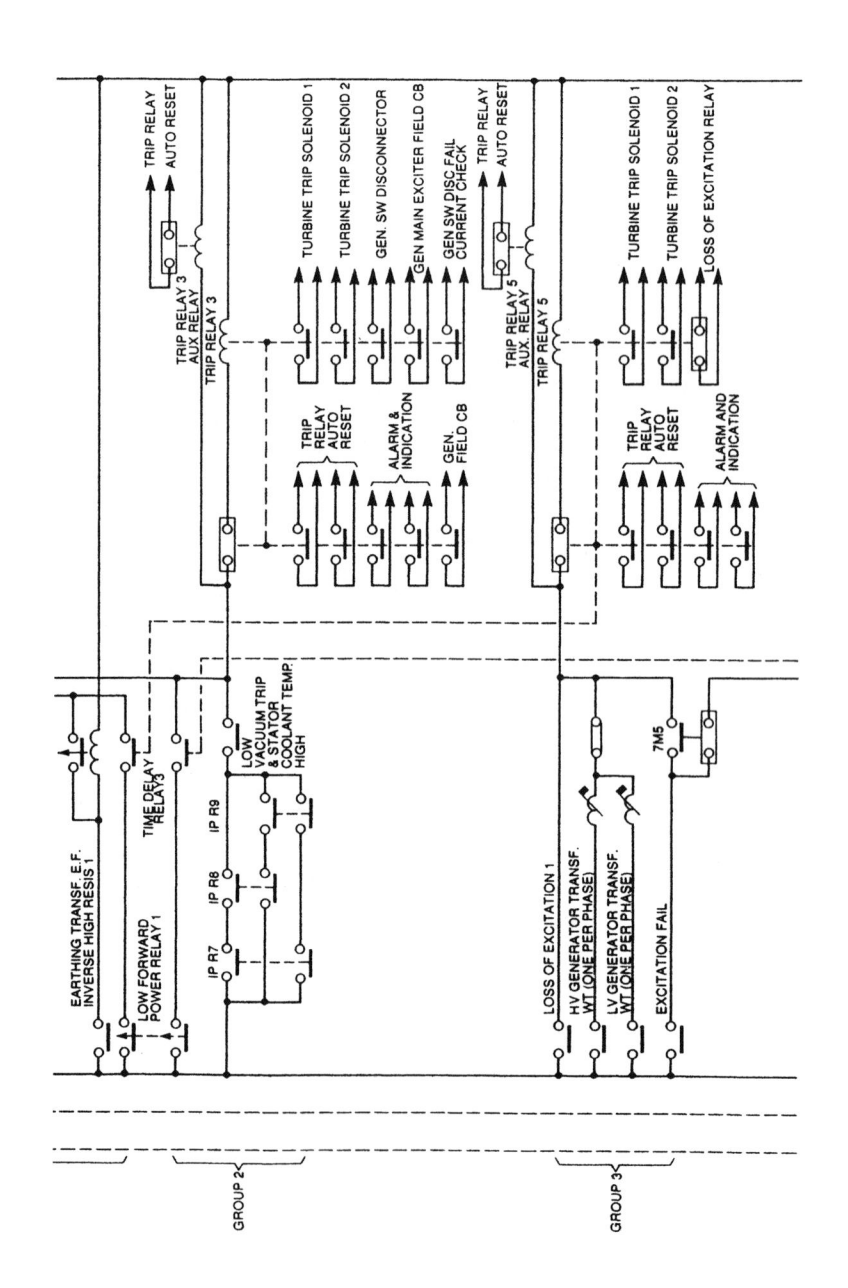

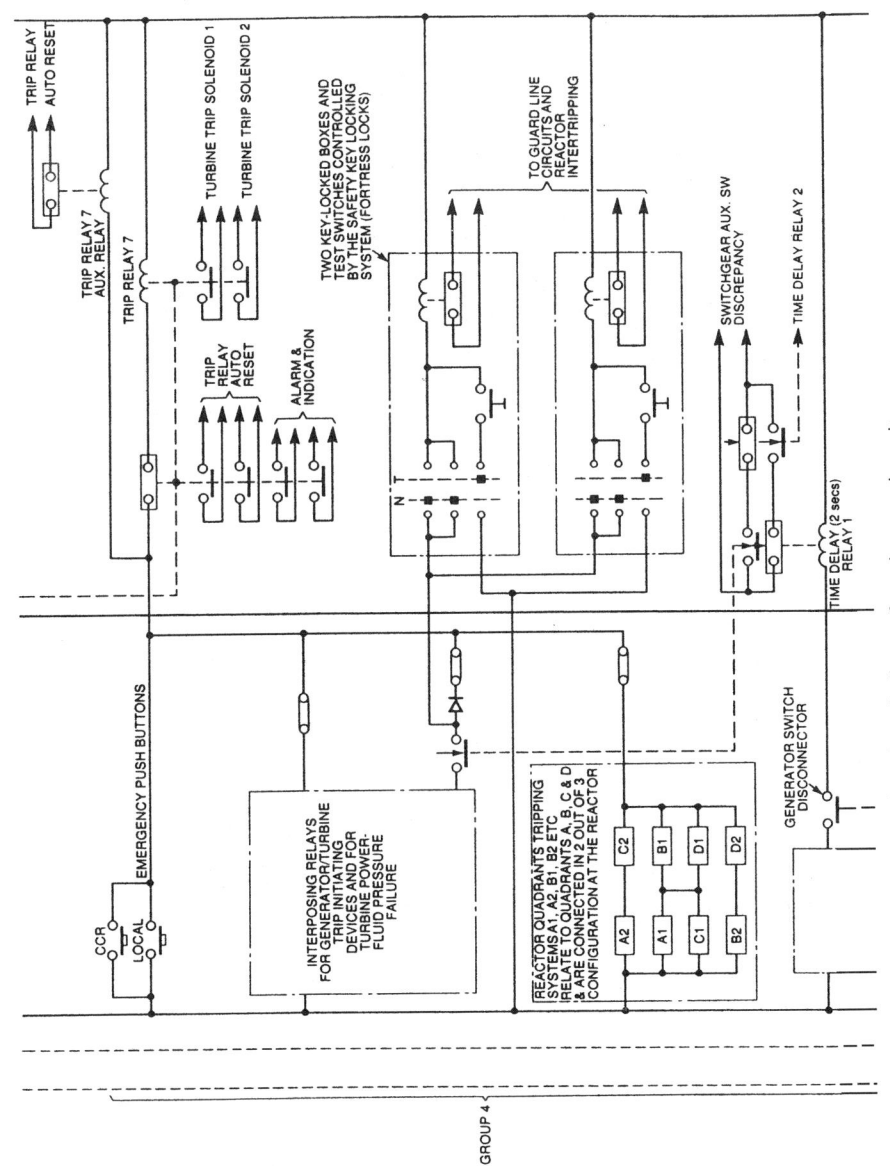

FIGURE 25.7 (*Continued*) Overall protection schematic diagram for main generating units.

25.17

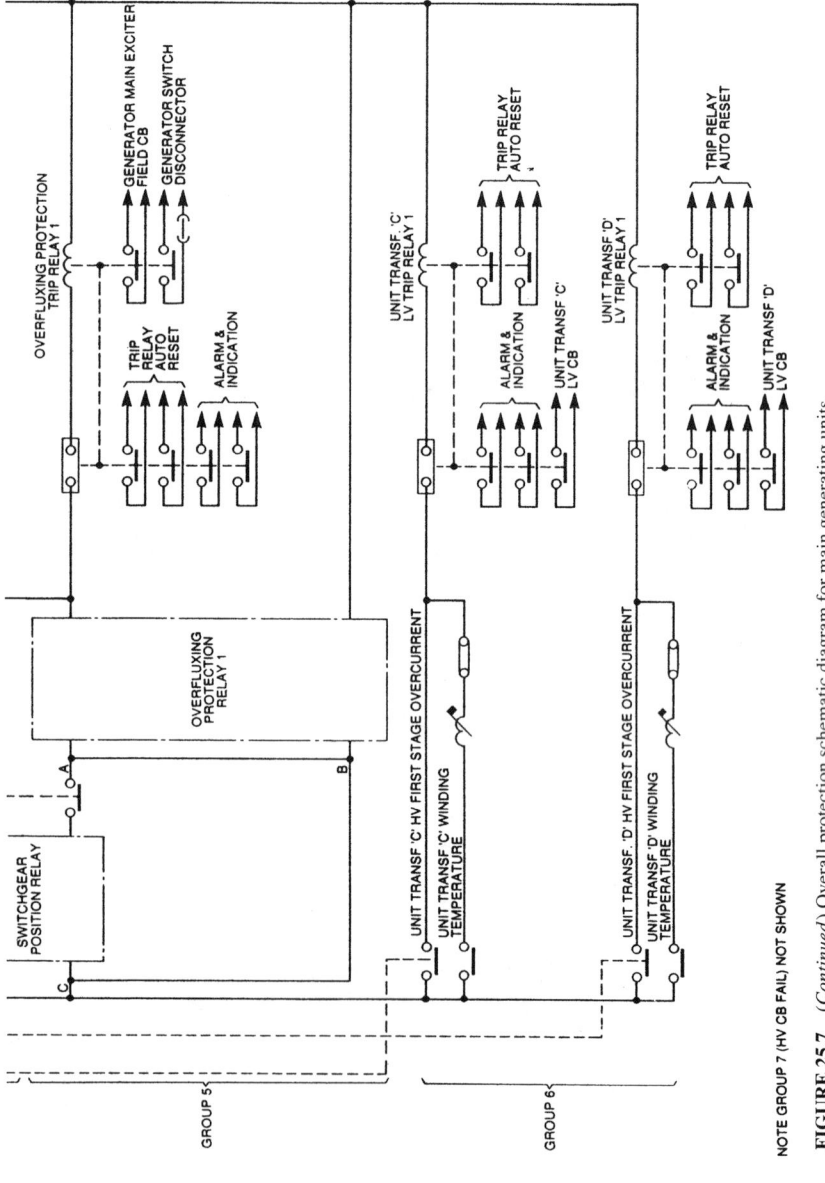

FIGURE 25.7 *(Continued)* Overall protection schematic diagram for main generating units.

INDEX

AC machine fundamentals, **5.**1–**5.**12
  induced torque in AC machine, **5.**9–**5.**10
    induced torque, 5.10
    magnetic field from rotor, **5.**9
    magnetic field from stator, **5.**9
    synchronous machine, **5.**9
  induced voltage in AC machines, **5.**6
    rotating magnetic field, **5.**6
  induced voltage in coil on two-pole stator, **5.**6–**5.**8
    right-hand rule, **5.**6
    stationary coil, **5.**6
    stationary magnetic field, **5.**6
  induced voltage in three-phase set of coils, **5.**8
    rotor magnetic field, **5.**8
  power flows and losses, **5.**11–**5.**12
    losses in AC machines, **5.**11
    mechanical losses, **5.**12
    no-load rotational losses, **5.**12
    rotor and stator copper losses, **5.**11
  proof of the rotating magnetic flux concept, **5.**3
    angular velocity **5.**3
  relationship between electrical frequency and speed of magnetic field rotation, **5.**3–**5.**6
    angular speed of rotation, **5.**3
    electrical frequency, **5.**4
    four-pole stator, **5.**4
    magnetic poles, **5.**3
    mechanical frequency, **5.**4
  reversing direction of magnetic field rotation, **5.**6
    direction of magnetic field, **5.**6
  rms voltage in three-phase stator, **5.**8–**5.**9
  rotating magnetic field, **5.**1–**5.**3
    direction of magnetic flux, **5.**3
    resulting magnetic flux densities, **5.**1
    rotating magnetic field of constant magnitude, **5.**1

AC machine fundamentals (*Cont.*):
  three-phase set of currents, **5.**1
  total magnetic field, **5.**2
  winding insulation in AC machines, **5.**10–**5.**11
    rule of thumb, **5.**10
    shorted insulation, **5.**10
    temperature rise above ambient, **5.**11
Alternating currents, **1.**15–**1.**21
  alternating current generator, **1.**15
  capacitive circuit, **1.**16–**1.**17
  inductive circuit, **1.**18–**1.**21
    apparent power, **1.**20
    inductive reactance, **1.**19
    inrush current, **1.**20
    instantaneous values, **1.**18
    phase angle of system, **1.**19
    reactive or inductive power, **1.**19
    real power, **1.**19
    synchronous motors, **1.**20
  power distribution systems, **1.**15
  resistive circuit, **1.**15–**1.**16
Ampere's law, **1.**7–**1.**8
  bubble chamber, **1.**7
  magnets, **1.**7
  permeability constant, **1.**8
Bearings:
  materials and finish, **21.**3
    hardness, **21.**3
    Rockwell, **21.**3
  sizes of bearings, **21.**3–**21.**5
    aircraft engines, **21.**3
    dimensions of rolling bearings, **21.**4
    fillet radius, **21.**4
    medium series, **21.**5
    pumping stations, **21.**3
    self-aligning ball bearing, **21.**5
  statistical nature of bearing life, **21.**3
    failure rate, **21.**3

Bearings (*Cont*):
  thrust bearings, **21**.7
    cylindrical rollers, **21**.7
    thrust bearing, **21**.7
  types of bearings, **21**.1–**21**.3
    ball and roller bearings, **21**.1–**21**.2
    inboard bearings, **21**.1
    stress during rolling contact, **21**.2–**21**.3
  types of roller bearings, **21**.5–**21**.7
    angular-contact bearing, **21**.6
    deep-groove ball bearing, **21**.5
    double-row ball bearings, **21**.6–**21**.7
    self-aligning ball bearings, **21**.15
  (*See also* Lubrication)
Capacitors, **1**.1–**1**.3
  applications, **1**.2
  capacitance, **1**.1
  dielectric constant, **1**.2
  dielectric strength, **1**.3
  flashover, **1**.3
Circuit breakers, **19**.1–**19**.15
  application, **19**.6
    power station auxiliaries, **19**.6
    arc splitter type, **19**.6
    short arcs, **19**.6
  conventional circuit breakers, **19**.4
    air-break circuit breakers, **19**.4
    automatic switch, **19**.4
  magnetic blowout type, **19**.4–**19**.6
    arc chutes, **19**.5
    blowout coils, **19**.5
    deionization, **19**.5
    magnetic blast, **19**.4
  methods for increasing arc resistance, **19**.4
    arc lengthening, **19**.4
  oil circuit breakers, **19**.6–**19**.9
    advantages of oil, **19**.7
    arc control circuit breakers, **19**.8–**19**.9
    disadvantages of oil circuit breakers, **19**.8
    generate hydrogen gas, **19**.6
    heat conductivity, **19**.6
    heavy short-circuit currents, **19**.7
    plain break oil circuit breakers, **19**.8
  physics of arc phenomena, **19**.1–**19**.3
    arc interruption theory, **19**.2–**19**.3
    breakdown voltage, **19**.1
    circuit emf, **19**.1
    degree of dissociation, **19**.2
    degree of ionization, **19**.2
    high-resistance interruption, **19**.3
    low resistance on current zero interruption, **19**.3
  plain break type, **19**.4

Circuit breakers (*Cont*):
  convection currents, **19**.4
  rating, **19**.3–**19**.4
    rated breaking capacities, **19**.3
    rated frequency, **19**.3
  recent developments, **19**.9–**19**.15
    application of vacuum switches, **19**.11
    construction of vacuum switch/circuit breakers, **19**.10
    sulfur hexafluoride (SF_6) circuit breakers, **19**.11–**19**.15
    theory of circuit interruption, **19**.1
    vacuum chamber, **19**.10
    vacuum circuit breakers, **19**.9
Current and resistance, **1**.4–**1**.6
  conductivity, **1**.6
  electric field, **1**.4
  resistivity, **1**.4
Faraday's law of induction, **1**.9–**1**.11
  electromagnetic force, **1**.10
  electromagnetism, **1**.9
Fuses, **20**.1–**20**.7
  advantages of fuses over circuit breakers, **20**.6
    blown fuse, **20**.6
    maintenance-free, **20**.6
  electrical system protection considerations, **20**.6–**20**.7
    core performance requirements, **20**.6–**20**.7
  features of current-limiting fuses, **20**.4–**20**.6
    fuseholder, **20**.4
    interrupting current, **20**.4
    moderate overloads, **20**.4
    time-current characteristics, **20**.5
  types of fuses, **20**.1
    current-limiting fuses, **20**.2
    dual element fuses, **20**.1
    single-element, **20**.1
Generator bearings and seals, **13**.13–**13**.14
  annular groove, **13**.13
  jacking oil taps, **13**.13
Generator brushless excitation systems, **13**.47
  rotating armature main exciter, **13**.47
Generator cooling systems, **13**.31–**13**.35
  hydrogen cooling components, **13**.29–**13**.31
    dew point of hydrogen, **13**.29
    explosive mixture, **13**.29
    resonance, **13**.29
  hydrogen cooling system, **13**.31–**13**.35
    gassing-up and degassing, **13**.31
    generator gas systems, **13**.34
    hygrometer, **13**.35
    katharometer-type purity monitor, **13**.35
    motor-driven blower, **13**.35

INDEX

Generator cooling systems: (*Cont*):
  other cooling systems, **13.**43
  hydrogen cooler, **13.**43
  hydrogen detectors, **13.**43
  rotating exciter, **13.**43
Generator core frame, **13.**17
  end plate flux shield, **13.**17
Generator electrical connections and terminals, **13.**26
  terminal bushing, **13.**26
Generator end winding support, **13.**26
  core ovalization, **13.**26
  electromagnetic forces, **13.**26
  molded glass fiber, **13.**26
  thermosetting resin, **13.**26
Generator excitation, **13.**43–**13.**47
  AC excitation system, **13.**43
  control, **13.**53
    over fluxing limit, **13.**53
    rotor current limiter, **13.**53
  exciter performance testing, **13.**46
  exciter transient performance, **13.**43
  main exciter, **13.**46
  pilot exciter, **13.**43
  pilot exciter protection, **13.**46
  salient-pole permanent generator, **13.**47
  system analysis, **13.**55
    dynamic stability, **13.**55
    transient stability, **13.**55
Generator exciter power system (GEP) characteristics, **13.**55
  pumped-storage plants, **13.**55
  rotor angle, **13.**55
Generator fans, **13.**9
  axial-type fans, **13.**9
  centrifugal-type fans, **13.**9
Generator flux probe text, **18.**11
  flux probe, **18.**11
  interturn fault, **18.**11
Generator high-speed balancing, **18.**9–**18.**10
  dynamic stability, **18.**9
Generator inspection and maintenance, **17.**1–**17.**12
  (*See also* Generator off-load maintenance; Generator on-load maintenance and monitoring)
Generator insulation, **18.**8–**18.**9
  all strip turn insulation, **18.**8
  mica mat tape, **18.**8
  Nomex, **18.**9
Generator main connections, **14.**1–**14.**5
  introduction, **14.**1

Generator main connections (*Cont*):
  isolated phase bus bar, **14.**1
  rating of bus bar installation, **14.**1
  isolated phase bus bar circulatory currents, **14.**1–**14.**3
  electrically continuous IPBB with short-circuit, **14.**1
  enclosures, **14.**1
  neutral connections, **14.**2
  unit transformer, **14.**2
  system description, **14.**3
  generator transformer, **14.**3
  high-voltage end, **14.**3
  synchronization, **14.**3
Generator off-load maintenance, **17.**4–**17.**7
  automatic voltage regulator, **17.**6
    motorized potentiometers, **17.**6
  exciter and pilot exciter, **17.**6
    inspection, **17.**6
  field switch, **17.**6
    overcurrent trip, **17.**6
  generator testing, **17.**7–**17.**12
    hydrogen loss test, **17.**12
    insulation testing, **17.**7–**17.**9
    rotor winding tests, **17.**12
    stator coolant circuit testing, **17.**10–**17.**12
    testing of stator core, **17.**9–**17.**10
  rectifier, **17.**6
    diodes, **17.**6
  rotor, **17.**5
    end rings, **17.**5
  slip rings and brush gear, **17.**5
    brush gear enclosure, **17.**5
    spring tension, **17.**5
    turning gear, **17.**5
  stator external work, **17.**4–**17.**5
    hydrogen coolers, **17.**4
    main connections, **17.**5
    stator water system, **17.**4
  stator internal work, **17.**4
    clean-conditions system, **17.**4
    hot spots, **17.**4
  supervisory and protection equipment, **17.**7
    protection system, **17.**7
Generator on-load maintenance and monitoring, **17.**1–**17.**3
  excitation system, **17.**3
    brush vibration, **17.**3
    brushless system, **17.**3
    slip ring systems, **17.**3
  rotor, **17.**2–**17.**3
    pedestal bearing insulation, **17.**2–**17.**3**

Generator on-load maintenance and monitoring (*Cont*):
    rotor ground fault detection, **17.2**
    shaft voltage measurement, **17.2**
    vibration monitoring, **17.2**
    stator, **17.1–17.2**
        core monitor, **17.2**
        end winding vibration, **17.1**
        hydrogen leakage, **17.2**
        radio-frequency monitor, **17.2**
        stator water flow, **17.1**
        temperature measurement, **17.1**
Generator operation, **13.55–13.58**
    application of a load, **13.57**
    capability chart, **13.57**
    neutral grounding, **13.57**
    open-circuit conditions and synchronizing, **13.56**
    rotor torque, **13.58**
    running up to speed, **13.55**
Generator operational problems, **18.1–18.7**
    collector, bore copper, and connection problems, **18.4**
    contamination, **18.4**
    forging concerns, **18.5–18.6**
        frequent start-stops, **18.6**
        moderate copper distortion, **18.5**
        negative sequence currents, **18.5**
        single forging, **18.5**
    misoperation, **18.6–18.7**
        abnormal system conditions, **18.7**
        catastrophic consequences, **18.6**
    retaining rings, **18.6**
        hoop stress, **18.6**
        nonmagnetic rings, **18.6**
    shorted turns and field grounds, **18.1**
        abnormal operating incidents, **18.2**
        contamination, **18.2**
        operation time, **18.1**
        type of operation, **18.2**
    thermal sensitivity, **18.3–18.4**
        blocked ventilation, **18.3**
        shorted turns, **18.3**
        uneven end winding insulation, **18.3**
        uneven insulation, **18.3**
        uneven wedge tightness, **18.3**
Generator power system stabilizer (PSS), **13.53–13.55**
    active power flow, **13.54**
    damper winding, **13.54**
    electromechanical oscillations, **13.53**
    flexible coupling, **13.53**
    objective of the PSS, **13.54**

Generator power system stabilizer (PSS) (*Cont*):
    power system oscillations, **13.53**
    reactive power flow, **13.54**
    synchronizing, torque, **13.54**
Generator rotor, **13.1**
    end rings, **13.6–13.8**
        balancing ring, **13.8**
        centrifugal forces, **13.6**
        nonmagnetic austenitic steel, **13.6**
        protective finish, **13.6**
        rotor and alignment threading, **13.9–13.13**
        shrink surface, **13.8**
        skid plate, **13.9**
        slip ring brush gear, **13.10**
        slip rings and connections, **13.9**
        stress-corrosion cracking, **13.6**
    hydrogen-initiated cracking, **13.1**
    modifications, upgrades, and uprates, **18.9**
        higher operating factor, **18.9**
        rotor rewind, **18.9**
    peak torque, **13.1**
    reliability and life expectancy, **18.7–18.8**
        generator experience, **18.7–18.8**
        refurbishment, **18.8**
        generator rotor rewind, **18.8**
    ultrasonic examination, **13.1**
    wedges and dampers, **13.8–13.9**
        brushes and holders, **13.9**
        D-leads, **13.9**
        damper winding, **13.9**
        slip rings, brush gear, and shaft grounding, **13.9**
        wetting current, **13.9**
    winding, **13.1–13.6**
        creep properties, **13.1**
        current-carrying conductors, **13.3**
        end winding, **13.1**
        epoxide glass strips, **13.4**
        flexible leads, **13.6**
        retaining wedge, **13.3**
        rotor body, **13.2**
        stiffness compensation, **13.5**
Generator shaft seals and seal oil system, **13.35–13.38**
    journal-type seal, **13.36**
    seal oil system, **13.37**
    thrust-type seal, **13.35**
Generator size and weight, **13.14**
    moisture absorbent, **13.14**
    water contamination, **13.14**
Generator stator casing, **13.31**
    annular rings, **13.31**
    end shields, **13.31**

Generator stator casing (*Cont*):
  fault torques, **13**.31
  hydrostatic pressure test, **13**.31
  liquid-leakage detectors, **13**.31
  pressure-tight enclosure, **13**.31
  shaft seal, **13**.31
  thermal stresses, **13**.31
Generator stator winding, **13**.17–**13**.26
  arrangement of stator conductors, **13**.22
  cooling components, **13**.26–**13**.29
    conductivity, **13**.27
    demineralized water, **13**.26
    dissolved oxygen, **13**.27
    flexible polytetrafluoroethylene (PTFE), **13**.27
    pH value, **13**.27
    two-pass design, **13**.28
  core end plate and screen, **13**.21
  core frame, **13**.20
  corrugated glass spring, **13**.24
  eddy currents in stator conductors, **13**.23
  Roebel method, **13**.24
  stator slot, **13**.25
  stray losses, **13**.24
  water cooling system, **13**.38–**13**.42
    demineralized water system, **13**.38
    demineralizer, **13**.38
    electrical flashover, **13**.38
    high-integrity insulation, **13**.38
    hydrogen pressure, **13**.38
Generator surveillance and testing, **16**.1–**16**.21
  diagnostic testing, **16**.2–**16**.21
    alternating current tests for stator windings, **16**.6
  direct current high-potential test, **16**.5
  direct current tests for stator and rotor windings, **16**.3
  dissipation factor and tip-up test, **16**.9
  generator operational checks (surveillance and monitoring), **16**.1
    brush gear inspection, **16**.1
    monitoring on-line partial discharge activity, **16**.1
  high-voltage step and ramp tests, **16**.5
  insulation resistance and polarization index, **16**.3–**16**.5
    interpretation, **16**.4
    test setup and performance, **16**.4
  low-core flux test (EL-CID), **16**.18
  major overhaul, **16**.1–**16**.2
    copper dusting, **16**.2
    electrical tests, **16**.1
    mechanical tests, **16**.1

Generator surveillance and testing (*Cont*):
    mechanical tests, **16**.18–**16**.21
      core lamination tightness check (knife test), **16**.20
      ground wall insulation, **16**.20–**16**.21
      rotor windings, **16**.21
      stator and rotor cores, **16**.21
      stator winding side clearance check, **16**.20
      stator winding tightness check, **16**.20
      visual techniques, **16**.20
  partial discharge tests, **16**.7
    off-line conventional partial discharge test, **16**.7
    on-line pd test, **16**.9
  stator insulation tests, **16**.2
  stator turn insulation surge test, **16**.10–**16**.11
  synchronous machine rotor windings, **16**.11–**16**.18
    air gap search coil for detecting shorted turns, **16**.13
    detecting the location of shorted turns with rotor removed, **16**.15
    impedance test with rotor installed, **16**.13
    open-circuit test for shorted turns, **16**.12
Generator vibration, **13**.13
  critical speeds, **13**.13
  equalization of stiffness, **13**.13
  overspeed, **13**.13
  rundown, **13**.13
  trim balancing, **13**.13
  two-plane balancing, **13**.13
Generator voltage regulator, **13**.49–**13**.53
  auto follow-up circuit, **13**.52
  AVR protection, **13**.52
  digital AVR, **13**.52
  dual-channel AVR, **13**.50
  manual follow-up, **13**.52
  proportional-integral-derivative (PID) algorithim, **13**.52
  system description, **13**.50
Inductance, **1**.13–**1**.14
  magnetic energy stored, **1**.14
  self-induction, **1**.13
Induction motors, **6**.1–**6**.21
  basic induction motor concepts, **6**.1–**6**.6
    concept of rotor slip, **6**.2–**6**.5
    electrical frequency of rotor, **6**.5–**6**.6
  construction, **6**.1
    squirrel-cage, **6**.1
    wound rotors, **6**.1
  control of motor characteristics by squirrel-cage rotor design, **6**.14–**6**.17

Induction motors (*Cont*):
  deep bar and double-cage rotor designs, **6.**16–**6.**17
  different rotor designs, **6.**16
  large, deep rotor bars, **6.**15
  leakage reactance, **6.**14
  NEMA design class A, **6.**15
  starting torque, **6.**14
  Thevenin equivalent, **6.**14
  equivalent circuit of an induction motor, **6.**6–**6.**9
    rotor circuit model, **6.**6–**6.**9
  losses and the power flow diagram, **6.**9
    hysteresis and eddy currents losses, **6.**9
    stator copper losses, **6.**9
  low-slip region, **6.**10
  magnetization current, **6.**9
  rotor slip, **6.**9
  starting circuits, **6.**19–**6.**21
    across-the-line starter, **6.**19
    disconnect switch, **6.**19
    on-time delay relays, **6.**21
    overload heater, **6.**19
    short circuit protection, **6.**20
    starting resistance, **6.**21
  starting induction motors, **6.**17–**6.**21
    across-the-line starting, **6.**17
    horsepower rating, **6.**18
    locked rotor, **6.**18
    starting circuit using auto transformers, **6.**18
    starting code letter, **6.**18
    starting current, **6.**17
    under voltage protection, **6.**20
  torque-speed characteristics, **6.**9–**6.**14
  torque-speed curve, **6.**11–**6.**12
  variation of the torque-speed characteristics, **6.**13
    complex automatic control circuit, **6.**13
    maximum torque, **6.**13
    poor efficiency, **6.**13
    slip rings, **6.**13
Inductive loads, effect of, on phase angle control, **9.**18–**9.**19
  inductive voltage, **9.**19
  magnetic energy, **9.**18
  (*See also* Speed control of induction motors)
Inverters, **9.**19–**9.**27
  external commutation inverters, **9.**20–**9.**21
    countervoltage, **9.**20
    rectifier circuit, **9.**21
    synchronous motor, **9.**20
    two commutation techniques, **9.**20

Inverters (*Cont*):
  pulse-width modulation inverters, **9.**22–**9.**27
    comparator, **9.**22
    derating, **9.**24
    fundamental frequency, **9.**24
    harmonic component, **9.**24
    output of PWM circuit, **9.**25
    peak comparator voltage, **9.**26
    pulse train, **9.**22
    reference voltage, **9.**23
    single-phase PWM circuit, **9.**23
    sinusoidal control voltage, **9.**27
  rectifier, **9.**19–**9.**20
    dc output voltage, **9.**19
    inductive load, **9.**20
    smoothing dc output, **9.**19
  self-commutated inverters, **9.**21–**9.**22
    current source inverters, **9.**21
    firing pulses, **9.**22
    output impedance, **9.**22
    self-commutation, **9.**21
    voltage source inverters, **9.**21
Lenz's law, **1.**11–**1.**13
  induced current, **1.**11
  induced voltage, **1.**12
  right-hand rule, **1.**11
  steady work, **1.**13
  thermal energy generation, **1.**13
Lubrication, **21.**7–**21.**15
  flow-through pipes, **21.**10
    laminar flow, **21.**10
    turbulent, **21.**10
  housing and lubrication, **21.**12–**21.**13
    airborne particles, **21.**12
    hypodermiclike needle, **21.**12
    prelubricated bearings, **21.**13
    self-aligning ball bearings, **21.**3
  lubrication of antifriction bearings, **21.**13–**21.**15
    constant-level oiler, **21.**15
    grease lubrication, **21.**13
    line bearings, **21.**15
    oil drain plug, **21.**15
    oil level, **21.**15
    oil rings, **21.**14
    seal oil, **21.**14
    vertical wet-pit condenser circulating pumps, **21.**13
  nonnewtonian fluids, **21.**11
    newtonian fluids, **21.**11
    nonlinear relationship, **21.**11
    oils at low temperatures, **21.**11
    pour point depressants, **21.**11

Lubrication (*Cont*):
  significance of viscosity, **21.**10
    elastohydraulic lubrication, **21.**10
    oil film, **21.**10
    variation of lubricant viscosity with use, **21.**11–**21.**12
      oxidation reactions, **21.**11
      physical reactions, **21.**12
    variation of viscosity with temperature and pressure, **21.**10
      MacCoull's equation, **21.**10
      temperature effect, **21.**10
      viscosity index, **21.**11
    VI improved oils, **21.**11
      permanent viscosity loss, **21.**11
      polymer, **21.**11
  viscosity of lubricants, **21.**7–**21.**9
    resistance of liquid to flow, **21.**17
    shear rate, **21.**8
    shear stress, **21.**8
  viscosity units, **21.**9–**21.**10
    centistokes, **21.**9
    dynamic viscosity, **21.**9
    kinematic viscosity, **21.**9
Machinery principles, **2.**1–**2.**15
  common terms and principles, **2.**1
    angular acceleration, **2.**1
    torque, **2.**1
  electric machines and transformers, **2.**1
    electric machines, **2.**1
    transformers, **2.**1
  energy losses in a ferromagnetic core, **2.**5–**2.**7
    coercive magnetomotive force, **2.**5
    domains, **2.**5
    hysteresis, **2.**5
    residual flux, **2.**5
  Faraday's law—induced voltage from a magnetic field changing with time, **2.**7–**2.**9
    eddy currents, **2.**9
    laminations, **2.**4
    rate of change of flux, **2.**7
  induced voltage on conductor moving in magnetic field, **2.**15
    induced voltage, **2.**15
  magnetic behavior of ferromagnetic materials, **2.**3–**2.**5
    magnetic permeability, **2.**3
    magnetizing intensity, **2.**4
    saturation curve, **2.**3
    saturation region, **2.**3
    unsaturated region, **2.**3
  magnetic field, **2.**2–**2.**3
    action of magnetic fields, **2.**2

Machinery principles (*Cont*):
  permanent magnets, **2.**9–**2.**14
    characteristics of permanent magnets, **2.**12
    core loss, **2.**11
    demagnetization and energy product curves, **2.**13
    energy product, **2.**10
    excitation source, **2.**9
    permeance ratio, **2.**13
  production of induced force on wire, **2.**14–**2.**15
    force induced in conductor, **2.**14
    magnitude of force, **2.**14
  production of magnetic field, **2.**2–**2.**3
    Ampere's law, **2.**2
    ferromagnetic materials, **2.**2
    relative permeability, **2.**3
Magnetic field, **1.**6–**1.**7
  flux, **1.**6
  lines of induction, **1.**6
Magnetic field in solenoid, **1.**9
  close-packed helix, **1.**9
Magnitude of charge, **1.**1
Maintenance of motors, **8.**1–**8.**28
  application data, **8.**3–**8.**4
    unusual service condition, **8.**3
    usual service conditions, **8.**3
  characteristics of motors, **8.**1
  design characteristics, **8.**4–**8.**5
    increased voltage, **8.**4
    reduced voltage, **8.**4
  diagnostic testing for motors, **8.**7–**8.**19
    dc high-potential testing, **8.**13–**8.**14
    dc tests for stator and rotor windings, **8.**9–**8.**12
    insulation resistance and polarization index, **8.**12
    interpretation, **8.**13
    manual rotation test, **8.**19
    stator current fluctuation test, **8.**18
    stator insulation tests, **8.**8–**8.**9
    surge testing, **8.**14
    terminal-to-terminal resistances, **8.**14–**8.**15
    test setup and performance, **8.**12–**8.**13
    tests for the detection of open circuits, **8.**15–**8.**18
  enclosures and cooling methods, **8.**1–**8.**2
    dust-ignition-proof, **8.**2
    explosion proof, **8.**2
    open drip proof, **8.**1
    totally enclosed, **8.**1
    totally enclosed water, air-cooled, **8.**2
  insulation of ac motors, **8.**5–**8.**6

I.8 INDEX

Maintenance of motors (*Cont*):
   lead insulation, **8.5**
   turn insulation, **8.5**
  motor troubleshooting, **8.7**
   motor overheats, **8.7**
   motor runs noisy, **8.7**
  predictive maintenance, **8.6–8.7**
   oil contamination, **8.6**
   repetitive problems, **8.7**
   thermography, **8.7**
   vibration, **8.6**
  repair and refurbishment of ac induction motors, **8.19–8.21**
   bearings, **8.20**
   motor repair, **8.21**
   motor rewind, **8.21**
   oil and water heat exchangers, **8.20**
   rotor work, **8.20**
   stator work, **8.19–8.20**
   temperature detectors, **8.20**
  winding failures in three-phase stators, **8.21–8.28**
   blown fuse, **8.22**
   damage caused by locked rotor, **8.27**
   phase damage due to unbalance voltage, **8.26**
   shorted connection, **8.26**
   undervoltage and overvoltage, **8.22**
   winding damaged by voltage surge, **8.28**
   winding damaged due to overload, **8.27**
   winding failures, **8.21**
   winding grounded at edge of slot, **8.25**
   winding grounded in slot, **8.25**
   winding shorted phase to phase, **8.23**
   winding shorted turn to turn, **8.24**
   winding single-phase, **8.22**
   winding with shorted coil, **8.24**
Performance and operation of generators, **15.1–15.18**
  condition monitoring, **15.3–15.6**
   hydrogen dew point monitoring and control, **15.4**
   hydrogen gas analysis, **15.3–15.4**
   temperature monitoring—thermocouples, **15.3**
   vibration monitoring, **15.4–15.6**
  fault conditions, **15.7–15.16**
   loss of generator excitation, **15.13**
   negative phase sequence currents, **15.10–15.13**
   pole slipping, **15.13–15.14**
   rotor faults, **15.14–15.16**
   stator ground (earth) faults, **15.7**

Performance and operation of generators (*Cont*):
   stator interturn faults, **15.10**
   stator phase-to-phase faults, **15.10**
  generator systems, **15.1–15.3**
   cooling of stator conductors, **15.2**
   excitation, **15.1**
   hydrogen cooling, **15.1**
   hydrogen seals, **15.2–15.3**
  operational limitations, **15.7**
   hydrogen leakage, **15.7**
   temperatures, **15.7**
Power electronics:
  components, **9.1–9.7**
   DIAC, **9.5**
   diode, **9.1–9.2**
   gate turnoff thyristor, **9.4**
   insulated gate bipolar transistor (IGBT), **9.7**
   power and speed comparison of power electronic components, **9.7**
   power transistor, **9.5–9.6**
  introduction, **9.1**
   high-power solid-state devices, **9.1**
   solid-state motor drive, **9.1**
  three-wire thyristor or SCR, **9.3–9.4**
   breakover or turn-on voltage, **9.3**
   rectification applications, **9.4**
   reverse voltage, **9.4**
   silicon controlled rectifier, **9.3**
  TRIAC, **9.5**
   positive or negative pulses, **9.5**
   two SCRs, **9.5**
  two-wire thyristor or PNPN diode, **9.2–9.3**
   conducting region, **9.3**
   forward-blocking region, **9.3**
   reverse-blocking diode-type thyristor, **9.2**
Power plant outages and faults, **24.15–24.16**
  forced outages, **24.15**
  planned outages and faults, **24.15**
Power station dc systems, **24.16–24.20**
  dc system functions, **24.18**
   emergency generator seal oil pump, **24.18**
   essential instruments, **24.18**
  introduction, **24.16–24.18**
   motor generators, **24.17**
   power source, **24.16**
   rectifiers, **24.17**
   static inverter scheme, **24.18**
   station service board, **24.17**
  mission time of dc systems, **24.18–24.20**
   boost (off load), **24.20**
   control equipment, **24.19**
   emergency lighting, **24.19**

INDEX

Power station dc systems (*Cont*):
  essential services board, **24.**20
  fire sirens, **24.**19
  float, **24.**8
  standby batteries, **24.**18
Power station electrical system description, **24.**3–**24.**4
  generator main output system, **24.**3–**24.**4
    electrical auxiliary systems, **24.**3
    fossil-fired power stations, **24.**4
    generator transformers, **24.**3
    main generator, **24.**3
    type of stations, **24.**3
Power station system performance, **24.**4–**24.**15
  effects of loss of grid supplies, **24.**15
    operating limits, **24.**15
    wind storm, **24.**15
  introduction, **24.**1
    degree of reliability, **24.**1
    station system transformer, **24.**1
    unit trip, **24.**1
  plant requirements, **24.**5
    oil pumps, **24.**5
  shutdown and power trip, **24.**14–**24.**15
    controlled shutdown, **24.**14
    power trip, **24.**14–**24.**15
  synchronizing unit to station, **24.**5
    coal-fired units, **24.**10
    condenser extraction pump, **24.**6
    general services, **24.**12
    generator service air compressor, **24.**9
    generator transformer, **24.**7
    LP heater pump, **24.**10
    lubricating pump, **24.**11
    manual check synchronizing, **24.**14
    standby boiler feed pump, **24.**11
    standby supplies, **24.**13
    station board, **24.**9
    station services boards, **24.**12
    station transformer, **24.**8
    stator coolant pump, **24.**11
    transmission services, **24.**7
    unit boiler services, **24.**8
    unit transformer, **24.**6
  synchronizing to grid, **24.**5
  turbine-generator, **24.**5
Power station systems requirements, **24.**1–**24.**3
  grid criteria, **24.**2
    frequency deadband, **24.**2
    overfrequency excursions, **24.**2
  safety requirements, **24.**3
    codes of practice, **24.**3

Power station systems requirements (*Cont*):
    standards, **24.**3
    switchgear, **24.**3
  station operation systems, **24.**1–**24.**2
    load rejection, **24.**2
    operating flexibility, **24.**2
    station technical particulars, **24.**1
  unit start-up, **24.**4
    gas turbine, **24.**4
    high-voltage (HV) circuit breaker, **24.**4
Power station uninterruptible power supply (UPS) systems, **24.**16
  black start, **24.**16
  guaranteed instrument supplies, **24.**16
  no-break supplies, **24.**16
Power station protective systems, **25.**1–**25.**18
  dc tripping systems, 25.-8–**25.**18
    logic diagram, **25.**8–**25.**18
  design criteria, **25.**1–**25.**7
    auxiliary trip circuit 1, **25.**3
    auxiliary trip circuit 2, **25.**4
    auxiliary trip circuit 3, **25.**5
    backup protection, **25.**1
    category A faults, **25.**2
    category B, **25.**5
    condenser level, **25.**3
    electromagnetic relays, **25.**1
    generator negative sequence protection, **25.**1
    governor fail, **25.**3
    lubricating oil, **25.**4
    relay fluid, **25.**3
    relay panels, **25.**2
    stator coolant, **25.**5
    transformers switchgear connections, **25.**2
    trip supervision alarm, **25.**7
    tripping system, **25.**2
    turbine trip solenoid, **25.**6
  generator protection, **25.**8
    stator ground (earth) faults—low impedance grounding, **25.**8
    stator ground faults—high-resistance grounding, **25.**8
    stator phase-to-phase faults, **25.**8
  introduction, **25.**1
    faulted systems, **25.**1
    overall protection scheme, **25.**1
Pulse circuits, **9.**12–**9.**13
  analog methods, **9.**12
  digital techniques, **9.**12
  relaxation oscillator, **9.**13
  read-only memory, **9.**13

I.10   INDEX

Pulse synchronization, **9.**15
  synchronizing the pulse, **9.**15
  triggering pulse, **9.**15
Rectifiers:
  basic rectifier circuits, **9.**7
    full-wave rectifier, **9.**8–**9.**9
    half-wave rectifier, **9.**8
    relative speeds, **9.**8
    ripple factor, **9.**7
    three-phase full-wave rectifier, **9.**9–**9.**10
    three-phase half-wave rectifier, **9.**9
  filtering rectifier output, **9.**10–**9.**12
    capacitors, **9.**10
    inductive filter, **9.**12
    low-pass filters, **9.**10
    output ripple, **9.**12
    output voltage of rectifier circuit, **9.**11
    three-phase half-wave rectifier circuit, **9.**11
Relaxation oscillator using PNPN diode, **9.**13–**9.**15
  extra transistor stage, **9.**14
  pulse-generating circuit, **9.**13
  SCR gate lead, **9.**14
  transistor amplifier, **9.**15
Speed control of induction motors, **7.**1–**7.**11
  by changing line frequency, **7.**1–**7.**3
    base speed, **7.**1
    derating, **7.**1
    family of torque-speed characteristic curves, **7.**2
    magnetization current, **7.**2
    power supplied to motor, **7.**3
    saturation, **7.**3
    saturated region, **7.**2
    solid state motor drives, **7.**3
    variable-frequency induction motor drive, **7.**1
  by changing line voltage, **7.**3–**7.**4
    speed control by varying rotor resistance, **7.**4
    variable-line voltage speed control, **7.**4
  by changing rotor resistance, **7.**5
    inserting extra resistances, **7.**5
  induction generator, **7.**5–**7.**8
    generator region, **7.**7
    induction generator operating alone, **7.**7–**7.**8
    pushover torque, **7.**7
    variable frequency control with a PWM waveform, **7.**6
    variable voltage control with a PWM waveform, **7.**7

Speed control of induction motors (*Cont*):
  induction motor ratings, **7.**8–**7.**11
    capacitor bank, **7.**9
    compounded induction generator, **7.**10
    current limit, **7.**11
    high-efficiency induction motor, **7.**8
    magnetization curve of an induction machine, **7.**9
    no-load terminal voltage, **7.**9
    terminal voltage-current characteristic, **7.**10
    voltage and current ratings, **7.**11
  motor protection, **7.**5
    excessive instantaneous currents, **7.**5
    overvoltage, **7.**5
  solid-state induction motor drives, **7.**5
    pulse-width modulation (PWM) technique, **7.**5
  (*See also* Induction motors)
Synchronous generators, **12.**1–**12.**29
  capability curves, **12.**26–**12.**28
    apparent power output, **12.**26
    capability diagram, **12.**26
    reactive power output, **12.**26
    real power output, **12.**26
    safe operating point for the generator, **12.**28
  conditions required for paralleling, **12.**15–**12.**19
  construction, **12.**1–**12.**3
    brushless exciter, **12.**1
    nonsalient poles, **12.**1
    permanent magnets, **12.**3
    rectifier circuit, **12.**3
    salient eight-pole synchronous machine rotor, **12.**2
    salient poles, **12.**1
    single salient pole, **12.**2
    slip rings and brushes, **12.**1
  effect of load changes, **12.**13–**12.**15
    effect of increase in generator loads, **12.**13
    magnitude of internal generated voltage, **12.**13
    power factor, **12.**13
    voltage regulation, **12.**14
  equivalent circuit, **12.**6–**12.**10
    armature reaction, **12.**6
    equivalent circuit of three-phase synchronous generator, **12.**8
    internal generated voltage, **12.**6
    magnetization curve, **12.**6
    rotor magnetic field, **12.**6
    stator magnetic field, **12.**6
    synchronous reactance, **12.**8
    windings self-inductance, **12.**8

Synchronous generators (*Cont*):
  frequency-power and voltage-reactive power characteristics, **12.**19–**12.**21
    no-load terminal voltage setpoint, **12.**20
    plot of terminal voltage versus reactive power, **12.**20
    prime mover, **12.**19
    speed-drooping characteristic, **12.**19
    speed-power curve, **12.**20
    voltage-reactive power characteristics, **12.**20
  general procedure for paralleling generators, **12.**16–**12.**19
    frequency of oncoming generator, **12.**16
    phase angles, **12.**17
    phase sequence, **12.**16
  internal generated voltage of synchronous generator, **12.**6
  operating alone, **12.**13
    power factor of the load, **12.**13
  operation of generators in parallel with large power systems, **12.**21–**12.**24
    infinite bus, **12.**21
    motoring of generator, **12.**22
    no-load frequency, **12.**23
    power-frequency characteristic, **12.**21
    reactive power output of generator, **12.**24
    reactive power-terminal voltage characteristic, **12.**21
    torque-speed characteristic, **12.**24
  parallel operation of AC generators, **12.**15
    major advantages, **12.**15
    reliability, **12.**15
  phasor diagram, **12.**10
    lagging power factor, **12.**10
    leading power factor, **12.**10
    per-phase equivalent circuit of a synchronous generator, **12.**8
    relationships between AC voltages, **12.**10
    winding resistance, **12.**10
  power and torque, **12.**11–**12.**13
    friction and windage losses, **12.**12
    induced torque, **12.**13
    power flow in synchronous generator, **12.**11
    rotational speed, **12.**11
    stray losses, **12.**12
    torque angle, **12.**13
  ratings, **12.**25–**12.**26
    apparent power and power-factor ratings, **12.**25
  short-time operation and service factor, **12.**28–**12.**29

Synchronous generators (*Cont*):
    insulation class, **12.**29
    momentary power surges, **12.**29
    service factor, **12.**29
  speed of rotation, **12.**23
    electrical frequency, **12.**3
    mechanical speed, **12.**3
    stator electrical frequency, **12.**3
  steps taken to synchronize incoming ac generator, **12.**18
  voltage, speed, and frequency ratings, **12.**25
Synchronous machines, **11.**1–**11.**12
  air gap and magnetic circuit, **11.**2–**11.**4
    air gap, **11.**2
    rotor, **11.**3
    stator, **11.**3
  excitation, **11.**9–**11.**10
    excitation characteristics, **11.**9
    leading power factor load, **11.**9
    synchronous motor V curves, **11.**10
  field excitation, **11.**5–**11.**6
    rotating rectifier excitation, **11.**5
    series excitation, **11.**6
  machine losses, **11.**10–**11.**12
    armature conductor loss, **11.**12
    core losses, **11.**12
    excitation loss, **11.**12
    explosion, **11.**11
    heat-transfer properties of hydrogen, **11.**11
    stray-load loss, **11.**12
    windage and friction losses, **11.**10
  no-load and short-circuit values, **11.**6–**11.**7
    armature windings, **11.**6
    no-load or open-circuit voltage, **11.**6
    short-circuit characteristics, **11.**7
  physical description, **11.**1–**11.**2
    armature reaction, **11.**1
    armature winding, **11.**1
    field winding, **11.**1
  pole pitch: electrical degrees, **11.**2
    number of poles, **11.**2
  torque tests, **11.**7–**11.**9
    effect of rotor bar material, **11.**9
    locked-rotor torque, **11.**7
    pull-in torque, **11.**8
    pull-out torque, **11.**7
    speed-torque characteristic, **11.**8
  windings, **11.**4–**11.**5
    damper windings, **11.**5
    distributed windings, **11.**4
    solenoid windings, **11.**5

INDEX

Three-phase systems, **1.21–1.25**
  power density, **1.21**
  power in three-phase systems, **1.23–1.25**
    instantaneous power, **1.24**
    power factor angle, **1.23**
    total power, **1.23**
  three-phase connectors, **1.22**
    line voltages, **1.22**
    phase voltages, **1.22**
  three-phase motor, **1.21**
Transformers, **3.1–3.28**
  autotransformer, **3.22–3.23**
    common winding, **3.22**
    series current, **3.23**
    series winding, **3.22**
  classification, **4.1–4.2**
    dry transformers, **4.1**
    harmonics, **4.2**
    oil-immersed transformers, **4.1**
  components and maintenance, **4.1–4.27**
  dot convention, **3.14**
    positive magnetomotive force, **3.14**
  efficiency, **3.21**
  equivalent circuit, **3.15–3.18**
    approximate equivalent circuits, **3.17–3.18**
    copper losses, **3.15**
    core-loss current, **3.17**
    hysteresis losses, **3.16**
    reluctance, **3.16**
  failures, cause of, **4.11–4.13**
    aging factor, **4.13**
    hoop force, **4.12**
    microscopic droplets, **4.12**
    percent moisture, **4.13**
  forces, **4.11**
    radial forms, **4.11**
    short circuit, **4.11**
  gas relay and collection systems, **4.22–4.24**
    gas relay, **4.22–4.24**
  ideal transformer, **3.–3.4**
    analysis of circuits containing, **3.5–3.9**
    power factor, **3.4**
    power in ideal transformer, **3.3**
  impedance transformation, **3.4–3.5**
    apparent impedance, **3.4**
    impedance, **3.4**
  importance, **3.1**
    transmission losses, **3.1**
    voltage level, **3.1**
  instrument transformer, **3.28**
    current transformer, **3.28**
    potential transformer, **3.28**

Transformers (*Cont*):
  insulation, types and features of, **4.9–4.11**
    laminated high-density boards, **4.9**
    low-density calendered board, **4.9**
    reasons for deterioration, **4.11**
    typical bushing construction, **4.10**
  interconnection with grid, **4.24–4.27**
    battery banks, **4.27**
    main output transformer, **4.24**
    power station or single-line diagram, **4.26**
    reliability of power supplies, **4.27**
    station service transformer, **4.24**
    utility switchyard, **4.25**
  magnetizing current in real transformer, **3.12–3.14**
    core-loss current, **3.12**
    excitation current, **3.14**
    magnetization current, **3.12**
    magnetization curve, **3.13**
    resulting flux, **3.14**
  phasor diagram, **3.19–3.20**
    negative voltage regulation, **3.20**
    series impedances, **3.19**
  power transformer, main components of, **4.2–4.9**
    breather, **4.7**
    bushing terminal connectors, **4.5**
    bushings, **4.9**
    conservator tank with air cell, **4.8**
    current transformers, **4.9**
    cutaway view, **4.2**
    gas detector relay, **4.7**
    H.V. bushings, **4.3**
    high-voltage winding, **4.6**
    L.V. bushings, **4.3**
    nitrogen demand system, **4.8**
    oil, **4.6**
    radiator, **4.7**
    rapid pressure rise relay, **4.4**
    safety harness anchor rod, **4.4**
    tank, **4.6**
    tap changers, **4.9**
    transformer core, **4.7**
    winding temperature indicator, **4.4**
    windings, **4.8**
  ratings, **3.25–3.28**
    apparent power rating of transformer, **3.27**
    derating, **3.26**
    inrush current, **3.27**
    saturation region, **3.26**
    transformer nameplate, **3.27**
    voltage and frequency ratings of transformer, **3.25**

INDEX

Transformers (*Cont*):
  relief devices, **4.**24
    diaphragm, **4.**24
    explosion vent, **4.**24
  simplified voltage regulation, **3.**20–**3.**21
    primary voltage, **3.**20
  taps and voltage regulators, **3.**21–**3.**22
    taps, **3.**21
    tap changing under load, **3.**22
  theory of operation of real single-phase transformers, **3.**9–**3.**10
    average flux per turn, **3.**10
    flux linkage, **3.**9
  three-phase transformer, **3.**23–**3.**25
    three-phase transformer connections, **3.**24
  transformer oil, **4.**13–**4.**21
    causes of deterioration, **4.**16
    gas-in-oil, **4.**20–**4.**21
    interfacial tension test, **4.**17
    methods of dealing with bad oil, **4.**19–**4.**20
    Myers index number, **4.**17–**4.**18
    neutralization number test, **4.**17
    testing transformer insulating oil, **4.**13–**4.**16
    (*See also* Used oil analysis)
  types and construction, **3.**1–**3.**2
    substation transformer, **3.**2
    thin laminations, **3.**1
  voltage ratio across, **3.**10–**3.**12
    average flux, **3.**10
    leakage flux, **3.**10
    mutual flux, **3.**11
  voltage regulation and efficiency, **3.**18–**3.**19
    full-load voltage regulation, **3.**18
Turbine-generator components—the stator, **13.**17
  circulating currents, **13.**17
  core laminations, **13.**17
  grain-oriented sheets of steel, **13.**17
  stator bore, **13.**17
Used-oil analysis, **22.**1–**22.**8
  proper lube oil sampling technique, **22.**1
    analysis laboratory, **22.**1
    gearbox, **22.**1
  test description and significance, **22.**1–**22.**7
    chemical and physical tests, **22.**2–**22.**3
    visual and sensory inspections, **22.**1–**22.**2
  summary, **22.**7
    continuing service, **22.**7
    optimum intervals, **22.**7
    trends, **22.**7
Variable-speed drives, **10.**1–**10.**24
  ac drive application issues, **10.**12–**10.**13
    diode source current unbalance, **10.**12
    introduction, **10.**12

Variable-speed drives (*Cont*):
    ac power factor, **10.**13
    ac input power changes with ac input voltage, **10.**13
    dc link reactor, **10.**13
    power circuit, **10.**13
  basic principles of ac variable-speed drives, **10.**1–**10.**3
    constant-power (extended speed) region, **10.**1
    constant-torque region, **10.**1
    insulated gate bipolar transistors, **10.**3
    inverters, **10.**1
    pulse-width-modulated inverters, **10.**2
    two-level pulse-width-modulated inverter (PWM-2), **10.**3
  cable details, **10.**16–**10.**17
    aluminum armor, **10.**16
    continuous corrugated aluminum cable, **10.**16
  cabling details for ac drives, **10.**16
    cable connections, **10.**16
    grounding practices, **10.**16
  common failure modes, **10.**23–**10.**24
    ac line transients, **10.**24
    motor insulation, **10.**24
  dc link energy, **10.**4
    bank of electrolytic capacitors, **10.**4
    harmonics, **10.**4
  IGBT switching transients, **10.**13–**10.**16
    cable terminating (matching) impedance, **10.**15
    extra insulation, **10.**15–**10.**16
    insulation voltage stress, **10.**13–**10.**14
    inverter output filter, **10.**15
    motor, cable, and power system grounding, **10.**17
    motor winding voltage distribution, **10.**14
    radiated electromagnetic interference (EMI), **10.**14–**10.**15
    voltage transients, **10.**13
  input power converter (rectifier), **10.**3
    regeneration, **10.**3
  input sources for regeneration or dynamic slowdown, **10.**5–**10.**7
    dynamic breaking, **10.**5–**10.**7
  maintenance, **10.**23
    electrolytic capacitors, **10.**23
    power supplies, **10.**23
  motor application guidelines, **10.**24
    constant-torque load, **10.**24
    electrostatic shield, **10.**24
    inverter-rated, **10.**24

INDEX

Variable-speed drives (*Cont*):
  nonsinusoidal PWM waveform, **10**.24
  motor bearing currents, **10**.17–**10**.20
    add motor shaft grounding brush, **10**.19
    common-mode voltage, **10**.17
    electrostatic discharge currents, **10**.19
    line to neutral sources, **10**.19
    motor cable wiring practices, **10**.20
    motor capacitances, **10**.18
    motor conduit box, **10**.18
    nonsymmetric switching pattern, **10**.17
    reduce the stator to rotor capacitance value, **10**.20
    transient switching voltages, **10**.17
    use conductive grease in motor bearings, **10**.20
  output IGBT inverter, **10**.4
    original sine wave reference, **10**.4
    power transistors, **10**.5
    SCR bridge, **10**.4
    storage capacitor, **10**.4–**10**.5
    three-phase voltage commands, **10**.5
  PWM-2 considerations, **10**.8–**10**.9
    bearing current, **10**.8
    fast switching transients, **10**.8
    output harmonics, **10**.8
    motor bearing currents, **10**.8
    reflections, **10**.9
  regeneration, **10**.7–**10**.8, **10**.21–**10**.23
    ac input current, **10**.8
    ac input harmonics, **10**.8
    diode bridge, **10**.7
    dynamic braking, **10**.23
    dynamometer, **10**.22
    inverter cost, **10**.7
    magnetic break, **10**.22
    paper machine winder, **10**.23
    power flow diagram, **10**.22
    PWM bridge, **10**.7
    rotating load, **10**.21
    synchronous speed, **10**.7
  selection criteria of VSDs, **10**.21
    compressors and pumps, **10**.21
    motor starting, **10**.21
    variable process speed, **10**.21
  summary of application rules for ac drives, **10**.20–**10**.21
    bearing currents, **10**.21
    dc link reactor, **10**.20
    radiated EMI, **10**.21
    voltage reflections, **10**.20
  thyristor failures and testing, **10**.11–**10**.12

Variable-speed drives (*Cont*):
    comments about failure rates, **10**.12
    recognizing failed SCR or diode, **10**.11
    testing of SCRs or diodes, **10**.11
  transients, harmonics power factor, and failures, **10**.9–**10**.11
    common failure modes, **10**.9–**10**.10
    device application, **10**.11
    device explosion rating, **10**.11
    fault current limit, **10**.10–**10**.11
    semiconductor failure rate, **10**.9
Vibration analysis, **23**.1–**23**.16
  application of sine waves to vibration, **23**.1
    accelerometer sensors, **23**.1
    spring mass system, **23**.1
    top dead center, **23**.1
  bearing defects, **23**.11
  blade or vane problems, **23**.11
  condensate pump misalignment, **23**.12–**23**.16
    bent shaft, **23**.14
    conclusion, **23**.16
    corrective actions, **23**.16
    coupling inspection, **23**.13
    eccentric journals, **23**.14
    final results, **23**.16
    oil or steam whirl, **23**.15
    problem, **23**.12
    reciprocating forces, **23**.15
    soft foot, **23**.14
    test data and observations, **23**.12–**23**.15
    velocity spectra, **23**.12
    vibration severity chart, **23**.13
  driveshaft torsion, **23**.12
  electric motor defects, **23**.11
  forcing-frequency causes, **23**.9
  frequency domain, **23**.8–**23**.9
    spectrum, **23**.8
    three-dimensional graph, **23**.8
  gear defects, **23**.11
  logarithms and decibels, **23**.4–**23**.5
    logarithmic scale, **23**.4
    real machine, **23**.5
    reference amplitude, **23**.4
    standard reference, **23**.4
  machinery example, **23**.9
    bearing defect, **23**.9
    frequency domain analysis, **23**.9
  mechanical looseness, **23**.11
  misalignment, **23**.10
  multimass systems, **23**.1–**23**.3
    maximum amplitude, **23**.2
    peak-to-peak, **23**.2

Vibration analysis (*Cont.*):
    total vibration level, **23.**1
    vibration of piston, **23.**2
    vibration velocity, **23.**3
  oil whirl, **23.**11
  resonance, **23.**3–**23.**4
    high internal forces, **23.**3
    machine-support-piping, **23.**3
    natural frequencies, **23.**3
    vibration acceleration, **23.**4
  resonant frequency, **23.**12
    running speed, **23.**12
    stiffness, **23.**12
  time domain, **23.**6–**23.**8
    core velocity, **23.**7
    damping, **23.**7
    housing velocity, **23.**7
    impulsive signal, **23.**6
    unbalanced rotor, **23.**6
  unbalance, **23.**10
  uneven loading, **23.**11

Vibration analysis (*Cont.*):
  use of filtering, **23.**5
    filters, **23.**5
    vibration causes, **23.**9
  vibration instrumentation, **23.**5–**23.**6
    displacement transducer (proximity probe), **23.**5
    transducer selection, **23.**6
    velocity transducer, **23.**6
  vibration severity, **23.**12
    vibration diagnostic chart, **23.**12
Voltage variation by ac phase control **9.**16–**9.**18
  for ac load, **9.**18
    power controller, **9.**18
  for dc load driven from ac source, **9.**16–**9.**18
    breakover voltage, **9.**16
    bridge circuit, **9.**16
    current waveforms, **9.**17
    phase angle power, **9.**16

ABOUT THE AUTHOR

Philip Kiameh, M.A.Sc., B. Eng., P. Eng., D. Eng., is a teacher at the University of Toronto. During the past twelve years, he taught certified courses in power generation, electrical and mechanical equipment, power plant engineering, and industrial instrumentation and modern control systems. In May 1996, he was awarded the first "Excellence in Teaching" award by the University of Toronto.

Mr. Kiameh wrote the following books: *Power Generation Handbook* and *Industrial Instrumentation and Modern Control Systems*. He received his engineering degree from Dal-Tech (formerly, Technical University of Nova Scotia) "with distinction" and completed his graduate studies at the University of Ottawa. He performed research about power generation equipment with Atomic Energy of Canada Limited (AECL) at their Chalk River and Whiteshell Laboratories. He also worked in the power plants of Ontario Power Generation (formerly, Ontario Hydro) for over nineteen years where he was responsible for the operation and maintenance of different power plant systems.